ABHANDLUNGEN
DER AKADEMIE DER WISSENSCHAFTEN DER DDR
Abteilung Mathematik — Naturwissenschaften — Technik

Jahrgang 1982 · Nr. 1 N

Vorträge und Poster
des VI. Internationalen Symposiums

SYSTEMISCHE FUNGIZIDE UND ANTIFUNGALE VERBINDUNGEN

veranstaltet durch
die Gesellschaft für Allgemeine und
Technische Mikrobiologie
in der Biologischen Gesellschaft der DDR und von
der Gesellschaft für Medizinische Mykologie der DDR

vom 4.—10. Mai 1980
im Schloß Reinhardsbrunn

Herausgegeben von
H. Lyr und C. Polter

AKADEMIE-VERLAG · BERLIN 1983

Herausgegeben im Auftrage des Präsidenten
der Akademie der Wissenschaften der DDR
von Vizepräsident Prof. Dr. Heinrich Scheel

Erschienen im Akademie-Verlag, DDR-1086 Berlin, Leipziger Straße 3—4

Lizenznummer: 202 · 100/520/83
Gesamtherstellung: VEB Druckerei „Thomas Müntzer", 5820 Bad Langensalza
Umschlag: Rolf Kunze
Bestellnummer: 763 014 9 (2001/82/1 N) · LSV 4275
Printed in the German Democratic Republic

DDR 72,- M

Inhaltsverzeichnis

Vorträge

Poster, Gruppe A

Poster, Gruppe B

H. A. Koch

Medizinische Akademie Erfurt, DDR

Mykotoxinbildende Pilze — Eine neue Zielgruppe für Fungizide

Mykotoxine sind sekundäre Metabolite von Schimmelpilzen, vor allem Vertreter der Gattungen *Aspergillus, Fusarium* und *Penicillium*. Bisher sind 104 Mykotoxine und ca. 200 toxinogene Schimmelpilze bekannt. Mykotoxine wirken nach peroraler Aufnahme für Mensch und Tier in höheren Dosen (ppm-Bereich) akut toxisch und können kurzfristig zum Tode führen, in niedrigen Dosen (ppb-Bereich) wirken sie kanzerogen, mutagen und/oder teratogen. Nahrungs- und Futtermittel werden oft von Schimmelpilzen befallen, wenn dabei Mykotoxine gebildet werden, kommt es nach Aufnahme zu Mykotoxikosen. Beim Menschen sind als Folge des Verzehrs mykotoxinhaltiger Nahrungsmittel schwere Erkrankungen und Todesfälle beschrieben, wie z.B. die Alimentary Toxic Aleukia oder Aflatoxikosen (Tab. 1). In Europa sind derartige Fälle

Tabelle 1
Wichtige Mykotoxine, Produzenten und Wirkungen

Mykotoxin	Produzenten	wichtige Substrate	Biol. Wirkungen
Aflatoxine (12 Derivate)	*Asp. flavus* *Asp. parasiticus*	Getreide*) Futtermittel, Obst	Hepatotoxin, karzinogen, mutagen, teratogen
Zearalenon (F 2-Toxin) (15 Derivate)	*Fus. graminearum* *Gib. zeae* u. a.	Getreide*) Futtermittel, Stroh, Pellets	Oestrogenismus, Abort, Infertilität, teratogen
Trichothecene (37 Derivate)	*Fusarium*arten, *Myrothecium*arten	Getreide*) Heu	Alimentary Toxic Aleukia, Knochenmarkzerstörungen, Dünndarm- und Hodendestruktion
Ochratoxin A	*Asp. ochraceus*, *P. viridicactum*, *P. cyclopium* u. a.	Getreide*) Futtermittel	Nephrotoxin (Balkan-Nephropathie)
Citrinin	*P. viridicactum*, *P. citrinum* u. a.	Getreide*) Futtermittel, Obst	Nephrotoxin, Chromosomenschäden
Patulin	*P. expansum* u. a.	Obst	zytotoxisch (karzinogen ?) Lebernekrosen, Lungen- u. ZNS-Ödeme

*) einschließlich Mais und Reis

selten, in einigen tropischen Gebieten müssen wir mit erheblichen Dunkelziffern rechnen.

Beim Tier — Nutz- und Haustiere — nehmen Mykotoxikosen besorgniserregend zu. Diese medizinisch und volkswirtschaftlich nicht vertretbare Zunahme ist Anlaß, den Mykotoxikosen zunehmende Aufmerksamkeit zu widmen.

Schimmelpilze bilden Mykotoxine nicht nur art- und stammspezifisch, sondern auch in Abhängigkeit vom Substrat und den umgebenden biophysikalischen Bedingungen. Die Anwesenheit eines toxinogenen Schimmelpilzes sagt deshalb nichts über eine mögliche Mykotoxinkontamination aus. Mykotoxikosen können exakt nur durch chemisch-physikalische oder biologische Mykotoxinnachweise diagnostiziert werden.

Der Befall einer Frucht durch Schimmelpilze kann bereits auf dem Feld eintreten, dies ist witterungsabhängig oft beim Mais der Fall, wobei im Anschluß an einen Befall mit Fusarien die Toxinbildung z. T. erst später (z.B. ausgelöst durch einen Kälteschock) eintreten kann. Futtermais erweist sich oft mit Zearalenon oder Trichothecenen kontaminiert; auf dieser Basis ist es bereits zu schwerwiegenden Zwischenfällen in Tierbeständen gekommen. Auch beim Obst (z.B. Äpfel) tritt der Befall mit Schimmelpilzen (u. a. *Penicillium expansum*) oft schon vor der Ernte ein. Bei unsachgemäßer Lagerung bis zur Verarbeitung kommt es zur Ausbreitung der Schimmelpilze im Stapel und die Praxis zeigt, daß z.B. zahlreiche Apfelsaftchargen das Mykotoxin Patulin enthalten. In der Mehrzahl der Fälle tritt die Verschimmelung von Erntegut erst während der Lagerung oder Zwischenlagerung ein. Dabei begünstigen Feuchtigkeit und höhere Temperaturen die mikrobielle Kontamination. Bei einer relativen Feuchte unter 13% unterbleibt im Getreide das Wachstum von Schimmelpilzen, in ölhaltigen Früchten muß diese Grenze noch niedriger gehalten werden (z.B. Leinsamen 9%, Sojabohnen 12%). Diese Bedingungen sind bei den oft notwendigen Zwischenlagerungen nicht einzuhalten. Da bei höherem Feuchtegehalt das Lagergut oft umgesetzt wird, um Temperaturanstiege zu unterbinden, kommt es regelmäßig zur Verbreitung von Schimmelpilzen im ganzen Stapel. Diese Erscheinung wird dadurch begünstigt, daß in der Regel feuchte Chargen und weniger feuchte Herkünfte nicht getrennt gelagert werden. Auch Stroh wird oft zu feucht gelagert und/oder vor der Verarbeitung werden die regelmäßig feuchten und pilzbefallenen Deckschichten der Mieten nicht entfernt. Die aus solchen Chargen hergestellten Futterpellets sind oft erheblich mit toxinogenen Pilzen kontaminiert. Lagerndes Getreide kann durch antimikrobielle Behandlung (z.B. Propionsäure oder Harnstoff) vor Verschimmelung relativ gut geschützt werden. Entsprechende Verfahren scheinen erfolgversprechend zu sein, sie sind jedoch noch nicht zur Praxisreife entwickelt.

Die beim Tier nach Mykotoxinaufnahme entstehenden akuten Schäden variieren nach dem spezifischen Angriffsort der einzelnen Mykotoxine — im allgemeinen handelt es sich um die Symptome einer akuten oder chronischen Toxikose. Diese Symptome können von einer Futterverweigerung, Verschlechterung der Futterverwertung bis zu drastischen Leistungsverminderungen und Massensterben reichen. Zu welchen Dauerfolgen bei Tieren subtoxische Mykotoxindosen führen, ist bisher kaum untersucht. Es muß befürchtet werden, daß diese chronischen Schäden für die Leistungszucht viel größere Bedeutung haben als wir gegenwärtig übersehen. In welchem Ausmaß ein „carry over", d. h. eine Übertragung in tierische Nahrungsmittel erfolgt, ist nur teilweise bekannt, nach Aufnahme größerer Dosen muß damit gerechnet werden. Aflatoxine z.B. gehen in die Kuhmilch und in Hühnereier über. Bei Mastbullen waren

Mykotoxinnachweise noch bis zu 12 Wochen nach dem Absetzen des kontaminierten Futters nachweisbar, und zwar sowohl im Muskelfleisch als auch in der Leber und Niere.
Mykotoxikosen sind in der Mehrzahl der Fälle therapeutisch nicht beeinflußbar, auch solche Schäden, die durch geringe Dosen angerichtet werden, sind wahrscheinlich irreversibel. Über Reparaturmechanismen an mykotoxingeschädigter DNA und RNA ist bisher kaum etwas bekannt.
Mykotoxindetoxifikationen sind im Labormaßstab möglich, in Futter- und Nahrungsgütern unter Praxisbedingungen aber nicht realisierbar. In Anbetracht dieser Umstände müssen sich alle Bemühungen auf die Prävention konzentrieren. Weder in Europa noch in feuchtwarmen Gebieten stehen entsprechende technische Voraussetzungen und Energievorräte zur Trocknung und trockenen Lagerung von Erntegütern, insbesondere Körnerfrüchten, Futtermitteln und Nahrungsgütern zur Verfügung. Der Einsatz von geeigneten antifungiellen oder antimikrobiellen Substanzen zur Unterdrückung des Wachstums von Schimmelpilzen könnte zur Lösung der Mykotoxinprobleme beitragen. Dabei wäre bei verschiedenen Früchten, wie z.B. Mais, aber auch Citrusfrüchten eine Behandlung auf dem Felde denkbar. Für Getreide wäre eine Behandlung während der Ernte oder unmittelbar danach überlegenswert, um das Wachstum von Schimmelpilzen schon im Zwischenlager zu unterbinden. Auch der Einsatz von antifungiellen Prinzipien während der Lagerung, wie z.B. Harnstoff, könnte außerordentlich nützlich sein. Die Entwicklung und/oder Vervollkommnung entsprechender Verfahren sollte von Phytopathologen, Veterinärmedizinern und Humanmedizinern gemeinsam vorangetrieben werden, um von Anfang an alle medizinischen und volkswirtschaftlichen Gesichtspunkte zu berücksichtigen. Da die dargestellten Probleme für alle Länder von großer Bedeutung sind, ist es dringend notwendig, sich der Mykotoxikosen-Prävention mit großer Kraft anzunehmen. Manche Länder exportieren Futtermittel zu günstigen Bedingungen, in vielen Fällen handelt es sich dabei um Chargen, die mit Mykotoxinen mehr oder weniger stark kontaminiert sind. Es erscheint dringend geboten, Exporteure und Importeure zu Qualitätskontrollen, die Mykotoxinnachweise einschließen, zu verpflichten und entsprechende Regelungen festzulegen, weil ohne diese Maßnahmen eine wirksame Prävention kaum möglich sein wird.

Diskussion

BURTH: Im Vortrag wurden Getreide, Obst und deren Verarbeitungsprodukte als besonders durch mykotoxinproduzierende Pilze gefährdet herausgestellt. Bei Speisekartoffeln ist in den letzten Jahren infolge der Zunahme der „Unter Vakuum Lagerung" und der Belüftung gegen bakterielle Naßfäulen ein deutlich erhöhtes Auftreten von *Fusarium*-Trockenfäulen zu beobachten. Ist dieses Problem schon im Hinblick auf Mykotoxinbildung untersucht worden?

KOCH: Konkrete Angaben aus Mitteleuropa sind mir bisher nicht bekannt.

MÜLLER: Welche Möglichkeiten sehen Sie zur Inaktivierung von Mykotoxinen, z.B. durch Komplexbindung mit dem Zweck einer Intoxikation?

KOCH: Für die Praxis keine.

MÜLLER: Gibt es beim Wirt (z.B. Apfel, aber auch anderen Kulturarten) ein sortenspezifisches Verhalten zur Mykotoxinbildung?

KOCH: Mykotoxinbildung ist substratspezifisch und stammspezifisch. Unterschiede bei Apfelsorten hinsichtlich *P. expansum*-Befall bestehen, sind aber wohl nicht von unterschiedlichen Inhaltsstoffen der Sorten, sondern von technologischen Bedingungen abhängig.

MÜLLER: Sie behandelten das Problem des „carry over". Vom Standpunkt des Nutzers ist ein „carry over" für die Tierproduktion durch Verschneiden uninteressant zu machen, d. h. die toxische Schwelle so stark zu unterschreiten, daß der Mykotoxingehalt unbedenklich ist.

KOCH: Dies ist technisch möglich und mitunter die einzige Möglichkeit Schäden in Grenzen zu halten, bedeutet aber, daß subtoxische Dosen und damit Folgeschäden unübersehbar mit gestreut werden.

MÜLLER: Sind Unterschiede in der Mykotoxinproduktion der Mikroorganismenart, auch das denkbare Fehlen der Mykotoxinbildung erkennbar?

KOCH: Dies kann beim gegenwärtigen Kenntnisstand über die Bildung stabiler Metabolite nicht entschieden werden.

ECKERT: Although the presence of patulin in juice of apple infected with *Penicillium expansum* may be a potential health hazzard, this problem does not appear to be a significant problem in apples that are consumed in their natural condition. Patulin does not diffuse from the disease lesion into the adjacent sound flesh of the apple. Therefore removal of the diseased lesion also remove virtually all of the patulin.

KOCH: Die Probleme treten bei der Produktion von Obstsäften und Konzentraten auf. Dazu werden die Äpfel lange gelagert und unsortiert und nicht ausgeschnitten verarbeitet.

BOCHOW: Welche Erfahrungen liegen bezüglich der Translokation von Mykotoxinen in den pilzbefallenen Substraten vor? Könnte man an Beeinflussungen der Translokation durch Antitranslokantien im Rahmen von Bekämpfungsmaßnahmen denken?

KOCH: Die Diffusion in den Medien ist sehr unterschiedlich, z.B. Patulin ist nur in der befallenen Stelle vorhanden, Aflatoxine diffundieren weit in das Medium. Für die Praxis der Probenahme ist dies sehr wichtig, für die Bekämpfung nicht.

BECKER: Wie ist ihre Meinung zur Bedeutung solcher Pilzprodukte, die wie Griseofulvin den Entwicklungsverlauf von Pathogenen hemmen und dabei die Bildung von Mykotoxinen und Antibiotika beeinflussen? Mit einer Kollegin aus Alma-Ata konnte ich zeigen, daß die Aflatoxinbildung vom Entwicklungsverlauf abhängig ist. Können Substanzen wie Griseofulvin als Schutzmittel gegen eine Mykotoxinproduktion dienen?

KOCH: Die Frage berührt ein hochinteressantes Problem, aber für eine Beantwortung ist unser gegenwärtiges Wissen zu gering.

Weiterführende Literatur

[1] WYLLIE, T. D. and L. G. MOREHOUSE,
Mycotoxic Fungi, Mycotoxins, Mycotoxicoses
An Encyclopedic Handbook
Vol. 1 Mycotoxic Fungi and Chemistry of Mycotoxins
Vol. 2 Mycotoxicoses of Domestic and Laboratory Animals, Poultry, and Aquatic Invertebrates and Vertebrates
Vol. 3 Mycotoxicoses of Man and Plants, Mycotoxin Control and Regulatory Practices
[2] MARCEL DEKKER, Inc.; New York and Basel, 1978

Joseph W. Eckert

University of California, Riverside, USA

Fungicide Treatment of Harvested Fruits and Vegetables: Opportunities and Problems

Introduction

Harvested fruits and vegetables are susceptible to attack by specific fungi and bacteria generally are distinct from the pathogens of the growing crop. For example, *Penicillium digitatum* is the most serious cause of postharvest losses in citrus fruits, but this fungus does not attack any other fruit, vegetables or growing plant. Postharvest diseases result in a substantial reduction in agricultural productivity. Fruits and vegetables are often produced on farms distant from population centers and may ripen at a time of the year when demand is weak or the market is glutted with the product. These circumstances often result in a delay of several weeks or months for storage and shipment before the product reaches the consumer. Significant deterioration may occur during this period if the product is not treated with an effective antimicrobial agent and stored in an environment unfavorable to disease development.

Diseases of fruits and vegetables observed in the market place can be traced chronologically to: 1) infection or infestation of the product during the period of its development in the field or 2) infection through wounds created during the harvesting operation and through physiological injuries that result from an unfavorable storage environment [13, 14, 16]. The subsequent progress of an infection depends upon the growth requirements and the enzymatic capabilities of the pathogen and upon the physiological resistance of the plant tissue to invasion. Thus, infections of several species of fungi, especially those initiated early in the growing season, enter a latent or quiescent state soon after penetration because of the resistance of the host tissues to further invasion. After harvest, the fruit gradually loses this resistance as it ripens and the pathogen resumes growth, giving rise to an active decay lesion on the host. *Diplodia Phomopsis*, and *Alternaria* establish latent infections in the stem-end of citrus fruits because this organ is vulnerable to attack when the fruit are very small. *Gloesporium* rots (*Pezicula* spp.) of apples arise from quiescent infections that are initiated in the lenticels of the fruit months before the rots are observed in storage. Anthracnose *(Colletotrichum)* diseases of bananas, papayas and mangoes arise from latent infections initiated randomly over the entire surface of these fruits during their development in the plantation. Gray mold *(Bortytis)* on grapes and brown rot *(Phytophthora)* of citrus fruits originate as infections that are initiated in the field shortly before harvest. These infections may not be truly quiescent, but they develop at a slow rate in mature fruit, depending upon the disease resistance of the host tissue and the ambient temperature. The control of a postharvest disease that arises from a latent or quiescent infection depends upon the penetration of an effective dosage of the fungicide to the infection site before the host is extensively colonized by the pathogen.

The second major class of postharvest diseases, from an etiological standpoint, originate by infection through injury sites that arise incidental to harvesting and preparing the product for market. These infection sites may be localized, e.g., the severed stem of a pineapple, avocado, or pear, or they may be randomly distributed over the entire surface as abrasions or cuts due to rough handling of the crop.

Methods of Postharvest Disease Control

Postharvest diseases may be controlled by maintaining the resistance of the host and by inhibiting the development of the pathogen by application of an antimicrobial agent. Practical methods for delaying the senescence of harvested fruits and vegetables, which is correlated with disease susceptibility, are: i) store at the lowest temperature that can be tolerated by the product; ii) reduce the concentration of oxygen in the storage atmosphere to reduce respiration of the product; iii) apply 2,4-D or gibberellic acid to retard senescence. Preservation of fresh fruits and vegetables by storage in an optimum environment is invariably combined with the use of antimicrobial agents in a program to minimize losses due to postharvest diseases. This approach minimizes the expense and complications associated with the use of excessive quantities of chemical preservatives on foods.

Properties and applications of postharvest fungicides

Fungicides and other chemical agents used for practical control of postharvest diseases of fresh fruits and vegetables are shown in Table 1 and Figure 1. These treatments perform one or more of the following functions: i) prevent the development of latent

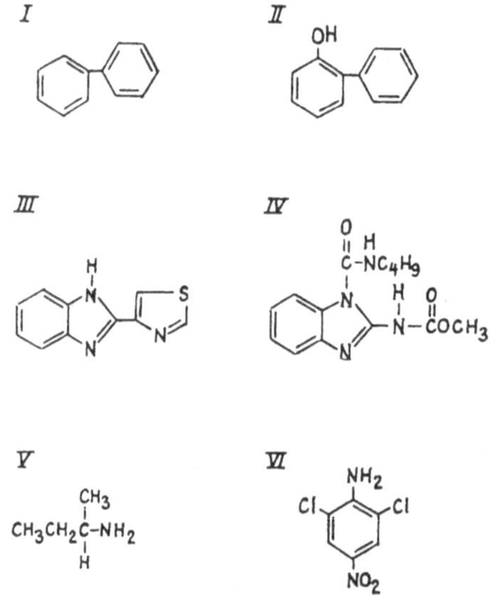

Fig. 1. Fungicides in practical use against postharvest diseases. I — biphenyl, II — *o*-phenylphenol, III — thiabendazole, IV — benomyl, V — *sec*-butylamine, VI — dichloran

Tabelle 1
Practical treatment for control postharvest disease of fruits and vegetables

Crop	Treatment
Apple and pear	sodium *o*-phenylphenate
	thiabendazole
	benomyl
Banana fruits	thiabendazole
	benomyl
	thiophanate methyl
	imazalil
Citrus fruits	sodium *o*-phenylphenate
	thiabendazole
	benomyl
	sec-butylamine
	biphenyl
	2,4-dichlorophenoxyacetic acid
	imazalil
Grape	sulfur dioxide fumigation
Mango and papaya	thiabendazole
	benomyl
Melons	sodium dimethyldithiocarbamate
Peach, nectarine, cherry	dichloran
	benomyl
Pineapple	sodium *o*-phenylphenate
	benomyl
Potato	thiabendazole
	benomyl
	sec-butylamine
Sweet potato	dichloran
	sodium *o*-phenylphenate
Tomato	dichloran

infections; ii) prevent the development of pathogenic fungi in harvest-related injuries; iii) protect the surface of the product against infection through injuries caused by handling the product after application of the fungicide; iv) inhibit the sporulation and spread of the pathogen from diseased fruit.

The selection of a fungicide for a specific application depends upon: i) the sensitivity of the pathogenic fungus to the chemical agent; ii) the ability of the chemical agent to penetrate to the infection site in the host tissue; iii) the tolerance of the agricultural product to the chemical agent, both from the standpoint of phytotoxicity and the effect of the treatment upon the sensory properties of the product. Few postharvest treatments have been developed that are effective against Phycomycetes, *Alternaria*, *Geotrichum* or bacteria; thus, diseases incited by these microorganisms are difficult to control today. Many compounds that exhibit in vitro activity against these pathogens do not prevent their development in the host, apparently because these fungicides are unable to penetrate to the sites of infection. Postharvest fungicides of the first generation, e.g., sodium o-phenylphenate, dichloran and *sec*-butylamine are effective in preventing decay by wound-invading pathogens, e.g., *Penicillium* and *Rhizopus*, but have little effect upon the development of deep-seated infections. Fungicides developed since 1965 have shown excellent protective, eradicant and antisporulation action.

Fungicides may be applied to harvested products either as gases or in water or wax formulations, depending upon the physical and chemical properties of the fungicide and the convenience of each method of application. Fumigation is advantageous for treatment of commodities such as strawberries and other highly perishable commodities that are not normally washed after harvest and may be adversely affected by contact with water. Today, most fungicide treatments are applied in water-base formulations that are sprayed or flooded onto the product as it is mechanically conveyed through the packing house. Formulations of fungicides in water-base waxes and in hydrocarbon-solvent-base waxes are in popular use also.

Fungicides applied in water-base formulations to prevent infection at superficial injuries may be classified into two categories according to their distribution on the surface of the treated product: i) non-ionic water insoluble compounds (e.g., benomyl) which are applied in a manner that produces a uniform coverage of the agent over the entire surface of the product; ii) water soluble salts such as sodium *o*-phenylphenate, *sec*-butylammonium phosphate, and sodium carbonate which are applied in relatively concentrated solution (0.5—3%) in water [13, 15]. The injuries on the fruit surface absorb the aqueous solution, whereas the intact cuticle is impermeable to the ionic compound. The treated product is rinsed lightly with fresh water following fungicide treatment and most of the chemical is removed from the surface of the fruit, except for a significant residue which remains active in the injury sites [13, 17, 21]. In the sodium *o*-phenylphenate treatment, the water insoluble *o*-phenylphenol is precipitated in the acidic environment of the injured tissues which have a pH considerably below the pK_a (10.01) of *o*-phenylphenol. An advantage of uniform, non-selective fungicide coverage of the fruit is that the entire surface of the product is protected to some degree from injuries incurred after the treatment is applied. Fungicides that may be phytotoxic must be applied selectivily to potential infection sites and the excess compound must be removed by rinsing the fruit with water to prevent injury to the surface cells of the fruit. Other fungicides, such as *sec*-butylamine and thiabendazole are not phytotoxic and substantial deposites may be left on the fruit without injury. Rinsing with water after treatment significantly reduces the fungicide residue on the product, and also reduces the antimicrobial effectiveness of the treatment as well.

o-Phenylphenol (Fig. 1, II) is a broad-spectrum antimicrobial agent that is toxic to bacteria, fungi and plant cells. However, the *o*-pehnylphenate anion does not readily penetrate the hydrophobic surface barriers of many fruits and therefore, may be safely applied to them for control of postharvest diseases [21]. Solutions of sodium *o*-phenylphenate have been used extensively for control of postharvest diseases of citrus fruits, apples, pears, peaches, sweet potatoes and other perishable fruits and vegetables [13, 14, 15, 16]. The benefits of treating fruits and vegetables with sodium *o*-phenylphenate are two-fold: i) fungus spores and bacteria on the surface of the fruit or in the cleaning solution are killed; ii) a residue of *o*-phenylphenol is deposited in superficial injuries on the product, preventing infection at these sites during storage and marketing of the crop. Krochta et al. [30] demonstrated that a foam prepared from a solution of 0.5% sodium *o*-phenylphenate and 0.3% sodium laurylsulfate was highly effective as a cushion for mechanically harvested tomatoes as they dropped into a bin and also prevented decay of fruit caused by *Geotrichum* and *Rhizopus*. These pathogens are not senitive to benzimidazole fungicides.

Biphenyl has been added to the packaging material for citrus fruits for the past 40 years to provide vapor-phase inhibition of sporulation of *Penicillium digitatum* and *P. italicum* during long distance shipment and marketing of the crop (Fig. 4). Despite these many years of successful use, biphenyl has three recognized problems that limit its value as a postharvest treatment [13, 15]: i) treated fruit may have a temporary "chemical" odor; ii) residues of biphenyl on treated fruit may exceed the 70 mg/kg tolerance in Europe and Japan, especially if the packages of fruit are not adequately cooled during shipment; iii) the biophenyl treatment is often ineffective when used on fruit that were previously treated with sodium *o*-phenylphenate. A low residue of *o*-phenylphenol on the fruit can suppress the growth *o*-phenylphenol/biphenyl-sensitive strains of *Penicillium* spp., permitting the proliferation of *o*-phenylphenol/biphenyl-resistant strains that exist at a low level in the natural population of *Penicillium* [27].

sec-Butylamine (Fig. 1, V) may be applied to harvested crops either as a fumigation treatment or as a solution of a salt of *sec*-butylamine [17, 18]. The fumigation treatment controls *Penicillium* on citrus fruits because: i) the injuries on the fruit surface become alkaline after absorption of the amine vapor and ii) fungistatic residues of neutral *sec*-butylamine salts persist in the injuries after the fumigation treatment is terminated [18]. The *sec*-butylammonium cation is uniquely fungistatic to several species of fungi [17, 22]. The optical isomer (—) *sec*-butylamine is substantially more active than the (+) isomer, although the racemic mixture is used in practice to control postharvest diseases. Fumigation of seed and ware potatoes in the UK with *sec*-butylamine controls *Oospora* and *Phoma* during storage [15, 26]. Gaseous *sec*-butylamine penetrates deeply into the potatoes so that the fumigation can be delayed for 14 days after harvest without diminishing the effectiveness of the treatment. However, the treatment is not effective against *Fusarium* dry rot.

In California, oranges are drenched with a solution of 1% *sec*-butylamine (phosphate salt, pH ca. 9) after harvest to protect the fruit injuries against infection by *Penicillium* spp. during the ethylene degreening period or during storage [16]. The degreening operation is highly conductive to the development of decay since it involves exposure of the fruit to a low concentration of ethylene gas (5 ppm) for several days in a warm humid environment. *Sec*-butylamine is also added to wax formulations applied to lemons before storage to control *Penicillium*, but the treatment has no effect upon diseases caused by *Alternaria* or *Geotrichum*, which are resistant to *sec*-butylamine as well as to the benzimidazole fungicides. The *sec*-butylamine treatment may be followed by a post-storage treatment with a benzimidazole fungicide or biphenyl because these three fungicides do not select for cross-resistant strains of *Penicillium*. *Penicillium expansum* and *Pezicula* spp. were controlled on apples treated with solutions of *sec*-butylamine salts [3]. The effectiveness against Penicillium was diminished, however, when treated fruit stored for several months at low temperatures [29].

Dichloran (Fig. 1, VI) is the most effective fungicide for control of *Rhizopus stolonifer*, a wound-invading pathogen of many fruits and vegetables. Cherries, peaches and other stone fruits are usually treated with this fungicide before shipment to distant markets [36, 49]. Dichloran is much less active against *Monilinia fructicola*, another major postharvest pathogen of stone fruits, and is only weakly active against other *Rhizopus* spp. and *Mucor* spp. [40]. Other fungicides must be combined with dichloran when these pathogens are a problem. Dichloran is the standard postharvest treatment for control of *Rhizopus* rot of sweet potatoes [45], but the treatment is ineffective against *Ceratocystis* black rot or *Penicillium* rots. Recent trials have demonstrated the value of dichloran applied to tomatoes after harvest to control *Botrytis* [8] and *Rhizopus* rots [30]. The treatment had no infleunce upon *Geotrichum* rot, however.

Fig. 2. Transformation of benomyl and thiophanate-methyl to carbendazim

Benzimidazole fungicides, thiabendazole, benomyl, carbendazim, and thiophanate methyl (Fig. 1-III, IV; Fig. 2) have been used extensively to control postharvest diseases since these compounds were introduced in the late 1960's [15]. All of these compounds and their derivatives share a common antifungal spectrum, although they differ quantitatively in activity [20, 48]. In general, the benzimidazole fungicides have been used interchangeably for certain postharvest applications while, in other cases, a specific compound is preferred because of its biological or chemical properties [3, 4, 15, 34]. The benzimidazole fungicides have been used world-wide to control *Penicillium* decay and stem-end rots of citrus fruits [6, 16, 34], *Penicillium* blue mold and *Gloeosporium (Pezicula)* lenticel rots of apples [2, 3, 7, 24, 29, 42, 43], Crown rot *(Fusarium, Ceratocystis)* and anthracnose on bananas, *Colletotrichum* and *Ceratocystis* on several tropical fruits and *Monilinia* rot on stone fruits (peaches, cherries, plums) [4, 15, 41, 49]. Benomyl is usually combined with dichloran for treatment of stone fruits in order to control both *Monilinia* and *Rhizopus*, the major postharvest diseases of these fruits. The success of the benzimidazole fungicides in the control of postharvest diseases is due to two factors: i) high intrinsic activity of these compounds against many of the fungi responsible for postharvest diseases and ii) ability of the benzimidazole fungicides to penetrate at least some degree through the hydrophobic surface barriers of the host, in order to reach the site of infection. Benomyl appears to penetrate the host surface more efficiently than thiabendazole, carbendazim, or thiophanate-methyl and may be more effective in disease situations in which the pathogen is situated below the surface of the host [1, 19, 23, 41].

Numerous trials conducted in the US and in Europe have demonstrated that the benzimidazole fungicides can control several diseases of potato tubers during storage-dry rot *(Fusarium)*, gangrene *(Phoma)*, skin spot *(Oospora)* and silver scurf *(Helminthosporium)* [5, 35, 38]. Thiabendazole appears to be somewhat more effective than benomyl for the control of *Fusarium* and *Phoma*. Thiabendazole is applied at 42 g/metric ton by means of an ultra low-volume mist applicator as the potatoes are transported into storage [35]. High-volume applications may be more effective, but the residual water on the potatoes in storage may increase the risk of bacterial soft rot. A similar treatment of sugar beets has been effective in reducing decay by *Penicillium* and other fungi that cause a sugar loss in the beets during storage in outdoor piles.

Treatment of citrus fruits with a wax formulation containing 4–6 g thiabendazole/liter results in a residue that inhibits sporulation of *Penicillium* spp. on decaying fruits [19, 23]. Benomyl produces this same effect at about one-half the dosage required for thiabendazole. The significance of this observation is that the benzimidazole fungicides offer an effective alternative to the biphenyl treatment for the control of *Penicillium* sporulation on citrus fruits during long-term storage and shipment (Fig. 4).

Thiabendazole and carbendazim are stable under most conditions that might be encountered in the treatment of fruits after harvest. Thiabendazole can even be applied as a "smoke" in fruit and vegetable storage rooms [12]. Benomyl and thiophanate-methyl, on the other hand, are unstable in aqueous solution and in organic solvents [9, 19]. Benomyl suspensions in pure water appear to be stable because of the very low solubility of benomyl in water (ca. 4 mg/liter). However, benomyl is not stable in very dilute aqueous solutions or in formulations of wax coatings applied to fruits and vegetables after harvest. Under neutral conditions, benomyl decomposes significantly within a few hours to methyl 2-benzimidazolecarbamate (carbendazim); under alkaline conditions benomyl is converted to 1,2,3,4-tetrahydro-3-butyl-2,4-dioxo-*s*-triazino(α)-benzimidazole (STB) within a few hours (Fig. 3) [19, 50]. Although carbenda-

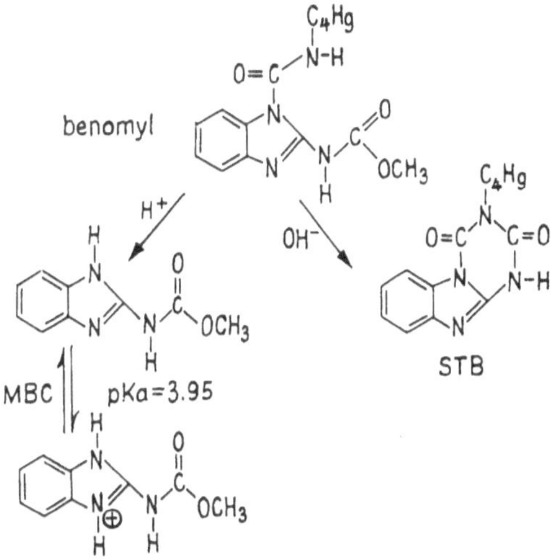

Fig. 3. Reactions of benomyl in aqueous formulations

zim is almost as active as benomyl in preventing infection of citrus fruits by *Penicillium* [34], STB is essentially devoid of antifungal properties. However, carbendazim is more polar than benomyl and is significantly less effective in controlling sporulation of *Penicillium* on citrus fruits because it does not penetrate into the peel of the fruit as efficiently as benomyl (Fig. 4) [19, 23]. Carbendazim and thiophanate-methyl, a progenitor of MBC [48], have been used as substitutes for benomyl. For some applications these compounds are equivalent in effectiveness to benomyl [3, 4, 31, 34], but for uses which require significant penetration into the host, MBC and thiophanate-methyl appear to be inferior.

All "benzimidazole" fungicides — thiabendazole, benomyl, carbendazim, and thiophanate-methyl have two serious biological limitations: i) all compounds in this group

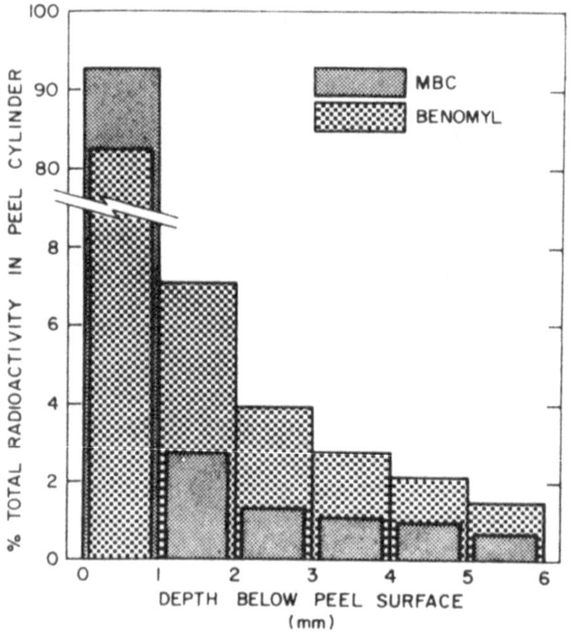

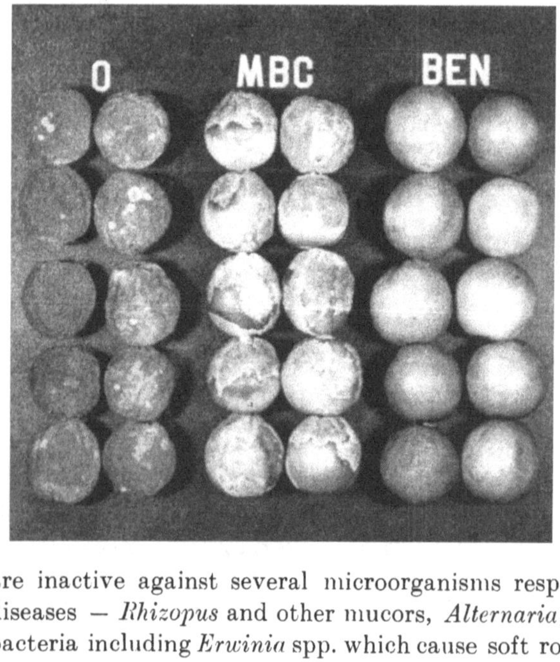

Fig. 4. Top. Penetration of ^{3}H-carbendazim (MBC) and ^{3}H-benomyl into the pericarp (peel) of an orange fruit. An equimolar amount of each compound was applied to 0.636 cm^2 areas of the fruit surface

Bottom. Growth and sporulation of *Penicillium digitatum* on orange fruits treated with equimolar concentrations of carbendazim and benomyl. 0 – untreated control; MBC – 1.25 g carbendazim per liter; BEN – 1.90 g benomyl per liter. The interiors of all fruits were extensively colonized by *P. digitatum*

are inactive against several microorganisms responsible for important postharvest diseases – *Rhizopus* and other mucors, *Alternaria*, *Geotrichum*, *Phytophthora*, and all bacteria including *Erwinia* spp. which cause soft rots of potatoes and vegetable crops. ii) resistant strains of pathogens exist at low levels in populations of "benzimidazole-sensitive" species of fungi. The resistant strains appear to be suppressed naturally by the competitive action of the normal sensitive strains, but under continuous use of the benzimidazole fungicide, the resistant strains may become the major component of the pathogen population and render the benzimidazole fungicide treatment useless.

Diseases of secondary importance caused by *Alternaria*, *Geotrichum* and other pathogens of citrus fruits that are resistant to benzimidazoles have become a limiting factor in the storage and handling of this crop after treatment with a benzimidazole fungicide [6, 15]. The resistance of strains of *Penicillium* to benzimidazole fungicides has seriously diminished the usefulness of these compounds as postharvest fungicides for citrus fruits and it now appears that this problem may also arise on other crops, following a period of continuous use of these fungicides.

Resistance to benzimidazole fungicides

The practices involved in preparing fresh lemons for market are diagramed in Figure 5, revealing the stages in this operation that favor the development of fungicide-

LEMONS
Harvest
↓
Fungicide
bath
↓
Emulsion wax
+2, 4-D +
fungicide
↓
Store 1—4 mo.
↓
Foam wash-
SOPP
↓
Organic solvent
wax + fungicide
↓
Pack in cartons
with biphenyl
↑
Ship to market

Fig. 5. Sequence of application of fungicides and other treatments to lemon fruits after harvest

resistant strains of *Penicillium*. Lemons may be washed in a bath of sodium *o*-phenylphenate and coated with a wax containing benomyl before storage. The selection pressure exerted by these treatments upon the *Penicillium* population results in the proliferation of strains that are resistant to either or both of the fungicides applied before storage. Consequently, the application of thiabendazole and biphenyl to these fruit immediately before shipment fails to control *Penicillium* decay during marketing because of the high frequency of fungicide-resistant strains.

Benzimidazole-resistance in *Penicillium* spp. was observed in California citrus packinghouses about 15 months after thiabendazole was incorporated in the wax formulation applied to lemons before storage. Harding [28] reported that thiabendazole-resistant isolates of both *P. digitatum* and *P. italicum* occurred at a high frequency in the atmosphere of lemon packinghouses that used thiabendazole intensively as a pre-storage treatment for lemons. These isolates were not controlled by treating inoculated fruit with a high dosage (3 g/liter) of thiabendazole. Benzimidazole-resistant strains of *P. digitatum* and *P. italicum* have been isolated also, in Florida, Israel, Australia and Japan. A survey of citrus fruits arriving in the port of Rotterdam revealed that approximately one-half of the isolates of *Penicillium digitatum* and *P. italicum* recovered from fruit from 20 countries was resistant to benzimidazole fungicides [37]. Benzimidazole-resistant strains of *P. expansum* have been reported on harvested apples and pears in Oregon [2], New York [42, 43] and Australia [29]. Most observations of benzimidazole-resistant strains of *Penicillium* spp. have followed intensive use of thiabendazole or benomyl over a period of some months, but KURAMOTO [31] reported a severe resistance problem on satsuma mandarins in Japan following application of a single annual spray of thiophanate-methyl immediately before harvest. Several investigators have recorded the isolation of benzimidazole-resistant strains from citrus groves and packinghouses where benzimidazole fungicides had never been used [28, 31].

Recent Developments in Postharvest Fungicides

The structures of six new fungicides that show promise against current disease problems are shown in Figure 6. Guazatine (Fig. 6, VII) is a broad spectrum fungicide that is effective against benzimidazole-resistant strains of *Penicillium* spp. and *Geotrichum* rot on citrus fruits [32]. Guazatine is a practical postharvest fungicide applied to citrus fruits in Australia.

Imazalil (Fig. 6, VIII) demonstrated, in large scale trials conducted throughout the world, high effectiveness against *Penicillium* decay and *Diplodia* stem end rot of

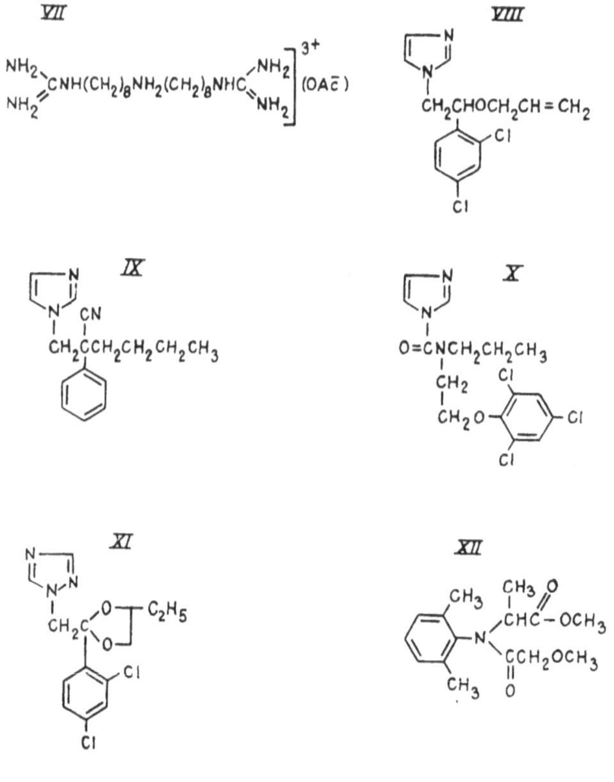

Fig. 6. New fungicides being evaluated for control of postharvest diseases. VII – guazatine, VIII – imazalil, IX – sisthane, X – prochloraz, XI – CGA-64251, XII – metalaxyl

citrus fruits [33]. Imazalil strongly inhibits sporulation of *Penicillium* spp. on citrus fruits and is fully active against strains of *Penicillium* that are resistant to benzimidazoles, *sec*-butylamine, and *o*-phenylphenol. Koffmann et al. [29] reported that imazalil was very effective in controlling benzimidazole-resistant strains of *Penicillium expansum* on apples during long term storage in a modified atmosphere. Our tests showed that imazalil was more effective than benomyl in controlling *Monilinia fructicola* (benzimidazole-sensitive strain) on sweet cherries, and others have reported good control of this pathogen on peaches [4]. Most remarkable, Spalding [44] reported the control of *Alternaria alternata* on tomatoes by treatment of inoculated fruit with imazalil.

Sisthane (Fig. 6, IX) and prochloraz (Fig. 6, X) have not been thoroughly tested against postharvest diseases, but these compounds appear to be highly effective against the same diseases as imazalil. A thorough evaluation may reveal unique applications for these compounds. A third compound in this class, CGA 64251 (Fig. 6, XI), controls benzimidazole-resistant strains of *Penicillium expansum* on apples [43] and we have shown that it is highly effective against *Penicillium* spp. on citrus as well as *Geotrichum* rot.

Metalaxyl (Fig. 6, XII) [47] has provided good control of *Phytophthora syringae* on apples [25] and we have controlled *P. citrophthora* on lemons. If applied early in the incubation period, metalaxyl can penetrate the waxy surface layer of the fruits to eradicate incipient infections of *Phytophthora* beneath. Local systemic action is essential for control of these diseases because *Phytophthora* infection is initiated in the orchard, but a postharvest fungicide treatment cannot be applied until several days later. Current practical control measures for *Phytophthora* fruit rots consist of protective fungicides (copper or captan) applied in the orchard before environmental conditions become conductive for infection.

Successes and Challenges in Postharvest Disease Control

Substantial progress has been made over the past decade in the chemical control of postharvest diseases. The benzimidazole fungicides provided a conceptual breakthrough in that they demonstrated for the first time that latent or incipient infections located within host tissue could be inactivated by chemical treatments applied to the surface of the product after harvest. Stem-end rots of citrus fruit, lenticel rots of apples, latent infections of *Colletotrichum* on bananas and other tropical fruits, and late-season infections of *Monilinia* on peaches have all been controlled by postharvest treatment of benomyl to an extent thought impossible in earlier times. Surface deposits of thiabendazole and benomyl on citrus fruits, within legal residue tolerance, can suppress the sporulation of *Penicillium* molds on the surface of citrus fruits more effectively than biphenyl, the standard fungicide for citrus shipment trade for four decades. Thiabendazole has equalled or surpassed the organomercurial fungicides in controlling certain important diseases of seed potatoes and this technology has been extended to the reduction of postharvest losses due to fungal pathogens in warehouse and fresh-pack potatoes as well. The outstanding successes of the benzimidazole fungicides in reducing storage losses in these crops has emphasized, by comparison, the inadequate control of diseases incited by microoganisms that are tolerant to this class of fungicides, i.e., *Alternaria*, *Geotrichum*, *Phytophthora*, *Mucor*, bacteria, and strains of fungal pathogens that have been selected for benzimidazole-resistance. More recent fungicide developments such as guazatine, sisthane, prochloraz and CGA 64251 have showed that diseases incited by these pathogens may also be controlled by postharvest fungicide treatment. In addition, these new fungicides have not yet selected virulent resistant strains from the pathogen population and it has been speculated that sensitive fungi do not possess the genetic capability to circumvent the biochemical lesion inflicted by these fungicides. Preliminary tests with metalaxyl have indicated that *Phytophthora* fruit rots may be successfully controlled by postharvest fungicide application of this fungicide when field conditions before harvest forecast the probability of an impending disease problem in storage. Although a few diseases of fresh fruits and vegetables cannot be satisfactorily controlled by postharvest treatments, existing fungicides have demonstrated the potential for eliminating the major pathological sources of postharvest losses. The challenge today is to develop these fungicides into truly practical and economical treatments.

Diskussion

GEORGOPOULOS: I would like to address two questions to Prof. ECKERT: First, why cross resistance to dichloran was not mentioned as one of the problems as it is in the case of cross-resistance between SOPP and biphenyl?
Second, is the effect of thiabendazole on the sugar content of sugarbeets an indirect one, i.e. does it act by prevention of sugar consumption during fungal growth?

ECKERT: SOPP-biphenyl cross resistance is a serious practical problem in citrus fruits because SOPP may be applied to these same fruits after storage. The SOPP-biphenyl resistant strains proliferate and these strains cannot be controlled with biphenyl. The treatment of fruits and vegetables with SOPP before storage and dichloran after storage is not a commercial practice; therefore cross resistance between these two fungicides is not a practical problem.
Thiabendazole reduces the loss of sugar from sugar beets during storage, presumably by reducing fungus utilisation of the beet sugar and by reducing host respiration that is stimulated by infection of the beets. I do not know which mechanism is the most important source of sugar loss in stored beets that become infected by fungi.

BOCHOW: We observed in the control of *Botrytis cinerea* on cabbage at postharvest time a difference between the efficiency of fungicides in vivo and in vitro. Some typical botryticides (thiram, glycophen, PCNB) were ineffective in vivo on cabbage leaves on the side and other fungicides, especially benzimidazoles, were effective both in vitro and in vivo. We found the only cause were not differences in penetration ability in the fungicides, but the presence or absence of the cytokinin side effect, shown in benomyl and related compounds. Cytokinins can stabilise the resistance of green plant tissue against the attack of pectolytic enzymes. In this connection we get in the case of benzimidazoles a combined effect on the resistance of plant tissue and against the pathogen, which was the real cause for the bad efficiency of benomyl in the practical control of *Botrytis* on cabbage leaves at postharvest time. Have You any indications too, in this line?

ECKERT: We have never observed this cytokinin effect of benomyl in citrus fruits treated after harvest, even in lemon fruits which were treated when they are in a green immature condition. On the other hand, it is a common commercial practice to spray oranges and lemons with gibberellic acid, when the fruit are developing on the tree for the purpose of delaying the harvest for several weeks. Fruit treated with gibberellic acid early in the season are more resistant to infection by *Penicillium* spp. after harvest. Cabbage leaf cells may be in a more juvenile state than the cells of citrus fruit pericarp and thus better respond to the action of anti-senescence agents.

I would like to point out, that, several independent investigations have shown that the pericarp of thiabendazole- and benomyl-treated grape fruit is more resistant to low-temperature (<10 °C) injury than untreated fruit. This effect of the benzimidazole fungicides must be mediated through an effect of these agents upon the plant cells.

GROSSMANN: You pointed out that benomyl penetrates into the fruits much better than carbendazim does. This may be of advantage so far as disease control is concerned. But it may cause residue patterns because the peel of oranges normally is not eaten in contrast to the other parts of the fruits. Could you comment on this situation?

ECKERT: Tolerances for pesticide residues on agricultural products are generally established on the basis of the weight of pesticide on or in the product at the time of potential consumption. No distinction is made between the pesticide which penetrates and that which remains on the surface. Thus, there is no penalty associated with the use of the pesticide that penetrates efficiently into the host for more effective disease control.

GROSSMANN: With regard to the restrictions of the use of benomyl e.g. in the Federal Republic of Germany, do You think that thiabendazol can replace benomyl in citrus disease control? How deep does it penetrate into citrus fruits?

ECKERT: I am not familiar with the exact limitations upon benomyl recently promulgated in the Federal Republic of Germany. None-the-less, thiabendazole is widely used today as a alternate fungicide to benomyl, despite the superior performance of benomyl in controlling many postharvest diseases of fruit crops. One of the reasons for the choice of thiabendazole for practical application are the limitations on benomyl residues on fruits imported into France, Belgium and Japan.

Thiabendazole appears to be more effective for control of *Fusarium* dry rot of potatoes during storage and, therefore, has been used more extensively for this application than benomyl. Both thiabendazole and carbendazim are inferior to benomyl in the their ability to penetrate into citrus fruits, presumably because the former fungicides are more polar than benomyl.

References

[1] BEN-ARIE, R., Benzimidazole penetration, distribution and persistence in post-harvest treated pears, Phytopathology **65** (1975), 1187–1189

[2] BERTRAND, P. F. and J. L. SAULIE-CARTER, The occurence of benomyl-tolerant strains of *Penicillium expansum* and *Botrytis cinerea* in the mid-Columbia region of Oregon and Washington, Plant Disease Reptr. **62** (1978), 302–305

[3] BOMPEIX, G. and F. MORGAT, Cires, anti-échaudures, fongicides et conservation des pommes, Fruits **32** (1977), 189–196

[4] BOMPEIX, G., M. COEFFIC and P. GREFFIER, Lutte contre les pourritures des pêches à *Monilia* sp., *Botrytis* sp. et *Rhizopus* sp., Fruits **34** (1979), 423–430

[5] BOYD, A. E. W., Fungicides for potato tubers, Proc. 8th British Insectic. Fungic. Conf. **3** (1975), 1035–1044

[6] Brown, G. E., Application of benzimidazole fungicides for citrus decay control, Proc. Int. Soc. Citriculture **1** (1977), 273–277
[7] Burchill, R. T. and K. L. Edney, An assessment of some new treatments for the control of rotting of stored apples, Ann. Appl. Biol. **72** (1972), 249–255
[8] Chastagner, G. A. and J. M. Ogawa, A fungicide-wax treatment to suppress *Botrytis cinerea* and protect fresh-market tomatoes, Phytopathology **69** (1979), 59–63
[9] Chiba, M. and F. Doornbos, Instability of benomyl in various conditions, Bull. Envir. Contam. Toxicol. **11** (1974), 273–274
[10] Cho, J. J., K. G. Rohrbach and W. G. Apt, Induction and chemical control of rot caused by *Ceratocystis paradoxa* on pineapples, Phytopathology **67** (1977), 700–703
[11] Couey, H. M. and G. Farias, Control of postharvest decay of papaya, HortSci. **14** (1979), 719–721
[12] Derbyshire, D. M. and M. R. Shipway, Control of post-harvest deterioration in vegetables in the UK, Outlook on Agric. **9** (1978), 246–252
[13] Eckert, J. W., Application and use of postharvest fungicides in D. C. Torgeson (ed.)/Fungicides – An Advanced Treatise, Academic Press, New York 1967, Vol. 1, 287–378, 13
[14] Eckert, J. W., Postharvest pathology – general principles in E. B. Panastico (ed.)/Postharvest Physiology, Handling and Utilization of Tropical and Subtropical Fruits and Vegetables, Avi Publ. Co., Westport, CT 1975, 393–414
[15] Eckert, J. W., Control of Postharvest diseases in Siegel, M. R. and H. D. Sisler (eds.)/Antifungal Compounds, Marcel Dekker, Inc., New York 1977, Vol. 1, 269–352
[16] Eckert, J. W., Post-harvest diseases of citrus fruits, Outlook on Agric. **9** (1978), 225–232
[17] Eckert, J. W. and Kolbezen, M. J., 2-Aminobutane salts for control of postharvest decay of citrus, apple, pear, peach, and banana fruits, Phytopathology **54** (1964), 978–986
[18] Eckert, J. W. and Kolbezen, M. J., Fumigation of fruits with 2-aminobutane to control certain postharvest diseases, Phytopathology **60** (1970), 545–550
[19] Eckert, J. W. and M. J. Kolbezen, Influence of formulation and application method on the effectiveness of benzimidazole fungicides for controlling postharvest diseases of citrus fruits, Neth. J. Pl. Path. **83**, Suppl. 1, (1977), 343–352
[20] Eckert, J. W. and M. L. Rahm, The antifungal activity of alkyl benzimidazole-2-ylcarbamates and related compounds, Pestic. Sci. **10** (1979), 473–477
[21] Eckert, J. W., M. J. Kolbezen and B. A. Kramer, Accumulation of o-phenylphenol by citrus fruits and pathogenic fungi in relation to decay control and residues, Proc. 1st Intern. Citrus Symp., Univ. Calif. (Riverside) **2** (1969), 1097–1103
[22] Eckert, J. W., M. L. Rahm and M. J. Kolbezen, Fungistatic activity of cations of non-aromatic amines, J. Agric. Food Chem. **20** (1972), 104–109
[23] Eckert, J. W., M. J. Kolbezen, M. L. Rahm and K. J. Eckard, Influence of benomyl and methyl-2-benzimidazole carbamate on the development of *Penicillium digitatum* in the pericarp of orange fruit, Phytopathology **69** (1979), 934–939
[24] Edney, K. L., Some experiments with thiabendazole and benomyl as postharvest treatments for the control of storage rots af apples, Plant Pathol. **19** (1970), 189–193
[25] Edney, K. L., Diseases caused by *Phytophthora* species. Postharvest fungicide trials Ann. Rept. East Malling Expt. Sta. 1979
[26] Graham, D., G. A. Hamilton, C. E. Quinn and A. D. Ruthven, Use of 2-aminobutane as a fumigant for control of gangrene, skin spot and silver scurf diseases of potato tubers, Potato Res. **16** (1973), 109–125
[27] Harding, P. R. Jr., Differential sensitivity to sodium o-phenylphenate by biphenyl-sensitive and biphenyl-resistant strains of *Penicillium digitatum*, Plant Disease Reptr. **46** (1962), 100–104
[28] Harding, P. J. Jr., Differential sensitivity to thiabendazole by strains of *Penicillium* italicum and *P. digitatum*, Plant Disease Reptr. **56** (1972), 256–260
[29] Koffmann, W., L. J. Penrose, A. R. Menzies, K. C. Davis and J. Kaldor, Control of benzimidazole tolerant *Penicillium expansum* in pome fruit, Scienta Hortic. **9** (1978), 31–39

[30] KROCHTA, J. M., R. A. CARLSON, J. M. OGAWA and B. T. MANJI, Harvesting into foam reduces tomato losses, Food Technol. **313** (1977), 42–46
[31] KURAMOTO, T., Resistance to benomyl and thiophanate-methyl in strains of *Penicillium digitatum* and *P. italicum* in Japan, Plant Disease Reptr. **60** (1976), 168–171
[32] KURAMOTO, T. and S. YAMADA, DF-125, a new experimental fungicide for control of satsuma mandarin postharvest decays, Plant Disease Reptr. **60** (1976), 809–812
[33] LAVILLE, E. Y., P. R. HARDING, Y. DAGAN, M. RAHAT, A. J. KRAGHT, and L. E. RIPPON, Studies on imazalil as a potential treatment for control of citrus fruits decay, Proc. Int. Soc. Citriculture **1** (1977), 269–273
[34] LÖCHER, F. and M. HAMPEL, Control of postharvest diseases of bananas, pineapples and citrus with carbendazim, Proc. 7th British Insecticide Fungicide Conf. **1** (1973), 301–308
[35] LOGAN, C., R. B. COPELAND and G. LITTLE, Potato gangrene control by ultra low volume sprays of thiabendazole, Ann. Appl. Biol. **80** (1975), 199–204
[36] LUEPSCHEN, N. S., K. G. ROHRBACH, A. C. JONES and C. L. PETERS, Methods of controlling *Rhizopus* decay and maintaining Colorado peach quality, Colorado State Univ. Expt. Sta. Bull. **547S** (1971)
[37] McDONALD, R. E., L. A. RISSE and B. M. HILLEBRAND, Resistance to thiabendazole and benomyl of *Penicillium digitatum* and *P. italicum* isolated from citrus fruit from several countries, J. Amer. Soc. Hort. Sci. **104** (1979), 333–335
[38] MEREDITH, D. S., Control of fungal diseases of seed potatoes with thiabendazole, Proc. 8th British Insectic. Fungic. Conf. **2** (1975), 581–587
[39] MUIRHEAD, I. F., Postharvest control of mango anthracnose with benomyl and hot water, Austral. J. Exptl. Agr. Animal Husbandry **16** (1976), 600–603
[40] OGAWA, J. M., J. H. MATHRE, D. J. WEBER, and S. D. LYDA, Effects of 2,6-dichloro-4-nitroaniline on *Rhizopus* species and comparison with other fungicides on control of *Rhizopus* rot of peaches, Phytopathology **53** (1963), 950–955
[41] PHILLIPS, D. J., Detection and translocation of benomyl in postharvest-treated peaches and nectarines, Phytopathology **65** (1975), 255–258
[42] ROSENBERGER, D. A. and F. W. MEYER, Benomyl-tolerant *Penicillium expansum* in apple packing houses in Eastern New York, Plant Disease Reptr. **63** (1979), 37–40
[43] ROSENBERGER, D. A., F. W. MEYER and C. V. CECILIA, Fungicide strategies for control of benomyl-tolerant *Penicillium expansum* in apple storages, Plant Disease Reptr. **63** (1979), 1033–1037
[44] SPALDING, D. H., Control of *Alternaria* rot of tomatoes by postharvest application of imazalil, Plant Disease **64** (1980), 169–171
[45] STEINBAUER, C. E. and KUSHMAN, L. J., Sweet potato culture and diseases, U.S. Dept. Agr. Handbook **388** (1971), 35–68
[46] STOVER, R. H., Banana, Plantain and Abaca Diseases, Commonwealth Mycological Institute, Kew, Surrey, England, 1972
[47] URECH, P. A., F. SCHWINN and T. STAUB, CGA 48988, a novel fungicide for the control of late blight, downey mildews, and related soil-borne diseases, Proc. 9th Brit. Insectic. Fungic. Conf. **2** (1977), 623–631
[48] VONK, J. W. and KAARS-SIJPESTEIJN, A., Methyl benzimidazol-2-yl-carbamate, the fungitoxic principle of thiophanate methyl, Pestic. Sci. **2** (1971), 160–164
[49] WADE, N. L. and P. G. GIPPS, Postharvest control of brown rot and *Rhizopus* rot in peaches with benomyl and dichloran, Australian J. Exptl. Agr. Animal Husbandry **13** (1973), 600–603
[50] WHITE, E. R., E. A. BOSE, J. M. OGAWA, B. T. MANJI, and W. W. KILGORE, Thermal and base-catalyzed hydrolysis products of the systemic fungicide, benomyl, J. Agr. Food Chem. **21** (1973), 616–618

M. Hrmová, L. Drobnica and M. Zemanová[1]

Department of Microbiology and Biochemistry of Slovak Polytechnical University and [1]Department of Microbiology of Comenius University, Bratislava, ČSSR

Induction of Mycelial Type of Development in *Candida albicans* by Fungicides and Nutritional Alternations

Besides the usually observable yeastlike (Y), i.e. round cells the true mycelial (M) forms can be present in *C. albicans*, the initial stage of which is termed germ tubes (GT). The propensity of dimorphic development in *C. albicans* is of great diagnostic value and it serves as a useful model for studying the regulation of fungal morphogenesis. The literature is very abundant in cultivation conditions conductive to M morphology (for a review see ref. [2]). Generally, the development of GT can be considered as a consequence of shift-down phenomenon (worsened nutrition). Nevertheless, till now mechanisms regulating morphogenesis in *C. albicans* remain unclear. In this connection our previous experimental results with *C. albicans* [3] stressed the importance of the growth phase of cells undergoing M morphology.

The purpose of this report was to point at the general importance of a certain physiological state of cells released for massive GT production in *C. albicans* as well as at three new GT induction techniques applicable at 28 °C. For all the methods published up to now the high cultivation temperature (37—43 °C) is the absolute prerequisite.

Figure 1 depicts the high GT production in *C. albicans*, clinical isolate No. 13 used throughout when transferring the cells into the modified glucose-mineral salts-vitamins liquid medium Vita [1] with "limited" glucose (250 mmol/l glucose concentration is in the original medium). The most pronounced conversion to M forms occurs within the use of cells from stationary growth phase (96 h cell culture). In this cultivation interval glucose was absolutely exhausted from the Vita liquid medium (residual P_i

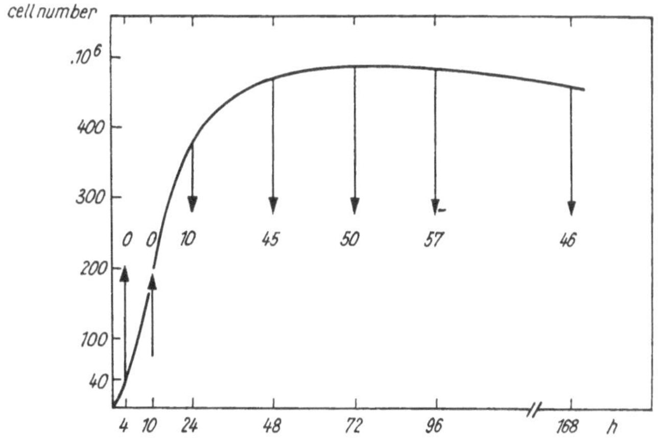

Fig. 1. Growth curve of *C. albicans* in the aerated medium Vita with 250 mmol/l glucose concentration and GT production in the same medium with "limited" glucose — 5 mmol/l. At appropriate time intervals/arrows/the cells were withdrawn, washed and transferred to medium Vita with low glucose reaching $18.2 \cdot 10^6$ cell concentration per ml. Numbers indicate the percentage of GT after 4 h cultivation at 28 °C

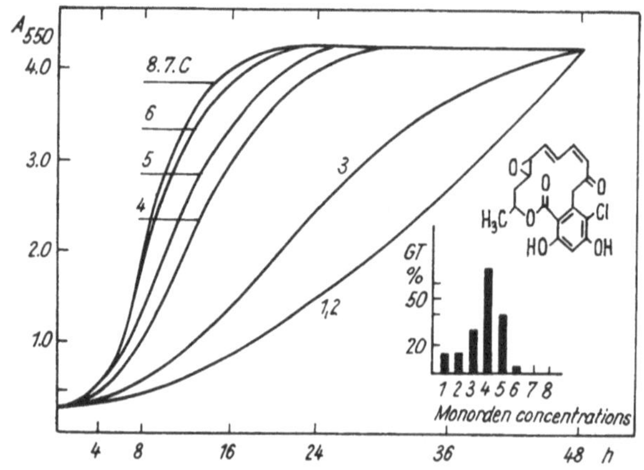

Fig. 2. Effect of monorden on the growth and GT production of *C. albicans* in the medium Vita with 100 mmol/l glucose concentration.
Monorden was added at zero hour during induction in the following concentrations: 500 [curve 1], 250 [2], 125 [3], 62.5 [4], 31.25 [5], 15.6 [6], 7.8 [7], 3.9 [8] μmol/l and without monorden being added [C]. Cells from stationary growth phase were used as inoculum [96 h culture]. The percentage of GT after 8 h shake cultivation at 28 °C in relation ot monorden concentrations is inserted

concentration was 13.8 μg per ml). However, with subsequent glucose addition regrowth appared until the deprivation of P_i from medium. It seems reasonable to consider glucose as the growth limiting factor. According to the same principle M morphology was induced in other clinical isolates and collection *C. albicans* strains (ATCC, CCY) as well. Nevertheless, for the pronounced frequency of GT formation (higher than 50%) in each strain the suitable "limited" glucose concentrations has to be determined.

In close connection with the explanation of GT induction mechanisms might have been important our finding that antibiotic monorden (radicicol) initiates M growth in *C. albicans*. Figure 2 indicates the relationship between monorden concentrations and GT formation. However, the highest dose of monorden did not fully suppress the growth of *Candida* cells. The mechanism of monorden attack is not clear at present; its effect on cell wall or cytoplasmatic membrane components syntheses is supposed. Recent results of our laboratories indicate that monorden is a thiol-combining agent.

The third new possibility for GT production resulted from experiments with N-acetyl-D-glucosamine as the sole carbon source in the synthetic medium Vita. Similarly as in previous techniques *Candida* cells from stationary growth phase were used as inoculum. With this aminosugar instead of glucose we observed at 28 °C in shake culture that on stationary phase cells, GT arose which later developed into true mycelia.

From the diagnostic standpoint of yeast and yeastlike microorganisms it is worth emphasizing the value of monorden especially for *C. albicans*. Table 1 summarizes that GT are produced only in *C. albicans* with the use of monorden and our method.

The results concerning the biochemical and physiological characterisation of Y and M forms in *C. albicans*, i.e. aerobic fermentation, respiration, RQ values, sensitivity against some specific inhibitors (iodoacetamide, uranyl nitrate, cycloheximide, carbonyl cyanide phenylhydrazones), dynamics of incorporation of radioactive precursors (glycine, leucine, adenine), radioactivity distribution in different cellular fractions after ^{14}C-glycine incorporation, chitin content in isolated cell walls and finally the results of microscopic and electronmicroscopic (ultrathin sections and freeze-etching technique) studies concerning Y and M forms in *C. albicans* are a part of Ph. D. dissertation (Hrmová, M. Morphogenesis of *C. albicans* and its regulation, Comenius University, Bratislava 1979) and are prepared for publication.

Tabelle 1
Effect of monorden on GT production in different yeast and yeastlike microorganisms. For detail see Figure 2. Monorden was applied in 62,5 µmol/l concentration and GT production was followed for 48 h

	GT		GT
S. cerevisiae, strain No. 7	—	*C. albicans*, strain Pn 10	+
S. cerevisiae, strain No. XII	—	*C. albicans*, isolate No. 1	+
S. cerevisiae, strain No. XIIa	—	*C. albicans*, isolate No. 3	+
S. cerevisiae, strain S	—	*C. albicans*, isolate No. 5	+
S. cerevisiae, strain PZ 43	—	*C. albicans*, isolate No. 8	+
S. cerevisiae, strain V_3	—	*C. albicans*, isolate No. 13	+
Debaryomyces hanseni	—	*C. utilis*	—
Hansenula anomala	—	*C. clausenii*	
		Brettanomyces anomalus	—
		Rhodotorula glutinis	—

In this short communication we found it necessary to mention the most important results concerning the time course of DNA biosynthesis and caryokinesis during bud and GT formation. In the process of budding DNA replicates at the same time as in the process of GT formation and then nuclear division occurs in both cases. The budding process is finished by septum formation between mother and newly formed daughter cell. But in the process of M development GT grows apically and with its elongation the nucleus of cylindrical outgrowth migrates to the proximity of apex. With subsequent GT elongation another caryokinesis takes place and a cross-wall is formed (division of a new terminal cell) as well as GT further elongates. At its cross-walls, according to the cultivation conditions, immediate or retarded blastospores bud off from the mycelium. The more complicated regulation of morphogenesis in *C. albicans* occurs in the case of monorden while the further M growth of cells is connected with another monorden addition.
According to us the regulation of morphogenesis in *C. albicans* depends upon the enzyme equipment of cells before transferring them into appropriate media governing Y and M development, i.e. before successive growth and DNA replication of released cells. This is emphasized mainly with biochemical and physiological differences of *Candida* cells obtained from exponential or stationary growth phases. The regulation of morphogenesis in *C. albicans* with various glucose concentrations found in our experiments we do not regard as a consequence of catabolite repression. This is in accordance with the results of SINGH and DATTA [4], that "glucose effect" or catabolite repression is absent in *C. albicans*.

Diskussion

LYR: Dimorphismus ist bei Pilzen weit verbreitet. Es scheint ein allgemein verbreitetes Regulationssystem zu geben, wodurch entweder Hefe- oder Myzelwachstum induziert wird. Ich werde in meinem Vortrag eine Arbeitshypothese dazu in einem Schema vorstellen. Es sieht so aus, daß ihre Ergebnisse sich gut einfügen. Unklar ist nur der Wirkungsmechanismus von Monorden. Hemmt das Antibiotikum die Glukose-Inkorporation oder den Glukose-Metabolismus? Dann wäre der Effekt leicht erklärbar.

HRMOVÁ: On the base of literature it is supposed monorden can interfere with synthesis of cell wall and membrane components. But we have found, that monorden reacts with low molecular thiols, i.e. monorden is a thiol binding agent. It can influence several metabolic processes, i.e. glycolysis and respiratory functions according its concentration. Its primary site of action is now investigated.

References

[1] DROBNICA, L., Adaptation of yeast and bacteria to antimicrobial substances during continuous flow cultivation, Folia Microbiol. 5 (1960), 257—263
[2] HASENCLEVER, H. F., The in vitro interactions of *Candida albicans* with nonspecific serum proteins, Mycopathologia **65** (1978), 169—176
[3] HRMOVÁ, M., L. DROBNICA and M. ZEMANOVÁ, Indukcia myceliarnych foriem *Candida albicans* a jej biochemicke pozadie, XI. Konferencia o kvasinkach, Kvasný průmysl **25** (1979), 161
[4] SINGH, B. R. and A. DATTA, Glucose repression of the inducible catabolic pathway for N-acetyl-D-glucosamine in yeast, Biochem. Biophys. Res. Commun. **84** (1978), 58—64

Gy. Josepovits

Forschungsinstitut für Pflanzenschutz, Budapest, Ungarn

Intermolekulares Transmissionsvermögen der Erregungsenergie bei Fungiziden und Konsequenzen

Einführung

Chemische Prozesse verlaufen unter Änderung der freien Energie und benötigen Energie zur Aktivierung der Substrate. Diese energetische Abhängigkeit der Reaktionsfähigkeit besteht auch bei biologisch aktiven Stoffen. Ein Sonderfall des Energie-Austausches zwischen zwei Molekülen ist die intermolekulare Transmission der Erregungsenergien, die ohne Transport von Substrat oder elektrischen Ladungen abläuft. Wir untersuchten diese Erscheinung bei fungiziden Wirkstoffen. Der Energie-Austausch zwischen verschiedenen Molekülarten kann nach Steinberg [9] wie folgt untersucht werden:

1.) durch Messung der Abnahme der Fluoreszenz des Energie-Donators
2.) durch Messen der Zunahme der Fluoreszenz des Akzeptors, sofern dieser fluoresziert
3.) durch Erfassen der Sensibilisierung des photochemischen Abbaues des Akzeptors.

Von der Möglichkeit zur Gewinnung derartiger Aussagen die Photolabilität von Fungiziden zu messen, wurde bereits mehrfach Gebrauch gemacht: Arbeiten von Buchenauer [1] über photochemische Umwandlungen verschiedener System-Fungizide, die von Clark und Watkins [2] über die Photolyse des Dichlofluanids, die von Clark u. a. [3] über die Photolabilität von Triadimephon, sowie die Mitteilung von Rouchard u. a. [7] über den Abbau der Piperazin-Fungizide durch UV-Licht.

Fluoreszenzlösung durch Fungizide

Wir haben uns zuerst die Frage gestellt, ob es System-Fungizide gibt, die die Fluoreszenz anderer Verbindungen vermindern können. Unsere Ergebnisse beweisen, daß Erregungsenergie fluoreszierender Stoffe auf manche Fungizide übertragen werden kann. Beispiele sind in Abbildung 1 zusammengestellt.

Unter den Akzeptoren ist mit ADP auch ein Naturstoff. Sein Fluoreszenzlöschungsvermögen wurde mit NADH im Vergleich zur Wirkung der Fungizide gemessen. Bemerkenswert ist, daß Benomyl und Carbendazim keine signifikante Verminderung der Fluoreszenz von Eosin, Chlorophyll oder Riboflavin verursachen, wohl aber Thiabendazol. Die Fluoreszenzlöschung durch Thiabendazol muß also dem Thiazol-Ring zugeschrieben werden.

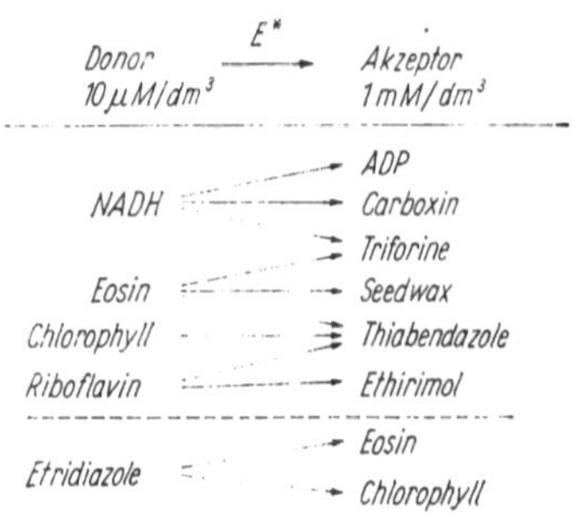

Abb. 1. Durch Fluoreszenzlöschung festgestellte Fälle intermolekularer Energie-Transmissionen

Etridiazol bildet unter den Energie-Donatoren eine Ausnahme. Dieser Wirkstoff kann die Fluoreszenz der untersuchten Verbindungen, wie Eosin etc. nicht verringern, weil sein Erregungsmaximum bei kürzeren Wellenlängen als die Maxima der Emissionsspektren dieser Verbindungen liegt. Entsprechend wurde durch spektrophotofluorimetrische Messungen nachgewiesen, daß seine Fluoreszenz durch Eosin oder Chlorophyll vermindert wird.

Zusammenhang zwischen der Fluoreszenzlöschung und Photosensibilisierung

Ein sensibilisierender Effekt von Riboflavin auf den photochemischen Abbau von Ethirimol ist schon lange bekannt. Auch SILBER u. a. [8] und BUCHENAUER [1] haben ähnliche photokatalytische Wirkungen bei Pestiziden erwähnt. Es blieb aber zu prüfen, ob Fluoreszenzlöschung und Photosensibilisierung bei einem Stoffpaar unter gleichen Umständen beobachtbar sind, ob also beide Effekte wirklich verschiedene Erscheinungsformen desselben Energie-Transportes sind.
Am Beispiel des Stoffpaares Riboflavin und Ethirimol konnten wir zeigen, daß bei bestimmten Konzentrationsverhältnissen eine Verminderung der Fluoreszenz des Riboflavins um 27% einer photochemischen Umwandlung des Ethirimols um 13% (5 Minuten) parallel geht (Tab. 1). Der Abbau von Ethirimol in der Kontrolle bleibt

Tabelle 1
Auslöschung der Fluoreszenz ($-\Delta I_f$) von Rhiboflavin und Sensibilisierung des photochemischen Abbaus von Ethirimol ($-\Delta C_e$) durch einander

Riboflavin	Ethirimol	$-\Delta I_f$	$-\Delta C_e$
10^{-5} M/dm³	10^{-4} M/dm³	27%	13%
$10^{-4,5}$ M/dm³	10^{-4} M/dm³		28%
10^{-4} M/dm³	10^{-4} M/dm³		35%
10^{-5} M/dm³	10^{-3} M/dm³	30%	

in dieser Zeit mit unter 1% unter der Nachweisgrenze. Durch Erhöhung der Riboflavin-Konzentration konnten diese Effekte noch gesteigert werden.
Einen ähnlichen Zusammenhang fanden wir in früheren Versuchen bei der Wechselwirkung von Eosin mit Alkylen-bis-dithiokarbamaten.

Kinetische Analyse der photokatalytischen Wirkungen

Eine Photosensibilisierung kommt nicht nur durch direkte Übertragung von Erregungsenergie des Donators auf den Akzeptor zustande, sondern auch durch Aktivierung des molekularen Sauerstoffs der Lösung oder durch Teilnahme des aktivierten Sensibilisators als chemischer Katalysator der Reaktion [8].
Hierzu wurden Versuche mit dem extrem photolabilen Carboxin und mit dem stabileren Ethirimol durchgeführt. Etridiazol diente in beiden Fällen als Sensibilisator, weil die UV-Absorption dieser Verbindung im kurzwelligen Bereich des Spektrums die photometrische Analyse beider anderen Substanzen nicht stört, und weil Etridiazol selbst, polarographisch gut bestimmbar ist.

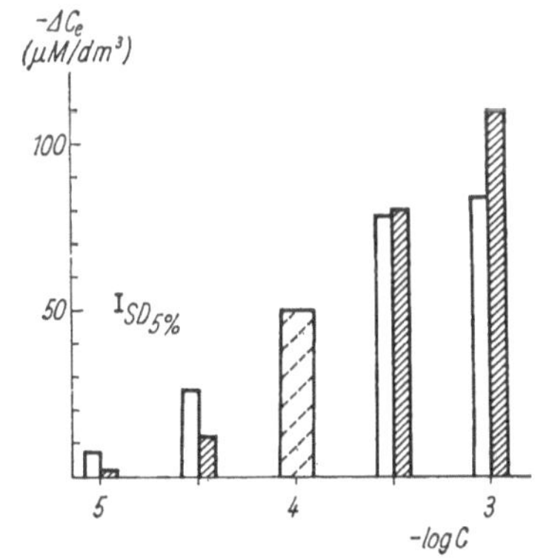

Abb. 2. Photochemische Umwandlung von Ethirimol bei Gegenwart von Etridiazol in verschiedenen Konzentrationen
Abszisse: —log Konzentration [molar] des Etridiazols
Ordinate: Verminderung der Konzentration von Ethiri mol, □ bei einer Ausgangskonzentration des Ethirimol- von 100 $\mu M/dm^3$; ▨ bei einer Ausgangskonzentrations des Ethirimols entsprechend der des Etridiazols

Abbildung 2 zeigt die Abhängigkeit der photochemischen Umwandlung des Ethirimols von der Konzentration des Sensibilisators bei konstanter bzw. bei äquimolarer Ethirimol-Konzentration. Die nach einer Bestrahlungszeit von 5 Minuten in 30 cm Abstand von der Quecksilberdampf-Lichtquelle abgebaute Ethirimol-Menge wächst mit steigenden Konzentrationen beider Verbindungen. Bei konstanter Ethirimol-Konzentration nähert sich die Umwandlung einem Grenzwert. Bei äquimolarem Anstieg der Konzentrationen findet man einen steilen Anstieg des photochemischen Effektes etwa bei 10^{-4} M/dm^3. Dieser Zusammenhang läßt sich besser auswerten, wenn die Konzentrationsskale durch eine Skale ersetzt wird, bei der die durchschnittlichen Molekülabstände abgetragen werden (Abb. 3). Eine starke Verminderung der photo-

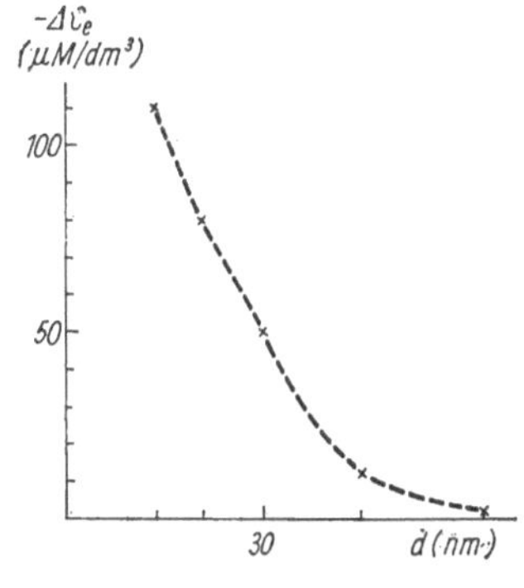

Abb. 3. Photochemische Umwandlung von Ethirimol bei Gegenwart von Etridiazol in verschiedenen äquimolaren Konzentrationen
Abszisse: Durchschnittlicher Molekül-Abstand errechnet aus den Konzentrationen
Ordinate: Verminderung der Konzentration von Ethirimol

chemischen Umwandlung ist bei einem Abstand von 30–40 nm zwischen den Donator und Akzeptor-Molekülen zu verzeichnen. Dies steht in guter Übereinstimmung mit der Theorie der Resonanz-Transmission der Erregungsenergie von FÖRSTER. Danach ist eine wirksame Energie-Transmission nur innerhalb des Abstandes von $R = \lambda/2\pi n$ gegeben [6].
Der photochemische Abbau von Carboxin in Gegenwart verschiedener Konzentrationen von Etridiazol als Sensibilisator zeigte einen ähnlichen Zusammenhang (Abb. 4). Zwischen den Konzentrationen 10^{-4} und 10^{-3} M/dm³ nimmt die Umwandlung stark zu. Es muß allerdings vermerkt werden, daß Carboxin auch ohne Sensibilisierung einen bedeutenden photochemischen Abbau erfährt.

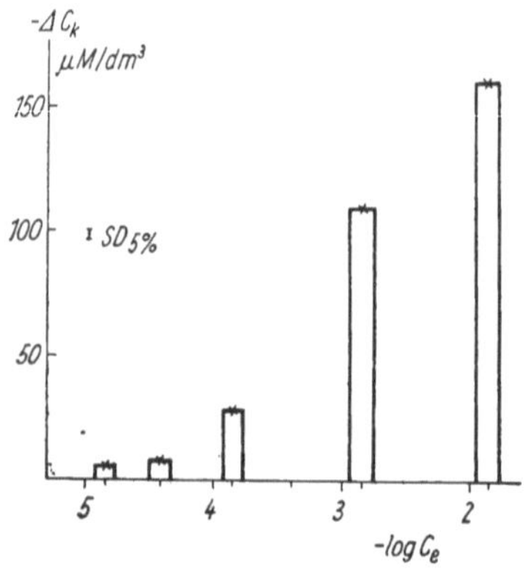

Abb. 4. Photochemischer Abbau von Carboxin bei Gegenwart von Etridiazol in verschiedenen Konzentrationen
Abszisse: –log Konzentration [molar] des Etridiazols
Ordinate: Verminderung der Konzentration des Carboxins, bei einer Ausgangskonzentration von 200 μM/dm³

Um den Einfluß des Mediums auf die Sensibilisierung überprüfen zu können, haben wir die photochemischen Umwandlungen in äthanolischer und in wässriger Lösung miteinander verglichen. Die Abnahme der Konzentration des Ethirimols war in wässrigem Milieu etwas größer als im alkoholischen, der Abbau des Carboxin zeigte umgekehrte Beeinflussung. Dies gibt einen Hinweis darauf, daß der Sauerstoff bei dieser photochemischen Umwandlung nicht mitwirkt, denn Äthanol enthält mehr gelösten Sauerstoff als Wasser.
Aus den kinetischen Versuchen geht hervor, daß der intermolekulare Energie-Austausch beim Prozeß der Photosensibilisierung der untersuchten Fungizide die entscheidende Rolle spielt. Der Sensibilisator nimmt chemisch an der Reaktion keinen Anteil. Die Photokatalyse tritt nur bei Molekül-Abständen entsprechend der Theorie von FÖRSTER auf.

Der Einfluß von Begleitstoffen auf die Sensibilisierung

Es ergibt sich nun die Frage, ob eine Photosensibilisierung auch innerhalb des Blattgewebes möglich ist. Im allgemeinen läßt sich sagen, daß weder der aufgenommene Wirkstoff noch eventuell vorhandene Sensibilisatoren im Innern des Blattgewebes eine solche Konzentration erreichen, daß die kritischen Molekülabstände zwischen Donator und Akzeptor erreicht werden. Trotzdem kann es zu Reaktionsmöglichkeiten kommen:

a) durch Komplex-Bildung zwischen Donator und Akzeptor
b) durch physikalische oder chemische Bindung beider Moleküle an eine dritte Komponente
c) durch Konzentration der Komponenten in Lipid-Phasen im Blattgewebe.

Mit der ersten Möglichkeit braucht man nur in Sonderfällen zu rechnen. Das Vorhandensein der zweiten Möglichkeit wurde an Naturstoffen nachgewiesen. Die Fluoreszenz von 10^{-5} M/dm³ NADH wurde durch 10^{-3} M/dm³ ADP stark vermindert. Zehnfache Verdünnung eliminierte diesen Effekt vollständig. Wenn nun Laktat-Dehydrogenase zu dieser verdünnten Mischung gegeben wurde, zeigte sich wieder eine Fluoreszenzlöschung. Das Enzym allein verursachte aber keine Verminderung der Fluoreszenz von NADH. Ähnlich wurde die Fluoreszenzlöschung von Eosin durch Seedwax untersucht. Oleyl-methyltaurid (Arkopon) wurde als dritte Komponente gewählt. Seine Bindung an Seedwax war früher schon nachgewiesen worden [5]. Ein Einfluß des Arkopons auf den Energie-Transport vom Eosin zum Seedwax war schon in einer Konzentration von 10^{-5} M/dm³ ausgeprägt vorhanden (Tab. 2).

Tabelle 2
Verminderung der Fluoreszenz ($-\Delta I_f$) von Eosin durch Seedwax und der Einfluß von Arkopon daran

Eosin	Seedwax	Arkopon	$-\Delta I_f$
10^{-5} M/dm³	10^{-3} M/dm³	–	40%
10^{-5} M/dm³	10^{-4} M/dm³	–	10%
10^{-5} M/dm³	10^{-4} M/dm³	10^{-5} M/dm³	42%

Wir wollten auch prüfen, ob ähnliche Einflüsse auf die Photosensibilisierung existieren. 3×10^{-5} M/dm³ Etridiazol sensibilisieren den photochemischen Abbau gleicher Ethirimol-Konzentrationen geringfügig. Wir haben dann Arkopon oder Emulsogen EL (einen nichtionogenen Emulgator vom Polyoxyäthylen-Typ) bzw. einen wässrigen Extrakt aus Weizen-Blättern zur wässrigen Ethirimol-Lösung gegeben. Die Detergentien wurden in Konzentrationen angewendet, die Micellenbildung und dadurch „Solubilisierung" der Fungizide ermöglichten. Entgegen unseren Erwartungen verminderten alle Begleitstoffe den Sensibilisierungseffekt (Tab. 3). In Versuchen von Tanaka u. a. [10] steigerte dagegen Micellenbildung den photochemischen Abbau von

Tabelle 3
Der Einfluß verschiedener Beimengungen auf den durch Ethridiazol sensibilisierten photochemischen Abbau von Ethirimol ($-\Delta C_e$).
Die Konzentration beider Fungizide = 3 × 10 μ M/dm³

Beimengungen	$-C_e$
–	21,5 μM/dm³
1 g/dm³ Arkopon	12 μM/dm³
0,5 g/dm³ Emulsogen EL	13 μM/dm³
5 g/dm³ Emulsogen EL	3 μM/dm³
3 g/dm³ Weizen-Extrakt	0 μM/dm³
$SD_{5\%}$	2 μM/dm³

Monuron. Unser negatives Ergebnis ist wahrscheinlich so zu erklären, daß die Detergenz-Micellen nur eines der beiden Fungizide“ solubilisiert“ haben. Dadurch wurden die Komponenten weiter getrennt statt angenähert. Dasselbe gilt wahrscheinlich für Weizenextrakte, die, wie wir früher bereits zeigten, Komponenten enthalten, die mit Etridiazol Addukte bilden [4].

Schlußfolgerungen

Unter den Systemfungiziden gibt es Wirkstoffe, die zu einem intermolekularen Energie-Austausch mit anderen Fungiziden oder mit anderen Verbindungen fähig sind. Das gibt die Möglichkeit eine photochemische Inaktivierung auch solcher Wirkstoffe zu sensibilisieren, die relativ stabil sind. Die praktische Bedeutung der hier näher untersuchten Beispiele besteht in der Verallgemeinerungsfähigkeit dieser Erkenntnisse und nicht so sehr in den konkreten Ergebnissen dieser Beispiele, die sich nicht auf Blatt-Fungizide beziehen.

Bei Anwendung von Wirkstoff-Kombinationen sollte man auch die Möglichkeit einer Sensibilisierung des photochemischen Abbaues des einen Wirkstoffes durch den anderen in Betracht ziehen. Auch Naturstoffe im Blattgewebe oder in Exudaten können einen photokatalytischen Effekt ausüben. Die Verhältnisse im Innern des Blattgewebes sind für die Sensibilisierung photochemischer Inaktivierungen weniger günstig als die an der Blattoberfläche. Anwesende Naturstoffe können die Sensibilisierung aber steigern oder verringern. Dies hängt vom chemischen Charakter des Wirkstoffes ab.

Die Sensibilisierung des photochemischen Abbaues bestimmter Naturstoffe (wie z.B. des Chlorophylls) durch ins Blatt aufgenommene Wirkstoffe kann eine der Ursachen der Phytotoxizität sein.

Unsere Ergebnisse beziehen sich auf die intermolekulare Übertragung von Erregungsenergien aus Lichtadsorptionen. Sie lassen die Frage offen, ob solche Wirkstoffe auch in Transmissionsprozesse von aus chemischen Quellen stammenden Erregungsenergien eingreifen können.

Danksagung

Ich danke Frau Dr. M. GASZTONYI für wertvolle Mitarbeit bei einem Teil der Versuche sowie Dr. K. SZÖKE für die Möglichkeit, spektrophotofluorimetrische Messungen im Institut für Gesundheitswesen durchführen zu können.

Literatur

[1] BUCHENAUER, H., Untersuchungen über die Stabilität verschiedener systemischer Fungizide gegenüber Einwirkung von UV- und Sonnenlicht auf Glas und Blättern von Bohnenpflanzen
Intern. Symp. Systemfungizide. Akad. Verlag, Berlin, 1975, 327

[2] CLARK, T., D. A. M. WATKINS, Photolysis of dichlofluanid, Pestic. Sci. 9 (1978), 225 to 228

[3] CLARK, T., u. a., Photolysis, metabolism and other factors influencing the performance of triadimephon as a powdery mildew fungicide, Pestic. Sci. 9 (1978), 497–506

[4] GASZTONYI, M., Faktoren, die die Selektivität der fungitoxischen Wirkung von Ethazol beeinflussen, Abhandl. d. Akad. Wiss. DDR 1979, N2, Intern. Symp. Systemfungizide, Akad. Verlag, Berlin, 1979, 207–213

[5] JOSEPOVITZ, GY., Versuche mit einigen heterozyklischen Stickstoffverbindungen hinsichtlich ihres Transportes durch Zell-Membranen
Intern. Symp. Systemfungizide. Akad. Verlag, Berlin, 1975, 279
[6] KNOX, R. S., Excitation energy transfer and migration: Theoretical considerations, in: GOVINDJEE (Ed.), Bioenergetics of photosynthesis. Acad. Press, New York, 1975, 183–221
[7] ROUCHARD, J., u. a., Photodecomposition of piperazine in water by "sunlight" ultraviolet radiation, Pestic. Sci. 9 (1978), 305–309
[8] SILBER, J., C. PREVITALI, N. SILBERA, Photoreactions of riboflavin in the presence of 2,4-dichloro-phenoxyacetic acid (2,4-D), J. Agric. Food Chem. **24** (1976), 679–680
[9] STEINBERG, I. Z., Long-range nonradiative transfer of electronic excitation energy in proteins and polypeptides, Ann. Rev. Biochem. **40** (1971), 83–114
[10] TANAKA, F. S., R. G. WIEN, E. R. MANSAGER, Effect of nonionic surfactants on the photochemistry of 3-/4-chlorophenyl-/1,1-dimethylurea in aqueous solution, J. Agric. Food Chem. 27 (1979), 774–779

D. J. Adams and G. W. Gooday

Department of Microbiology, University of Aberdeen, U. K.

Chitin Synthesis as a Target — Current Progress

There is a requirement for wide-spectrum, non-toxic antifungal compounds active against human and animal pathogens, as contemporary anti-fungal drugs are variably toxic in therapeutic dosage. Chemotherapy is a problem of comparative biochemistry in that the effectivness of a chemotherapeutic agent is dependent upon there being an exploitable difference between the pathogen and the host. The chemical nature of present day anti-fungal antibiotics, for example the polyenes, and so their effects, lack total specificity in that in addition to destroying fungal cells they also have detrimental effects on the cells of the host.

Chitin synthase is the enzyme responsible for the catalysis of the last step in chitin biosynthesis. It is found in fungi but not in mammals and so a specific inhibitor might be expected to prevent growth of fungi, but should lack host toxicity. The analogy is with the inhibition of bacterial cell wall synthesis by penicillin. There is one group of compounds with this mode of action, the polyoxins. They are powerful specific anti-fungal antibiotics that have been used widely in Japan as agricultural fungicides. They were discovered in 1963 (Suzuki et al. [11]) and the producing strain classified as *Streptomyces cacoi* var. *asoensis*. More recently nikkomycin, a closely related antibiotic, was described from a strain of *Streptomyces tendae* (Dähn et al. [6]). There is strong evidence to suggest that the primary mode of action of the polyoxins involves chitin biosynthesis. Endo et al. [7] showed that polyoxin D competitively inhibited the synthesis of chitin by hyphae of *Neurospora crassa*. All subsequent work has borne out their suggestion that the anti-fungal properties of the polyoxins are dependent upon the structural resemblence of these compounds to the substrate molecule for chitin biosynthesis, UDP-N-Acetylglucosamine (UDP-GlcNAc) (Gooday [8]).

A group of compounds with structures resembling part or all of the UDP-GlcNAc molecule was selected for study. Their inhibitory properties were tested against a solubilized chitin synthase preparation from *Coprinus cinereus* and compared with those of polyoxins. The synthesis of a polyoxin-type compound was also investigated and the inhibitory properties of the compound derived, 5'-*O*-octanoyluridine, were tested against a solubilized enzyme preparation. The anti-fungal activity of nikkomycin, which has already been described as a potent inhibitor of chitin synthase from *C. cinereus* (Brillinger [4]) was tested against a pathogenic fungus. In addition, the toxicity of nikkomycin to mammalian cells was compared with the toxicities of a range of anti-fungal drugs.

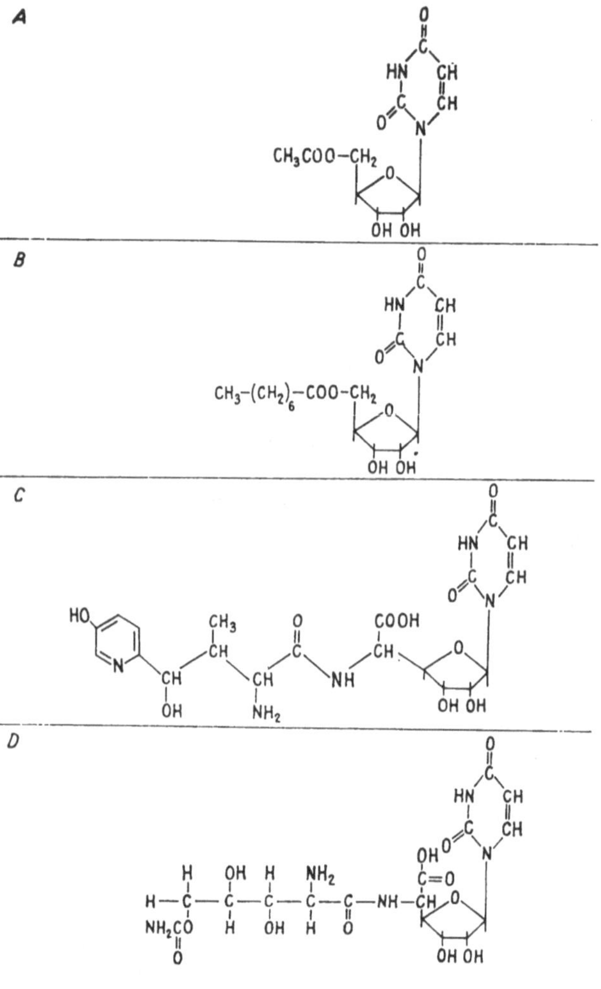

Fig. 1. Structures of uridine derivatives. A, 5′-*O*-acetyluridine; B, 5′-*O*-octanoyluridine; C, nikkomycin; D, piomycin B/polyoxin L

Materials and Methods

Preparation of chitin synthase and enzyme assays were as described by ADAMS & GOODAY [2]. Tunicamycin was dissolved in 1 M-NaOH before addition to the assay mixture and assays with 5′-*O*-Octanoyluridine contained 10% dimethylformamide. 5′-*O*-octanoyluridine was synthesized as described by ADAMS [1]. To test the effect of nikkomycin on *Candida albicans*, a suspension of yeast cells (Roche Products, Basle, Switzerland; strain 109) containing approximately 10^6 cells $ml.^{-1}$ was used to inoculate a 9 cm Petri dish of Bacto-SABOURAUD dextrose agar (Difco, Detroit, Michigan, USA). Plates were surface dried in a laminar flow cabinet and antibiotic assay discs placed in the centre of the agar. 10 or 100 μg of nikkomycin, dissolved in 20 μl of distilled water, were added to each disc. Plates were incubated at 30 °C.

For cytotoxicity testing in vitro, V79 Chinese hamster fibroblast cells (Royal Marsden Hospital, Surrey, U.K.) were used under sterile conditions. A monolayer of cells was grown until just sub-confluent in a 200 ml tissue culture flask using medium I (Dulbecco's minimal essential medium (Gibco/Biocult) containing 15% (w/v) foetal calf serum (Gibco/

Biocult) and gentamycin (Sigma; 200 µg ml^{-1})), incubated at 37 °C and 90% humidity. The medium was removed and the monolayer rinsed with saline (1.0%, w/v).
Trypsin (Difco, 0.25% w/v) in buffer (0.05 M-tri-sodium citrate; 0.5 M-NaCl; pH 7.6) was added to the flasks so that it just covered the cells. Flasks were incubated at 37 °C for 10 min. A cell suspension was prepared by adding approximately 40 ml of medium I. The suspension was transferred to two plastic universal bottles and centrifuged (50 g, 5 min). The pellet was resuspended in medium I. A single-cell suspension was ensured by vigorous pumping with a PASTEUR pipette. Cells were counted using a COULTER counter (Coulter Electronics Ltd., Dunstable, Bedfordshire, U.K.) and diluted with medium I to give a suspension containing 5 × 10^4 cells min^{-1}.
Drugs were dissolved in dimethylsulphoxide (DMSO) and added to 2 ml portions of cell suspension in the well of a COSTAR culture plate 3524 (L.H. Engineering Ltd., Stoke Poges, Buckinghamshire, UK) so that a 0.5% (v/v) solution of DMSO was prepared. Control incubations contained only DMSO. The plates were incubated at 37 °C and 90% humidity for 4–5 days or until there was adequate growth in the control wells. The medium was discarded and the wells were washed four times with saline (1.0%, w/v). 1 ml of a solution containing 0.19 M-Na_2CO_3, 0.1 M-NaOH, 0.4 mM-$CuSO_4 \cdot 5\,H_2O$, 0.9 mM-potassium sodium tartrate and 0.3 mM-sodium dodecyl sulphate in distilled water was dispensed into each well. Plates were incubated at 37 °C for 30 min or until all the cell layers were digested. FOLIN and CIOCALTEU's phenol reagent (0.1 ml) was added to each well with thorough mixing and plates incubated at 37 °C for 60 min. $E_{660\,nm}$ was read in a Pye Unicam SP 1800 spectrophotometer. To compensate for any residual protein originating from the growth medium, the instrument was adjusted to zero against digests from cell-free controls. A graph was drawn of $E_{660\,nm}$ against drug concentration (see Fig. 3). The amount of drug required to reduce cell protein concentration by half was called the I_{50} of that drug.

Results and Discussion

Structural analogues of UDP-GlcNAc significantly inhibited the chitin synthase activity from *C. cinereus* (Table 1). The natural products uridine, UDP, polyoxin L

Table 1
Effect of natural and synthetic derivaties of uridine on chitin synthase activity

Compound	Concentration (mM)	% Inhibition
5′-*O*-Acetyluridine	3.5	60
5′-Amino-5′-deoxy-5-fluouridine	4.0	41
5′-Amino-2′,5′-dideoxy-5-fluorouridine	3.8	0
1-(5′-Deoxy-α-*L*-lyxofuranosyl)-5-fluorouracil	4.0	19
1-(5′-Deoxy-β-*D*-arabinofuranosyl)-5-fluorouracil	4.0	63
1-(5′-Deoxy-β-*D*-lyxofuranosyl)-5-fluorouracil	4.0	18
2′-Deoxy-5-fluoro-3-methyl-α-uridine	3.8	0
2′-Deoxy-5-fluorouridine	4.0	45
5′-Deoxy-5-fluorouridine	4.0	56
2′-Deoxyuridine	4.4	60
5′-Deoxyuridine	4.4	64
5-Fluorouridine	4.0	87
Nikkomycin	2.0	99
5′-*O*-Octanoyluridine	2.7	80
Piomycin B/Polyoxin L	4.0	99
Tunicamycin	0.2	50
Uridine	4.0	93
Uridine-5′-diphosphate	2.1	99
Uridine-5′-monocarboxymethyl disodium phosphate	4.0	44

and nikkomycin were the most inhibitory, but considerable inhibition was observed with the synthetic derivatives 2′ and 5′-deoxyuridine, 5-fluorouridine, 5′-*O*-acetyluridine and 5′-*O*-octanoyluridine.

The effectiveness of uridine and its derivatives in inhibiting enzyme activity is in keeping with the results of HORI et al. [9]. Using a particulate chitin synthase preparation from *Piricularia oryzae* they investigated the active site of the enzyme by examining the effect of various derivatives of the polyoxins and uridine analogues on chitin synthase activity, assuming them to be competitive inhibitors. They suggested that the oxygen on the C-4 position of the uracil of UDP-GlcNAc forms a hydrogen bond with an amino group on the protein molecule, and that the hydroxyl groups at C-2′ and C-3′ of the ribose moiety of UDP-GlcNAc form hydrogen bonds to/with the enzyme molecule. This correlates with our results from affinity chromatography which suggest that the initial interaction of the enzyme with UDP-GlcNAc involves the uridine moiety of the substrate molecule (ADAMS & GOODAY [3]).

HORI et al. also suggested that in the catalytic process an unprotonated imidazole group of a histidine residue on the enzyme interacts with the pyrophosphate moiety of UDP-GlcNAc. The specificity of this interaction was investigated by synthesizing 5′-*O*-octanoyluridine, a structure with an inert chain of unsubstituted carbon atoms to replace the pyrophosphate moiety of UDP. This compound imposed a level of inhibition similar to that of uridine and less than that of UDP, nikkomycin and polyoxin L (Table 1). The Ki for polyoxin L was 10 μM (Fig. 2). It appears that the analogue of the pyrophosphate moiety of UDP-GlcNAc in nikkomycin and polyoxin more closely resembles that component of the substrate molecule than does the acyl side chain of the octanoyluridine.

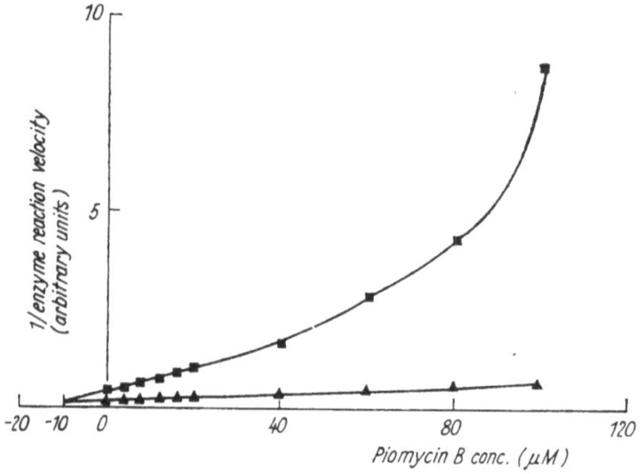

Fig. 2. Dixon plot of inhibition of chitin synthase by piomycin B/polyoxin L. ■, 0.6 and ▲ 2 mM-UDP-GlcNAc

The antibiotic tunicamycin also inhibited chitin synthase activity. SELITRENNIKOFF [10] showed that tunicamycin was an apparent competitive inhibitor of chitin synthase from *Neurospora crassa* with a Ki of approximately 480 μM.

Nikkomycin inhibited the growth of *Candida albicans*. One day after inoculation, growth of the fungus was inhibited around the assay discs (Table 2). If the inhibition imposed is directed at chitin synthesis, then this is of particular significance since the

Table 2
Effect of nikkomycin on growth of Candida albicans

Nikkomycin (µg/disc)	Diameter of inhibition zone (mm)
0	0
10	13
100	25

cell wall of this yeast form of *C. albicans* has a relatively low chitin content (CHATTAWAY et al. [5]).

The toxicity tests against mammalian cells showed that nikkomycin was effectively non-toxic in comparison with the other anti-fungal drugs tested (Fig. 3 and Table 3).

Table 3
Cytotoxicity of Antifungal Compunds

Drug	I_{50} (µg ml^{-1})
Griseofulvin	5.2
Amphotericin B	9.4
Miconazole	10.2
Nystatin	92
Nikkomycin	>900

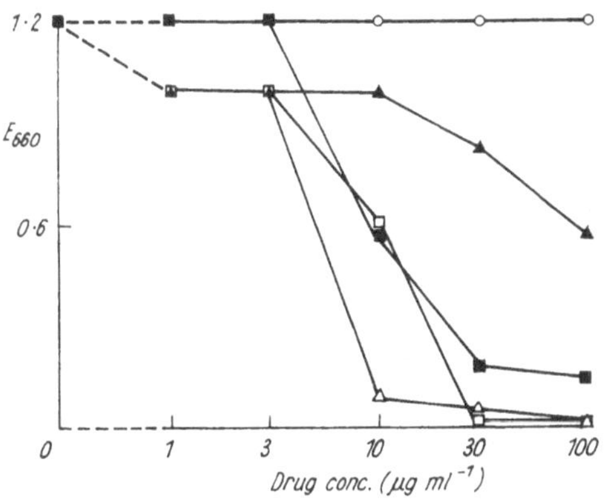

Fig. 3. Cytotoxicity tests of antifungal drugs. With control value here of 1.2 units, I_{50} is concentration at E_{660} of 0.6 O, nikkomycin; ▲ nystatin; □, miconazole; ■, amphotericin B; △ griseofulvin

Although the synthetic analogues tested here have not proved as inhibitory to the enzyme as the antibiotics polyoxin and nikkomycin, nevertheless the results show that this line of research is worth pursuing. The results of the inhibition of enzyme activity and fungal growth, and the non-toxicity to mammalian cells of the compounds discussed here shows us that the goal — of the devising of specific non-toxic antifungal agents — should be attainable in the forseeable future.

Diskussion

PFLIEGEL: Why have dimilin and its analogues no fungicidal activity? They are said to be inhibitors of chitin synthase.

ADAMS: I do not know the mode of action of dimilin although I have been told that it inhibits chitin synthase activity in insects.

LYR: Dimilin und ähnliche Verbindungen haben keine Wirkung auf die Chitinsynthetase. Der Wirkungsmechanismus ist offensichtlich ganz verschieden von Polyoxin.

ADAMS: On consideration of the structure of dimilin it appears unlikely, that it inhibits chitin synthase by acting as a substrate analogue. Indeed the mode of inhibition of dimilin is probably non-specific .As we have discussed it can inhibit chitinase activity and perhaps its in vivo inhibitory properties manifest themselves as a result of this alteration in the balance between chitin synthesis and lysis during growth.

SISLER: A recent paper in Pesticide Biochemistry and Physiology shows that polyoxin prevents chitin synthesis in insects. The effect is similar to that produced by diflubenzuron. Do these compounds act in the same manner?

ADAMS: So far I am aware, the mode of inhibition of chitin synthesis by diflubenzuron is unknown. However it seems unlikely that it imposes competitive inhibition in the manner of the polyoxin range of compounds.

HRMOVÁ: Have You investigated some physico-chemical properties of your compounds. If the substance is lipophylic, it can penetrate very quickly through the cell membrane, i.e. it is more effective than the hydrophylic one.

ADAMS: We have not examined the physico-chemical properties of our compounds in detail. It would certainly be very interesting to establish the affinity of, for example 5'-octanoyluridine, for cell membranes.

GEORGOPOULOS: In view of the work done in Japan which showed that polyoxins compete with dipeptides for the same permease, I would like to comment on the possibility of chitin synthase inhibition by a structure, the uptake of which would not be inhibited by dipeptides?

LYR: Wie Dr. GEORGOPOULOS erwähnte, fanden japanische Forscher, daß Polyoxin eine spezifische Permease braucht, um in die Zelle aufgenommen zu werden. Das ist aber ein großer Nachteil für eine Verbindung, da Resistenzmutationen leicht eintreten können. Es würde aber gut die Selektivität erklären. Die Permease müßte artspezifisch unterschiedlich auftreten.

ADAMS: As Professor LYR said, it does seem a disadvantage for an inhibitor to require transport into the cell. Perhaps a group of chitin synthase inhibitors which readily diffuse into the fungal cell should be saught. However, it is difficult to envisage how such compounds could impose the specific inhibitors achieved by polyoxins.

ECKERT: Is polyoxin-resistance in fungi dependent upon an alteration in the polyoxin-transport system?

What is the evidence that the active site of chitin synthetase is located within the cell permeability barrier. Presumably the enzyme must be closely associated with the elongating chitin microfibril.

ADAMS: At the last Reinhardsbrunn symposium Dr. GOODAY suggested that polyoxin resistance was dependent upon alteration in the polyoxin transport system for two reasons: Polyoxin-resistant strains of fungi that have arisen amongst polyoxin-treated populations in the field have chitin synthases just as susceptible to inhibition by polyoxin; secondly, simple peptides strongly antagonise the action of polyoxin in vivo, but have no effect on the activity of chitin synthase in vitro. That is, if the entry of the polyoxin into the cells is the site of antagonism by the peptides, polyoxin resistant strains may achieve as a result of an alteration in their peptide transport system.

On extraction, chitin synthase is generally associated with plasmamembrane fractions. Experiments including the effects of toxic compounds on chitin synthesis in protoplasts suggest that the active site of the enzyme is located on the inner surface of the plasmamembrane. The evidence suggests therefore that chitin synthase is an integral protein of the cell membrane, interacting with UDP-GlcNAc in the cytosol and forming the chitin microfibril on the outer face of the membrane.

LYR: Gibt es Unterschiede im Abbau von Polyoxin und Nikkomycin innerhalb einer Pilzzelle? Ist Nikkomycin stabiler?

ADAMS: We have no information on the comparative degradation of polyoxin and nikkomycin. Perhaps nikkomycin is an effective inhibitor of *Candida albicans* in vitro because it is a comparatively stable compound.

Literature

[1] ADAMS, D. J., Chitin synthase of *Coprinus cinereus*, Ph. D. Thesis, University of Aberdeen, U.K. 1980

[2] ADAMS, D. J. & G. W. GOODAY, A rapid chitin synthase preparation for the assay of potential fungicides and insecticides, Biotechnology Letters **2** (1980), 75–78

[3] ADAMS, D. J. & G. W. GOODAY, Partial purification and characterisation of chitin synthase from *Coprinus cinereus*, Abs. S7–9, Special FEBS Meeting on Enzymes, Dubrovnik, Yugoslavia, April 1979

[4] BRILLINGER, G. U., Metabolic products of microorganisms 181. Chitin synthase from fungi: a test model for substances with insecticidal properties, Archives of Microbiology **121** (1979), 71–74

[5] CHATTAWAY, F. W., M. R. HOLMES, & J. E. BARTOW, Cell wall composition of the mycelial and blastopore forms of *Candida albicans*, Journal of General Microbiology **51** (1968), 367–376

[6] DÄHN, U., H. HAGENMAIER, H. HOHNE, W. A. KONIG, G. WOLF, & H. ZÄHNER, Stoffwechselprodukte von Mikroorganismen 154 Mitteilung. Nikkomycin, ein neuer Hemmstoff der Chitinsynthese bei Pilzen, Archives of Microbiology **107** (1976), 143–160

[7] ENDO, A., K. KAKIKI, & T. MISATO, Mechanism of action of the antifungal agent polyoxin D , Journal of Bacteriology **104** (1970), 189–196

[8] GOODAY, G. W., The action of polyoxin on fungi, Abhandlungen Der Akademie Der DDR Nr. 2N "Systemfungizide" (1979), 159–168

[9] HORI, M., K. KAKIKI, & T. MISATO, Interaction between polyoxin and the active centre of chitin synthase, Agricultural Biological Chemistry **38** (1974), 699–705

[10] SELITRENNIKOFF, C. P., Competitive inhibition of *Neurospora crassa* chitin synthase activity by tunicamycin, Archives of Biochemistry and Biophysics **195** (1979), 234–244

[11] SUZUKI, S., K. ISONO, J. NAGUTSA, T. MIZUTANI, Y. KAWASHIMA, & T. MIZUNO, A new antibiotic, polyoxin A, Journal of Antibiotics **18** (1975), 131–132

Acknowledgements

We thank Mrs. JANE HAWKES, Mr. R. MIDDLETON and Mr. E. BARBER for their advice and able technical assistance, and Roche Products for financial support.

St. Neumann und F. Jacob

Sektion Biowissenschaften der Martin-Luther-Universität Halle-Wittenberg, DDR

Neue Ergebnisse zur Carboxylhypothese des Phloemtransports von Xenobiotika

Der gezielte Einsatz von Pflanzenschutzmitteln erfordert in Abhängigkeit von der erwünschten Wirkung bestimmte Mobilitätseigenschaften der eingesetzten Verbindungen. Der Erfolg einer Wirkstoffbehandlung hängt in entscheidenem Maße davon ab, daß die Angriffsorte von den Substanzen zum geeigneten Zeitpunkt in ausreichender Konzentration erreicht werden. Besondere Bedeutung besitzen in diesem Zusammenhang Verbindungen, die gemeinsam mit den Assimilaten über das Phloem in der Pflanze verteilt werden und alle wachsenden und speichernden Teile oberhalb und unterhalb der Fütterungsstelle erreichen können. Sie gelangen auch in die nach der Applikation hinzuwachsenden Blätter, die ebenso wie andere wenig transpirierende Organe von xylemmobilen Substanzen nach Wurzelapplikation nicht erreicht werden. Die Zahl der phloemmobilen Xenobiotika ist jedoch außerordentlich gering. Die bisher bekannten Fungizide, Insektizide und Nematizide sind, wie Übersichten zeigen, fast ausschließlich xylemmobil, nur bei den Herbiziden finden wir einige Gruppen mit phloemmobilen bzw. ambimobilen Eigenschaften [7, 8, 1, 3]. Alle diese Verbindungen besitzen als gemeinsames strukturelles Merkmal eine Carboxylgruppe im Molekül. Offensichtlich erleichtert diese Struktur den entsprechenden Verbindungen den Eintritt in das Phloem. Dieser von unserer Arbeitsgruppe auf Grund einer Literaturrecherche gefundene Zusammenhang führte zur Formulierung der Carboxylhypothese des Phloemtransportes von Xenobiotika, einer Arbeitshyptohese, die sich in den letzten Jahren als außerordentlich fruchtbar erwiesen hat. Alle von uns getesteten Substanzen zeigten nur dann einen Transport im Phloem, wenn sie eine -COOH-Gruppe im Molekül besaßen bzw. Metabolite mit dieser Gruppierung bilden konnten. Unabhängig von uns stellte Crisp [1] ähnliche Zusammenhänge zwischen chemischer Struktur und Phloemmobilität bei Insektiziden fest. In Analogie zu der von Mitchell u. a. [5] vorgenommenen Umwandlung des xylemmobilen Chlorprophams in ein phloemmobiles Milchsäurekonjugat gelang es ihm, durch Konjugation des xylemmobilen Insektizids Matazil mit Milchsäure eine phloemmobile Verbindung zu erhalten. Eine wesentliche Stütze der Hypothese stellen Experimente dar, bei denen die Mobilität bekannter Wirkstoffe mit der ihrer carboxylgruppen-freien bzw. carboxylgruppenhaltigen Derivate verglichen wurde.

Material und Methoden

Sinapis alba wurde nach Keimung auf Filterpapier in 1/10 konzentrierter Nährlösung nach Eschrich [2] unter konstanten Bedingungen (15°, rel. Luftfeuchtigkeit 55—60%, 16 Stunden Licht) in einer Klimakammer angezogen. Zum Versuchszeitpunkt waren die Keimblätter fast ausgewachsen, die Primärblätter hatten ungefähr ein Drittel ihrer Endgröße erreicht. Die Applikation der zu untersuchenden Substanzen erfolgte durch Einstellen der Keimpflanzen in die radioaktive Lösung (Wurzelapplikation) bzw. durch Auflegen eines Keimblattes mit der Oberseite auf die Testlösung (Blattapplikation). Nach 3 Stunden wurden die Pflanzen in einzelne Abschnitte zerlegt, diese in Methanol extrahiert und die Radioaktivität im Flüssigkeitsszintillationszähler gemessen. Zur Charakterisierung des Mobilitätsverhaltens wurden folgende Größen berechnet:

$$\text{Translokationsrate } (\%) = \frac{\text{dmp außerhalb des Fütterungsorgans}}{\text{dmp in der Gesamtpflanze}} \cdot 100$$

Translokationsquotient

$$Q_{tr} = \frac{\text{proz. Anteil an der Gesamtradioaktivität in Wurzel und Hypokotyl nach Blattapplikation}}{\text{proz. Anteil an der Gesamtradioaktivität in Keimblättern und Spitze nach Wurzelapplikation}}$$

Für die *Yucca*-Versuche wurden 20—30 cm lange Abschnitte der Blütenstandsachsen von im Freiland kultivierten *Yucca filamentosa*-Pflanzen über die untere Schnittfläche mit den Testsubstanzen gefüttert. Das aus der oberen Schnittfläche austretende Exsudat wurde über 12 Stunden gesammelt, anschließend zur Trockne eingedampft, mit geeigneten Lösungsmitteln aufgenommen und dünnschichtchromatografisch auf Silufol getrennt. Die Auswertung der Chromatogramme erfolgte mit dem Dünnschichtscanner bzw. durch Messung der Radioaktivität der einzelnen Folienabschnitte im TRICARB. Die Achsen wurden in 1 cm lange Abschnitte zerlegt, mit geeigneten Lösungsmitteln extrahiert und die Radioaktivität der Extrakte gemessen.

Isolierte Leitgewebe erhielten wir durch Präparation des zentralen Leitgewebsstranges aus den Blattstielen von *Cyclamen* spec. Nach 1 Stunde Wässerung in 0,25 mmol/l $CaSO_4$ wurden kleine Abschnitte 1 h in inaktiver Substratlösung vorinkubiert, anschließend 1 h radioaktiv gefüttert (Substratansatz 10 mmol/l Saccharose in 0,25 mmol/l $CaSO_4$ bzw. 100 µmol/l MCPA in 0,25 mmol/l $CaSO_4$, MES-Puffer pH = 5, Temperatur 25 °C). Nach Auswaschen des AFS in 0,25 mmol/l $CaSO_4$ für 10 min. wurde in Methanol extrahiert und die Radioaktivität gemessen.

Eingesetze Verbindungen, spez. Aktivitäten und verwendete Laufmittel: U-^{13}C-Saccharose, 540 mCi/mmol, Papier, n-Butanol: Aceton: H_2O = 2:7:1; U-^{14}C-Glukose, 100 mCi/mmol, Papier, n-Butanol:Aceton:H_2O = 2:7:1; 2-^{14}C-2,4-D, 28 mCi/mmol, Silufol, Benzol:Dioxan:Essigsäure = 90:25:4; 1-^{14}C-MCPA, 60 mCi/mmol, Silufol, n-Butanol:Äthanol:3N NH_3 = 120:30:50;

U-^{3}H-Dichloranisol, 18,4 mCi/mmol, Silufol, Chloroform:Aceton = 7:3; Ring-^{14}C-Defenuron, 2,06 mCi/mmol, Silufol, Chloroform:Aceton = 7:3; U-^{3}H-Phenylureidoessigsäure, 294 mCi/mmol, Silufol, Äthanol:H_2O = 7:3, U-^{3}H-Atrazin, 2,5 mCi/mmol, Silufol, Chloroform:Aceton = 7:3, U-^{3}H-Alanyl-s-triazin, 2,9 mCi/mmol, Silufol, Äthanol:H_2O = 7:3

Ergebnisse und Diskussion

Zur Kennzeichnung des Mobilitätsverhaltens benutzten wir den von unserer Arbeitsgruppe entwickelten *Sinapis*-Mobilitätstest. Mit ihm ist es möglich, mit relativ einfacher Versuchsanordnung und bei einer Versuchsdauer von nur 3 Stunden, das Mobilitätsverhalten einer Verbindung zu erkennen und quantitativ durch eine Größe, den Translokationsquotienten Q_{tr}, zu beschreiben. Der Q_{tr} drückt das Verhältnis zwischen der Phloem- und der Xylemmobilität einer gegebenen Verbindung nach Blatt- und Wurzelapplikation aus. Die für die einzelnen Mobilitätsgruppen charakteristischen Werte unterscheiden sich deutlich voneinander, so daß aus dem im Test für die untersuchte Verbindung erhaltenen Ergebnis eine klare Zuordnung eines bestimmten Mobilitätsverhaltens erfolgen kann. Q_{tr}-Werte >1 besitzen phloemmobile Substanzen, Q_{tr}-Werte <0,2 kennzeichnen xylemmobile Verbindungen, Q_{tr}-Werte zwischen 0,3 und 0,9 weisen Ambimobilität der Verbindungen aus. Man kann also erwarten, daß eine durch Variation in der Molekülstruktur bedingte qualitative Veränderung des Mobilitäsverhaltens sich auch in einer Veränderung des Q_{tr} widerspiegelt.

2,4-D gilt als Standardbeispiel für phloemmobile Verbindungen, sie besitzt im Senftest einen Q_{tr} von 6,4. Ausgehend von unserer Vorstellung, daß die Mobilität im Phloem durch das Vorhandensein der Carboxylgruppe bedingt ist, müßte mit dem Verlust dieser Gruppe auch die Fähigkeit im Phloem geleitet zu werden, verloren gehen. Durch Eliminierung der -COOH-Gruppe erhält man 2,4-Dichloranisol (Abb. 1). Diese

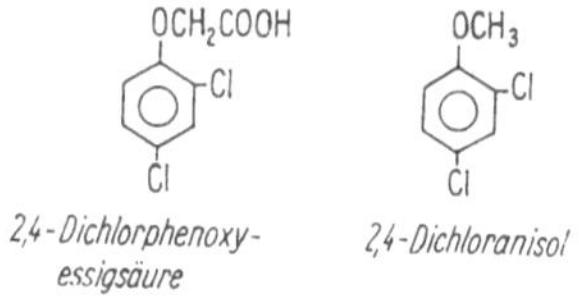

Abb. 1

Verbindung zeigt nach Blattapplikation einen äußerst geringen Transport aus dem Fütterungsblatt. Nur 1,5% der aufgenommenen Radioaktivität sind in Wurzel und Hypokotyl nachzuweisen. Für 2,4-D liegt dieser Wert fast zehnmal so hoch. In zwei weiteren Beispielen versuchten wir durch Einführung von Carboxylgruppen in typisch xylemmobile Verbindungen Phloemmobilität zu erreichen. Das Harnstoffherbizid Defenuron verbleibt nach Blattapplikation zum größten Teil im gefütterten Keimblatt. In den durch den Phloemstrom erreichten Abschnitten Wurzel und Hypokotyl finden wir nur 0,9% der durch die Pflanze aufgenommenen Radioaktivität, eine Menge, die vernachlässigt werden kann. Nach Wurzelapplikation beträgt die Translokationsrate 71%, in den Keimblättern und der Spitze finden wir 52,6%. Der Q_{tr} liegt bei 0,03. Die durch Einführung einer -COOH-Gruppe erhaltene Verbindung zeigt ein völlig anderes Mobilitätsverhalten. Im Gegensatz zum Defenuron beträgt die Translokationsrate nach Blattapplikation 35,0%, der prozentuale Anteil an der Gesamtradioaktivität in der Wurzel und im Hypokotyl 18,2%. Der Q_{tr} von 0,62 kennzeichnet die Verbindung als ambimobil (Abb. 2). Die Triazine sind ebenfalls ausgesprochen xylemmobile Substanzen. Atrazin z.B. zeigt im Sinapistest ein ähnliches Verteilungs-

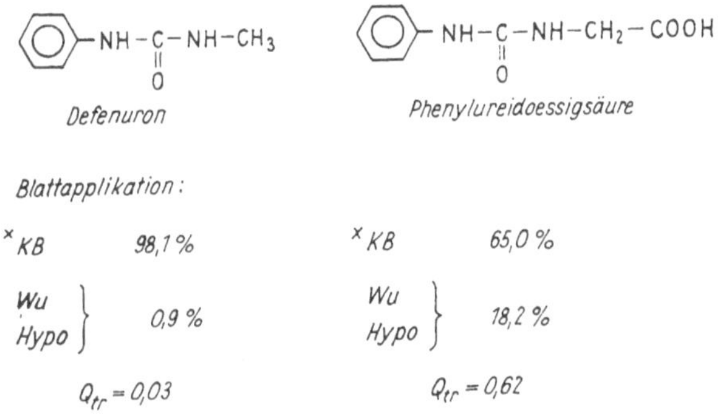

Abb. 2

Atrazin

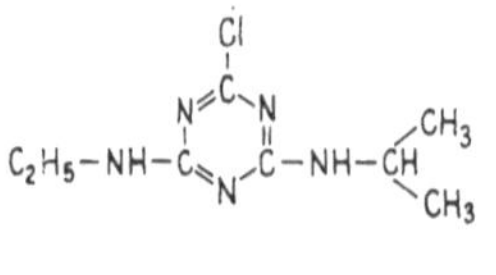

Alanyl-s-Triazin

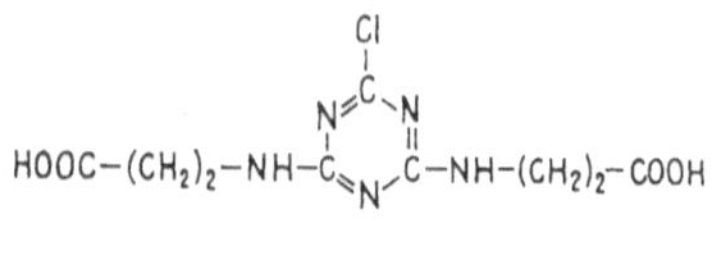

Abb. 3. *Sinapis alba* Verteilungsmuster von ^{14}C-Defenuron und ^{3}H-Phenylureidoessigsäure nach Blattapplikation

Tabelle 1
Sinapis alba
Verteilungsmuster von ^{3}H-Atrazin und ^{3}H-Alanyl-s-Triazin nach Blatt- und Wurzelapplikation. Versuchsdauer: 3 h; 15°, rel. F. 60%

	^{3}H-Atrazin Blattapplikation	Wurzelapplikation	^{3}H-Alanyl-s-Triazin Blattapplikation	Wurzelapplikation
Aufnahme (nmole) Außenlösung 5 × 10^{-5} mol/l	6,9	22,9	0,75	2,3
Translokationsrate (%)	3,1	53,9	13,4	25,1
% der Gesamtradioaktivität in				
Wurzel }		46,1		74,9
Hypokotyl }	1,7	20,2	7,6	12,5
gefüttertem Keimblatt	96,6	}	86,6	}
ungefüttertem Keimblatt	0,6	33,7	2,6	12,6
Primärblättern	0,8	}	3,2	}
Q_{tr}	0,05		0,60	

muster wie Defenuron (Q_{tr} = 0,05). Bei dieser Stoffgruppe verknüpften wir den Grundkörper mit Aminopropansäure und gelangten zu einer Verbindung mit ambimobilen Eigenschaften (Q_{tr} = 0,60) (Abb. 3; Tab. 1). Um die im Phloem transportierten Substanzen näher zu charakterisieren und auszuschließen, daß Stoffwechselprodukte der applizierten Verbindungen bei der Radioaktivitätsmessung im Senftest erfaßt wurden, führten wir eine Reihe von Experimenten mit Blütenstandsachsen von *Yucca filamentosa* durch. Wie Tammes u. a. [8] zeigten, handelt es sich bei dem aus der Schnittfläche austretenden Exsudat um reinen Phloemsaft. Wir applizierten an abgeschnittene 20—30 cm lange Blütenstandsachsen über die untere Schnittfläche die zu prüfenden Substanzen und sammelten über einen Zeitraum bis zu 12 Stunden das aus der oberen Schnittfläche austretende Exsudat. Anschließend wurde die Achse in 1 cm lange Abschnitte zerlegt und zur Radioaktivitätsmessung aufgearbeitet. Am Exsudat bestimmten wir ebenfalls Radioaktivität und trennten es dünnschichtchromatografisch auf. Tabelle 2 zeigt eine Zusammenstellung der Ergebnisse.

Im Exsudat tritt nur bei Applikation von Verbindungen mit -COOH-Gruppen Radioaktivität auf. Mit Ausnahme von Glukose ist die applizierte Verbindung in allen Fällen unverändert im Exsudat nachzuweisen. Die carboxylgruppen-freien Verbindungen werden an der Schnittfläche zwar in die Achse aufgenommen und im Xylem transportiert — eine Aufarbeitung der Achse zeigt eine durchgehende radioaktive Markierung — können aber offensichtlich nicht in das Phloem eintreten.

Es ergibt sich die Frage nach dem Mechanismus, mit dem die Aufnahme der phloemmobilen, carboxylgruppen-haltigen Xenobiotika in das Phloem erfolgt. Nach unseren

Tabelle 2
Yucca filamentosa
Applikation von ausgewählten organischen Verbindungen an isolierte Blütenstandsachsen

Applizierte Verbindung	Carboxyl-gruppen-haltig	Radioaktivität im Exsudat	im Exsudat nachgewiesene radioaktive Verbindung
^{14}C-Defenuron	–	–	–
^{3}H-Phenylureidoessigsäure	+	+	^{3}H-Phenylureidoessigsäure
^{3}H-Atrazin	–	–	–
^{3}H-Alanyl-s-Triazin	+	+	^{3}H-Alanyl-s-Triazin
^{14}C-Glukose		+	^{14}C-Saccharose

heutigen Kenntnissen muß angenommen werden, daß die Aufnahme in den Symplasten nicht den limitierenden Schritt darstellt. Experimentelle Beobachtungen zeigten, daß auch viele Substanzen, die nicht phloemmobil sind, wie z.B. Harnstoffe, gut in den Symplasten aufgenommen werden. Diese Aufnahme ist Voraussetzung für die entsprechende Wirkung dieser Verbindungen. Auch beim Transport durch die Wurzel muß für xylemmobile Substanzen spätestens an der Endodermis eine symplastische Passage angenommen werden. Als entscheidender Schritt muß die Aufnahme in den Siebröhren-Geleitzellen-Komplex angesehen werden. Untersuchungen der Aufnahmekinetik von MCPA in isoliertem Leitgewebe der Blattstiele von *Cyclamen* ergaben keinen Anhaltspunkt für eine Carrier-vermittelte Aufnahme. Die erhaltenen Kurven sprechen für eine Aufnahme durch Diffusion und unterscheiden sich deutlich von denjenigen, die für Saccharose erhalten wurden. Diese weisen eine Michaelis-Menten-Kinetik auf mit einem K_M-Wert von 11,1 mmol/l und V_{max} von 12,6 mol/gxh (Abb. 4). Wir müssen für Xenobiotika einen anderen Aufnahmemechanismus

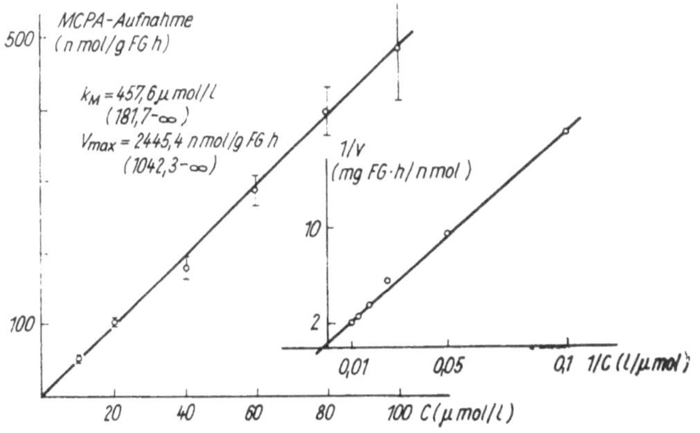

Abb. 4. Isoliertes Leitgewebe aus dem Blattstiel von *Cyclamen* spec., Aufnahme von ^{14}C – MCPA

annehmen als für Zucker. Im Apoplasten des SR-GZ-Komplexes beträgt der pH-Wert 5, im Inneren der Siebröhre 8 [4]. Alle von uns untersuchten phloemmobilen Xenobiotika zeigen in den beiden pH-Bereichen deutliche Unterschiede in ihren Verteilungskoeffizienten, bedingt durch den Übergang von der undissoziierten in die dissozierte Form. Im sauren Bereich sind sie wesentlich lipophiler als im basischen. Es ist denkbar, daß die im Apoplasten vorliegende undissoziierte, lipophile Form leicht in das

Phloem eintreten kann, dort zur hydrophileren Form dissoziert, deren Austritt in den Apoplasten aber nur sehr langsam erfolgt. Die Akkumulation der phloemmobilen Xenobiotika würde nach dem Prinzip einer Ionenfalle erfolgen. Unter diesem Gesichtspunkt wäre die Rolle der Carboxylgruppe beim Phloemtransport von Xenobiotika als die Fähigkeit der betreffenden Verbindung aufzufassen, bei unterschiedlichen pH-Werten, die den physiologischen Bedingungen von Apoplast- und Siebröhre entsprechen, in 2 Formen aufzutreten, die sich in ihrer Lipophilie und damit in ihrem Permeationsvermögen deutlich voneinander unterscheiden. Es bleibt zu prüfen, inwieweit das auch von anderen chemischen Strukturen erreichbar ist.

Diskussion

Zanke: Wir kennen nur sehr wenige Fungizide mit freier Carboxylgruppe, z.B. einige Salizylsäure derivate. Inder Regel ist mit der Einführung einer -COOH-Gruppe in ein aktives Molekül ein Verlust der fungiziden Wirksamkeit um mehrere Größenordnungen verbunden. Haben Sie bei bestimmten Verbindungsklassen eine Korrelation zwischen der Größe des Q_{zv}-Wertes und der Fungitoxizität gefunden?

Neumann: Wir haben bei allen unseren Untersuchungen keine Prüfung der biologischen Aktivität vorgenommen. Es ist aber zu erwarten, daß die von uns vorgestellten -COOH-gruppenhaltigen Derivate der Wirkstoffe nicht mehr biologisch aktiv sind. Es wäre von großer Bedeutung, wenn es gelänge, Substanzen zu synthetisieren, die von der Pflanze aufgrund bestimmter Molekülstrukturen gut aufgenommen und transportiert werden und am Wirkort durch metabolische Prozesse in die biologisch aktiven Verbindungen umgewandelt würden. Ähnliche Überlegungen wurden von Crisp für Insektizide angestellt.

Lyr: Eine Frage an Prof. Schwinn: Wird Metalaxyl, ein ambimobiler Carboxylsäuremethylester, in der Pflanze gespalten, sodaß die freie Säure für den in der Literatur beschriebenen Phloem-Transport verantwortlich sein könnte?

Schwinn: Metalaxyl wird zwar in Xylem und Phloem transloziert, doch ist der Xylemtransport sehr viel stärker ausgeprägt. Soweit wir wissen, wird das Original-Molekül transloziert; die Säure tritt zwar als Metabolit auf, doch ist ihre fungizide Wirkung wesentlich kleiner als die des Original-Moleküls. Vielleicht kann die Tatsache, daß die Carboxylgruppe von Metalaxyl verestert ist, die relativ geringe Phloemmobilität erklären.

Josepovits: Haben Sie versucht eine Korrelation zwischen den gemessenen Q_{tr}-Werten und den p_{Ka}-Werten der Carboxylsäure zu finden?

Neumann: Alle bisher von uns untersuchten Verbindungen besitzen p_{Ka}-Werte, die die Dissoziation zwischen pH 5 und pH 8 möglich machen. Eine Korrelation zwischen der Lage des p_{Ka}-Wertes und der in das Phloem aufgenommenen Menge an Xenobiotikum ist zu erwarten. Wir werden diese Frage prüfen, z.B. durch eine vergleichende Betrachtung der Aufnahme von Essigsäure und Trichloressigsäure.
Für den Eintritt in das Phloem genügt offensichtlich nicht, daß die Substanzen schwache Säuren schlechthin sind. Phenol z.B. ist auch eine schwache Säure, es ist aber nicht phloemmobil. Voraussetzung ist, daß die Säure bei pH 5 und pH 8 in zwei unterschiedlichen Formen, undissoziiert und dissoziiert auftreten kann. Für Phenol liegt der p_{Ka}-Wert aber über 8.

Josepovits: Es ist bekannt, daß einige anionische dissoziierbare Verbindungen zur Komplexbildung mit kationogenen Stoffen fähig sind. Es könnte interessant sein zu untersuchen, ob eine solche Komplexbildung die Phloemmobilität beeinflussen könnte.

NEUMANN: Darüber haben wir keine Untersuchungen. CRISP stellte auf dem IUPAC-Kongreß in Zürich einige Derivate von 2,4-D vor und fand, daß nur die Verbindungen zur Translokation fähig waren, die in der Pflanze in die freie Säure umgewandelt wurden.

MÜLLER: Die von Ihnen vorgestellten Ergebnisse beziehen sich auf den *Sinapis alba*-Test. Ist es denkbar, daß holzige Pflanzen anders reagieren und der von Ihnen für das Testregime ausgeschlossene „carrier Effekt" in holzigen Pflanzen wirken kann? Haben Sie vergleichende Untersuchungen machen können?

NEUMANN: Mit holzigen Pflanzen haben wir nicht gearbeitet. Vergleichende Mobilitätsuntersuchungen an anderen Testobjekten, Keimpflanzen von *Cucumis sativus*, *Helianthus annuus* und *Linum usitatissimum* haben aber, ebenso wie Experimente am isolierten Maisblatt, gezeigt, daß das im *Sinapis*-Test für eine bestimmte Substanz ermittelte Mobilitätsverhalten auch für alle anderen Objekte gilt. Eine Substanz, die im *Sinapis*-Test z.B. phloemmobil ist, ist auch in anderen Pflanzen phloemmobil. Unterschiede ergeben sich lediglich in quantitativer Hinsicht, z.B. in der Größe der beobachteten Translokationsraten. Hier spielen Stoffwechselvorgänge im Fütterungsorgan eine Rolle.

Literatur

[1] CRISP, C. E., Molecular design of systemic insecticides and functional groups in translocation. Pesticid. Chem. Proc. Int. Congr. Pestic. Chem. 2nd **1** (1972), 211–264

[2] ESCHRICH, W., Translocation ^{13}C-markierter Assimilate im Licht und im Dunkeln bei *Vicia faba* Planta **70** (1966), 99–124

[3] JACOB, F., ST. NEUMANN and U. STROBEL, Studies on mobility of exogen applied substances in plants, Transactions of the 3rd symposium on accumulation of nutrients and regulators in plant organism, Warszawa, in Proceedings of the Research Inst. of Pomology, Skierniewice, Poland, Ser. E. **5** (1973), 315–330

[4] MALEK, F. and D. A. BAKER, Proton Co-transport of sugar in phloem loading, Planta **135** (1977), 297–299

[5] MITCHEL, J. W., P. C. MARTH and W. H. PRESTON, Structural modification that increases translocability of some plant regulating carbamates, Science **120** (1954), 263–265

[6] MITCHEL, J. W., B. C. SMALE and R. C. METCALF, Absorption and translocation of regulators and compounds used to control plant diseases and insects., in METCALF, R. C. (Hsg.), Adv. Pest. control. Res., Interscience publishers, New York, 1960, Vol. **3**, 359–436

[7] WAIN, R. L. and G. A. CARTER, in TORGESON, D. C. (Hsg.), Fungicides, Academic press, New York, 1967 Vol. **1**, 561–611

[8] TAMMES, P. M. L., C. R. VONK and J. VAN DIE, Studies on phloem exudation from *Yucca flaccida* HAW. XI. Xylem feeding of ^{14}C-sugars and some other compounds, their conversion and recovery from the phloem exsudate, Acta Bot. Neerl. **23** (1973), 233–327

М. Н. Колосов, В. В. Оноприенко, Л. Н. Гладкова,
М. Г. Карапетян, Ф. А. Мамедов, А. Л. Никонов

Институт биооранической химии имени М. М. Шемякина
Академии наук СССР, Москва, СССР

Химическое Изучение Антибиотика Имбрицина

Среди антифунгальных антибиотиков особую и мало изученную в химическом отношении группу составляют гуанидинсодержащие неполиеновые макролиды: азаломицины F_1—F_5 [1, 2], копиамицин [3], нифимицины [4, 7, 8], имбрицин [9], примицин [6] и некоторые другие. Они обладают широким спектром действия, угнетая рост патогенных и фитопатогенных грибов, дрожжей, простейших и, в меньшей степени, грамположительных бактерий и микобактерий. Описано клиническое применение азаломицина F для вагинального лечения кандидозов и трихомонадоза, а нифимицин выпускается промышенностью Болгарии для использования в медицине и ветеринарии.

Имбрицин, выделенный в СССР в 1965—1970 г.г. из нескольких штаммов *Actinomyces imbricatus*, по своему компонентному составу и некоторым физико-химическим свойствам близок азаломицину F [9]. Идентичность компонентов этих антибиотических комплексов, однако, строго не доказана, а их компонентный состав весьма различен: по данным M. Arai [2] соотношение азаломицинов F_3:F_4:F_5 составляет в разных препаратах 2—3:5—6:1—3, тогда как в имбрицине содержится не менее 85—90% главного компонента 4.

После очистки хроматографией на силикагеле в системе метанол-вода-хлороформ 9:1:35 и/или перекристаллизацией из 30% метанола имбрицин состоит практически из одного компонента 4 и имеет т. пл. 132—133°, $[\alpha]_D^{20}$ + 39° (c 1, MeOH), $\lambda_{макс}$ (MeOH) 236 п, 240 и 269 нм (ε 39700, 42000 и 24300), $\nu_{макс}$(KBr) 3350 (OH, HH), 1708 (COOR), 1632 (гуанидиновая группа), 1605 (COO^-) $см^{-1}$. В его спектре ^1H-ЯМР проявляются сигналы 6 $C{-}\overset{|}{\underset{|}{C}}{-}Me$ групп (0,8—1,2 м. д.), двух N—Me групп (6H-синглет при 2,86 м.д.), 13—14 протонов NH, олефиновых и RCOOC**H**-группировок (5—7 м.д.). По данным осмометрии в метаноле имбридин имеет молекулярный вес 1100—1300, а при титровании $HClO_4$ в AcOH обнаруживает эквивалентный вес 1150 + 20. Поскольку в водно-метанольном растворе он обладает очень слабой основностью (pK_a 3,5—4), но титруется MeONa в придине как монокислота, антибиотик является цвиттер-ионом. Элементный анализ указывает на его вероятную эмпирическую формулу $C_{57}H_{103}N_3O_{20}$.

Наличие гуанидиновой группировки в имбрицине было доказано флуоресцентной реакцией Кона и Дэвиса, реакция Сакагучи, как и следовало ожидать для трехзамещенного гуанидина, была отрицательной. Поскольку при жестком щелочном гидролизе имбрицина образовывался только один летучий амин — метиламин, а аммиака и диметиламина обнаружить не удалось, из трех возможных частичных формул (1—3, R = $C_{53}H_{94}O_{18}$) антибиотику была приписана формула (3). Она была расширена до формулы (4), так как единственным легкокипящим продуктом кислотного метанолиза имбрицина оказался диметилмалонат, идентифицированный хромато-масс-спектрометрией, и логично было предположить, что малоновая кислота связана в молекуле антибиотика в форме кислого эфира, свободный карбоксил которого ионизован и является противоионом гуанидиновой группы. Моносахаридов или их метилгликозидов при мягком и жестком кислотном гидролизе и метанолизе имбрицина обнаружить не удалось.

$$\underset{COO^-}{\overset{|}{R}}-NMe-\underset{NH_2^+}{\overset{\|}{C}}-NHMe \quad \underset{COO^-}{\overset{|}{R}}-NH-\underset{NH_2^+}{\overset{\|}{C}}-NMe_2 \quad \underset{COO^-}{\overset{|}{R}}-NH-\underset{NHMe^+}{\overset{\|}{C}}-NHMe$$

(1))2) (3)

$$\underset{OOCCH_2COO^-}{\overset{|}{C_{51}H_{92}O_{16}}}-NH-\underset{NHMe^+}{\overset{\|}{C}}-NHMe$$

(4)

Метилирование имбрицина действием MeI и Ag_2O в DMF приводит к перметильному производному, в котором по данным ^1H-ЯМР содержится 14—15 MeO- и MeNгрупп. При каталитическом гидрировании антибиотик образует декагидропроизводное, $C_{57}H_{113}N_3O_{20}$, которое обладает только концевой абсорбцией в УФ-спектре, а по спектру ЯМР отличается от исходного имбрицина отсутствием сигналов 6 олефиновых протонов. При обработке йодметилатом трифенилфосфита и дальнейшем восстановлении Mg + i—PrOH и H_2/Pt декагидроимбрицин образует полидезоксипроизводное, которое при действии оснований отщепляет I моль HI и превращается в соединение с эквивалентным весом 880 ± 60 ($HClO_4$ в AcOH), молекулярным весом 935 (MS) и вероятной эмпирической формулой $C_{57}H_{113}N_3O_6$. Судя по ИК-спектру это соединение сохраняет гуанидиновую, COO^- и COOR группы, и его образование связано, очевидно, с потерей декагидроимбрицином 14 OH-групп. При дальнейшем восстановлении $LiAlH_4$ оно теряет также COOH и COOR группы и дает сильное основание с эквивалентным весом 935 ± 50 (0,05 N HC_8 в 50% MeOH), образование которого ясно показывает, что имбрицин содержит непрерывную углеродную цепь не менее, чем из 50 C-атомов.

Из пяти гидрируемых двойных связей имбрицина одна является изолированной, а четыре других входят в состав двух диеновых группировок (5) и (6), отвечающих двум хромофорам антибиотика. В более длинноволновом из них диеновая группа должна быть сопряжена с карбалкоксильной (лактонной) или, менее вероятно, кетогруппой, а в коротковолновом, судя по его $\lambda_{макс}$, она должна быть три- или даже тетразамещенной.

Действительно, при озонолизе перметилимбрицина с последующим восстановлением $NaBH_4$ и бензоилированием мы выделили бензаоты этиленгликоля и 1,2-пропандиола, которые, очевидно, образуются при расщеплении этих диеновых группировок. Кроме того, при этой деградации мы получили N-метил-N, O-дибензоил-δ-аминобутанол (7), структуру которого подтвердили контрсинтезом. Выделение соединения (7) показало, что метилирование антибиотика в присутствии Ag_2O сопровождается гидролизом гуанидиновой группировки.

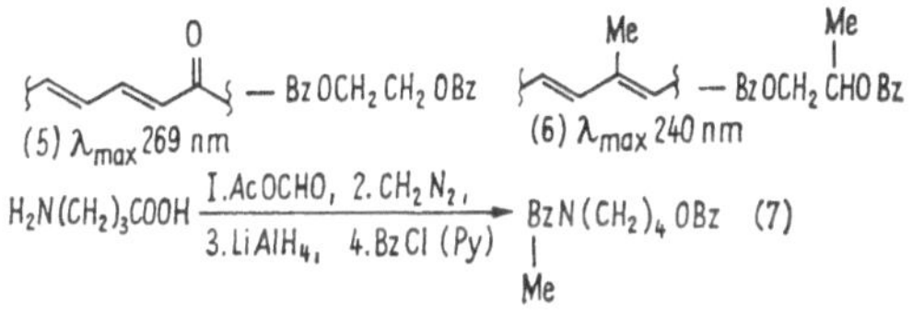

При озонолизе немодифицированного имбрицина с последующим восстановлением $NaBH_4$ и ацетилированием помимо ацетатов этиленгликоля и пропандиола было выделено три вещества. Строение первого из них (8), установленное на основании, главным образом, данных ЯМР и масс-спектров, приведенных на формуле (ионы, состав которых опеределен HRMS отмечены звездочкой), дополнительно подтверждает сделанный ранее вывод о наличии заместителей у всех трех N-атомов гуанидиновой группы. Строение второго из выделенных веществ (9) также вытекает из его спектральных данных, причем характер спин-спинового взаимодействия протона при вторичной ацетоксильной группе однозначно доказывает расположение обеих метильных групп. Для подтверждения структуры соединение (9) было окислено $KMnO_4$ и образовавшиеся кислоты исследованы в виде метиловых эфиров GC—MS. Основным продуктом оказался при этом эфир α-метиладипиновой кислоты (10), образующийся также при аналогичном окислении имбрицина и декагидроимбрицина.

Самым сложным из продуктов озонолиза имбрицина является третий. По данным HRMS наиболее тяжелый ион в его масс-спектре имеет состав $C_{56}H_{88}O_{26}$ (m/e 1176), а так как, судя по спектру ^1H-ЯМР, он содержит 13 AcO-групп, но 13—14 протонов CHOAc и CHOR групп и одну группу CH—CH_2OAc, ему была приписана эмпирическая формула $C_{58}H_{90}O_{27}$. Спектр ЯМР указывает также на наличие в этом соединении 3 C—CHMe—C и одной C—CHMe—O группы. Для установления характера углеродного скелета оно было подвергнуто гидрированию над Pd/$BaCO_3$ при 250 до-300°, в результате чего, по-видимому, с потерей 2 C-атомов скелета образовался углеводород $C_{30}H_{62}$, имеющий формулу (II) или (12).

Сопоставление имеющихся данных по структуре нифимицина A_1, имбрицина и примицина показывает, что общим для этих антибиотиков является значительная величина углеродного скелета и, вероятно, макролактонного цикла, наличие 12—14 гидроксилов и гуанидиновой группы. Представляется очевидным, что количество и распределение в молекуле двойных связей мало влияет на антимикробные свойства этих антибиотиков, и их необычная для неполиеновых макролидов высокая анти-

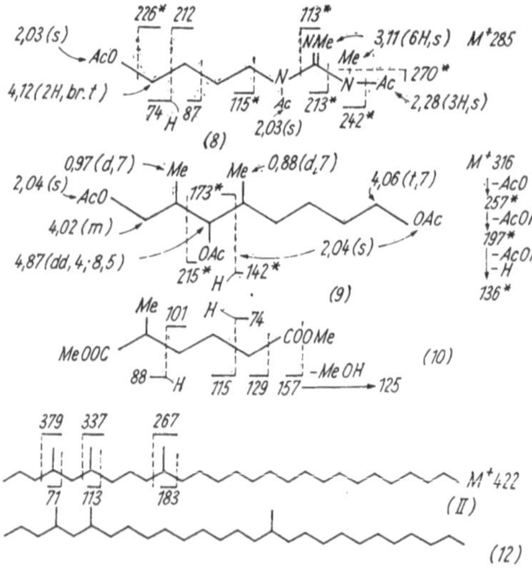

фунгальная активность и довольно значительная токсичность обусловлены присутвием неметилированного (примицин) или частично метилированного (нифимицины и имбрицин) гуанидинового остатка.

Diskussion

GEORGOPOULOS: Is there any information available on the mechanism of action of this group of antibiotics?

ONOPRIENKO: There are some papers published about azalomycin F (S. SUGASAWA "Mode of action of azalomycin F. Effect of azalomycin F on *C. albicans*", J. Antibiotics 20 (1967), 93) and copiamycin (see, for instance, K. SEIGA, V. SAMAZI, Appl. Microbiol. 21 (1971) 986). The data are contradictory. Some workers point out to the membran activity of the antibiotics, the others to their action on the protein biosynthesis.

SISLER: Does the antibiotic belong to the polyene group?

ONOPRIENKO: No, it belongs to the little known group of nonpolyenic guanidine containing macrolides.

LACHOWICZ: On what yeast this antifungal action of imbricin was tested?

ONOPRIENKO: Imbricin has a very broad spectrum of activity. It is active against such yeasts as *Saccharomyces cerevisiae* and many others in concentrations 1—5 mkg/ml; the data are published in "Antibiotiki" (USSR), 1970, 208, 1974, 68. Activity of imbricin is the same, as for azalomycin F, see M. AZAI, K. HAMANO, J. Antibiotics, 23 107, and references cited there.

Литература

[1[ARAI, M., Azalomycin F — an antibiotic against fungi and *Trichomonas*, Arzneimittel Forschung **18** (1968) 1396—1399

[2] ARAI, M., K. HAMANO, Isolation of three main components, F_3, F_4 and F_5, from azalomycin F-complex, J. Antibiotics **23** (1970) 107—112

[3] Arai, T., S. Kuroda, H. Ohara, Y. Katoh, H. Kaji, Copiamycin, a new antifungal antibiotic derived from *Str. hygroscopicus* var. *crystallogenes*, J. Antibiotics **A18** (1965) 63—67

[4] Блинов, Н. О., В. В. Оноприенко, Т. М. Родина, С. Й. Агайна, Е. И. Хлебарова, Химическая характеристика нифимицинов A_1 и A_2 Антибиотики (1974) 579—581

[5] Fehr, T., R. C. Jain, P. de Mayo, O. Motl, I. Szilagyi, L. Baczynskyj, D. E. F. Gracey, H. L. Holland, D. B. MacLean, The constitution of primycin. Part III. Degradation of methylated primycin and the structure of primycin, J. Chem. Soc. Perkin I (1974) 836—847

[6] Оноприенко В. В., С. Й. Агайна, А. Д. Кузовков, Нифимицин A_1. Функциональный анализ и восстановительные перврващения Реферативная информация, сер. "Хим.-фарм. пром.-сть" (1977) Москва, ЦБНТИ-медпром, № 1, 7—11

[7] Оноприенко, В. В., С. Й. Агайна, А. Д. Кузовков, Нифимицин A_1. Окислительная и гидролитическая деградация Реферативная информация, сер. ,,Хим.-фарм. пром-сть (1977) Москва, ЦБНТИ-медпром, № 3, 5—9

[8] Петрова, Л. Я., В. Г. Митрофанова, Ю. Ф. Свешников, Выделение и физико-химическая характеристика основных компонентов имбрицина и препарата ЛИА-ОIII Антибиотики (1974) 68—71

P. WERNER und H. LYR

Institut für Pflanzenschutzforschung der Akademie der Landwirtschaftswissenschaften, Kleinmachnow, DDR

Zum Wirkungsmechanismus des systemischen Fungizids Chloroneb und den möglichen Ursachen erworbener Resistenz bei *Mucor mucedo* (L.) FRES.

Auf dem Reinhardsbrunner Symposium im Jahre 1977 berichteten wir über erste Ergebnisse bei der Erforschung der Wirkung von Chloroneb und dessen Metaboliten auf Wachstum und Ultrastruktur verschiedener Pilzarten.

Im Mittelpunkt weiterer Untersuchungen standen folgende Fragestellungen:

1. Ist Chloroneb selbst das fungizide Wirkprinzip oder ein Metabolit?
2. Welches ist der Wirkungsmechanismus von Chloroneb?
3. Welches sind die Ursachen der selektiven Wirkung von Chloroneb und einer Resistenz gegen das Fungizid?

Mehrere Autoren [4, 5, 9] haben nachgewiesen, daß Chloroneb von verschiedenen Pilzarten oxidativ zu 2,5-Dichlor-4-methoxyphenol (DCMP) und 2,5-Dichlorhydrochinon (DCHQ) metabolisiert wird, wobei jedoch nicht eindeutig geklärt ist, ob diese Metaboliten für die fungizide Wirkung bedeutsam sind. Unsere Wachstumsuntersuchungen ließen erkennen, daß DCMP eine ähnlich fungizide Wirkung besitzt wie Chloroneb selbst [8]. Chloroneb-unempfindliche Pilzarten reagieren auch unempfindlich auf 2,5-Dichlor-4-methoxyphenol. Vergleichende Untersuchungen an einem Chloroneb-sensiblen und einem in unserem Institut isolierten Chloroneb-resistenten Stamm von *M. mucedo* ergaben, daß die Fähigkeit der Pilze, Chloroneb zu metabolisieren, nichts mit ihrer Empfindlichkeit gegen diesen Wirkstoff zu tun hat. Sowohl der Chloroneb-sensible als auch der Chloroneb-resistente Stamm von *M. mucedo* bauen Chloroneb nur in sehr geringen Mengen ab (Abb. 1). Nur ca. 1% des eingesetzten ^{14}C-Chloroneb wurde in Form von Metaboliten nachgewiesen.

Ein von uns untersuchter Stamm von *Neurospora crassa* zeigt mittlere Empfindlichkeit gegen Chloroneb ($ED_{50} = 35$ ppm) und baut ebenfalls Chloroneb nur in sehr geringem Umfang ab. Es konnte ausgeschlossen werden, daß es in den Mitochondrien, die nach Chloroneb-Einwirkung schwere Schädigungen aufweisen, zu einer möglichen selektiven Anreicherung von Chloroneb-Metaboliten kommt. In isolierten Mitochondrien aus Myzel von *Neurospora crassa*, das mit ^{14}C-Chloroneb inkubiert worden war, wurde nur der unveränderte Wirkstoff Chloroneb nachgewiesen. Auch die Hemmung der Sauerstoffaufnahme isolierter Mitochondrien aus *Neurospora crassa* wird durch Chloroneb selbst hervorgerufen. Das zeigen die Messungen mit der Platinelektrode im Minutenbereich, einem Zeitraum, in welchem noch keine nennenswerte Metabolisierung zu erwarten ist.

Wie wir bereits 1977 berichtet hatten [8], verändert Chloroneb die Ultrastruktur der Pilzzelle bereits 1 Stunde nach Zusatz des Wirkstoffes. Dabei ist von diesen charakteristischen Schädigungen in besonderem Maße die Innenmembran der Mitochon-

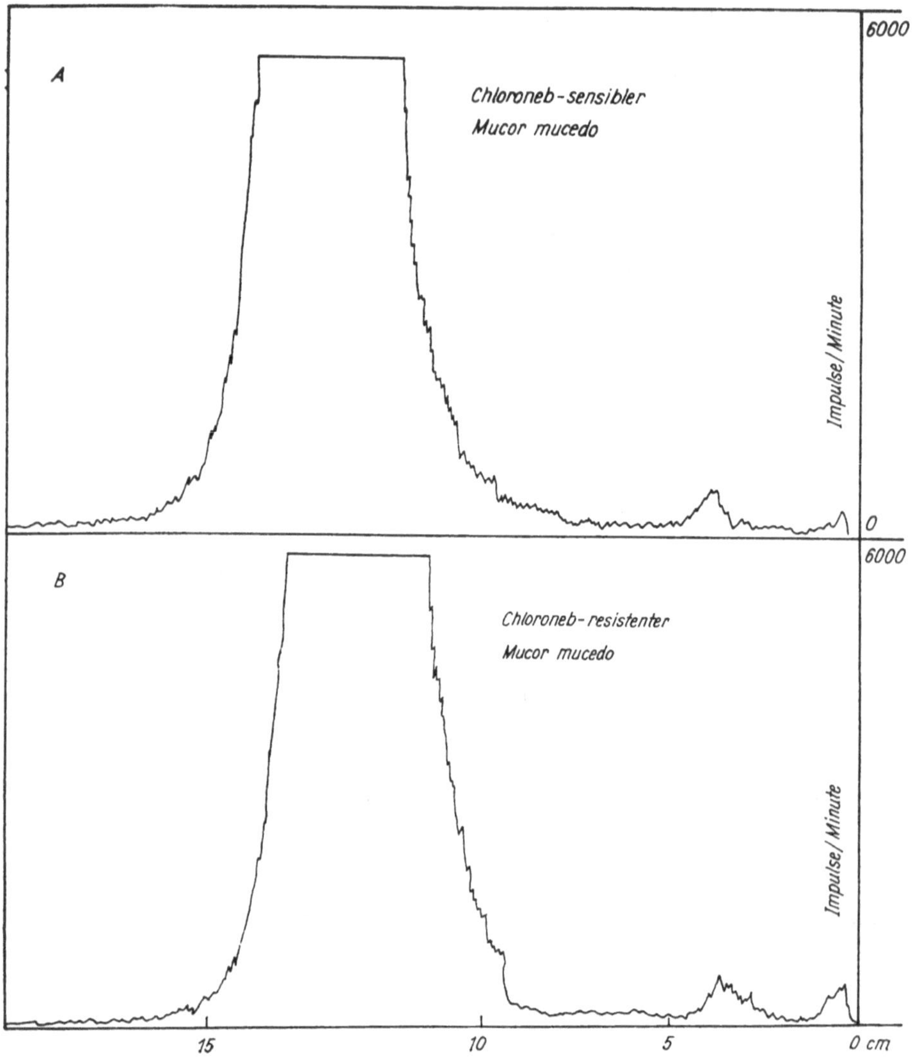

Abb. 1. Dünnschicht-Scannogramme
A — Methanol-Extrakt aus Myzel von chloronebsensiblem *Mucor mucedo* (S-CHl-P3) 24 Stunden nach Zugabe von 1,2 μCi $O^{14}CH_3$-markiertem Chloroneb in das Inkubationsmedium
B — Methanol-Extrakt aus Myzel von chloronebresistentem *Mucor mucedo* 24 Stunden nach Zugabe von 1,2 μCi $O^{14}CH_3$ markiertem Chloroneb in das Anzuchtmedium

drien betroffen, daneben auch andere Membransysteme. Der O_2-Verbrauch junger Keimmyzelien von *M. mucedo* wird durch stark wachstumshemmende Konzentrationen von Chloroneb im Warburg-Versuch jedoch nur schwach gehemmt. Auch die Hemmung der Sauerstoff-Aufnahme isolierter Mitochondrien aus Chloroneb-empfindlichen Stämmen von *Mucor mucedo* und *Neurospora crassa* erreicht selbst bei sehr hohen Chloroneb-Konzentrationen ein relativ geringes Ausmaß, wobei bemerkenswerterweise der Respirationsquotient konstant bleibt. Isolierte Mitochondrien aus

einem Chloroneb-resistenten Stamm von *Mucor mucedo* und dem unempfindlichen Pilz *Neurospora sitophila* ($ED_{50} > 100$ ppm) erwiesen sich bezüglich der Hemmung des O_2-Verbrauches als nahezu völlig Chloroneb-unempfindlich.
Die Zusammenstellung von Ergebnissen polarographischer Messungen läßt erkennen, daß eine Beziehung zwischen Chloroneb-Empfindlichkeit bzw. -Unempfindlichkeit im Wachstums-Test der untersuchten Pilzarten (in vivo) und dem Ausmaß der Hemmung der Sauerstoff-Aufnahme bei isolierten Mitochondrien (in vitro) besteht (Tab. 1). Das bereits erwähnte schwache Ausmaß der Atmungshemmung, das bei Mitochondrien aus sensiblen Pilzen nach Chloronebbehandlung registriert wurde, deutet darauf hin, daß Chloroneb kein spezifischer Hemmstoff der Atmungskette ist.

Tabelle 1
Vergleich der Wirkung von Chloroneb auf Wachstum in vivo und O_2-Konsumption isolierter Mitochondrien sensitiver und resistenter Pilz-Stämme

Organismus	ED_{50} Wachstum µm	ED_{50} O_2-Verbrauch isolierter Mitochondrien (5 min Werte)
N. crassa	175	180
N. sitophila	1000	$\gg$ 200
M. mucedo (sensibel)	10	70
M. mucedo (resistent)	$>$1000	$\gg$200

Das steht in guter Übereinstimmung mit Ergebnissen, die von MÜLLER (persönl. Mitteilung) im Rinderherz-ETP-Test gewonnen wurden. Hier wurde festgestellt, daß die Halbhemm-Konzentration für das NADH-Oxidase-System ca. 100 µM Chloroneb beträgt, während sich das Succinat-Dehydrogenase-System bei diesen Konzentrationen als unempfindlich erweist. Diese bei relativ hohen Chloroneb-Konzentrationen einsetzende Hemmung des NADH-Reduktase-Systems ließe sich auf unspezifische Wirkstoff-Membran-Protein Wechselwirkungen zurückführen.
Tatsächlich konnten wir nachweisen, daß Chloroneb relativ intensiv an isolierte Mitochondrien-Membranen gebunden wird. Es wurde die Fähigkeit isolierter Mito-

Tabelle 2
% fest gebundene $O^{14}CH_3$-Chloroneb-Aktivität bezogen auf 100 mg lösliches Protein isolierter Mitochondrien von Mucor mucedo (10 Versuche)

Chloroneb-sensibler Stamm	Chloroneb-resistenter Stamm
20,0	7,6
20,0	7,6
10,3	4,6
10,3	5,6
9,0	6,0
12,0	6,4
9,0	6,1
9,0	6,4
10,3	6,4
10,3	6,1
∅ 9,9	∅6,2

Unterschiede statistisch gesichert

chondrien-Membranen eines Chloroneb-sensiblen und eines resistenten *Mucor mucedo* Stammes -$O^{14}CH_3$-markiertes Chloroneb zu binden, miteinander verglichen (Tab. 2). Dabei zeigte sich, daß Mitochondrien aus dem sensiblen Stamm signifikant mehr markiertes Chloroneb binden als solche aus dem resistenten Stamm. Auch Mitochondrienmembranen aus *Neurospora crassa* zeigten eine vergleichsweise höhere Chloroneb-Bindungsfähigkeit als Mitochondrienmembranen aus dem Chloroneb-resistenten Pilz *Neurospora sitophila* (Tab. 3). Somit besteht interessanterweise ein Zusammenhang

Tabelle 3
% fest gebundenes $O^{14}CH_3$-Chloroneb bezogen auf 100 mg lösliches Protein isolierter Mitochondrien (6 Versuche)

Neurospora crassa (sensibel)	*Neurospora sitophila* (unempfindlich)
5,1	2,1
3,2	2,1
2,1	1,0
2,7	1,7
2,5	1,1
2,2	2,2
∅2,9	∅1,7

zwischen der Empfindlichkeit der Pilze in Wachstumsuntersuchungen in vivo sowie der Empfindlichkeit der isolierten Mitochondrien bezüglich O_2-Verbrauch in vitro mit der Bindung von Chloroneb an Mitochondrienmembranen. Nach Behandlung mit Chloroneb wurde — parallel zu den Lysiserscheinungen in der Pilzzelle — bei dem Chloroneb-sensiblen Stamm von *M. mucedo* ein vorübergehender Anstieg der Gesamt-Protease-Aktivität gemessen. Beim Chloroneb-resistenten Stamm von *Mucor mucedo* dagegen blieb der Gesamt-Protease-Spiegel nach Behandlung mit Chloroneb unverändert (Abb. 2).

Im Ergebnis unserer Untersuchungen zum Wirkungsmechanismus von Chloroneb kommen wir zu folgenden Schlußfolgerungen: Chloroneb wird von sensiblen Pilzen stärker als von resistenten an Membranen gebunden, wobei Mitochondrienmembranen in besonderem Maße Wirkstoff binden. Folge dieser Wechselwirkungen von Chloroneb mit Membranen, deren Art bisher unbekannt ist, könnte die Freisetzung oder Aktivierung membrangebundener Proteasen sein, diese dürften die Ursache der Lysis der Mitochondrien Innenmembranen sein, wobei auch andere lytische Enzyme (Phospholipasen) beteiligt sein können, so daß wegen mangelhafter Resynthesen der Membranproteine ein Katastrophenzyklus in Gang kommt, der zu der festgestellten fungiziden Wirkung führt, wobei die unspezifische Hemmung der O_2-Verbrauch und damit der ATP-Produktion sukzessive zunimmt.

Wir haben auch einige Untersuchungen zur Wirkung des Chloroneb-Metaboliten 2,5-Dichlor-4-methoxyphenol durchgeführt. DCMP hemmt die Atmung von jungen Keimmyzelien in ähnlich schwachem Maße wie Chloroneb (Abb. 3). Auch die Sauerstoff-Aufnahme isolierter phosphorylierender Mitochondrien aus *Neurospora crassa* wird durch DCMP verringert. Jedoch entkoppelt DCMP im Gegensatz zu Chloroneb den Elektronentransport von der oxidativen Phosphorylierung. Da jedoch, wie das

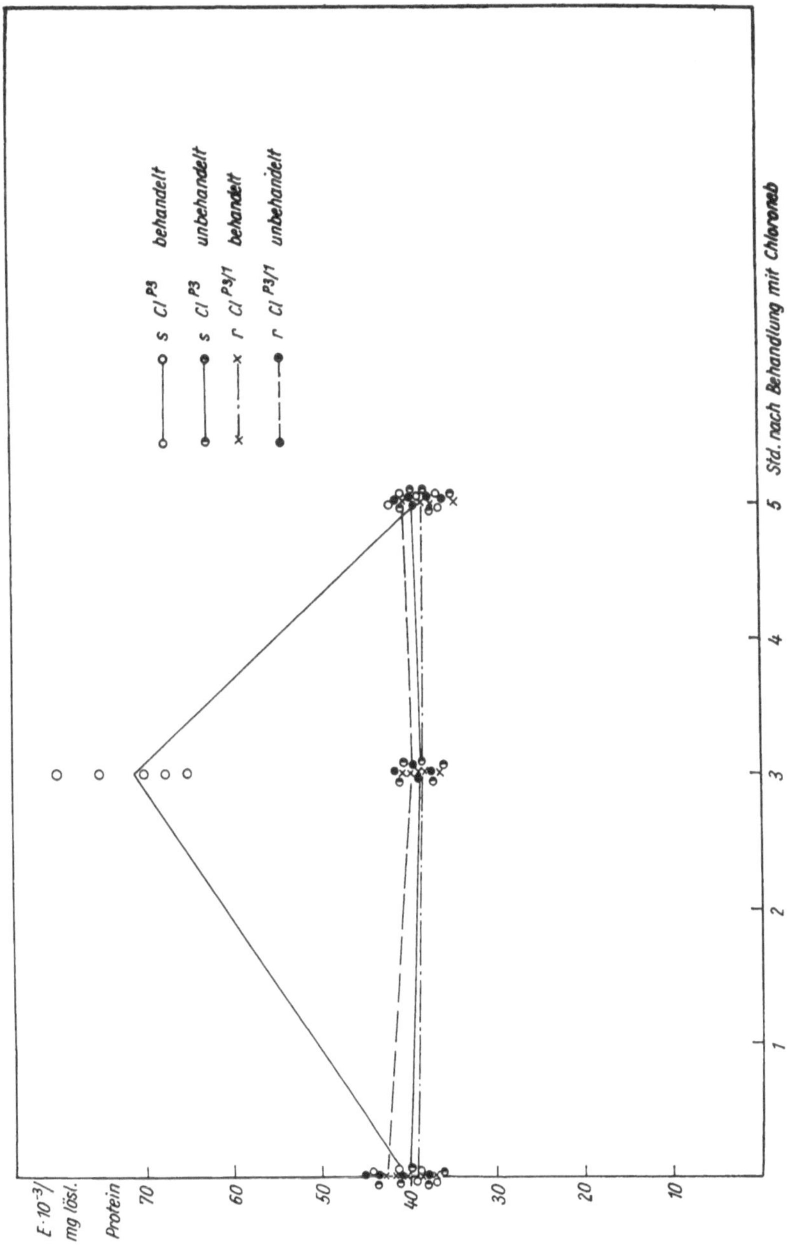

Abb. 2. Gesamtaktivität von Proteasen im Myzel von chloronebsensiblem (S-Chl-P3) und Chloroneb-resistentem *Mucor mucedo* (R-Chl-P3/1) 0,3 und 5 Stunden nach Chloroneb-Zugabe. Im Vergleich dazu die unbehandelten Kontrollen

Wirkungsspektrum zeigt, im Vergleich zum Chloroneb alle biologischen Wirkungen nahezu Chloroneb-identisch sind, scheint es sich hierbei um einen nachgeordneten Effekt zu handeln.

Nach WEINBACH und GARBUS [7] ist für chlorierte Phenole neben der Entkopplerwirkung die Bindung an Mitochondrienproteine von essentieller Bedeutung. Chloroneb und DCMP unterscheiden sich hierin offenbar nur geringfügig, so daß ihr primärer

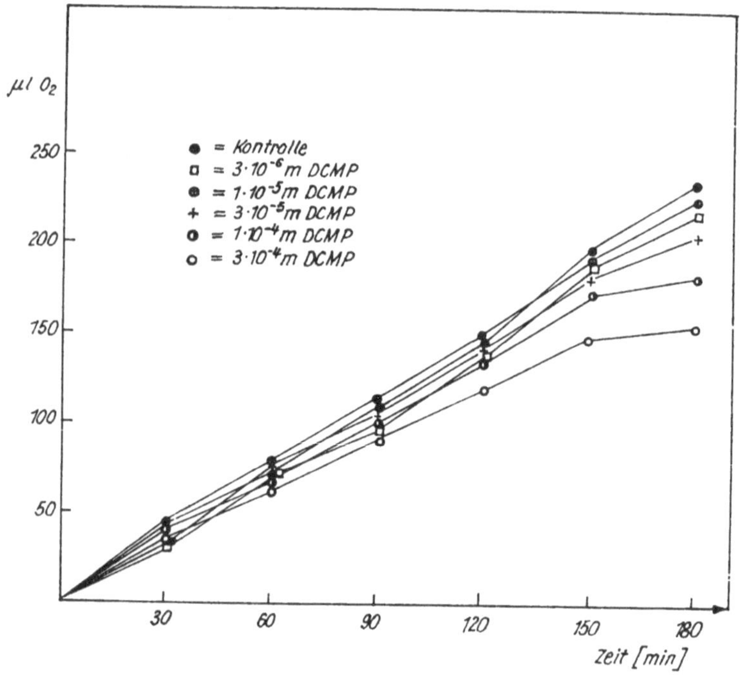

Abb. 3. Sauerstoffaufnahme durch Keimmyzelien von *Mucor mucedo* im Warburg-Atmungstest nach Inkubation verschiedener Konzentrationen von 2,5-Dichlor-4-methoxyphenol (nach LYR, unveröffentlicht)

Wirkungsmechanismus identisch ist. DCHQ hat dagegen einen stärkeren Entkopplereffekt, aber keine Chloroneb-analoge Wirkung.

Chloroneb ist als Fungizid besonders dadurch interessant, daß es eine ausgeprägt selektive Wirkung gegenüber einigen Pilzarten, darunter den wirtschaftlich bedeutsamen Oomycetengattungen *Pythium* und *Phytophthora* besitzt. Unsere Ergebnisse deuten die Möglichkeit an, daß die artspezifisch unterschiedliche Affinität von Mitochondrienmembranen eine Ursache für diese Selektivität ist.

Vergleichende Untersuchungen der Aminosäure-Zusammensetzung von Mitochondrienmembranen-Protein aus einem sensiblen und einem resistenten *Mucor mucedo* Stamm ergaben eine mutative Veränderung der Aminosäurezusammensetzung. So wurde beim resistenten Stamm, der aus dem sensiblen über mehrere Passagen isoliert worden war (BISCHOFF und LYR [2]), nur noch ein Tyrosin-Anteil von 0,2% festgestellt, während der Tyrosin-Gehalt des sensiblen Stammes 2,9% betrug. Alle anderen Aminosäuren waren bei R- und S-Stämmen in absolut gleichen Proportionen vorhanden (Tab. 4). Modellvorstellungen von AESCHBACH et al. [1] zufolge ragen die hydrophoben Tyrosinreste aus den Peptidketten heraus und tragen durch Wasserstoffbrückenbildung wesentlich zur Stabilisierung der Proteinstruktur bei. Denkbar wäre, daß Chloroneb auf bisher unbekannte Weise mit diesen hydrophoben Resten in Wechselwirkung tritt und entweder Protein-Bindungen oder die spezifischen und anfälligen Protein-Phospholipid-Bindungen stört. Diese Wechselwirkungen könnten zu Konfigurationsänderungen, allosterischen Aktivierungen oder Inaktivierungen der an die Mitochondrienmembran gebundenen Enzyme führen, was eine unspezifische Re-

Tabelle 4
Aminosäure-Zusammensetzung der Mitochondrienproteine von R- und S-Stämmen von Mucor mucedo (Auswahl)

Aminosäure	R 3/1	S 3
Lysin	7,1	6,9
Histidin	2,8	2,7
Threonin	6,3	5,7
Serin	7,1	6,8
Alanin	10,5	10,1
Valin	7,5	7,5
Methionin	1,3	1,9
Leucin	9,2	8,8
Tyrosin	0,2	2,9
Phenylalanin	4,8	4,7

duktion des O_2-Verbrauches erklären würde. Denbkare Folge solcher Wechselwirkungen könnte auch die Freisetzung bzw. Aktivierung membrangebundener Proteasen sein, die Membranzerstörungen bewirken und dadurch einen Katastrophenzyklus einleiten, ähnlich wie er für Terrazol [6] beschrieben wurde. Als Ursachen der Selektivität sind wahrscheinlich auch neben unterschiedlichen Proteinstrukturen Unterschiede in der Lipid-Zusammensetzung z. B. die Phospholipid-Sterol-Relation bedeutsam, die die räumliche Struktur der Membran und ihre Stabilität beeinflußt. Dem Problem der selektiven Wirkung von Chloroneb soll in weiteren Untersuchungen nachgegangen werden. Bei einzelnen Pilzarten können unterschiedliche Ursachen vorliegen, wie der Nachweis von 2 bis 5 differenten R-Genen bei aromatischer Hydrokarbon-Fungizid-Resistenz zeigt [3]. Neben der Mitochondrienmembran binden offenbar auch andere Membranen das EPR Chloroneb, was Effekte im Kernbereich und Anomalien der Chromosomen-Verteilung zur Folge haben könnte (Georgopoulos [3]).

Diskussion

Schwabe: Wurden auch die Strukturproteine der Mitochondrien isoliert und die Bindung an diese bei sensiblen und resistenten Stämmen untersucht?

Werner: Es wurden keine gereinigten Proteine untersucht.

Grossmann: Sie erwähnten die gute Wirkung von Chloroneb gegen Oomyceten (*Pythium, Phytophthora*). Haben Sie geprüft, ob auch die Mitochondrien-Membranen solcher Pilze eine hohe Affinität zu Chloroneb aufweisen?

Werner: Affinitätsuntersuchungen an Oomyceten-Membranen wurden bisher nicht durchgeführt.

Josepovits: Sie haben gezeigt, daß Chloroneb und sein Metabolit Dichloro-methoxy-phenol eine ähnliche fungitoxische Wirkung haben. Haben Sie geprüft, ob dieser Metabolit gleiche Unterschiede in der Bindung an Mitochondrien-Membranen des empfindlichen und resistenten Stammes zeigt wie Chloroneb und ob Kreuzresistenz zwischen beiden Verbindungen besteht?

Werner: An markierten Metaboliten haben wir keine Bindungsuntersuchungen durchgeführt. Zwischen Chloroneb und DCMP besteht in hohem Maße Kreuzresistenz.

DAVIDSE: Did you look at the kinetics of the binding of chloroneb to the mitochondrial protein? It should be interesting to know whether you can saturate the binding site, indicating a specific binding.

WERNER: Wir haben keine kinetischen Untersuchungen zur Bindung von Chloroneb an Mitochondrienprotein durchgeführt. Wir haben bisher kein gereinigtes Protein hergestellt, um solche hochspezifischen Bindungsuntersuchungen durchführen zu können.

SCHEWE: Aufgrund der hier vorgestellten ultrastrukturellen und biochemischen Veränderungen der Membranen muß man mit einer möglichen Wirkung des Chloroneb über den Radikalstoffwechsel rechnen. Xenobiotika, die in situ Radikale bilden können, führen in tierischen Zellen zu ähnlichen Veränderungen. Daher wäre es interessant, die zelleigenen Schutzsysteme gegen Radikalschädigungen, wie z.B. Superoxid-Dismutase und Glutathion-Peroxidase, sowie den SH-Gehalt der Mitochondrien von resistenten und empfindlichen Pilzen vor und nach Chloroneb-Behandlung zu untersuchen. Auch die auffälligen Unterschiede im Tyrosingehalt ließen sich in dieser Hinsicht erklären, da Tyronsinreste bei Radikalschädigungen besonders leicht angegriffen werden.

Literatur

[1] AESCHBACH, R., R. AMADO, H. NEUKOM, Formation of Dityrosine Cross-Links in Proteine by Oxydation of Tyrosine Residues, Biochem. et. Biophys. Acta **439** (1976), 292

[2] BISCHOFF, G. und H. LYR, Das Resistenzverhalten von *Mucor mucedo* L. FRES. gegenüber Chloroneb und anderen Fungiziden, Arch. f. Phytopathol. u. Pflanzenschutz 1980, im Druck

[3] GEORGOPOULOS, S. G., Development of fungal resistance to fungicides, In: Antifungal Compounds, Vol. II, M. R. SIEGEL and H. D. SISLER, DEKKER, New York, 1977

[4] HOCK, W. K. and H. D. SISLER, Specifity and Mechanism of Antifungal Action of Chloroneb, Phytopathology **59** (1969a), 627

[5] HOCK, W. K. and H. D. SISLER, Metabolism of Chloroneb by *Rhizoctonia solani* and other Fungi, J. Agr. Food Chem. **17** (1969b), 123

[6] RADZUHN, B., Zum Wirkungsmechanismus des systemischen Fungizide Terrazol Dissertationsschrift, Humboldt-Universität zu Berlin 1978

[7] WEINBACH, E. and J. GARBUS, The Interaction of uncoupling Phenols with Mitochondria and with Mitochondria Protein, J. Biol. Chem. 240 (1965), 1811

[8] WERNER, P., H. LYR und G. CASPERSON, Die Wirkung von Chloroneb, einigen Abbauprodukten und chlorierten Phenolen auf das Wachstum und die Ultrastruktur einiger Pilzarten, Vortrag Intern. Symposium „Systemfungizide" Reinhardsbrunn 1977, Akademie-Verlag 1979, 299

[9] WIESE, M. V. and J. M. VARGAS, Interconversion of Chloroneb and 2,5-Dichlor-4-methoxyphenol by soil Microorganisms, Pesticide Biochem. and Physiology **3** (1973), 214

H. Lyr und G. Casperson

Institut für Pflanzenschutzforschung der Akademie der Landwirtschaftswissenschaften, Kleinmachnow, DDR

Pathologische Zellwandsynthese bei *Mucor mucedo* (L.) Fres. unter Einwirkung einiger Fungizide und anderer Verbindungen

Eine anomale Zellwandverdickung unter der Einwirkung wachstumshemmender Konzentrationen einiger Fungizide, z.B. Terrazol, Chloroneb und Pentachlornitrobenzol (PCNB) konnte bei *Mucor mucedo* elektronenoptisch nachgewiesen werden (Lyr u. a. [11], Casperson und Lyr [5, 6], Werner u. a. [24, 25], Radzuhn und Casperson [19]). Hierbei wird das normale Spitzenwachstum mehr oder weniger stark blockiert, es entsteht ein dickwandiges Myzel mit oft bizarren Verzweigungen (Abb. 1). Bei anderen *Mucorales*-Arten wird ein hefeartiges Wachstum (Y-Form) mit gleichzeitiger Zellwandverdickung ausgelöst, wobei die Neigung zum M (mycelial) — Y (yeast) — Dimorphismus bei den einzelnen Arten unterschiedlich stark ausgeprägt ist.

Während man ursprünglich in einer Umschaltung des Stoffwechsels auf Glykolyse (Anaerobiose) oder Schädigung der Atmungsfunktion eine hinreichende Erklärung für den Übergang von der M- zur Y-Form sah (Bartnicki-Garcia [1], Terenzi und Stock [22], Friedenthal u. a. [8], Schulz u. a. [21], zeigen neuere Befunde, daß der Prozeß komplizierter ist. Zwar führt Anaerobiose zur Y-Form, aber nur, wenn verwertbare Hexosen (Glucose > Fructose > Mannose > Galaktose) in ausreichender Konzentration vorhanden sind (Bartnicki-Garcia [1]). Wahrscheinlich ist der endogene cAMP-Spiegel in noch ungeklärter Weise an der morphogenetischen Regulation beteiligt (Paznakow and Sypherd [17]), während GMP keine derartigen Beziehungen erkennen läßt (Orlowski and Sypherd [16]). Anaerob ist ein 4-fach höherer Spiegel an cAMP in der Zelle vorhanden als anerob (Larsen and Sypherd [9]). Bei Zugabe von 1 mM cAMP bildet sich auch aerob die Y-Form aus, jedoch nur in Gegenwart von Hexosen, nicht jedoch mit Glycerin, Succinat oder Maltose (Orlowski and Sypherd [15]).

Interessanterweise kann die Y-Form auch durch einige Fungizide bei den Mucorales induziert werden (Fisher [7]). Es soll in diesem Beitrag versucht werden, die eigenen Ergebnisse in Zusammenhang mit denen aus der Literatur zu diskutieren, zumal sie auch von allgemeinem Interesse für die Zellregulation sind.

Ergebnisse

In den in Tabelle 1 zusammengestellten Ergebnissen fällt auf, daß keine volle Übereinstimmung zwischen Schädigung des Atmungssystems und der Ausbildung der Y-Form besteht. Über die Schädigung der Mitochondrienstruktur liegen nicht von

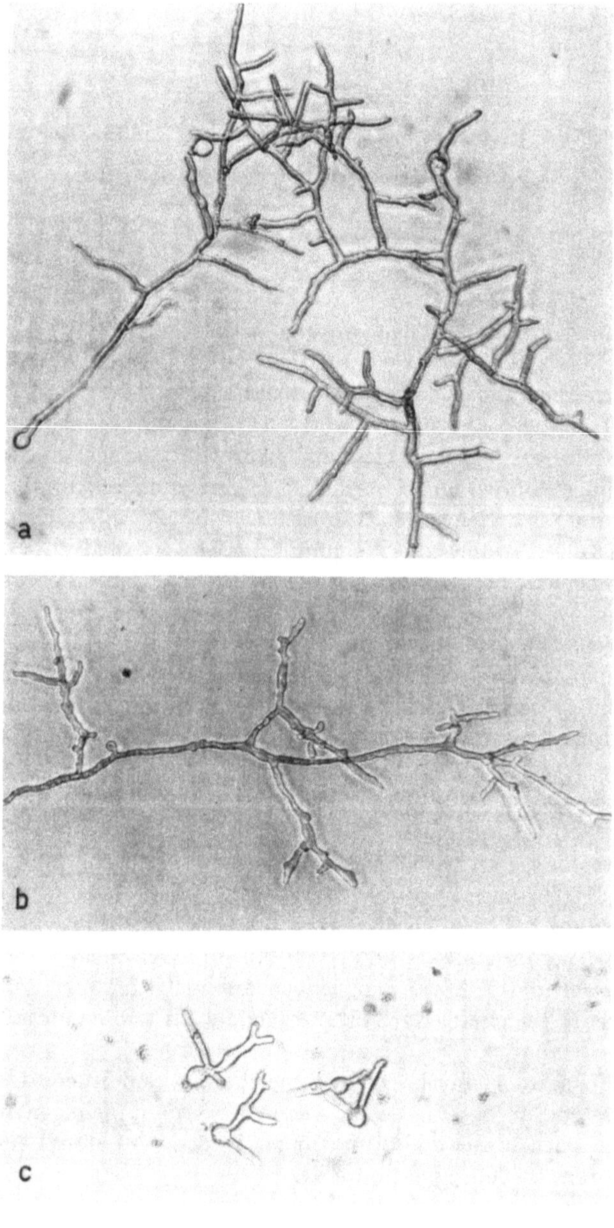

Abb. 1. *Mucor mucedo*, 12 h Kultur mit a) 0 ppm, b) 2 ppm, c) 10 ppm Terrazol. Vergr. 200:1

allen Verbindungen Daten vor. Offensichtlich reicht eine Blockierung der Atmungskette und die damit verbundene Umschaltung auf aerobe Glykolyse nicht aus, um die Morphogenese umzuschalten, was besonders für Amytal, Rotenon, Theonyltrifluorazeton, sowie den Entkoppler 2,4-Dinitropehnol zutrifft (Schulz u. a. [21]).

Die eigenen Versuchsergebnisse sind in Tabelle 2 dargestellt. Die Inkubationszeit betrug, wenn nicht anders vermerkt, 2 Stunden nach einer etwa 4-stündigen Keimzeit der Sporen in 100 ml Flüssigkeitskultur auf Schütteltischen. Nähere Angaben siehe

Tabelle 1
Zusammenstellung einiger Verbindungen, die hefeartiges Wachstum (Y-Form) bei Mucorales auslösen

Verbindung	Konzentration	Wirkung auf Atmungssystem			Schädigung der Mitochondrienstruktur		Y-Form	Organismus	Referenz
β-Phenyläthylalkohol	$1.4 \cdot 10^{-2}$ M	+	+	+	+	+	+	*Mycotypha microspora*	21
	0,22%	+	+	+			+	*Mucor rouxii*	22
Chloramphenicol	$1 \cdot 10^{3}$ M $1 \cdot 10^{3+}$M	+	+		+	+	+	*Mycotypha microspora*	21
Glukose	0,22 M						+	*Mycotypha microspora*	30
Cycloheximid	$1 \cdot 10^{-4}$ M		—			—	—	*Mycotypha microspora*	21
Rotenone	$1 \cdot 10^{-3}$ M	+	+	+			—	*Mycotypha microspora*	21
Theonyltrifluoraceton	$1 \cdot 10^{-4}$ M	+	+				—	*Mycotypha microspora*	21
Antimycin A	$1 \cdot 10^{-5}$ M	+	+				—	*Mycotypha microspora*	21
	$3 \cdot 10^{-4}$ M	+	+	+			+	*Mycotypha microspora*	21 30
2,4 DMP	$1 \cdot 10^{-4}$ M	+	+	+			—	*Mycotypha microspora*	21

Tabelle 2
Einige Versuchsergebnisse an Mucor mucedo.
Dickenzunahme der Zellwand nach 2 Stunden Inkubation in % der Kontrolle. Glukose-Konzentration im Medium 3 %

Agens	Konzentration µg/ml	Zellwanddicke in %	Schädigungen an Mitoch.	Zellkern
Kontrolle	0	100	0	0
Chloroneb	20	225	++	++
3,5-Dichlor-4-methoxyphenol	20	245	++	++
Dichloran	20	218	++	++
Diphenyl	5	182	+	+
	15	95	++	+
Drazoxolon	20	154	+	+
PCNB	20	280	++	++
PCNB + N_2	20	175	++	++
Terrazol	10	344	+++	++
Terrazol + 10% Glukose	10	200	++	++

Casperson und Lyr [5]. Die elektronenmikroskopischen Aufnahmen wurden auf Zellwanddicke hin vermessen und die Veränderungen an Mitochondrien und Zellkern insgesamt erfaßt (Abb. 2 u. 3).

Fungizide, die zu der Gruppe der sogenannten „Aromatischen Hydrokarbon-Fungizide" zusammengefaßt werden, wie PCNB, Chloroneb, Dichloran und Diphenyl ver-

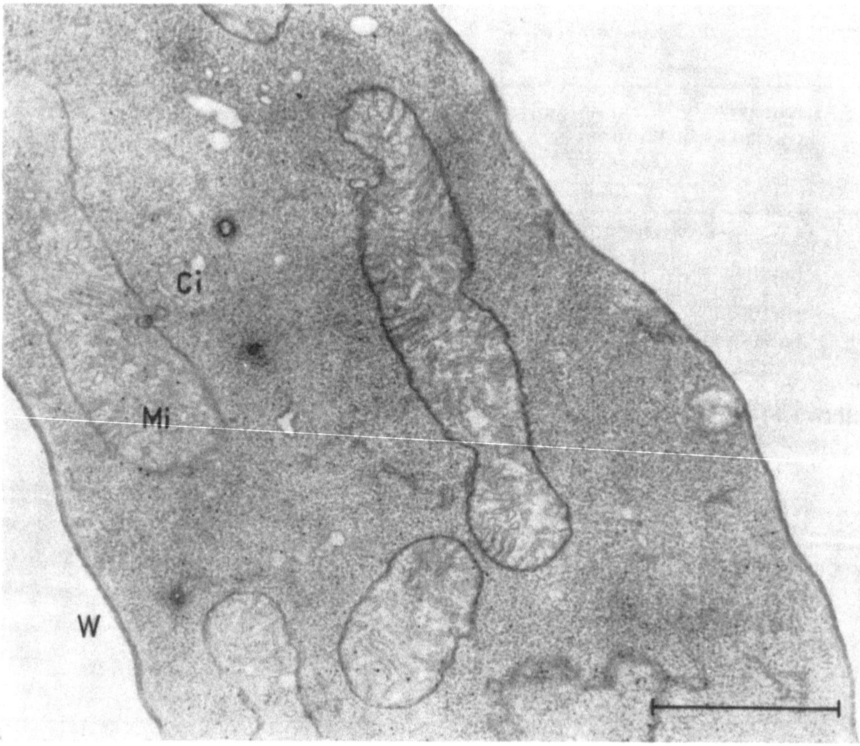

Abb. 2. *Mucor mucedo* ohne Behandlung. Dünne Zellwand (W), Mitochondrien(Mi), Cisternaler Ring (Ci)

halten sich bezüglich der Zellwandverdickung und den morphologischen Veränderungen der Zellstruktur sehr ähnlich (Werner u. a. [24, 25], Casperson und Lyr [6]), so daß man einen gleichen oder sehr ähnlichen Wirkungsmechanismus annehmen kann, was durch den Nachweis von Kreuzresistenz bestärkt wird (Bischoff und Lyr [4]). Nach Werner und Lyr [26] wird Chloroneb an die Mitochondrienmembran gebunden, wobei hydrophobe Aminosäure-Reste, wie der aromatische Ring des Tyrosins, mögliche Bindungsstellen sind. Das gleiche könnte auch für andere Fungizide zutreffen. Obwohl Terrazol (Ethridiazol)einen anderen Wirkungsmechanismus hat als Chloroneb besteht eine Kreuzresistenz zu Chloroneb, deren Deutung allerdings noch aussteht (Bischoff und Lyr [4]). Die Zellwandverdickung zeigt eine deutliche Abhängigkeit zur Zeitdauer des Wirkstoffeinflusses (Abb. 4), wie zur eingesetzten Konzentration der Wirkstoffe (Abb. 5). Während die Kurve der Zellwanddicke bei Terrazol, PCNB und Chloroneb nach einem Anstieg abflacht, durchläuft die Kurve bei Diphenyl ein Maximum bei relativ niedrigen Konzentrationen und fällt dann unter den Kontrollwert ab. Typische Entkoppler, wie 2,4-Dinitrophenol, Pentachlorphenol und 2,4,5- und 2,4,6-Trichlorphenol ergaben bezüglich einer Zellwandverdickung in Übereinstimmung mit den Resultaten von Schulz u. a. [21] keine Effekte, schädigten teilweise aber die Mitochondrienstruktur, wobei Pentachlorphenol eine starke Lysis verursachte. Antimycin A ergab in unseren Versuchen bis zu einer Konzentration von 2 ppm, die die Atmung von *Mucor* völlig blockiert, keine Zellwandverdickung.

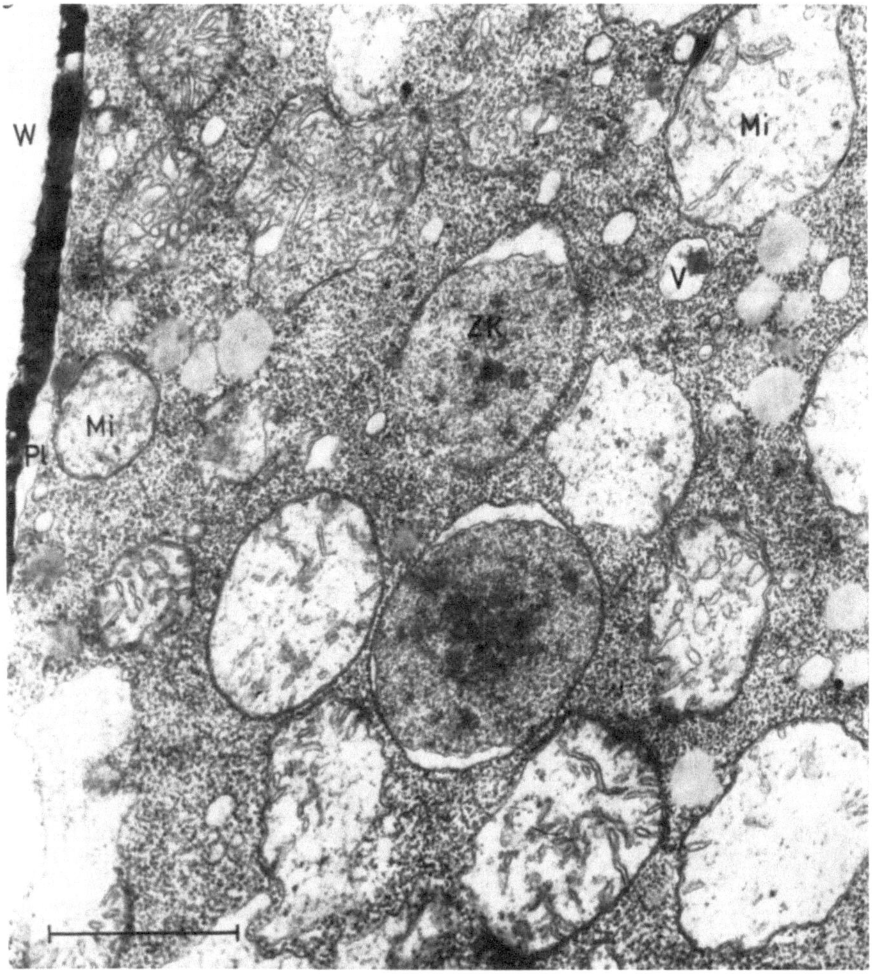

Abb. 3. *Mucor mucedo* nach 2 h Behandlung mit 20 ppm Pentachlornitrobenzol (PCNB). Stark verdickte Zellwand (W), Mitochondrien (Mi) geschwollen, Cristae teilweise in Lysis, Zellkern (ZK) mit Erweiterung des perinukleären Raumes, Plasmalemma (Pl) hebt sich stellenweise von der Zellwand ab, Zelle ist stark vakuolisier (V)

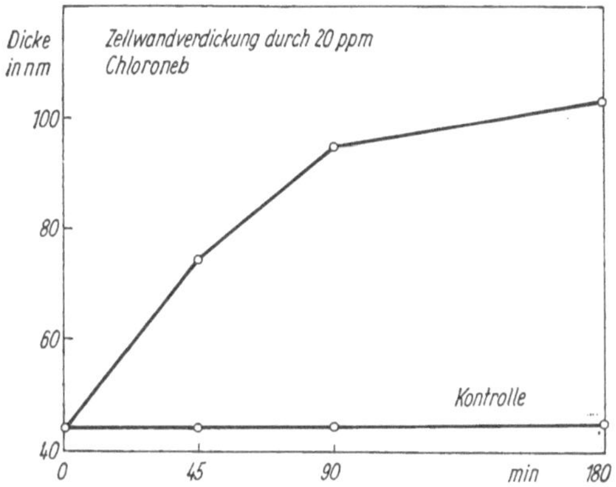

Abb. 4. Zellwandverdickung bei *Mucor mucedo* in Abhängigkeit von der Einwirkzeit von 20 ppm Chloroneb

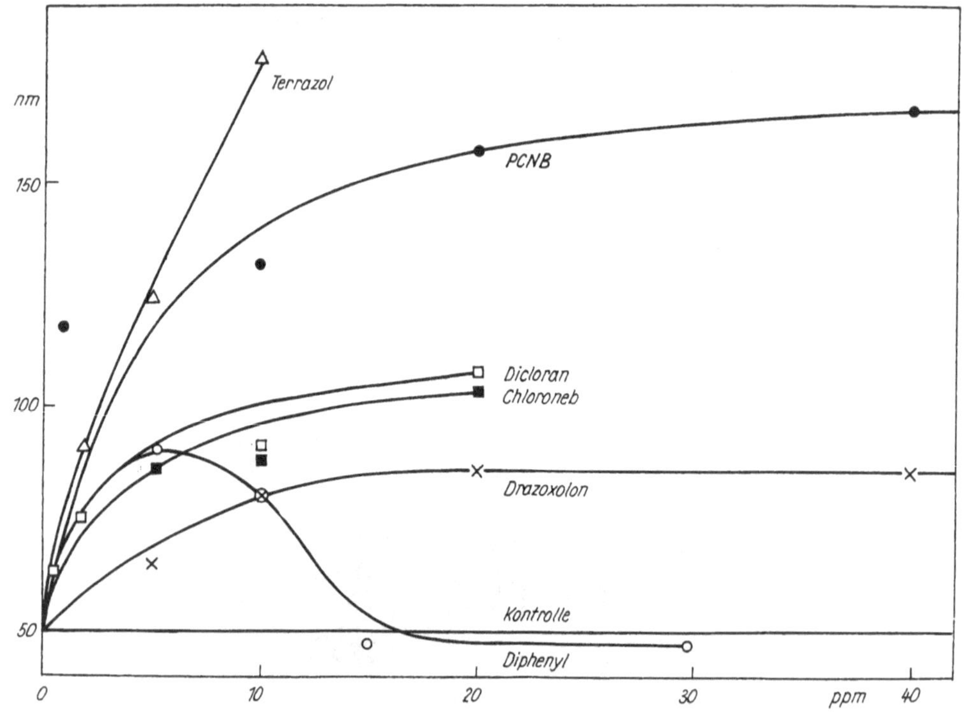

Abb. 5. Zellwandverdickung bei *Mucor mucedo* nach 2 h Behandlung mit verschiedenen Wirkstoffen und Wirkstoffkonzentrationen

Unter anaeroben Kulturbedingungen (Durchströmen der Kultur mit reinem N_2) erhielten wir Zellwandverdickungen nach 2, 4, 8 und 20 Stunden von 90, 100, 116 und 134% bezogen auf aerobe Kontrolle = 100%. Somit läßt sich ein leichter Effekt bei ausschließlicher Glykolyse nachweisen, wie er für die Bildung der Y-Form beschrieben wurde. Gleichzeitiger Zusatz zur anaeroben Kultur von 20 ppm PCNB ergab jedoch Zellwandverdickungen von 133, 138, 240 und 300%. Der 2 Stunden Wert für PCNB ohne N_2 lag in diesem Versuh bei 175%, woraus eine leichte Reduktion der Zellwandverdickung unter anaeroben Bedingungen zu erkennen ist, jedoch keine prinzipielle Blockierung der Fungizidwirkung.

Diskussion

Die Ergebnisse zeigen, daß unter Einwirkung bestimmter Fungizide eine extrem rasche Zellwandsynthese initiiert wird, so daß bereits nach 2 Stunden je nach Wirkstoffart und -konzentration eine Zunahme der Zellwanddicke auf 200 bis 300% erreicht wird. Es ist bisher nicht analysiert, wie sich hierbei das Verhältnis der einzelnen Zellwandkomponenten verschiebt. Macris und Georgopoulos [12] erhielten nach PCNB-Behandlung bei Neurospora crassa eine Erhöhung der Glukansynthese und

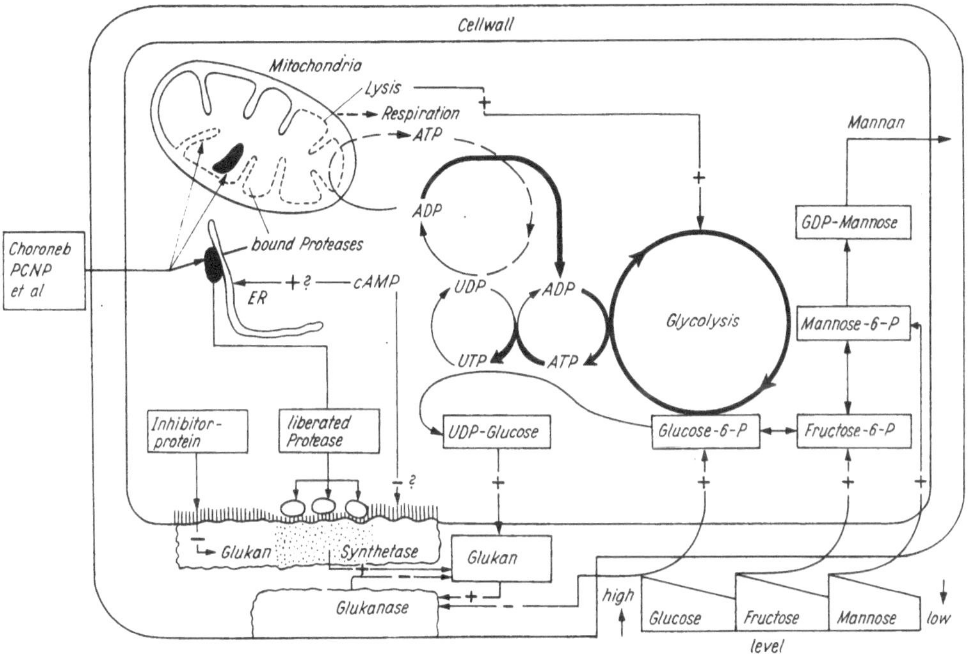

Abb. 6

eine Hemmung der Chitinsynthese. Auch bei der Umschaltung auf Hefewachstum bei *Mucor rouxii* wird die Glucan- und Mannansynthese gegenüber der Chitinsynthese erhöht (Bartnicki-Garcia und Nickerson [3]). Dieses massive Appositionswachstum, das die gesamte Zellwand betrifft, ist mit dem Verlust der Apikaldominanz der Hyphenspitze bezüglich der Wandsynthese verknüpft, wobei je nach Pilzart hefeartiges Wachstum oder abnormes Myzelwachstum mit stark verdickten Zellwänden auftritt. Für die massive Zellwandsynthese, wie sie sowohl unter Fungizid-Einwirkung als auch in abgeschwächtem Maße unter anaeroben Bedingungen abläuft, ist ein Überschuß an Vorstufen (UDP-Glukose, UDP-Azetylglukosamin, GDP-Mannose) sowie an ATP zur Regeneration von UTP bzw. GTP erforderlich. Hierzu ist ein Überschuß an Glukose, Fruktose oder Mannose im Medium erforderlich, die über Fruktose-6-Phosphat als gemeinsames Intermediat konzentrierbar sind. Im Stadium areober Glykolyse ist bei *Saccharomyces cerevisiae* der ATP-Spiegel der Zelle im Maximum bei gleichzeitiger hoher Glukose-Aufnahmerate (v. Meyenburg [13]). Die Glukan-Synthese bei *Candida* wird zu etwa 60% direkt aus der Glykolyse und zu 40% durch Resynthese von Pentose aus dem Pentose-Phosphat-Stoffwechselweg realisiert (Röber und Reuter [20]). Fungizide, wie Chloroneb, Terrazol, PCNB führen zu einer Strukturzerstörung der Mitochondrien und dadurch zu einem begrenzten Absinkten der oxidativen ATP-Bereitstellung und einer Erhöhung des ADP-Spiegels, was zur aeroben Glykolyse führt (Pasteur-Effekt). Allein damit ist jedoch noch nicht die starke Wirksamkeit erklärbar, denn PCNB stimuliert auch unter anaeroben Bedingungen die Zellwandsynthese (Tab. 2), was nicht durch Schädigung der Mitochondrienfunktion erklärt werden kann, die bei den zellwand-aktiven Wirkstoffen typisch ist. Offensichtlich

tritt zugleich eine allgemeine Depression der zellwandgebundenen Chitin- bzw. Glukansynthethasen ein, die nach Ulane und Cabib [23] durch Anwesenheit eines Inhibitorproteins in inaktiver Form an den Zellwänden vorhanden sind.
Aufgrund unserer Untersuchungen zum Wirkungsmechanismus von Terrazol und Chloroneb kamen wir zur Auffassung, daß diese Verbindungen kryptische, membrangebundene Phospholipasen und Proteasen aktivieren, was zur Destruktion der Mitochondrienmembran aber auch zu einem Angriff auf andere Membransysteme führt (Radzuhn [18], Werner u. a. [26]). Da begrenzte Proteolyse ein offenbar weit verbreiteter rasch und irreversibel wirkender Regulationsmechanismus in der Morphogenese ist (Neurath und Walsh [14]), liegt der Gedanke nahe, daß bestimmte Fungizide durch induzierte Protetolyse das Inhibitorprotein der Chitin- bzw. Glukan-Synthetase eliminieren und dadurcheine anomale Zellwandsynthese auslösen, die im Zustand einer aeroben Glykolyse sehr intensiv verläuft, da dann sowohl der ATP-Spiegel als auch der Glukose-6-Phosphat-Spiegel besonders hoch sind. Entkoppler der oxidativen Phosphorylierung und Atmungskettenblocker sind deshalb nicht oder nur sehr gering wirksam, da sie den ATP-Spiegel drastisch senken und wahrscheinlich keine direkte Protease-Aktivierung auslösen.
Inwieweit Fungizide, wie PCNB, die in vitro die Chitin-Synthetase stark stimulieren, auch in vivo einen zusätzlichen Effekt auslösen, ist unklar. Eine direkte Interaktion der Fungizide mit der UDP-N-Acetylglukosamin: Chitin-N-Acetylglukosaminyl-Transferase als Ursache der Zellwandverdickung ist wahrscheinlich, da bei Chloroneb und Terrazol ein solcher Effekt in vitro nicht gefunden wurde (Lyr und Seyd [10]). Es ist denkbar, daß bei Glukose-Zusatz oder beim Übergang zur Anaerobiose ebenfalls spezifische Proteasen aktiviert werden, die zu einer Derepression der Chitin-Synthetase führen. cAMP durfte nach Paznokas und Sypherd [17] durch die Phosphorylierung von spezifischen Proteasen ein wichtiger Regulationsfaktor sein. N^6, $O^{2'}$-Dibutyryl-Adenosin-3,5-cyclo-monophosphat verursacht ähnlich wie ein hoher Spiegel an cAMP Stillstand des Apikalwachstums und Hyphenveränderungen an *Mucor, mucedo* und *Mucor hiemalis* [27].
Andererseits erfolgt bei niedrigem Glukose-Spiegel eine Derepression von Glukanasen die in latenter Form Zellwand gebunden vorliegen [29], die einer anomalen Zellwandsynthese vorbeugt [28]. Die komplexen Beziehungen lassen sich in einem Übersichtsschema darstellen (Abb. 6).

Diskussion

Delp (Comment): Chloroneb unlike the other fungicides discussed here, is a systemic fungicide and has relatively poor activity in vitro against *Pythium* and *Rhizoctonia* compared to rates which control disease in treated seedling.
Seed must be protected from seed rot pathogen with other fungicides like thiram or captan. But once the seedling has emerged, the chloroneb accumulates in the hypocotyl and protects against damping off. Additional mode of action studies may find an effect on the host (perhaps cell wall) to make the host more resistant to the pathogen.

Georgopoulos: I would like to address two questions to Prof. Lyr First, since biphenyl is highly volatile, how did you control concentration in your experiment which seems to indicate a difference in the effect on wall thickness depending on biphenyl concentration?

Second, would you like to comment on the ability of apparently all members of the aromatic hydrocarbon group to disturb the hereditary process in fungi?
LYR: These experiments were performed within 2 hours in nutrition solution where the loss of biphenyl can be neglected.
Second question: This property depends on the mechanism of action of this group of fungicides, which at present is not yet exactly known. It may be, that the observed interference with the nuclear membrane is an indication for this effect, because membrane activity is a general feature of these fungicides, but secondary effects can not be excluded at present.

Literatur

[1] BARTNICKI-GARCIA, S., Control of dimorphism in *Mucor* by hexoses: inhibition of hyphal morphogenesis, J. Bact. **96** (1968), 1586–1594

[2] BARTNICKI-GARCIA, S. and E. LIPPMAN, Fungal morphogenesis: Cell wall construction in *Mucor rouxii*, Science **165** (1969), 302–304

[3] BARTNICKI-GARCIA, S. and W. J. NICKERSON, Isolation, composition and structure of cell walls of filamentous and yeast-like forms of *Mucor rouxii*, Biochim. Biophys. Acta **58** (1962), 102–119

[4] BISCHOFF, G. und H. LYR, Das Resistenzverhalten von *Mucor mucedo* gegenüber Chloroneb und anderen Fungiziden, Arch. f. Phytopathol. u. Pflanzenschutz **16** (1980), 111–118

[5] CASPERSON, G. und H. LYR, Wirkung von Terrazol auf die Ultrastruktur von *Mucor mucedo*, Z. allg. Mikrobiol. **15** (1975), 481–493

[6] CASPERSON, G. und H. LYR, Die Wirkung von Pentachlornitrobenzol (PCNB) auf die Ultrastruktur von *Mucor mucedo* und *Phytophthora cactorum*, Z. allg. Mikrobiol. im Druck

[7] FISHER, D. J., Induction of yeast-like growth in Mucorales by systemic fungicides and other compounds, Trans. Br. mycol. Soc. **68** (1977), 397–402

[8] FRIEDENTHAL, A., A. EPSTEIN, S. PASSERON, Effect of potassium cyanide, glucose and anaerobiosis on morphogenesis of *Mucor rouxii*, J. gen. Microbiol. **82** (1974), 15–24

[9] LARSEN, A. D. and P. S. SYPHERD, Cyclic adenosine 3′,5′-monophosphate and morphogenesis in *Mucor racemosus*, J. Bact. **117** (1974), 432–438

[10] LYR, H. und W. SEYD, Hemmung der Chitin-Synthetase von *Mucor rouxii* in vitro durch Fungizide und andere Werkstoffe, Z. allg. Mikrobiol. **18** (1978), 721–729

[11] LYR, H., B. LAUSSMANN und G. CASPERSON, Wirkungsmechanismus von Terrazol, Z. allg. Mikrobiol. **15** (1975), 345–355

[12] MACRIS, B. and S. G. GEORGOPOULOS, Apparent effect of pentachlornitrobenzene on the hyphal wall polysaccharides of two strains of *Neurospora crassa*, Z. allg. Mikrobiol. **13** (1973), 415–423

[13] MEYENBURG, K., von Katobolit-Repression und der Sprossungszyklus von *Saccharomyces cerevisiae*, Vierteljahresschr. Naturf. Ges. Zürich, **114** (1969), 113–222

[14] NEURATH, H. and K. A. WALSH, Role of proteolytic enzymes in biological regulation. A review, Proc. Natl. Acad. Sci. USA **73** (1976), 3825–3832

[15] ORLOWSKI, M. and P. S. SYPHERD, Cyclic guanosine 3′,5′-monophosphate in the dimorphic fungus *Mucor racemosus*, J. Bact. **125** (1976), 1226–1228

[16] ORLOWSKI, M. and P. S. SYPHERD, Cyclic guanosine 3′,5′-monophosphate in the dimorphic fungus *Mucor racemosus*, J. Bact. **125** (1976), 1226–1228

[17] PAZNOKAS, L. J. and P. S. SYPHERD, Respiratory capacity, 3′,5′-cyclic adenosine monophosphate and morphogenesis of *Mucor racemosus*, J. Bact. **124** (1975), 134 to 139

[18] RADZUHN, B., Zum Wirkungsmechanismus des systemischen Fungizids Terrazol, Dissertation, Universität Berlin, 1978

[19] RADZUHN, B. und G. CASPERSON, Zum Wirkungsmechanismus von Terrazol, In: LYR, H. und C. POLTER (Hsg) Systemfungizide Abh., Akad. Wiss. DDR Nr. 2 (1979), 195–206

[20] RÖBER, B. und G. REUTER, Biosynthese der Zellwand-Polysaccharide Mannan und Glucan als Spiegelbild unterschiedlicher Abbauwege der Glucose durch *Candida* spec., Z. allg. Mikrobiol. **19** (1979), 187–194

[21] SCHULZ, B. E., G. A. KRAEPELIN and W. HINKELMANN, Factors affecting dimorphism in Mycotypha (Mucorales): a correlation with the fermentation/respiration equilibrium, J. gen. Microbiol. **82** (1974), 1–13

[22] TERENZI, H. F. and R. STOCK, Stimulation of fermentation and yeast-like morphogenesis in *Mucor rouxii* by phenethyl alkohol, J. Bact. **97** (1969), 1248–1261

[23] ULANE, R. F. and E. CABIB, The activating system of chitin synthetase from *Saccharomyces cerevisiae*, J. Biol. Chem. **249** (1974), 3418–3422

[24] WERNER, P., H. LYR und G. CASPERSON, Die Wirkung von Chloroneb, seinen Abbauprodukten, sowie von chlorierten Phenolen auf das Wachstum und die Ultrastruktur verschiedener Pilzarten, Arch. Phytopath. u. Pflanzenschutz **14** (1978), 301–312

[25] WERNER, P., H. LYR und G. CASPERSON, Die Wirkung von Chloroneb, einigen Abbauprodukten und chlorierten Phenolen auf das Wachstum und die Ultrastruktur einiger Pilzarten, In: LYR, H. und C. POLTER (Hsg) Systemfungizide, Abh. Akad. Wiss. DDR Nr. 2 (1979), 299–310

[26] WERNER, P. und H. LYR, Der Wirkungsmechanismus von Chloroneb und die möglichen Ursachen der Resistenz gegenüber dieser Verbindung, Symposium Reinhardsbrunn 1980

[27] JONES, B. E. and J. D. BU'LOCK, Die Wirkung von $N^6,O^{2'}$-Dibutyryl-Adenosin-3,5-cyclomonophosphat auf die Morphogenese der Mucorales, J. gen. Microbiol. **103** (1977), 29–36

[28] SANTOS, T., J. R. VILLANUEVA and G. NOMBELA, Regulation of β-1,3-glucanase synthesis in *Penicillium italicum*, J. Bact. **133** (1978), 542–548

[29] KRITZMAN, G., I. CHET and Y. HENIS, Localization of β-(1,3)-glucanate in the mycelium of *Sclerotium rolfsii*, J. Bact. **134** (1978), 470–475

[30] HALL, M. G. und N. KOLANKAYA, The physiology of mould-yeast dimorphism in the genus *Mycotypha* (Mycorales), J. gen. Microbiol. **82** (1974), 25–34

P. Leroux, R. Fritz and M. Gredt

Institut National de la Recherche Agronomique, Laboratoire de Phytopharmacie, CNRA, 78000 Versailles — France

Cross-Resistance between 3,5-Dichlorophenyl Cyclic Imide Fungicides (Dichlozolin, Iprodione, Procymidone, Vinchlozolin) and Various Aromatic Compounds

Introduction

The 3,5-dichlorophenyl cyclic imides (dichlozolin, iprodione, procymidone, vinchlozolin, ...) are new fungicides. They are active against various phytopathogenic fungi and they are used especially for control of grew mould when resistance to benzimidazole fungicides occurs.

Resistance to iprodione, procymidone or vinchlozolin is readily obtained in *Botrytis cinerea*, *Penicillium expansum* and *Ustilago maydis* under laboratory conditions. Cross-resistance always occurs between the 3,5-dichlorophenyl cyclic imides. This phenomenon is also observed with chloroneb, dichloran, biphenyl and quintozene (Leroux et al. [6, 7]).

In this work we investigated the effects of various other compounds (especially aromatics) on wild and mutant strains of *Aspergillus nidulans*, *Botrytis cinerea*, *Penicillium expansum* and *Ustilago maydis*.

Material and Methods

Fungal strains

Aspergillus nidulans strain 003 was generously supplied by J. M. van Tuyl (Laboratory of Phytopathology, Wageningen, the Netherlands). Mutant strains AC, AD and AI descended from strain 003 were selected on nutrient medium containing chloroneb, dichlobenil or iprodione (100 µg pesticide/ml medium). Strain CHL-4, tolerant to chloroneb was provided par J. M. van Tuyl.

The sensitive strains of *Botrytis cinerea*, (S) *Penicillium expansum* (P) and *Ustilago maydis* (M) were the same as those described previously (Leroux et al. [7]). *Botrytis cinerea* strains SUG 2 and SOV 2 were selected respectively on iprodione and vinchlozolin (30 µg/ml). *Penicillium expansum* strain POG 2 was selected on iprodione (100 µg/ml) and *Ustilago maydis* strain MUC 1 on chloroneb (100 µg/ml).

Chemicals

Technical or pure fungicides (dichlozolin, iprodivne, procymidone, vinchlozolin, chloroneb, dichloran, hexachlorobenzene, quintozene and tecnazene) were kindly provided by BASF (Ludwigshafen-am-Rhein, Germany), Du Pont de Nemours (Welmington, USA), Hoechst (Frankfurt, Germany), Rhône Poulenc (Vitry, France) or Sumitomo (Hyogo, Japan). Technical chlorthiamid and dichlobenil were generously supplied by Shell (Sittingbourne, England) and Philips-Duphar (The Netherlands). The other chemicals were obtained from Ega-Chemie (Germany), Eastman (USA), Fluka-Buchs (Switzerland), Interchim (France), Pfalz and Bayer (USA) and Prolabo (France).

Growth tests

Inhibitory effects of toxicants on mycelial growth were measured as described previously (Leroux et al. [6]). All tests were carried out on solid medium and chemicals were added in methanol to molten agar; the final solvant concentration was 1%. Inoculation was realized by placing inverted agar disc with young mycelium on the agar surface. With *Ustilago maydis* similar experiments were conducted, using disc containing sporidies instead of mycelium. Incubation took place at 35 °C for *Aspergillus nidulans* and 25 °C for the other fungi. Radial growth was measured every day during one week. For each condition, the concentration (ED 50) which gave a growth rate inhibition of 50% was calculated.
The resistance factor (R) was defined as the ratio between the ED-50 value of resistant strain and the ED 50-value of sensitive strain.

Results

Dichlozolin, iprodione, procymidone and vinchlozolin were very effective against wild strains of fungi. ED 50-values between 1 and 10 µg/ml were obtained with *Aspergillus nidulans*, *Penicillium expansum* or *Ustilago maydis* and they were below 1 µg/ml with *Botrytis cinerea*. Strains selected either on chloroneb, dichlobenil, iprodione or vinchlozolin were tolerant to all 3,5-dichlorophenyl cyclic imide fungicides. Resistance factors were always greater than 100 (Fig. 1, Table 1).
High levels of resistance were also observed with several polynuclear aromatics especially acenaphthene, fluorene and phenanthrene (Fig. 1, Table 1). Resistance factors were generally below 10 with naphthalene and related compounds (Table 1). With biphenyl derivatives, resistance often occured but isomers generally had different comportments. Two- substitution was usually found to impart more toxicity towards sensitive strains and higher level of resistance, than four-substitution. This phenomenon was specially clear with hydroxybiphenyls (Fig. 2).
In our experiments the most effective compound among the chlorinated benzenes was pentachlorobenzene. The increasing number of chlorine atoms from two to five enhanced the fungitoxicity (Table 2 and 3). The resistance level was always greather than 25 with pentachlorobenzene. A mutant of *Botrytis cinerea* selected on this toxicant was also resistant to the 3,5-dichlorophenyl cyclic imides. Chlorinated anilines were generally more active than chlorinated benzenes. Resistance factors greater than 100 were obtained with 2,3,4,6-tetrachloroaniline and pentachloroaniline (Table 2). High levels of resistance were also obtained with some chlorinated nitrobenzenes, especially with the fungicides tecnazene and quintozene (Table 2).
Among the aromatic fungicides mentioned in Table 4, resistance occured to dicloran and chloroneb but not to chlorothalonil and pentachlorophenol. This phenomenon was also observed with the herbicides chlorthiamid and dichlobenil.
Mutant strains of *Botryis cinerea* had a normal growth on medium containing 1% of methanol whereas the wild strain was partially inhibited. Experiments conducted with higher concentrations of methanol indicated that ED 50-values for resistant and sensitive strain were 1. 1% and 5%. This phenomenon was also observed with the other fungi (Fig. 3).

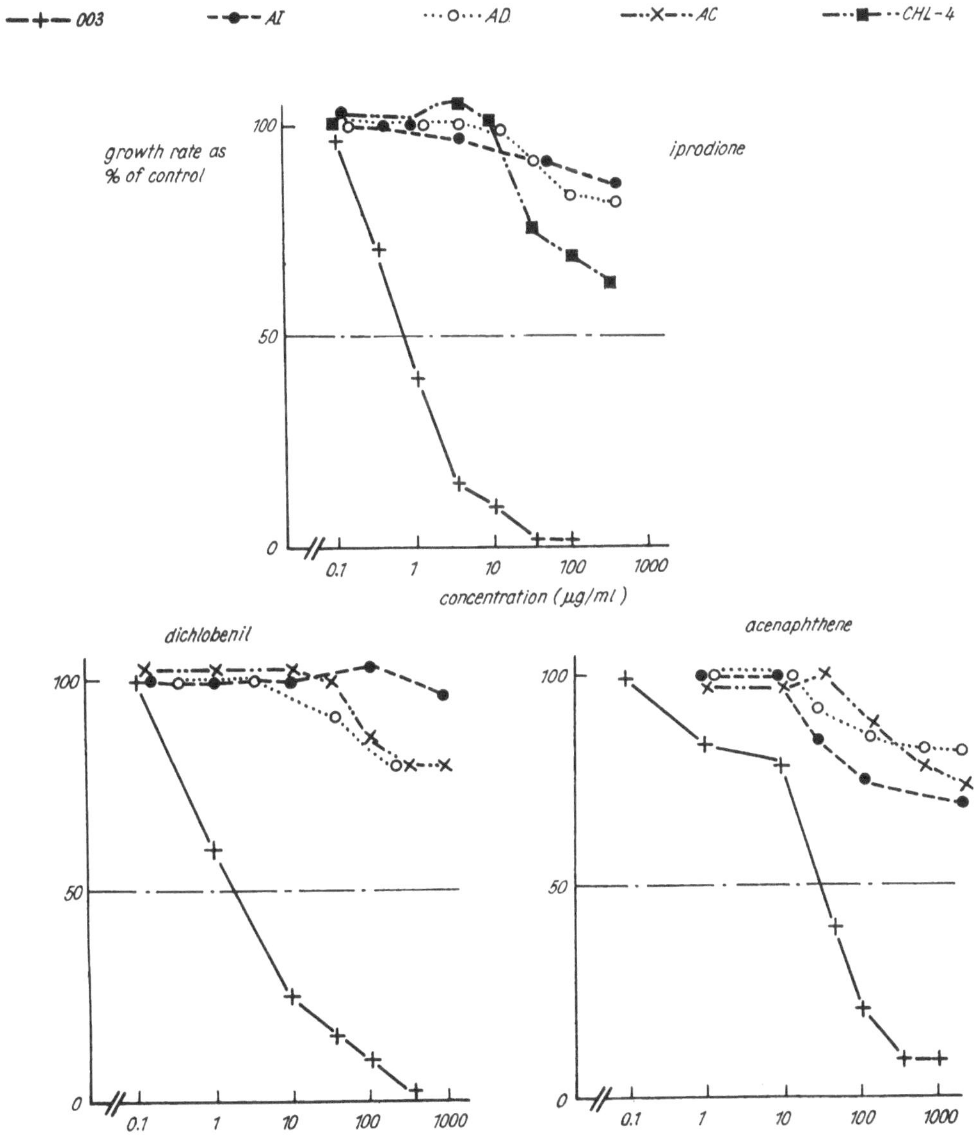

Fig. 1. Effects of iprodione, acenaphthene and dichlobenil on mycelial growth of *Aspergillus nidulans*

Discussion

Strains of *Aspergillus nidulans, Botrytis cinera, Penicillium expansum* and *Ustilago maydis* tolerant to 3,5-dichlorophenyl cyclic imide fungicides are also resistant towards toxicants belonging to the "aromatic hydrocarbon group" defined by Georgopoulos and Zaracovitis [3]. Our results indicated that the herbicides chlorthiamid and dichlobenil can be included in this group.

Table 1
Effects of 3,5-dichlorophenyl cyclic imides and polynuclear aromatics on mycelial growth of fungi

Toxicants	*A. nidulans*		*B. cinerea*		*P. expansum*		*U. maydis*	
	ED 50(a)	R(b)	ED 50(a)	R(b)	ED 50(a)	R(b)	ED 50(a)	R(b)
	[003]	(c)	[S]	(d)	[P]	[POG 2]	[M]	[MUC 1]
3,5-DICHLOROPHENYL CYCLIC IMIDES								
dichlozolin			4	4	3	4		
iprodine	4	4	4	4	3	4	3	≥3
procymidone	3	4	4	4	3	4	4	≥3
vinchlozolin	3	4	4	4	3	4	3	≥3
POLYNUCLEAR AROMATICS								
acenaphthene	2	≥3	3	4	3	4	3	≥3
fluorene	2	≥3	3	4	3	4	3	≥3
phenanthrene	3	4	3	4	3	4	3	4
naphthalene			2	1	1	1	1	≥1
α-naphthol			1	0	2	0	2	1
β-naphthol			2	0/1	2	0	2	1
α-naphthylamine	2	0	1	0	1	0	0	1
biphenyl	2	1	2	≥3	2	>2	2	2
2-bromobiphenyl	2	1/2	3	2	3	1	3	2
4-bromobiphenyl	0	—	2	1/2	1	>1	1	≥1
2-chlorobiphenyl	2	1	2	1/2	2	2	3	3
3-chlorobiphenyl	2	0/1	3	3/4	3	3	3	4
4-chlorobiphenyl	0	—	1	≥1	1/2	>1	1	≥1
arochlor 1232	2	1	3	3	2	>2	2	≥2
arochlor 1242	1/2	1	1	≥1	1	>1	2	≥2
2-aminobiphenyl	2	1	3	2/3	3	2	3	1
2-nitrobiphenyl	2	1	3	2/3	2	2	2	2
2-hydroxybiphenyl	3	1/2	4	3	3	2	3	2
3-hydroxybiphenyl	3	0/1	3	1/2	2	0	3	1
4-hydroxybiphenyl	3	0/1	3	1	2	0	3	1

(a) ED 50-values for sensitive strains; 4 = ED 50 < 1 μg/ml, 3 = 1 μg/ml < ED 50 < 10 μg/ml; 2 = 10 μg/ml < ED 50 < 100 μg/ml; 1 = 100 μg/ml < ED 50 < 1000 μg/ml; 0 = ED 50 < 1000 μg/ml.
(b) resistance factors (see text); 0 = 0 < R < 2.5, 1 = 2.5 < R 10, 2 = 10 < R < 25 3 = 25 < R < 100, 4 = R < 100.
(c) resistance factors were calculated with strains AC, AD and AI.
(d) resistance factors were calculated with strains SUG 2 and SOV 2.

We did not realize genetic analysis, however, the same chromosomal genes probably control resistance to these various compounds.
The biochemical mode of action of either 3,5-dichlorophenyl cyclic imide or aromatic hydrocarbon fungicides ist no eluciated. Some of them inhibit the nucleic acid biosynthesis; this effect can be responsible for the increasing sectoring in diploid strains of *Aspergillus nidulans* (Fritz et al. [2], Georgopoulos et al. [4, 5]).
Some of the tested polynuclear aromatics are procarcinogens which bind, after metabolic activation, to different macromolecules, especially nucleic acids. This activation requires cytochrom P-450 mediated monoxygenase or mixed function

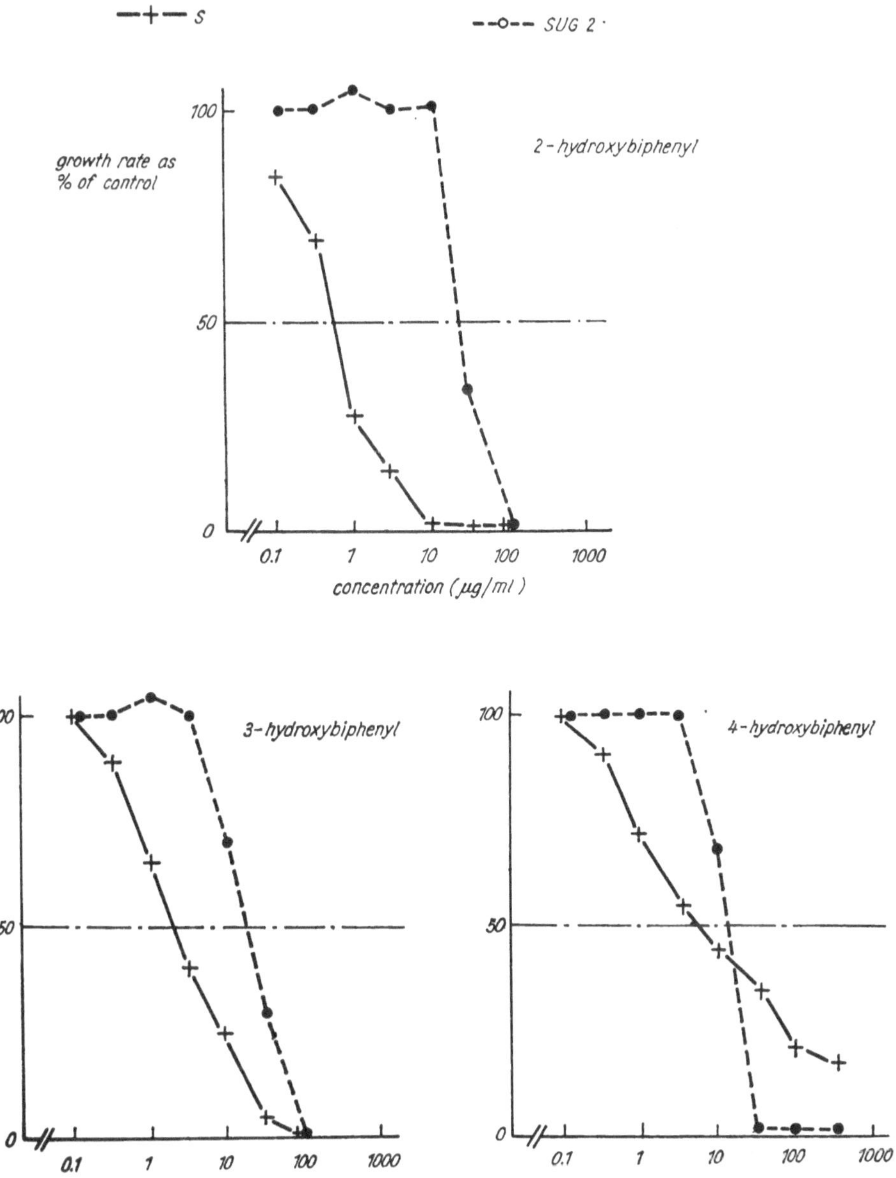

Fig. 2. Effects of hydroxybiphenyls on mycelial growth of *Botrytis cinerea*

oxidase system. In animals this types of cytochrome are induced by various products including aromatic compounds AHOKAS [1].
A special form of cytochrome P-450 (cytochrome P_1-450 or P 448) is induced by polynuclear aromatics in "responsive" mouse strains but the induction is absent in "non responsive" strains. This "responsiveness" to aromatic hydrocarbons is genetically controlled by locus Ah. NEBERT [8].

Table 2
Effects of chlorinated benzenes, chlorinated anilines or chlorinated nitrobenzenes on mycelila growth of fungi (a)

Toxicants	*A. nidulans* ED 50	R	*B. cinerea* ED 50	R	*P. expansum* ED 50	R	*U. maydis* ED 50	R
CHLORINATED BENZENES								
1,2-dichlorobenzene			1	≥2			1	≥1
1,3-dichlorobenzene	0	–	1	≥1	0	–	1	≥1
1,4-dichlorobenzene	0	–	1	≥1	0	–	1	≥1
1,2,3-trichlorobenzene	1	0/1	1	0/1	1	0	1	0
1,2,4-trichlorobenzene	1	0	1	1	1	0	1	1
1,3,5-trichlorobenzene	1	≥1	1	≥1	1	≥1	1	≥1
1,2,3,4-tetrachlorobenzene	1	≥1	2	2	2	1	2	≥2
1,2,3,5-tetrachlorobenzene	1	≥1	2	≥3	2	≥2	2	≥2
1,2,4,5-tetrachlorzene	1	≥1	2	≥3	1/2	≥1	1	≥1
pentachlorobenzene	3	3/4	3	4	3	4	3	≥3
hexachlorobenzene	0	–	0	–	0	–	2	≥2
CHLORINATED ANILINES								
3-chloroaniline	2	1	1	0/1	2	0/1	2	1
4-chloroaniline	1	0	1	0	1/2	0	2	1
3,4-dichloroaniline	2	0/1	2	0/1	2	0	2	1
2,6-dichloroaniline	1	1	2	1	2	1	2	1
3,5-dichloroaniline	3	0/1	3	2	2	1	2/3	1
2,4,6-trichloroaniline	2	1	3	2	3	1	3	1
3,4,5-trichloroaniline	3	1	3	1	3	0	3	4
2,3,5,6-tetrachloroaniline	3	1	4	4	3	3	4	4
pentachloroaniline	2/3	4	3	4	1	≥1	4	4
CHLORINATED NITROBENZENES								
nitrobenzene	1	0/1	1	0/1	0	–	1	1
2-chloronitrobenzene			3	2			2	2
4-chloronitrobenzene			2	1/2	1	0	2	1
3,5-dichloronitrobenzene	2	0/1	3	2	2	1	3	1
2,3,4-trichloronitrobenzene	3	1	3	1/2	2	0	3	0
2,4,5-trichloronitrobenzene	3	1	3	2	3	1	3	1
tecnazene	3	2	3	4	3	4	4	4
quintozene	3	2/3	3	4	3	4	4	4

(a) notations the same as in Table 1.

Table 3
Effects of chloronitated benzenes on mycelial growth of Botrytis cinerea strains S and SUG 2

Chlorinated benzenes	ED 50-values (µg/ml) strain S	strain SUG 2
1,2-dichlorobenzene	950	>1000
1,3-dichlorobenzene	850	>1000
1,4-dichlorobenzene	950	>1000
1,2,3-trichlorobenzene	200	500
1,2,4-trichlorobenzene	300	1000
1,3,5-trichlorobenzene	250	>1000
1,2,3,4-tetrachlorobenzene	25	350
1,2,3,5-tetrachlorobenzene	25	>1000
1,2,4,5-tetrachlorobenzene	30	>1000
pentachlorobenzene	3	>1000
hexachlorobenzene	>1000	>1000

Table 4
Effects of nitroaniline derivatives and various aromatic compounds on mycelial growth of fungi (a)

Toxicants	*A. nidulans*		*B. cinerea*		*P. expansum*		*U. maydis*	
	ED 50	R	ED 50	R	ED 50	R	ED 50	R
NITROANILINE DERIVATIVES								
2-nitroaniline	2	0	2	1/2	1/2	0	2	1
3-nitroaniline	1	0	1	0/1	1	0	1	0
4-nitroaniline	1	0	1/2	0/1	2	0	1	1
2-chloro-4-nitro-aniline	2	1	2	1	2	0	2	0
2,4-dichloro-6-nitro-aniline	3	1/2	3	3	3	1	3	3
dichloran	3	≥3	4	4	3	≥3	3	≥3
VARIOUS AROMATIC COMPOUNDS								
chloroneb	3	3	3	4	3	≥3	3	≥3
chlorthiamid	2	1	3	2	2	1	2	2
dichlobenil	3	4	3	4	3	4	3	≥3
1,3,5-trimethyl-benzene	0	–	0	–	0	–	0	–
1,2,4,5-tetramethyl-benzene	0	0	1	≥1	1	>1	1	≥1
pentamethylbenzene	1	0/1	2	2	2	2	1	≥1
chlorothalonil	4	0	3	0	1	0	1	0
pentachlorophenol			3	0	3	0	3	0

(a) notations the same as in Table 1.

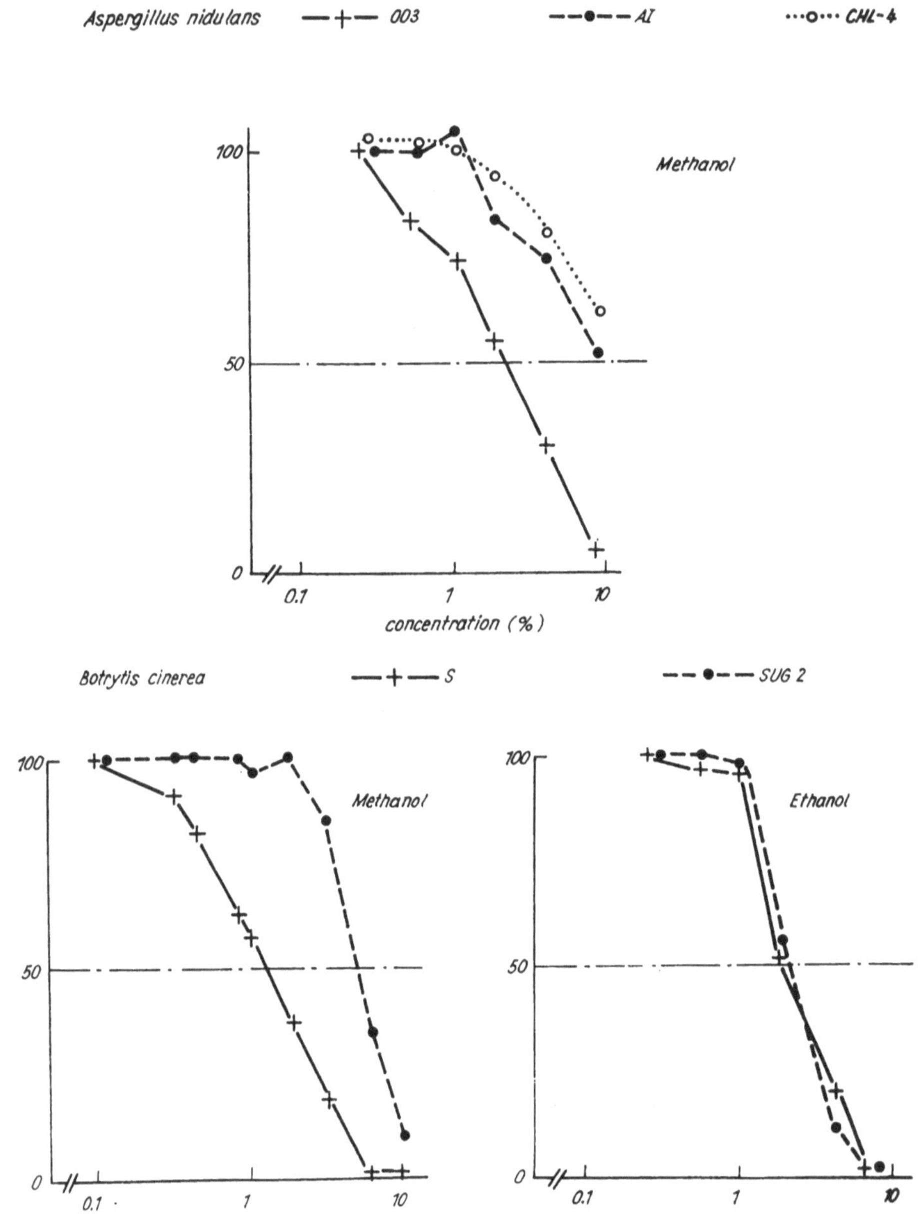

Fig. 3. Effects of ethanol and methanol on mycelial growth of *Aspergillus nidulans* or *Botrytis cinerea*

It is possible that a similar phenomenon occurs with fungi. In other words, resistance can be due to a deficiency in certain forms of cytochrome P-450. The antagonistic effects of methylene dioxyphenyl compounds (piperonylbutoxide, sulfoxide, ...) towards 3,5-dichlorophenyl cyclic imides in sensitive strains is an indirect proof of the interference of these fungicides with cytochrome P-450.

It ist necessary to do new experiments to discover the biochemical mode of action of these various antifungal compounds and to elucidate the mechanism of resistance.

Diskussion

LANCAKE: Is it possible that cross resistance to all these unrelated compounds is due to a common mechanism, for reduced permeability of fungal cells to the fungicides?

LEROUX: We observed, that absorption and "exsorption" of either iprodione or vinchlozoline is similar with resistant and sensitive strains.

LYR: The common structure in all new *Botrytis* active compounds you investigated, seems to be the dichlorobenzene-group. The results on mechanism of chloroneb (see WERNER and LYR) are in agreement with this. If you take pentachlorophenol you have another dominating mechanism of action (uncoupler of oxidative phosphorylation) and you get no cross resistance. Of course some other properties can modify the activity (penetration, inactivation).

LEROUX: As mentioned by other research workers cross resistance between pentachlorophenol or chlorothalonil (this compound reacts with thiol groups) and dichlorophenyl cyclic imides and other aromatics tested does not exist. This observation indicates that these last fungicides are not uncouplers of oxidative phosphorylation and do not react with thiol groups.

SISLER: Does sensitivity to methanol reflect a membrane sensitivity to alternation by hydrocarbon?

LEROUX: It is proved that methanol can be oxidized into formaldehyde by an enzyme requiring cytochrome P_{450}. It is possible that if resistance to aromatic fungicides is due to a deficiency in cytochrome P_{450}, the activation of methanol does not occur.

References

[1] AHOKAS, J. T., Cytochrome P-450 in fish liver microsomes and carcinogen activation in KHAN M. A. Q., J. J. LECH, and J. J. MENN (eds)/Pesticide and xenobiotic metabolism in aquatic system, ACS symposium series **99**, Washington 1979, 279–296

[2] FRITZ, R., P. LEROUX et M. GREDT, Mechanisme de l'action fongitoxique de la promidione (260 19 RP ou glycophène), de la vinchlozoline et du dicloran sur *Botrytis cinerea* Phytopath. Z. **90** (1977), 152–163

[3] GEORGOPOULOS, S. G. and C. ZARACOVITIS, Tolerance of fungi to organic fungicides, Ann. Rev. Phytopathol. **5** (1967), 109–130

[4] GEORGOPOULOS, S. G., A. KAPPAS and A. C. HASTIE, Induced sectoring in diploid *Aspergillus nidulans* as a criterion on fungitoxicity by interference with hereditary processes, Phytopathology **66** (1976), 217–220

[5] GEORGOPOULOS, S. G., M. SARRIS and B. N. ZIOGAS, Mitotic instability in *Aspergillus nidulans caused* by the fungicides iprodione, procymidone and vinchlozolin, Pestic. Sci. **10** (1979), 389–392

[6] LEROUX, P., R. FRITZ et M. GREDT, Etudes en laboratoire de souches de *Botrytis cinerea* résistantes à la dichlozoline, au dichloran, au quintozène, a la vinchlozoline et au 260 19 RP (ou glycophène), Phytopath. Z. **89** (1977), 347–358

[7] Leroux, P., M. Gredt et R. Fritz, Etudes en laboratoire de souches de quelques champignons phytopathogenes résistantes à la dichlozoline, à la dicyclidine, a l'iprodione, à la vinchlozoline et à divers fongicides aromatiques, Med. Fac. Landbouww. Rijskuniv. Gent **43** (1978), 881–889
[8] Nebert, D. J., Genetic differences in microsomal electron transport: the Ah locus. In Fleisher S. and L. Packer (eds.)/Methods in enzymology, volume L II: biomenbranes, Academic Press, New York (1978), 226–240

MAGDA KOVÁCS and M. TÜSKE

Research Institut for Plant Protection, Budapest, Hungary

The Possibility of Development of Resistance to Triadimefon in Obligate Phytoparasitic Fungi

Triadimefon (Bayleton[R]) is one of the most important systemic fungicides of the last five years, which effectively control among others the obligate parasitic fungi of cereals. Earlier benomyl was used against powdery mildews, but lately this excellent fungicide has lost its earlier importance in this field, because of the appearence of benomyl resistant strains. In our study we wanted to find out, whether we have to reckon with similar problems following the use of triadimefon. This problem has been theoretically treated by DELP [2].
Literature data on the problems of fungicide resistance of phytopathogenic fungi are abounding, but informations on the resistance of the obligate parasite powdery mildews are scarcely available even in the reviews of GEORGOPOULOS [6] or FEHRMAN [3].
When we restrict our investigations to fungicides which inhibit sterol-biosynthesis, the field is relatively easy to survey. In their experiments SHERALD et al. [8] and LEROUX and GREDT [7] used a *Ustilago maydis* strain, which was resistant against triarimol and triforine. The problem of resistance against sterol-biosynthesis inhibitors was extensively studied by FUCHS and DRANDAREVSKI [4] and FUCHS et al. [5], but they worked with facultative parasites, as did DE WARD and VAN NISTELROOY [9].

As to the above authors the fungi which are resistant against sterol-biosynthesis inhibitors, do not represent real danger in the agricultural practice because of their decreased fitness. CORBETT [1] does not share this opinion, but he does not elaborate. WALMSLEY-WOODWARD et al. [10] experimented with barley powdery mildew in glasshouse and in the field, but they used ethirimol and tridemorph, and these fungicides are not sterol-biosynthesis inhibitors.

As we could not find in the literature appropriate method which would have best suited our aims, we worked out the following experimental arrangement (Fig. 1).

The circles indicate the pots, which were treated with increasing doses of fungicides. In the arrangement "basic sensitivity" we checked the constancy of the fungicide sensitivity of the fungi used in our experiments; the inoculums were taken always from untreated plants. In the arrangement "forced" the inoculums were taken from plants, where the rate of infection was $50 \pm 10\%$. In the arrangement "one way" the passage was made between the plants treated with the same doses, when the rate of infection was higher than 50%. Plants treated with higher doses were infected with inoculums from plants with an 50% infection rate. In the arrangement "steps forward" the passage was made always to a plant treated the next higher dose.

The experimental conditions are shown on Figure 2.

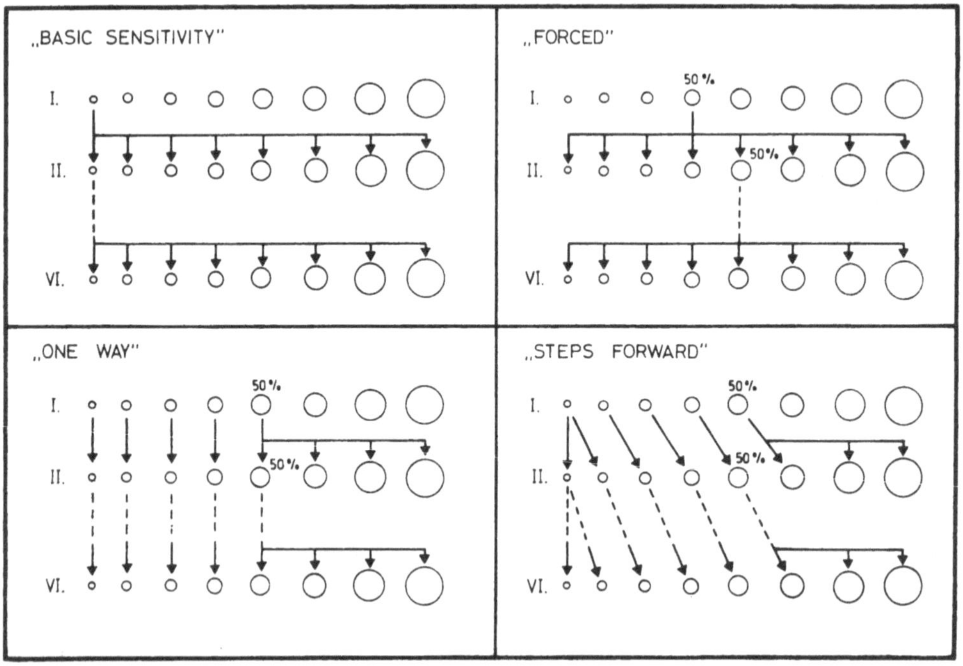

Fig. 1. Scheme of the experimental arrangements

FUNGI	PLANTS	plants per pot	TIME-TABLE (days)						
			sowing	treatm.	inocul.	evaluation			
						1st	2nd	3rd	4th
Erysiphe graminis f.sp. tritici	winter wheat cv. MV-4	50	0.	5.	7.	13.	15.	17.	19.
Erysiphe graminis f.sp. hordei	winter barley cv. MV-37	50	0.	5.	7.	13.	15.	17.	19.
Uromyces appendiculatus	bean cv. Pinto	4	0.	8.	10.	22.	—	—	—
Uromyces vicia-fabae	broad bean cv. Bonus 10	4	0.	11.	13.	25.	—	—	—

Rearing: in plastic pots (200 ml), Vermiculite, nutrient solution

Fungicides: TRIADIMEFON (Bayleton 25 WP), BENOMYL (Chinoin-Fundazol 50 WP)

Treatment: 10 ml via roots; 5 parallel runs per dose

	powdery mildews	rusts
Mode of inoculation:	wind-chamber	pinsel
Mode of evaluation:	number of infected plants number of pustules on the 1st leaf	number of pustules per cm^2 per leaf

Fig. 2. Experimental conditions

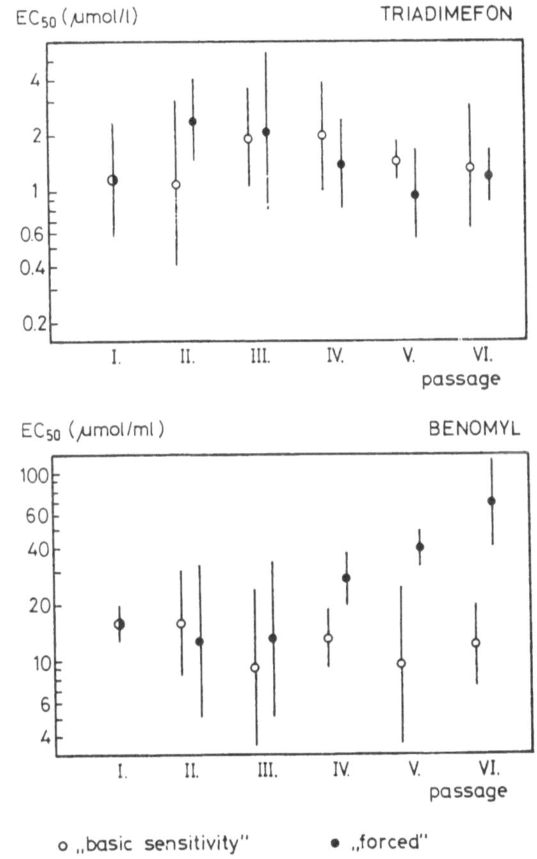

Fig. 3. Changes of the sensitivity of wheat powdery mildew as a result of passages

In our experiments with rust fungi unforseeable phenomena appeared, which made impossible to interpret our results. We found, that the symptoms of disease differred in case of two fungicides, and changed in case of the same fungicides depending on the dose.

The basic sensitivity of wheat powdery mildew had not changed in six passages (Fig. 3). The sensitivities (EC_{50} values with the 95% confidence interwalls) have been calculated from dose-activity data pairs of the 2. evaluation, using the least square method. It can be seen, that in the "forced" arrangement, where the inoculum was obtained from plants with about 50% infection rate, in case of benomyl the EC_{50} values increase, indicating, that the strain more and more tolerates the fungicide. In case of triadimefon this phenomena does not appear, and the sensitivities measured in the "forced" arrangement correspond with the basic sensitivity within the error limit.

In the "one way" arrangement we wanted to find out, whether the rate of infection decreases or remains the same during the passage between plants treated with same doses. The activities of the two fungicides are seen on Figure 4A, where the rates of infection are shown at doses resulting in about 50% infection. No significant changes appear.

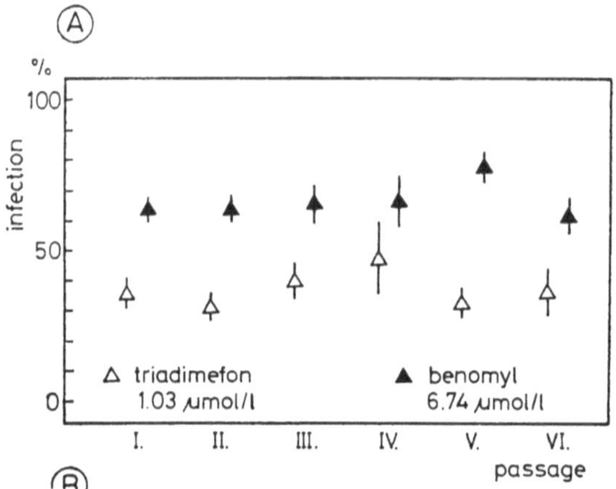

(B)

TRIADIMEFON — dose (μmol/l)

	0	0.59	1.03	1.81	3.16	5.55
I.	100					
II.	100	45				
III.	100	80	38			
IV.	100	90	74	39		
V.	100	64	31	28	1	
VI.		73	47	18	0	×

BENOMYL — dose (μmol/l)

	0	6.74	13.5	26.9	53.9
I.	100				
II.	100	97			
III.	100	69	32		
IV.	100	86	57	23	
V.	100	41	28	7	0
VI.		57	38	19	0

Fig. 4. Wheat powdery mildew
A. Changes of the infection rates of the plants treated with a given dose of benomyl and triadimefon ["one way"]
B. Changes of the infection rates ["steps forward"]

In the "steps forward" arrangement the passages were made from plants to an other plant, treated with the next higher dose in the concentration series. By this way we wanted to see, whether the fungi can be adapted to stepwise increasing doses of fungicides. Figure 4B shows, that rate of infection decreases, as the doses of fungicides increases. The experiment "steps forward" has not revealed any difference between the two fungicides.

The data obtained from the experiments with barley powdery mildew were treated by the similar way. Figure 5 shows, that the sensitivity of barley powdery mildew does not change after the treatment with triadimefon, contrary to the treatments with benomyl, where in the "forced" experiment the VI. passage resulted in EC_{50} values, which were five times higher than the basic sensitivity.

Figure 6 shows, that similary to the experiments with wheat powdery mildew, we have not found significant changes. It is intersting to mention, that in the arrangement "steps forward", the rate of infection temporarily increased with higher doses, in the low concentration region.

In our experiments we wanted to obtain information about the extent of danger of the appearance of resistant strains against triadimefon. In order to check the applicability of our experimental arrangement, we included benomyl as control, which is known to produce resistant strains.
In the "forced" arrangement, we could induce the tolerance of both powdery mildews against benomyl, while the "one way" and "steps forward" experiments have not revealed significant differences between the two fungicides. We concluded, that the "forced" arrangement can be used for the study of the development of resistance.
As in the "forced" treatment with triadimefon the sensitivity of the fungi has not changed after six passages, we concluded in accordance with FUCHS et al. [5], that in the agricultural practice the appearence of resistant strains against sterol-biosyn-, thesis inhibitor fungicides is unlikely. Our conclusion is confirmed by our experience, as to which in each of four arrangements, after treatment with higher doses of benomyl we could often find heavily infected pots, next to the healthy ones. This phenomena have never occured with triadimefon in our experiments, using up 1500 pots and lasting one and a half year.

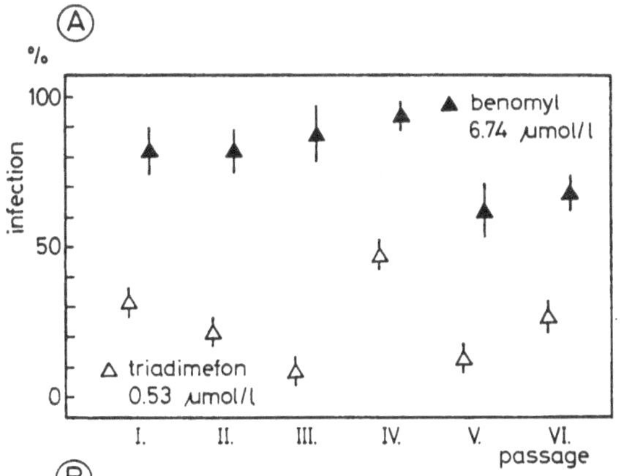

B

TRIADIMEFON — dose (μmol/l)

	0	0.13	0.27	0.53	1.07	2.13
I.	100					
II.	100	139				
III.	100	95	135			
IV.	100	96	94	31		
V.	100	100	105	17	0	
VI.		102	97	26	0	X

BENOMYL — dose (μmol/l)

	0	3.37	6.74	13.5	26.9	53.9
I.	100					
II.	100	98				
III	100	92	67			
IV.	100	100	109	91		
V.	100	91	75	18	23	
VI.		100	63	39	0	0

Fig. 5. Changes of the sensitivity of barley powdery mildew as a result of passages

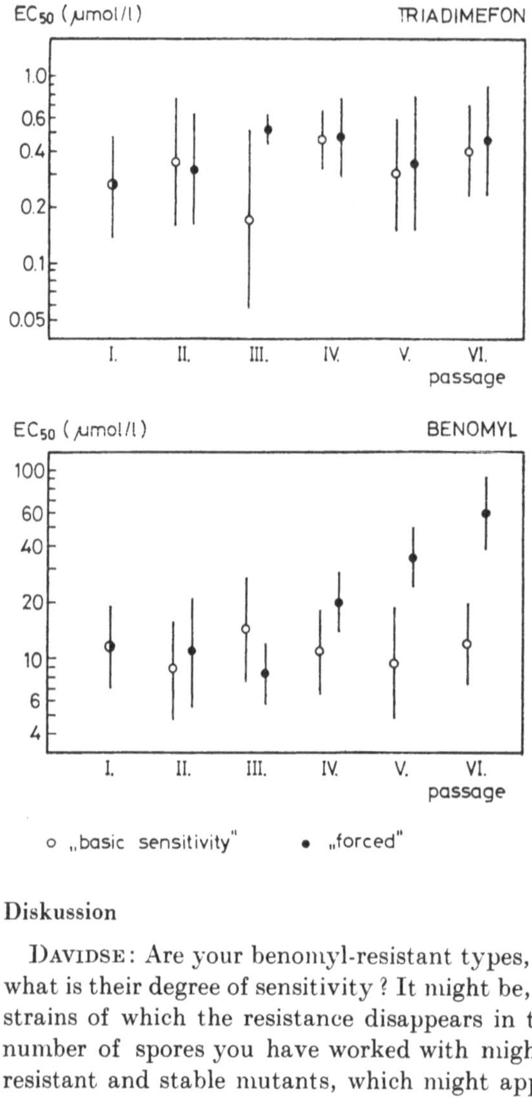

Fig. 6. Barley powdery mildew
A. Changes of the infection rates of the plants treated with a given dose of benomyl and triadimefon ["one way"]
B. Changes of the infection rates ["steps forward"]

Diskussion

Davidse: Are your benomyl-resistant types, you got in the experiments, stable and what is their degree of sensitivity? It might be, that you only got moderately-resistant strains of which the resistance disappears in the absence of benomyl. When so, the number of spores you have worked with might have been too low to select highly-resistant and stable mutants, which might appear at much lower mutation frequencies.
When the mutation frequencies at which triadimefon-resistant mutants might appear are even lower, you cannot select these types in your experiments.
Although the results of your model-system agree with the experience in practice with respect to the absence of development of resistance to triadimefon, they do not necessarily indicate that never resistance will be found after many years of use of triadimefon.

Tüske: In the presence of benomyl the basic sensitivity of wheat and barley powdery mildew lowered to 20 and 12 percent, respectively. We have not tested, whether the strains regain their sensitivity. We think, that we have not obtained highly-resistant strains because the number of passages was not enough, and not because of the low spore population. We agree with the last research.

GEORGOPOULOS: Was the size of the spore population the same whith both fungicides? Can you exclude the possibility of a much lower mutation frequency in the case of triadimefon?

TÜSKE: The size of the spore population was the same. As to the other question: triadimefon is not mutagenic on the basis of available informations.

Literature

[1] CORBETT, J. R., Resistance and selectivity: biochemical basis and practical problems, Proc. 1979 Brit. CropProtect Conf. — Pests and Diseases, Vol. 3, 717—730

[2] DELP, CH. J., Resistance to plant disease control agents; how to cope with it, IX. Intern. Congr. of Plant Protect., Washington (USA), 1979

[3] FEHRMANN, H., Systemische Fungizide — ein Überblick II. Fungizidresistenz phytopathogener Pilze, Phytopath. Z. **86** (1976), 144—185

[4] FUCHS, A. and C. A. DRANDAREVSKI, The likelihood of development of resistance to systemic fungicides which inhibit ergosterol biosynthesis, Neth. J. Plant Pathol. **82** (1976), 85—87

[5] FUCHS, A., S. P. DERUIG, J. M. VAN TUYL and F. W. DEWRIES, Resistance to triforine: a nonexistent problem?, Neth. J. Plant Pathol. **83** (Suppl. 1) (1977), 189—205

[6] GEORGOPOULOS, S. G., Development of fungal resistance to fungicides in SIEGEL, M. R. and SISLER, H. D. (eds.): Antifungal Compounds, MARCEL DEKKER Inc., New York 1977, Vol. 2, 439—495

[7] LEROUX, P. and M. GREDT, Comparaisons des modes d'action fongitoxique du Triadiméfon (Meb 6447), du Triarimol et de la Triforine, Phytopath. Z. **86** (1976), 276—279

[8] SHERALD, J. L., N. N. RAGSDALE and H. D. SISLER, Similarities between the systemic fungicides triforine and triarimol, Pestic. Sci. **4** (1973), 719—727

[9] DEWAARD, M. A. and J. G. M. VAN NISTELROOY, Mechanism of resistance to fenarimol in *Aspergillus nidulans*, Pestic. Biochem. Physiol. **10** (1979), 219—229

[10] WALMSLEY-WOODWARD, D. J., F. A. LAWS and W. J. WHITTINGTON, Studies on the tolerance of *Erysiphe graminis* f. sp. *hordei* to *systemic fungicides*, Ann. appl. Biol. **92** (1979), 199—209

Charles J. Delp, Juan M. Serres and Bruse A. Hadley

E. I. du Pont de Nemours & Co., Inc. Wilmington, DE, USA

Calculations of Buildup of Resistant Fungi against Benomyl in Combinations with Other Fungicides

This paper identifies and relates some of the key factors which influence the development of resistance to benomyl. It deals primarily with the experiences and theoretical calculations for the use of benomyl to avoid or delay problems from resistance. Two mathematical models based on different principles are used to demonstrate the rate of buildup of resistance under various spray strategies. The Kable/Jeffery model points to spray coverage as a major factor in the rate of resistance and shows that mixtures with stronger companions and/or reduced effects from benomyl have the greatest influence to delay resistance. The Varner/Du Pont model demonstrates the value of starting the use of mixtures as soon as possible. The strategies which involve the use of strong unrelated fungicides in combination with benomyl are clearly superior to alternate sprays.

Introduction

Resistance in plant pathogens to chemicals used to help control diseases has, in recent years, been a serious problem. There is general agreement that the most promising strategies for coping with resistance problems are designed to prevent the sole, long-term exposure of the pathogen to a single disease control agent. But a controversy over the best strategies centers on the relative merits of combinations vs. single component in rotation schedules. This paper deals primarily with the experiences and theoretical calculations for the use of benomyl. Where benomyl has been combined with an effective companion fungicide in the absence of any exclusive benomyl use, resistance problems have been delayed and, in some cases, completely avoided [2]. Although some of these concepts and experiences with benomyl may be applied directly to other disease control agents, it is quite possible that other agents are sufficiently different to result in misleading interpretations.

This paper identifies and relates some of the key factors which impact on the development of resistance, especially as they concern benomyl usage. One factor is the mode-of-action and/or mode-of-resistance at biochemical and genetic levels. For instance, multi-site inhibitors like maneb and captan interfere with numerous vital metabolic processes of the pathogen and allow little chance for the development of resistance since multiple changes in the pathogen's genome are required to circumvent the action. These multi-site inhibitors are useful components in strategies designed to prolong the usefulness of compounds like benomyl with a propensity for resistance.

Benomyl, a specific-site inhibitor, acts when methyl benzimidazolecarbamate binds to fungal tubulin, thus disrupting the microtubule functions mainly during mitosis and mycelial development [1]. Fungal strains resistant to benomyl may have only a single gene difference and can be found in very low natural proportions before any exposure to benomyl. Benomyl is so effective for the control of a broad number of diseases that in some cases it has been used extensively and exclusively. When benomyl is used in situations where most of the sensitive population is controlled, resistant strains are selected and in time may dominate the population. The rate at which this process occurs differs due to several biological and environmental factors. Some can be quantitatively measured, but others must be estimated. Various attempts have been made to model this selection process and to anticipate the population dynamics. Two of those models will be discussed and exemplified.
The words resistance and tolerance are used interchangeably by plant pathologists. But exclusive use of the word resistance is recommended by the United Nations FAO Panel of Experts on Pest Resistance to Pesticides [3].

Background

Some factors influencing resistance development are outlined as follows:

A. Characteristics of Resistant Strains

1. Genetics [5]

   The natural occurrence of benomyl resistant fungal mutants appears to be 1×10^{-6} to 10^{-9}.

   In some fungi, only one gene change is necessary for benomyl resistance.

2. Fitness

   Although pathogenicity, reproductive ability and overwintering survival ability of some benomyl-resistant strains is equal to wild-type sensitive strains, their extremely low numbers in natural populations and the gradual reduction in mixed populations where benomyl usage has ceased, indicates some disadvantage for resistant strains.

3. Chemical Resistance

   Disease control is possible until the resistance level in the population approaches a threshold concentration. In most cases, this level is 5 to 20X the concentration lethal to benomyl-sensitive strains. The level of resistance to benomyl is frequently greater than 50X.

   Cross-resistance to other benzimidazole fungicides makes them ineffective substitutes or companions. Non-benzimidazole fungicides, on the other hand, are equally effective on benomyl resistant or sensitive strains. No practical field demonstration has been made of negative cross-resistance, but it is worth the search for agents that will be more effective against resistant strains.

B. Characteristics of Disease

1. Area

   As a crop area increases with a corresponding increase in the proportion of that area diseased and treated, the selection for resistance increases.

2. Intensity

   Resistance selection increases in a treated area as various conducive environmental factors intensify the disease.

3. Dilution
   A reduction in resistance selection is possible through fungal population dilution from spore migration, crop rotation, sanitation, etc.

C. Characteristics of Chemicals
1. Mode-of-Action
   The mode-of-action determines cross-resistance and a propensity for pathogens to develop resistance. Multi-site inhibitors have a lower propensity than specific-site inhibitors such as benomyl.
2. Persistence
   Chemicals that remain on the host will select for resistance over a longer period than those with a short half-life.
3. Application
   Spray coverage is influenced by application technique and timing. More complete coverage (such as dip treatment) selects resistance faster than poor coverage.

D. Control Strategies
1. Integrated Pest Management
   The use of a variety of different methods to lessen the level of disease and number of reproductive cycles of the pathogen will reduce the selective pressure.
2. Rotation (alternate)
   The reduction of selection pressure by using alternative unrelated chemicals can delay resistance development. This use of chemicals alone increases the risk of loss if resistance does develop.
3. Mixtures
   The addition of another unrelated agent in mixture with benomyl has been shown to delay or prevent the emergence of resistance problems. The choice of companion and the rates and schedules of sprays are critical.

Experience has proven that the use of mixtures is the best single method, but resistance is delayed most effectively and with least risk when mixtures are used in combination with rotation and integrated pest management strategies.

Theoretical Mathematical Models

Experience and logic point to the usefulness of mixtures; therefore, field experiences are supplemented with mathematical models for verification of this concept. Models are bound to oversimplify the complexities of population dynamics, but are valuable to highlight interactions that otherwise are obscured by less important complications. Two models based on different principles are used to demonstrate the rate of buildup of resistance under various spray strategies. They are in remarkable agreement as to the value mixtures to avoid or delay problems from resistance.

Kable/Jeffery Model

This simple mathematical model (Tab. 1) published in *Phytopathology* last January [4], was constructed to simulate the process of selection that occurs when a population of organisms with a component resistant to a particular biocide is repeatedly exposed to that biocides when combined with a second biocide to which no resistance exists. The model determines the change in the proportions of the subpopulations (% resistance) with successive spray applications. The variables included in the model are (1) the efficacy of the first biocide against the sensitive subpopulation, (2) the efficacy of the second biocide against both resistant and sensitive subpopulations (equal), (3) the degree of completeness of spray coverage (the proportion of the population contacted at each spray application), and (4) the proportion of resistant to sensitive spores. No consideration is given to factors of persistence (half-life), systemic effects, inoculum survival, fitness, etc.

Kable/Jeffery [4] express results as the number of spray applications until resistance dominates. This effectively, although indirectly, considers disease intensity and reproductive cycles or generations which can vary greatly depending on the disease, weather, etc. For instance, when condiciions are unfavorable and only a few spray applications are needed in a season, the selective pressure is lower and resistance will not develop for a longer time than under conditions where multiple disease cycles make multiple sprays necessary.

Table 1
Kable/Jefferey *Mathematical model for rate of resistance buildup*

$$P_1^m = \frac{n^1[1 + (1 - E)(f_{1A} - f_{1B} - f_{1A} - f_{1B})]^m}{n_1[1 + (1 - E)(f_{1A}f_{1B} - f_{1A} - f_{1B})]^m + n_2[1 + (1 - E)(f_{2A}f_{2B} - f_{2A} - f_{2B})]^m}$$

P_1^m	= Percent (decimal −0 to 1) of resistant/sensitive spores	
n_1	= Initial number of spores resistant to product A (1×10^{-9})	
n_2	= Initial number of spores sensitive to product A ($1 - 1 \times 10^{-9}$)	
E	= Escaping parameter (decimal −0 to 1) due to imperfect coverage	
f_{1A}	= Efficacy (0 to 1) of product A on subpopulation 1 (resistant)	
f_{2A}	= Efficacy (0 to 1) of product A on subpopulation 2 (sensitive)	
f_{1B}	= Efficacy (0 to 1) of product B on subpopulation 1	assumed equal
f_{2B}	= Efficacy (0 to 1) of product B on subpopulation 2	
m	= Number of sprays	

The model (Tab. 1) is based on the following assumptions:

- Pre-existance of subpopulations of the target organism which are resistant. For benomyl, 1 resistant spore for every 10^9 sensitive spores is assumed to be found in a wild population before selective spray application.
- The population will be challenged repeatedly by a mixture of unrelated biocides.
- Biocides exert independent joint action.
- The proportions of each subpopulation change with each challenge but not between sprays. There is no dilution from outside sources.
- Efficacies (f) of the biocides against the subpopulations are constant. f = proportion (0 to 1) of individuals killed when contacted (dependent on concentration).
- Genetic recombination among the two subpopulations is absent, whether through sexual or asexual means.
- For the sake of simplicity, fungicide degradation and residual deposits are not considered.

Conditions selected for demonstration Figure 1

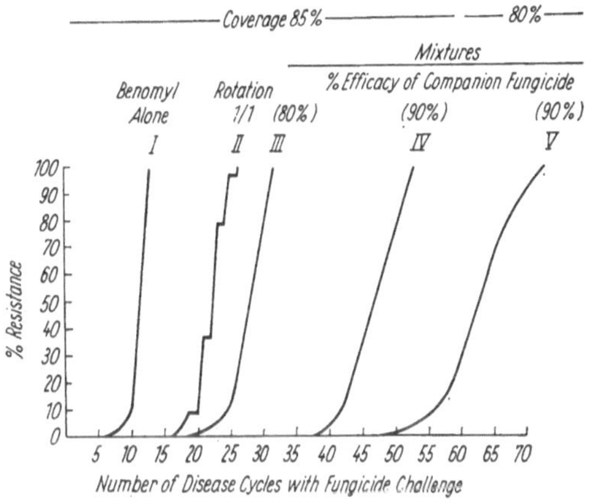

Fig. 1. Kable/Jeffery Model for rate of resistance buildup

Specialized equipment and high-volume application is necessary to attain more than 90% coverage under field conditions. It is our observation that 85% coverage is common on crops such as fruit trees and peanuts, about 80% on sugar beets, rice and cereals and about 75% on soybeans.

E $= 0.15 =$ 85% host covered by spray treatments.
E $= (0.20 =$ 80% in example V only)
$f_{1A} = 0.01 =$ 1% efficacy of product A (benomyl) on subpopulation (resistant)
$f_{2A} = 0.99 =$ 99% efficacy of product A (benomyl) on subpopulation 2 (sensitive)

Variable factor

$f_B =$ Companion fungicide efficacy equal on benomyl sensitive and resistant.
I $f_B = 0$ when benomyl is used alone
II $f_B =$ any number below 10 when rotation strategy is used.
III $f_B = 0.8$ when companion gives 80% efficacy on fungi contacted.
IV $f_B = 0.9$ when companion gives 90% efficacy on fungi contacted.
V $f_B = 0.9$ as in IV, but E is reduced to 0.2 (80% host covered), and thus benomyl exerts less selective pressure.

Conclusions from the use of the Kable/Jeffery model under various conditions are as follows:

- When the initial population of resistant strains is low (1 in 10^9) there is a relatively slow buildup to 1 percent. But once above 1 percent, the proportion rises rapidly and resistance may dominate within a few cycles.
- The greatest effect on the rate of selection is spray coverage (E). Excellent coverage (99%) can result in a dominance of resistance in less than 20 sprays, but poor coverage (50%) delays resistance to beyond 500 sprays.
- Increasing the strength of the companion fungicide or decreasing the strength of benomyl has a delaying effect on resistance buildup.
- To calculate the buildup of resistance from the alternate use of two fungicides, double the values obtained from using benomyl alone since the companion fungicide used alone has no effect on the proportion of resistance and sensitive.

Varner/Du Pont Model

R. W. Varner [2 and 6] of Du Pont has developed a theoretical mathematical model to help understand the development of benomyl-resistant fungus strains under various field conditions (Table 2). The major factors which contribute to the selection and development of resistance are considered in this model as follows:

Table 2
Varner/Du Pont *Mathematical model for rate of resistance buildup*

$A[([F \times G \times B]^D \times C)(F \times G \times H)]^E + ['']^{E-1} + ['']^{E-2}$ etc. = Resistant lesions per 10,000 HA.

A	=	Initial resistance level (A = 1 if 1.5×10^{-8})
B	=	Efficacy of companion fungicides as decimal percent *survival* of resistant spores
C	=	Decimal percent of inoculum survial to next season (5%)
D	=	Infection cycles per season (3)
E	=	Number years continuous fungicide use
F	=	Number spores produced per lesion (10,000)
G	=	Decimal percent of spores produced that infect new host tissue (0.0015)
H	=	Net dilution effect (0.9 = 10%)

A. In the natural state, it has been estimated that the frequency of resistant spores is 1.5×10^{-8}. This, of course, varies with the fungus and the previous selection pressure, but represents the well-established concept that resistant strains are a natural occurence.

B. The percentage of resistant spores which survive is dependent upon the fitness of the strains and the kinds of control measures employed.

C. The amount of inoculum which survives to infect the next season is also a critical factor in the rate of development.

D. Organisms with only one or two infection cycles per season could be expected to develop resistance much slower than those with multiple infection cycles.

E. Each year of continuous use of a particular treatment program has a cumulative effect.

H. Net dilution effects are influenced by import and export of spores into and out of treated areas.

The numerical values assigned to these factors vary widely among particular combinations of hosts, pathogens, and environments; therefore, the time intensity of resistance development will also change.

Various situations and experiences have been computed on this model and one example is presented in Figure 2. Under conditions conductive for development of a *Cercospora* leaf spot of an annual crop like peanuts, the model shows that the frequent use of benomyl alone can result in an unacceptable level of resistance after five or six disease cycles. It also shows that when benomyl plus fungicide combinations are used from the start (companion at a level providing at least 80 percent control of the resistant strain) the development of significant resistance is delayed more than 15 cycles. By using a companion to provide at least 90 percent control of the resistant strain, resistance is delayed more than 55 disease cycles. When diseases with few reproductive generations and requiring fewer chemical treatments are computed on the model, resistance develops more slowly.

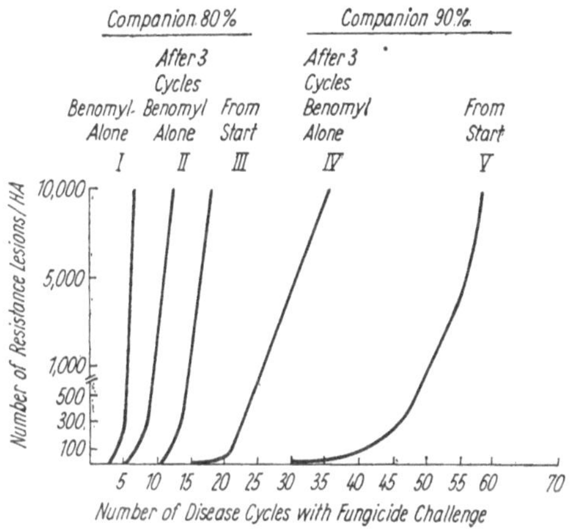

Fig. 2. VARNER/DU PONT Model for rate of resistance buildup

These relatively simple models have been helpful, but each has shortcomings. For example, neither computes factors for fitness of various strains or considers the many epidemiological factors with the sophistication being developed in many research centers and described by ZADOKS and SCHEIN [8]. Although these models are not designed to predict the emergence of resistance in a specific grower's situation, they are useful to understand the theory associated with various spray regimes and to detect general trends in population dynamics. They definitely demonstrates the advantages of the timely use of mixtures.

Future prospects

Disease control now and in the future will be much more complex than during the days when only a few multi-site agents were available. But the rewards from greatly improved disease control with the more diverse agents are worth the efforts toward enlightened use. As a part of the integrated pest management concept, the most effective use of all chemicals includes ways to cope with resistance. Several new classes of disease control agents entering the marketplace will be confronted with the possibility of resistance. Notable are the new generation of chemicals described by SCHWINN [7], which are highly effective and specific against the Phycomycetes. This specificity calls for mixtures to broaden the spectrum of disease control, but at the same time mixtures could help prevent the development of resistant strains. Conventional disease control agents are needed alone and in combinations with the specific new chemicals, including those yet to be discovered. There is an urgent need for sufficient information and then its rational use in registering agents for safe and effective use. Restrictions to control abuse should be balanced to allow for necessary use flexibility. Resistance problems compound the need to maintain effective use registrations for the older agents and to discover new ones for a full arsenal of chemicals to control plant diseases. The loss of proven fungicides and delays in the availability of new control agents, both multi-site and site-specific, through government regulatory action clearly complicate the necessary job of producing healthy crops.

Diskussion

LYR: You prefer combinations before rotation as you demonstrated by your model. Have you already good recommendations for combinants besides maneb?

Delp: In some cases mixtures with captan have been satisfactory. The use of potent sterol biosynthesis inhibitors in combination with benomyl has shown promise and should be good mixtures for the future.

Lyr: Would you agree, that as combinants for benomyl only can be used fungicides which give no resistance? Because you mentioned the multiple resistance, than you could spoil the whole population. This is well known from insecticides.

Delp: No, I think it is possible to find an advantage in combining benomyl with other fungicides as long as there is no detectible resistance in the population to either. One can help prolong the effective life of the other. The use of each alone will result in resistance problems just as soon or sooner than combinations we shall make maximum use of combinations with multi-site agents because they offer many advantages, but they are not our only choice.

Lyr: There is an invention for combination of benomyl and chloroneb. Would not this be very dangerous way, because of the broad cross resistance of hydrocarbon fungicides. This would restrict very much combinations of systemic fungicides.

Delp: There is little advantage in combining chloroneb with benomyl, because chloroneb is most active against Pythium and benomyl has none. I think the patent you refer to is a combination of MBC with Chlormequat for *Cercosporella* (eye spot) and lodging control on cereals. There is no synergism here on chemical, only an additive (expected) effect of a growth modifier and a fungicide (see my other discussion comments).

Eckert: It would seem that the combination of benomyl with protective fungicides (maneb, captan) should be a more effective strategy for delaying the emergence of resistant strains in these diseases that can be efficiently controlled by purely protective treatments. Alternately this strategy should be less effective for controlling resistant strains in those diseases that are best controlled by treatment with post-infective (local systemic) acting, e.g. rots, apple scab, postharvest wound pathogens.

Delp: The companion fungicide will have no effect on fungi not contacted. Protectant fungicides fail to contact some of the fungi which have penetrated deeply into the host. But there are many fungi which have not penetrated and which may serve as inoculum for later infection and these can be contacted. There are situations like the ones you describe where a preventive companion is less than ideal, but a little effect against possible selection of resistance may be of economic value.

Schwinn: The factor which in practice will decide on the chance of mixture programs, is the acceptance by the farmer who first of all is attracted by the striking activity of a compound like benomyl. To what extent is the farmer prepared to use mixtures instead of the single product in situations when resistance has not yet developped?

Delp: Many growers can understand the risk of resistance in future years, but elect to continue exclusive use until the problem is with them. The greatest success for implementing the use of mixtures has been in situations where a companion fungicide is necessary to control an additional disease. The economics of the mixture is of high considerations to the grower and so is convenience. In some cases regulations requiring mixtures or special programs are necessary to persuad growers to use these methods.

SCHWINN: Additional question: Does your answer not indicate that in order to implant such strategic use programs into the agricultural practice the commercial companies depend on the support of officials, extension services and advisers?

DELP: Yes very much so. But in addition companies have technical representatives who can provide growers with practical information.

MÜLLER: Do you think, that the IPM-system will give a base for different spraying-programs by the farmers in your country for the future too?

DELP: Yes, but it will take time (many years) because good IPM programs have been in effect for plant disease control of many crops. The integrated use of cultural procedures, resistant varieties, spray practice based on forcasting systems and the changing idea that 100% disease control is not necessary will make some spray program change a bit in the future. It will be a gradual growth, but efficient methods for high production will continue to have high priority as the demand for food for a growing world population increase.

Literature

[1] DAVIDSE, L. C., In: *Microtubules and Microtubular Inhibitors* (M. BORGERS and M. DE BRANDER, eds.), North Holland Publ. Co., Amsterdam 1975, 483–495

[2] DELP, C. J., Resistance to plant disease control agents — how to cope with it. Presented at the IX. International Congress of Plant Protection. Reprinted by E. I. du Pont de Nemours & Co., Inc., Wilmington, DE USA (1979)

[3] FAO Plant Production and Protection Paper 6/2 Pest Resistance to Pesticides AGP: 1979/M/2

[4] KABLE, P. F., and H. JEFFERY, Selection for tolerance in organism exposed to sprays of biocide mixtures: A theoretical model, Phytopathology **70** (1980), 8–12

[5] TUYL, J. M. VAN, Genetics of fungal resistance to systemic fungicides, Mededelingen Landbouwhogeshool, Wageningen 77 **(2)** (1977), 1–136

[6] VARNER, R. W., Mathematical model — Development of Benlate®-tolerant strains. Internal Du Pont Report (1976). Summary published in reference by DELP (2)

[7] SCHWINN, F. J., Control of Phycomycetes: A Changing Scene, Proc. 1979 British Crop Prot. Conf. 791–802

[8] ZADOKS, J. C., and R. D. SCHEIN, Epidemiology and Plant Disease Management, Oxford Univ., N. Y., 1979

Ferenczy L.*, Frankó, A., N. Hamar G., R. Hegedüs E., Pfliegel T.

CHINOIN, Pharmaceutical and Chemical Works, Budapest, Hungary
*József Attila University, Szeged, Hungary**

Interactions between Benzimidazole Fungicides and Other Types of Systemic Fungicides

Although it must by no means be regarded as a natural law, systemic fungicides introduced till now into plant protection practice are without exception site-specific to target organisms (Dekker [7], Georgopoulos [10], Fritz [9], Ragsdale [13]). This, in comparison with the site non-specific contact fungicides, increases the probability of developing tolerant or resistant fungal strains (Lacy [12]). Especially in the case of benzimidazole fungicides (carbendazim and its precursors: benomyl, thiophanate-methyl, cypendazol and mecarbinzid; fuberidazol; thiabendazol) which possess systemic activity to a wide range of fungi, including members of the classes Ascomycetes, Deuteromycetes, and Basidiomycetes (Delp [9]), many fungal species of real or potential commercial importance are reported to have developed resistance in the laboratory (Schroeder [16]) as well as in the field or greenhouse (Berger [1], Bollen [2], [3], Georgopoulos [10], Vargas [18], Wicks [19].

Fungicide-resistant strains might well be less virulent or less aggressive than wild strains without the selection pressure of the fungicides (Day [6]). This was really found with benomyl in the case of *Colletotrichum musae* (Griffe [11]) and *Botrytis cinerea* (Bollen [3]). Wild and benomyl-resistant strains of *Cercospora apii* (Berger [1]) and *Cercospora beticola* (Ruppel [15]), however, exhibited similar virulence and aggressivity in the field.

A simple and logical means to retard or even to prevent the occurence of resistance to site-specific systemic fungicides is to mix two or more of them which are active against a given parasite but possess different sites of action. The only limitations for the practical use of such fungicide mixtures which may occur are antagonism between the fungicides combined and/or instability of the mixture.

Knowing both, the spectra of activity and the primary mechanisms of action of the different groups of systemic fungicides, we concentrated our efforts to combine carbendazim and its precursors — benomyl and thiophanate-methyl — with two other groups of systemic fungicides: promidione and vinchlozolin on the one hand, and triadimefon and fenapronil on the other. The activity spectra (with the exception of triadimefon whose scope is substancially narrower) are more or less identical, and the mechanisms of action are distinctly different.

The benzimidazoles inhibit mitosis (Richmond [14]). Promidione and vinchlozolin repress DNA and RNA synthesis and cause marked accumulation of fatty acids and triglycerides (Fritz [9]). Triadimefon, triarimol, and presumably the structurally closely related fenapronil are inhibitors of ergosterol synthesis (Buchenauer [4], Ragsdale [13]).

With non-obligate fungi we used the standard serial dilution method on yeast extract media containing agar at pH 7.2 to determine the minimal inhibitory concentrations (MIC) of different fungicides by themselves, and of their dual combinations in different ratios.

Of the obligate parasites, activity to *Erysiphe graminis* on wheat ("Bezostaya") was only examined in a phytotrone, applying different concentrations and different ratios of the fungicides.

The degree of infection was compared with that of the untreated, infected control (100%).

Using COLBY's known formula [5] and inserting for biological response the reciproke of the MIC:

$$\frac{1}{\text{MIC (calc.)}} = \frac{m_A}{\text{MIC}_A} + \frac{m_B}{\text{MIC}_B}$$

for calculating the expected minimal inhibitory concentration of the binary mixture of fungicides A and B, taken in the ratio $m_A : m_B$ ($m_A + m_B = 1$) we expressed the degree of synergism (D. S.) by the ratio

$$\text{D. S.} = \frac{\text{MIC calc.}}{\text{MIC found.}}$$

If D. S. was smaller than unity, the interaction of the fungicides was regarded antagonistic, if D. S. was around unity it was regarded additive, and if D. S. was greater than unity it was regarded synergistic.

In many cases we controlled the results obtained by the method described above by graphic representation of TAMMES' isoboles [17]. These gave identical results.

Our experimental results were very unexpected indeed.

As it can be seen from Table 1, the combination of carbendazim or benomyl with the lipid-accumulating fungicides promidione and vinclozolin showed strong antagonism in all our tests.

Table 1

Antagonism between benzimidazole and lipid accumulating fungicides

Fungal species	Fungicide	MIC (μg/ml)		D.S.
		calc.	found.	
A Carbendazim (I) and Promidione (II)				
Botrytis cinerea	I	—	0,07	—
	II	—	25	—
	I + II (1:1)	0,14	0,70	0,2
Fusarium moniliforme	I	—	3	—
	II	—	50	—
	I + II (1:1)	5,66	12	0,5
Epidermophyton floccosum	I	—	—	0,15
	II	—	50	—
	I + II (1:1)	0,30	1,5	0,2
B Benomyl (III) and Vinclozolin (IV)				
Botrytis cinerea	III	—	0,07	—
	IV	—	50	—
	III + IV (1:1)	0,14	0,70	0,2
Sclerotinia scler.	III	—	0,03	—
	IV	—	0,3	—
	III + IV (1:1)	0,05	0,07	0,8
Trichophyton	III	—	0,7	—
mentagrophytes	IV	—	50,0	—
	III + IV (1:1	1,38	3,0	0,5

Additivity of fungicidal action of benomyl and triadimefon could be demonstrated in all cases examined. This practically means that triadimefon does not interfere with the action of benomyl on fungi towards which it is ineffective.
Most interesting results were obtained by combining benomyl with fenapronil (α-Butyl-α-phenyl-lH-imidazole-l-propanenitrile, DOS 2604047).
Strong synergism could be observed in all experiments, especially in those in which the ratio of benomyl to fenapronil was 2:1 or 1:1. With the ratio of 1:2 synergistic, additive, and even antagonistic effects could equally be registered (Table 2).

Table 2
Synergism between benomyl (III) and fenapronil (V)

Fungal species	Fungicide	MIC (µg/ml) calc.	MIC (µg/ml) found	D.S.
Alternaria tenuis	III	—	300	—
	V	—	0,07	—
	III + V (2:1)	0,22	0,11	1,9
	(1:1)	0,14	0,06	2,3
	(1:2)	0,10	0,05	2,0
Stemphylium radicinum	III	—	300	—
	V	—	0,7	—
	III + V (2:1)	2,12	0,05	42,4
	(1:1)	1,40	0,06	29,3
	(1:2)	1,04	1	1
Cercospora beticola	III	—	0,07	—
	V	—	0,6	—
	III + V (2:1)	0,10	0,05	2,0
	(1:1)	0,13	0,06	2,1
	(1:2)	0,17	0,22	0,8
Phytophthora infestans	III	—	300	—
	V	—	12	—
	III + V (2:1)	33,6	24	1,4
	(1:1)	23,1	12	1,9
	(1:2)	17,6	9	2,0
Trichophyton rubrum	III	—	3	—
	V	—	0,03	—
	III + V (2:1)	0,09	0,05	1,8
	(1:1)	0,06	0,06	1,0
	(1:2)	0,04	0,05	0,9

Of greatest practical significance perhaps is the fact that benomyl-tolerant *Botrytis cinerea* strains isolated from the field whose MIC were above 200 µg/ml (the MIC of benomyl-sensitive *Botrytis cinerea* is 0.07 µg/ml) could be re-sensibilized to benomyl by the addition of fenapronil whose MIC towards these strains was 15 and 3.7 µg/ml, resp.
Synergism was 6 to 54-fold, the greater value was always found with mixtures in which the quantity of benomyl prevailed over that of fenapronil. We have no explanation for the experimental fact that smallest degree of synergism or even antagonism was found with the ratio 1:1 of the two fungicides. The only exception was strain "C" in which initial tolerance had probably proceeded onto genetical resistance. It is interesting to note that in this case a distinct antagonsim manifested itself, since addition of benomyl increased the MIC of fenapronil (Table 3).
The in vivo experiments using mixtures of benomyl with triadimefon or fenapronil on

Table 3
Combined action of benomyl (III) and fenapronil (V) against four benomyl-resistent Botrytis cinerea strains

Fungal strain	Fungicide	MIC (μg/ml) calc.	MIC (μg/ml) found	D.S.
A	III	—	200	—
	V	—	15	—
	III + V (2:1)	39,5	3,7	10,7
	(1:1)	27,9	15	1,9
	(1:2)	21,6	3,7	5,8
B	III	—	200	—
	V	—	3,7	—
	III + V (2:1)	10,8	0,2	54,0
	(1:1)	7,3	15	0,5
	(1:2)	5,4	0,4	13,7
C	III	—	200	—
	V	—	3,7	—
	III + V (2:1)	10,8	60	0,2
	(1:1)	7,3	200	0,0
	(1:2)	5,5	30	0,2
D	III	—	200	—
	V	—	15	—
	III + V (2:1)	39,5	7,5	5,3
	(1:1)	27,9	30	0,9
	(1:2)	21,6	3,7	5,8

wheat infected with conidia of *Erisyphe graminis* in both cases showed additivity or a small degree of synergism.

In conclusion it may be stated that fungicidal mixtures of mitosis inhibiting benzimidazoles with ergosterol biosynthesis inhibiting systemic fungicides, — especially with fenapronil, — can be of great practical value for three reasons: 1) in most cases they can retard or even prevent the occurence of tolerant or resistant fungal strains; 2) because of synergism between them, the applied doses may often be reduced; 3) in cases where one of them fills up a gap in the activity spectrum of the other, even activity spectra may be broadened.

The combination of benzimidazoles with promidione or vinchlozolin is contraindicated because of antogonism between theses two groups of systemic fungicides.

Diskussion

GEORGOPOULOS: I would like to make one comment on the difference observed between the benzimidazole resistant strains of *Botrytis* obtained from the field and the one selected in the laboratory.

Botrytis may be heterocaryotic and that the presence of the fungicide may change the ratio of nuclei in a heterocaryon is well known. It is, perhaps, possible for the companion fungicide to change the nuclear ratio to the opposite direction. In the laboratory-selected strain the nuclear ratio might be very much in favor of the benzimidazole resistant gene, so that the effect of the companion, if any, may take a long time to recognise.

PFLIEGEL: Thank you for this comment, professor GEORGOPOULOS, which most probably is the one reason of the differences found among different *Botrytis* strains.

But still the explanation why in all cases the ratio I:I of benomyl to fenapronil is so much worse than either the 2:1 or the 1:2 ratio is missing.

LEROUX: Synergistic effects are observed between prodione and benzimidazole fungicides in cereal seed treatments.

PFLIEGEL: Results between in vitro experiments with differing kinds of fungi may differ from those obtained in seed dressing. Anyhow, I must confess that after having observed a strong antagonism between benzimidazoles and vinchlozolin or iprodione (D.S. = 0.2) at the ratio 1:1 we discontinued work on this kind of dual compositions. Thanks to your comment, we shall immediately check the situation with other ratios as well.

LEROUX: Fenapronil is an imidazole fungicide related to imazalil. It inhibits ergosterol biosynthesis in the same manner as triadimefon. We observed also in *Ustilago maydis* cross-resistance between fenapronil, triadimefon, imazalil and related compounds (LEROUX, unpublished results).

PFLIEGEL: I suppose that cross-resistance should be expected generally between fungicides with the same mechanism of action. According to our laboratory experiments, however, tolerance to fenapronil can be provoked onyl with difficulty, and it is of transient character.

DELP: The patent claiming synergestic action of MBC + Chlormequat for *Cercosporella* control (yield improvement) of cereals is not valid because data from all sources including the patent show only expected additive effects.

I am impressed with Dr. PFLIEGEL in results which show synergistic effects with 1/2 and 2/1 ratios of benomyl/fenapronil but little effect at 1/1 ratio against *Botrytis* strain. We have done some work on combination and have negative results which 1/1 ratios but with the mixtures we did not try other ratios and may have missed possible synergism.

Most demonstrations of synergism in specific situation in culture have not proved to be of practical value under use condition in the field. New information which will lead to practical use of synergism is needed and welcome.

PFLIEGEL: Allow to thank you most cordially, professor DELP, for your words of acknowlegement, and for your comments. It goes without saying that our laboratory results can be of real practical value only if they can be proven in the field.

KERKENAAR: Some inhibitors of ergosterol biosynthesis also led to accumulation of fatty acids like triadimefon. How do you explain in this case the synergestic effect because you mentioned the antagonistic effect of compounds causing accumulation of fatty acids?

PFLIEGEL: The leading idea of our experiments was to find systemic fungicides complementary to the benzimidazoles having distinctly different mechanisms of action which, if possible, are synergistic with them. On basis of your experimental results we are convinced that we have found such a group in the ergosterol biosynthesis inhibiting compounds, but we do not want to deduce theoretical hypothesis from these experiments. Anyhow, your comment may be an explanation of the fact that interaction between triadimefon and the benzimidazoles is only additive while the same between fenapronil and the benzimidazoles is synergistic.

BECKER: Es ist sehr interessant, was Sie über die Einhaltung der Mischungsverhältnisse bei Fungizidmischungen dargelegt haben. Etwas Ähnliches haben wir bei der Abhängigkeit der Pilzanfälligkeit der Baumwolle von Mineralstoffen, insbesondere von

Zink und Kobalt gefunden. Auch hier ist ein bestimmtes Verhältnis dieser Ionen für hohe Resistenz erforderlich. Meinen Sie nicht, daß dies dafür spricht, daß es bei Pilzen einen spezifischen „Parasitär-Stoffwechsel" gibt, der durch Fungizide neben dem Normalstoffwechsel mit erfaßt werden muß?

PFLIEGEL: Ich bin vollkommen mit Ihrer Beantwortung einverstanden, bitte aber unsere bisherigen Ergebnisse als viel versprechende, aber große Vorversuche zu betrachten. Die Vereinfachung und Optimierung der Fungizidmengen-Verhältnisse in vitro und in vivo stehen noch vor uns.

References

[1] BERGER, P. D., Disease progress of *Cercospora apii* resistant to benomyl, Plant Dis. Rep. **57** (1973), 837–860

[2] BOLLEN, G. J., Resistance of benomyl and some chemically related compounds in strains of *Penicillium* species, Neth. J. Plant Pathol. **77** (1971), 187–193

[3] BOLLEN, G. J., and G. SCHOLTEN, Acquired resistance to benomyl and some other systemic fungicides in a strain of *Botrytis cinerea* in cyclamen, Neth. J. Plant Pathol. **77** (1970), 83–90

[4] BUCHENAUER, H., Inhibition of ergosterol biosynthesis in *Ustilago avenae* by triadimefon and fluotrimazole, Mitt. Biol. Bundesanst. Land- u. Forstwirtschaft, Berlin-Dahlem, Heft 165, 154-5 (CA. **1975**, **85**, 138417)

[5] COLBY, S. R., Calculating synergistic and antagonistic responses of herbicide combinations, Weeds, **16** (1968), 20–31

[6] DAY, P. R., Recent developments in the genetics of the hostparasite system, Ann. Rev. Phytopathology, **4** (1966), 245–268

[7] DEKKER, J., L. C. DAVIDSE, Acquired resistance to benzimidazole derivatives. In: "Pesticides"/IUPAC 3rd Int. Congr., Helsinki, 1974/, 410-3. Thieme Verlag, Stuttgart (1975)

[8] DELP, C. J., and H. L. KLOPPING, Performance attributes of a new fungicide and mite ovicide candidate, Plant Dis. Rep. **52** (1968), 95–99

[9] FRITZ, R., P. LEROUX, and GREDT, Mechanism of antifungal activity of promidione, vinchlozolin and dichloran on *Botrytis cinerea* PERS, Phytophatol **90**/2 (1977) 152–163

[10] GEORGOPOULOS, S. G., and C. DOVAS, A serious outbreak of strains of *Cercospora beticola* resistant to benzimidazole fungicides in northern Greece, Plant Dis. Rep. **57** (1973), 321–4

[11] GRIFFE, P. J., Resistance to benomyl and related fungicides in *Colletotrichum musae*. Trans. Brit. Mycol. Soc. **63** (1974), 137–141

[12] LACY, M. L., and J. M. VARGAS, Jr, Resistance of fungi to fungicides. In: Chemie der Pflanzenschutz- und Schädlingsbekämpfungsmittel (ed. R. WEGLER), Vol. 4, 239–256, Springer Verlag, Berlin, 1977

[13] RAGSDALE, N. N., Specific effects of triarimol on sterol biosynthesis in *Ustilago maydis*, Biochim. Biophys. Acta **380** (1) (1975), 81–96

[14] RICHMOND, D. V., and A. PHILIPS, Effect of benomyl and carbendazim on mitosis in hyphae of *Botrytis cinerea* and roots of *Allium cepa*, Pesticide Biochem. Physiol. 5 (1975), 367–9

[15] RUPPEL, E. G., Biology of benomyl-tolerant strains of *Cercospora beticola* from sugar beet, Phytopathology, **65** (1975), 785–9

[16] SCHROEDER, W. T., and R. PROVVIDENTI, Resistance to benomyl in powdery mildew of cucurbits, Plant Dis. Rep. **53** (1969), 271–5

[17] TAMMES, P. M. L., Isoboles, a graphic representation of synergism in pesticides, Neth. J. Plant Pathol. p. 73 (1964)

[18] VARGAS, J. M., Jr, A benzimidazole resistant strain of *Erysiphe graminis*, Phytopathology, **63** (1973) 1366–8

[19] WICKS, T., Tolerance of the apple scab fungus to benzimidazole fungicides, Plant Dis. Rep. 58 (1974), 886–9

Maya Gasztonyi

Research Institute for Plant Protection, Budapest

The Possible Role of Different Enzymes in the Activation of Triadimefon

Introduction

It has been found that in beans, barley and soil triadimefon, 1-(4-chlorophenoxy)-3,3 dimethyl-1-(1,2,4-triazol-1-yl)-butanone, systemic fungicide is transformed into triadimenol (a reduced derivative of the above compound), also having a fungicide effect [1, 8]. We have proved similar transformation in other plants and phytopathogenic fungi [3, 4]. At the same time we found that the degree of transformation varies in species and it is in correlation with the sensitivity of fungal species to triadimefon. Triadimenol is more active against most fungi and is never less active than triadimefon [5], (Fig. 1). Thus the transformation may be considered as activation in terms of fungicidal activity.

Through the increase in the number of asymetrical centers, the two stereoisomers of triadimefon are transformed into four isomers of triadimenol, in the form of two enantiomer pairs. Figure 2 shows one mirror image model from each enantiomer pair and the diastereomer relationship between these two. Vogeler [8] and Specht [7] determined that the two enantiomer pairs are not produced in the same proportion in soil and plants. Others have found, however, the proportions in marrow and *Aspergillus niger* to be equal [2]. It was later shown that their fungitoxicities differ, and Bayer researches worked out a method for production of an enriched form of the more active product [6].

In view of these facts, we set out to answer these questions:

1. Is the biological reduction of triadimefon reversible?
2. Is only one enzyme or enzyme-complex responsible for the transformation, or are there separate enzymes involved?
3. Is the proportion of enantiomer pairs produced dependent upon the species of fungus, and does this have a role in selectivity?

The Investigation of the Irreversibility of Reduction

In our previous experiments, we found the *Cladosporium cucumerinum* mycelia to reduce triadimefon the most intensively, therefore we once again included this species in our experiments. For our other test object we chose *Saccharomyces cerevisiae*, which is advantageous due to the ease in method of biochemical assay in cell suspensions. After incubation for varying lengths of time in media containing triadimefon or tria-

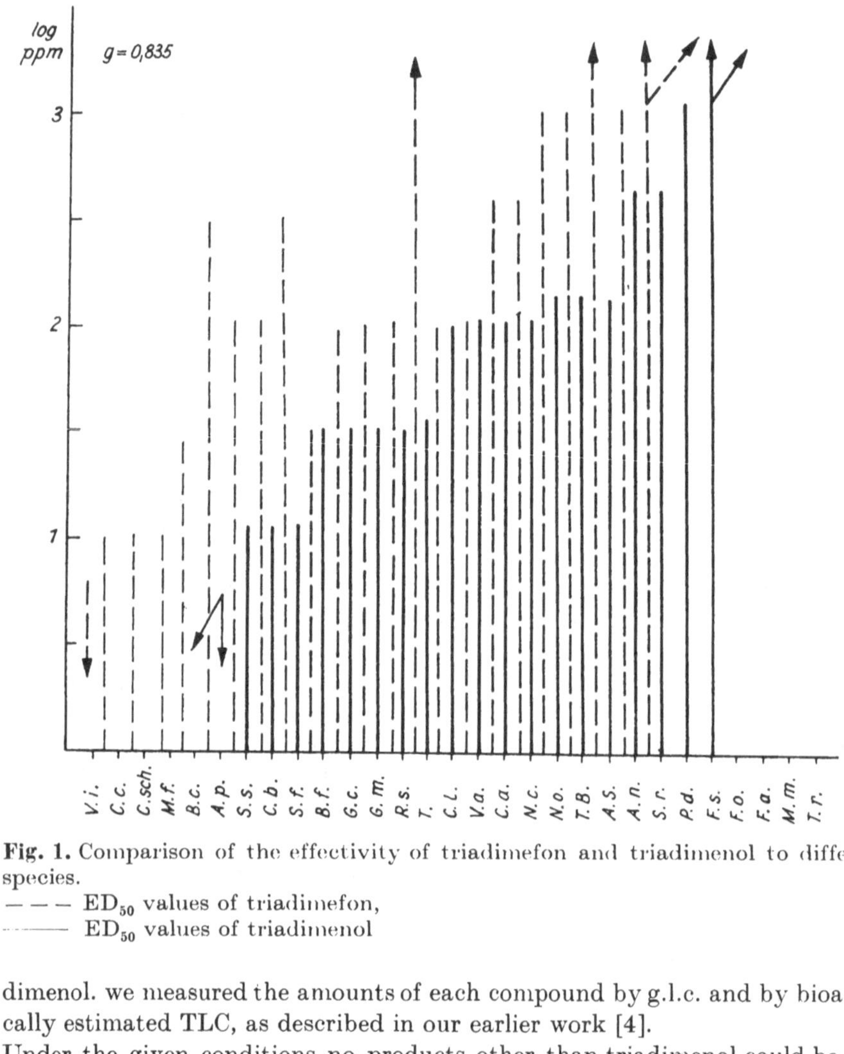

Fig. 1. Comparison of the effectivity of triadimefon and triadimenol to different fungal species.
— — — ED_{50} values of triadimefon,
——— ED_{50} values of triadimenol

dimenol. we measured the amounts of each compound by g.l.c. and by bioautographycally estimated TLC, as described in our earlier work [4].

Under the given conditions no products other than triadimenol could be found. The amounts of triadimefon and triadimenol present in *C. cucumerinum* show a slight tendency to decrease during incubation. This deviation was found to be not significant. Therefore, all calculations expressed the produced amount of triadimenol as a percentage of the combined amounts of the two compounds, thereby reducing the error due to loss in recovery Figure 3 shows the percentage amount of produced triadimenol from incubated triadimefon, that is the degree of reduction as a function of incubation time. One gram fresh weight of yeast was capable of transforming 25—30% of the 2 mg of triadimefon added to it, while *C. cucumerinum* mycelia in similar proportions were capable of reducing 70%, and in one case almost all, of the added triadimefon. In the reverse reaction, no trace of triadimefon could be found in the *C. cucumerinum* incubation medium. The same was observed in previous experiment with *Fusarium oxysporum* [5]. The yeast only oxidized 0.5% of the triadimenol to triadimefon after 5 days. Therefore the slowing down of reduction as seen on the graph cannot be attri-

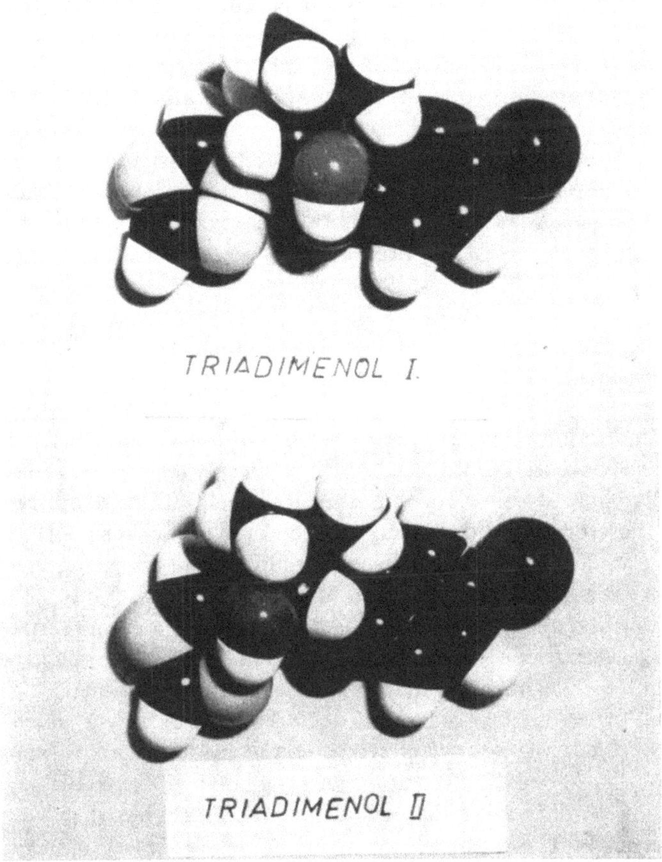

Fig. 2. Diastereoisomeric forms of triadimenol

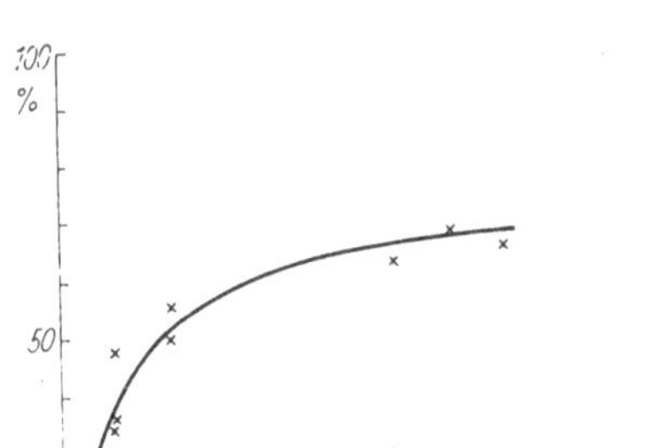

Fig. 3. Time-curve of triadimenol formation — × — in *Cladosporium cucumerinum*, --○-- in *Saccharomyces cerevisiae*

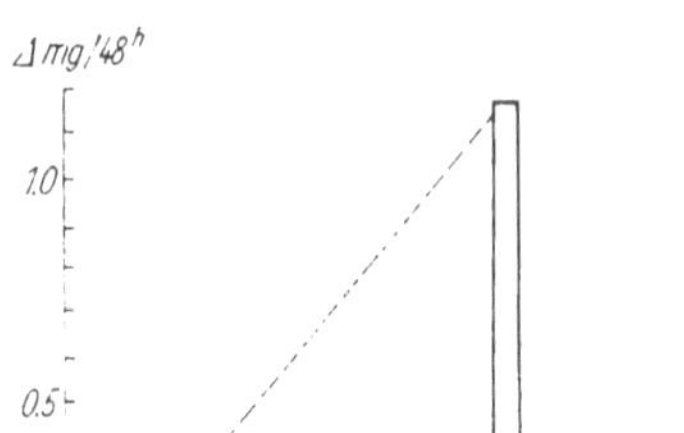

Fig. 4. Rate of triadimefon reduction by *Cladosporium cucumerinum* depending on initial concentration in incubation media

buted to the onset of equilibrium. Damage to the fungi's metabolism over a prolonged period must be a cause of this phenomenon, although a decrease in triadimefon concentration is a first factor, too. The latter's role is illuminated by the results of an incubation of *C. cucumerinum* in a triadimefon medium of different concentration. As seen in Figure 4 in the 10 to 100 mg/dm^3 concentration range, the amount of triadimefon reduced after 48 hours varied directly with regard to the initial concentration. Therefore, the reduction of triadimefon in fungi can in effect be considered irreversible; for example, in the case of *Saccharomyces cerevisiae* the oxidation reaction constant is two orders of magnitude smaller than the reduction reaction, and in the case of *C. cucumerinum* equilibrium is shifted in the direction of the reduced product.

Experiments for Determining the Localisation of Enzyme Reduction Using Authentic Enzyme Preparations

We attempted to follow the reduction reaction in cell fractions of both organisms, but achieved results only with the more actively reducing *C. cucumerinum*. We were unable to desintegrate the cells completely or to separate intact cells from membrane fragments and cell nuclei, therefore we didn't examine the enzyme activity of the first centrifuge sediment. Figure 5 compares the degree of triadimefon reduction in the intact mycelia, total homogenate, cell-free supernatant, mitochondrial fraction or microsomal and plasma fraction. Ignoring slight losses naturally occuring in the course of procedures, the cell-free supernatant's microsomal fraction measured reductive activity was three times that of the mitochondrium fraction. This experiment also proved this to be an enzyme process, since reduction doesn't begin in a heat-inactivated system; it also proved that extracellular enzymes in the media aren't responsible for the transformation. (These negative results aren't included in the Figure). The relatively strong reduction potential of triadimefon makes it incapable of serving as an aspecific electron acceptor on the terminal oxidation end of the electron trans-

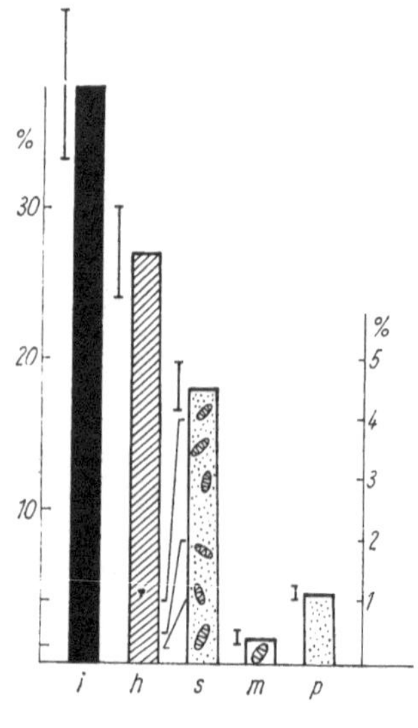

Fig. 5. Comparison of the degree of triadimefon reduction in the intact mycelia [i], total homoganate [h], cell-free supernatant [s], mitochondrial fraction [m], or microsomal and plasma fraction [p]. The values of i and h refer to the left axis of ordinates, the values of s, m and p refer to the right axis of ordinates

port system. Therefore, to find the enzyme responsible for the reduction, we had to look among oxidoreductases capable of reducing a keton group. Among the enzyme preparations available for our use, alcohol dehydrogenase (1.1.1.1.), glycerol-3-phosphate dehydrogenase (1.1.1.8.), and lactate dehydrogenase (1.1.1.27.) didn't reduce triadimefon under conditions optimal to them. Incubation with malate dehydrogenase (1.1.1.37.) produced measurable triadimenol (Fig. 6). The reduction of pyruvate, which stands closer to the natural substrate, proceeded at a rate 25 times greater than the reduction of triadimefon by malate dehydrogenase.

Malate dehydrogenase is among those enzymes which are found in both mitochondrium and plasma. This agrees with the results obtained in the cell fraction experiment.

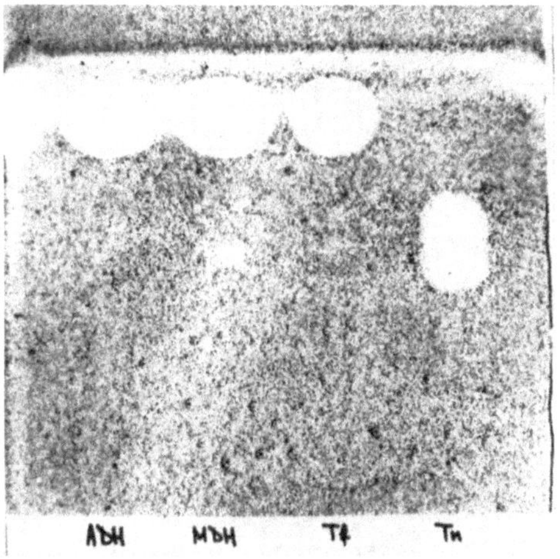

Fig. 6. Thin-layer chromatogram of extracts of incubation media, prepared after 24^h incubation of triadimefon with alcohol-dehydrogenase [ADH], malate-dehydrogenase [MDH]. Tf — triadimefon standard, Tn — triadimenol standard

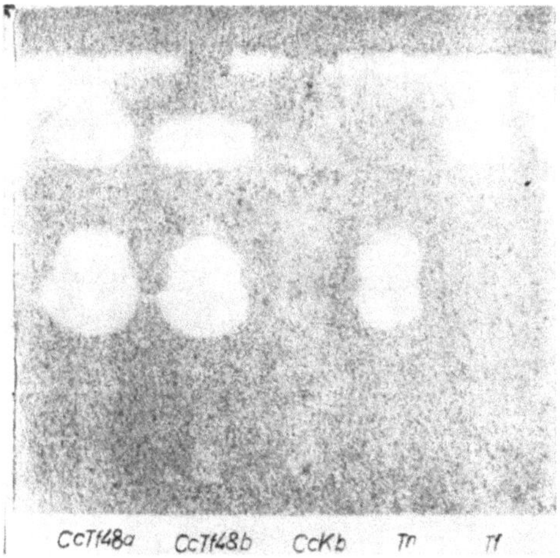

Fig. 7. Thin-layer chromatogram of extracts of mycelia [CcTf48b] or incubation media [CcTf48a] prepared after incubation of *C. cucumerinum* mycelial mats for 48^h with triadimefon. CcKb — mycelia, incubated without triadimefon, Tn — triadimenol standard, Tf — triadimefon standard

We must also point out that we compared the amount of triadimefon reduction by *C. cucumerinum* mycelia in an oxygen-poor system to the amount in an aerated system and found no differences after 24 hours.

In the TLC experiments, using the same solvent utilized by VOGELER [8], the triadimenol's two enantiomer pairs appeared as two spots (Rf = 0.43 and Rf = 0.54). Figure 7 shows that the isomer with the smaller Rf value (from here on refered to as triadimenol A) is produced in greater amounts in *C. cucumerinum* mycelia than the other (B) enantiomer pair. These results were in agreement with the values obtained by g.l.c., which were easier to interpret quantitatively. According to the results, incubated intact mycelia produced A and B triadimenol in a 2:1 ratio. In the cellfree filtrate of the homogenate this ratio was 10:1, and in the malate dehydrogenase experiment only triadimenol A was found. This indicates that the reduction of triadimefon in *C. cucumerinum* is not accomplished by one single enzyme, that instead, in addition to malate dehydrogenase there is another enzyme with a differing stereospecificity attached to its structure taking in the reaction.

The Enantiomer-Pair Ratio Differs among Varying Fungi

In the following experiments we measured the proportion of triadimenol A to triadimenol B produced. In addition to using *C. cucumerinum* and *S. cerevisiae* we tested the *Stemphylium radicinum* species, which on the basis of previous experiments [4] is slow to reduce triadimefon and its therefore considered to belong to the group of triadimefon-resistant species of fungi. Figure 8 shows that according to a thin-layer-chromatogram in *S. cerevisiae* cell suspension, in contrast to that of *C. cucumerinum*, triadimenol B is produced in greater quantity. The enantiomer-pair ratios for all three fungi species and their respective degrees of reduction are given in Table 1. *S. cerevisiae* and *St. radicinum* not only reduce triadimefon slower than *C. cucumerinum*, but the enantiomer-pair ratios are reversed.

Fig. 8. Thin-layer chromatogram of extracts prepared after incubation for 72[h] with triadimefons.
CcTf72a — incubation media of *C. cucumerinum*, CcTf72b — mycelia of *C. cucumerinum*, ScTf72a — incubation media of *S. cerevisiae*, ScTf72b — cells of *S. cerevisiae*, Tn — triadimenol standard
a The lone spot with Rf = 0.22 on the TLC of extract from ScTf72a probably origins from high content of sugar [0.65 M] in the media used in this case. The same spot was also observed on /the TLC of extract of empty media

Table 1
Grade of triadimefon reduction in fungi after 24 h incubation and the ratio of diastereomers of triadimenol (A/B) formed

	C. cucumerinum	*St. radicinum*	*S. cerevisiae*
reduction (%)[a]	38 ± 5%	6.4 ± 0.8%	9.1 ± 1%
A/B	1.9 ± 0.3	0.23 ± 0.06	0.14 ± 0.04

a. Reduced triadimefon expressed as a percentage of the combined amounts of the two compounds (triadimefon and triadimenol).

This is a significant finding, since we have established that the A form is a more active fungicide than the B form, and that their activity relative to each other varies among fungi species. *C. cucumerinum* is about 3—4 times more sensitive to the A form than to triadimenol B, in the case of *St. radicinum* the ratio is at least 10:1 (Figures 9 and 10).

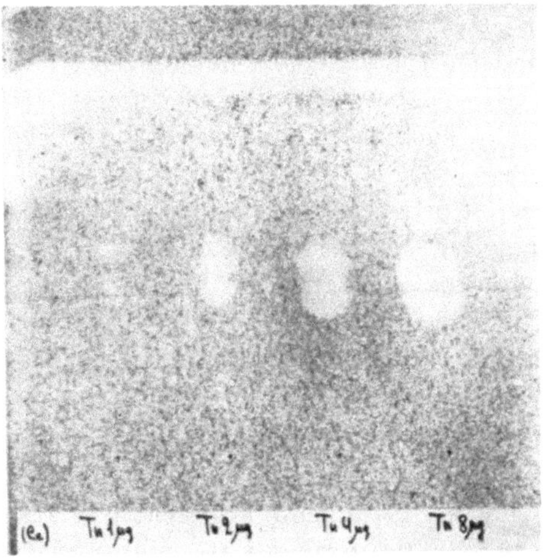

Fig. 9. Thin-layer chromatogram of different amounts of triadimenol standard solution, detected by bioassay with *C. cucumerinum*

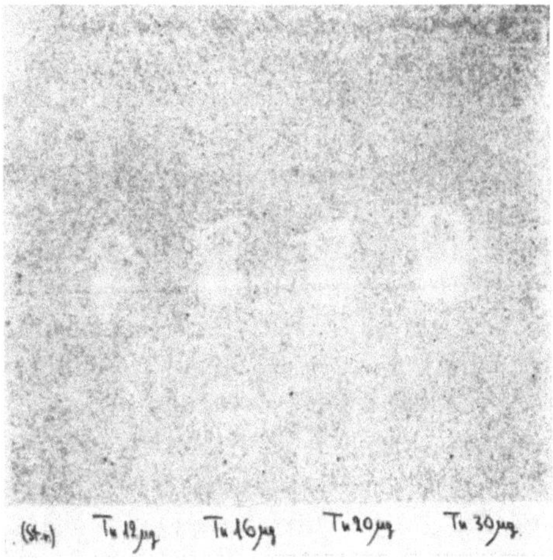

Fig. 10. Thin-layer chromatogramm of different amounts of triadimenol standard solution, detected by bioassay with *St. radicinum*

These findings give us greater insight into the reasons for *St. radicinum*'s resistance to triadimefon. In part, it reduces triadimefon more slowly than other fungi, and further in the course of reduction it produces the less active diastereoisomer in greater proportion, to which it is less sensitive.

Conclusions

On the basis of our experiments, we found the following answers to the questions put forth at the introduction:

In fungal organisms the triadimefon reduction to triadimenol is an enzyme reaction whose equilibrium is shifted greatly in the direction of the reduced product.

It seems that both structure-bound and free enzymes take part in the reation, with malate dehydrogenase also taking a role.

Triadimenol's two possible enantiomer pairs are produced in unequal amounts, and their proportion is characteristic for each of different fungal species. The triadimefon-sensitive *C. cucumerinum* produces the more active form of triadimenol in greater quantity, while the resistant *St. radicinum* produces the much less active diastereoisomer form.

It can therefore be concluded that the selectivity of triadimefon's fungicidal activity depends on two factors: its speed of reduction to triadimenol and the stereochemical direction of its reduction.

Acknowledgements

Thanks are due to Dr. Gy. Josepovits for statistical analysis and for helpful discussion, and to J. Toth for technical assistance at g.l.c.

Diskussion

BANASIAK: From the chemical viewpoint it must be possible that besides reduction products the metabolic pathway gives also oxidation or dealkylation products, for instance carboxylic acid or isopropyl analogs. Can such reactions play a role in the toxification or detoxification of triadimefon?

GASZTONYI: It is known from literature data that other metabolites may be formed in long term experiments. Among them the demethylated derivative of triadimefon, detected by CLARK et al., is known to be less fungitoxic.
In our experiments no products other than triadimenol could be detected by glc. and other methods, although the combined amounts of triadimefon and triadimenol show a slight tendency to decrease during the incubation. This tendency may be due to the other metabolic pathway, but it was not significant in our experiments.

GROSSMANN: Your studies on the transformation of triadimefon were done in vitro. Since, as we found in field experiments with cereals, triadimefon apparently is also transformed in higher plants, the picture may be even more complicated under practical conditions. Do you think, that the fungi are able to change the ratio between triadimenol A and triadimenol B which they may find inside the plant?

GASZTONYI: We also investigated in our experiments the reduction of triadimefon to triadimenol in plants. The method of detection used in those experiments weren't able to distinguish the diastereomeric forms of triadimenol, but we established that the rate of reduction was dependent on plant species. It may be assumed that by the analogy with fungi species the ratio of triadimenol A and B, produced in plants is characteristic for plant species. As it has been shown by our present experiments, fungi do not change the ratio of triadimenol diastereomers previously existed in media.

SISLER: Why not start with triadimenol as a fungicide instead of triadimefon?

GASZTONYI: We are not competent to answer the question, although we had also put it ourselves. An argument for using triadimefon may be the higher selectivity of this compound, which may be in some circumstances advantageous from the point of view of epiphytic fungal species.

SISLER: Which isomer is more active as an antigibberellin agent in plant?

GASZTONYI: The antigibberellin activity of diastereomers of triadimenol has not been estimated by us.

BUCHENAUER: Triadimenol A ist wirksamer gegen Pilze, B hat einen stärkeren Antigibberellineffekt.

References

[1] BUCHENAUER, H., Systemisch-fungicide Wirkung und Wirkungsmechanismus von Triadimefon (MEB 6447), Mitt. Biol. Bundesanst. Land- u. Forstwirtsch., 165 (1975), 154–155

[2] CLARK, T., CLIFFORD, D. R., DEAS, A. H. B., GENDLE, P., WATKINS, D. A. M., Photolysis, metabolism and other factors influencing the performance of triadimefon as a powdery mildew fungicide, Pestic, Sci., 9 (1978), 497–506

[3] GASZTONYI, M. and JOSEPOVITS, GY., Translocacion and metabolism of triadimefon in different plant species, Acta Phytopathol. Acad. Sci. Hung., 13 (1978), 403–415

[4] GASZTONYI, M. and JOSEPOVITS, GY., The activation of triadimefon and its role in the selectivity of fungicide action, Pestic. Sci., 10 (1979), 57—65
[5] GASZTONYI, N. and JOSEPOVITS, GY., Triadimefon és triadimenol fungicid hatása és ezt befolyásoló tulajdonságaik, Növényvédelem, 15 (1979), 529—535
[6] KRÄMER, W., BÜCHEL, K. H., PFLUGEL, W. D., FROHBERGER, D. E., BRANDES, W., Diastereomere Triazolyl-O,N-acetale. Verfahren zu ihrer Herstellung sowie ihre Verwendung als Fungicide (Bayer AG), DE 2743767 (BRD), (1979)
[7] SPECHT, W., Methode zur gaschromatografischen Bestimmung der Fungicide Fuberidazol, Fluotrimazol, und Triadimefon in Pflanzen und Boden, Pflanzenschutz-Nachrichten Bayer, 30 (1977), 55—71
[8] VOGELER, K., Der Metabolismus von Triadimefon in Gerste unter Gewächshausbedingungen und im Boden, Leverkusen-Bayerwerk, (1976), RA-272

F. J. Schwinn and T. Staub

CIBA-GEIGY LTD., Agricultural Division, Research and Development Department, Basle, Switzerland

Biological Properties of Metalaxyl

1. Introduction

Over the past six years a number of new chemicals for the control of Oomycetes has been introduced (Schwinn [12]). They belong to four different chemical classes, as presented in Table 1.

Table 1
New chemicals for control of Oomycetes

Chemical group	Common name	Trade marks+ Formulation (s)	First report
carbamate	prothiocarb	PREVICUR S 70; SCW	1974
cyanoacetamideoxime	(DPX 3217)	CURZATE; WP	1977
acylalanine	furalaxyl*	FONGARID; WP, G	1977
O-alkyl-phosphonate	phosetyl Al**	ALIETTE; WP	1977
acylalanine	metalaxyl*	RIDOMIL; WP, G ACYLON; WP APRON; SD	1977
carbamate	propamocarb	PREVICUR N; SCW	1978

* ISO draft recommendation
** ISO draft proposal
+ In this paper, trade marks are given in capital letters.

Ciba-Geigy Ltd., Basle ,Switzerland, has contributed to this progress by the detection of the acylalanines as a new class of highly active protective and curative fungicides with systemic properties, and the commercial development of two molecules of this class, namely metalaxyl for agricultural crops and furalaxyl for ornamentals. For furalaxyl the trade mark FONGARID has been chosen. The trade marks and formulations for metalaxyl are shown in the following table (Table 2).

2. General Properties and Use Concept

The basic chemical, physical-chemical, toxicological and biological properties of the two compounds have been described earlier (Schwinn et al. [11], Wiersema and Wis-

Table 2
Trade marks, formulations and uses for metalaxyl.

Trade mark	Formu-lation	Content of active ingredient	Main use
RIDOMIL	25 WP	25% metalaxyl	foliar application
RIDOMIL PLUS = ACYLON BLEU	50 WP	15% metalaxyl plus 35% copper	foliar application
RIDOMIL	5 G	5% metalaxyl	soil application
APRON	35 SD	35% metalaxyl	seed dressing application

sink [18], Urech et al. [17]). Therefore, this paper summarizes the major aspects only and reports on some newer findings, mainly on metalaxyl.
The spectrum of activity of metalaxyl centers around the order of Peronosporales (Table 3). Here, the in vitro-ED 50-values for the inhibition of mycelial growth are in

Table 3
In vitro activity of metalaxyl on mycelial growth

Pathogen	ED_{50} in ppm
Oomycetes	
Saprolegniales	
Aphanomyces euteiches Drechs.	>100
Aphanomyces laevis de Bary	>100
Peronosporales	
Phytophthora cactorum (Leb. et Cohn) Schroet.	0,4
Phytophthora cinnamomi Rands	0,2
Phytophthora citrophthora (Sm. et Sm.) Leon.	0,2
Phytophthora infestans de Bary	0,3
Phytophthora megasperma Drechs.	>0,1
Phytophthora nicotianae var. parasitica (Dastur) Waterhouse	0,6
Pythium debaryanum Hesse	0,5
Pythium ultimum Trow	0,2
Zygomycetes	
Mucor hiemalis Wehmer	>100
Rhizopus nigricans Ehrenb.	>100
Ascomycetes	
Sclerotinia fructigena Aderh. et Ruhl.	>100
Basidiomycetes	
Sclerotium rolfsii Sacch.	>100
Rhizoctonia solani Kühn	>100
Deuteromycetes	
Alternaria solani Sor.	>100
Botrytis cinerea Pers. ex Fr.	>100
Fusarium oxysporum Schlecht. ex Fr.	>100
Gloeosporium musarum Cooke et Massee	>100
Verticillium albo-atrum Reinke et Berth.	>100

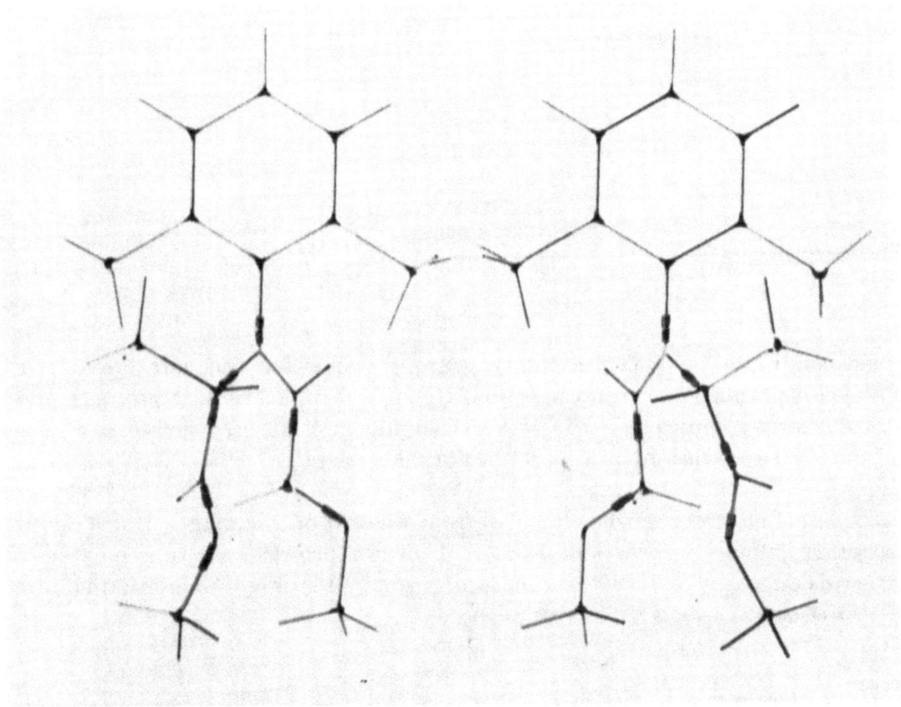

Fig. 1. Dreiding-model of the two enantiomers of metalaxyl
left: L(+)-Form (= CGA 76538)
right: D(−)-Form (= CGA 76539)

the range of 0.01 to 1 ppm. In laboratory studies with *Phytophthora* species, metalaxyl showed a peculiar dose-response curve characterized by an unusually flat part over a broad dosage range (STAUB et al. [14]).

Metalaxyl is an alanine derivative with a centre of chirality. The optically pure enantiomers (Fig. 1) were prepared separately and tested for biological activity (MOSER and VOGEL [10]). In tests carried out in petridishes as well as in greenhouse tests the D(−)-form (code number = CGA 76539) was by far more active than the L(+)-form code number = CGA 76538), as shown in Table 4 and Figure 2.

Table 4
In vitro fungicidal activity of metalaxyl and its enantiomers against Peronosporales (Inhibition of mycelial growth, agar incorporation technique)

Species	ED_{50}-Values in ppm a. i.		
	metalaxyl	CGA 76539 D(−)-Form	CGA 76538 L(+)-Form
Phytophthora infestans	0.4	<0.1	>100
Phytophthora megasperma	<0.1	<0.1	>100
Pythium debaryanum	0.7	0.3	>100
Pythium ultimum	0.4	0.1	48

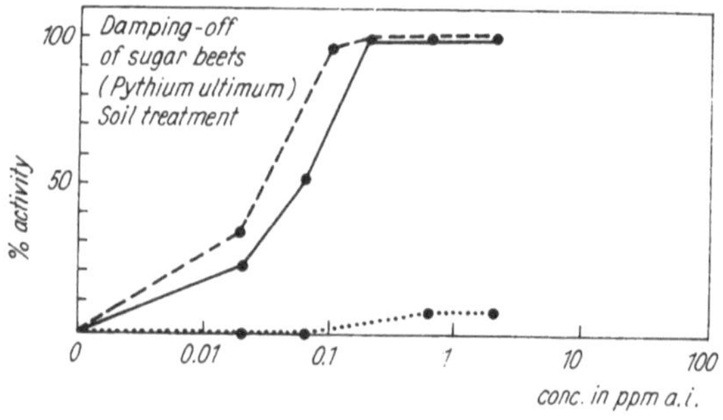

Fig. 2. In vivo fungicidal activity of metalaxyl and its enantiomers —·— metalaxyl, – – – CGA 76539 (D), ... CGA 76538 (L)

Under field conditions metalaxyl exhibits a strong protective and curative activity against all plant pathogenic Peronosporales. After foliar application, it provides good control of air-borne Oomycete pathogens, even under high disease pressure, at dose rates of 200–250 g a.i./ha in spray volumes of 200–500 l/ha, or at 20–30 g a.i./hl, respectively.

Its rapid penetration prevents the chemical from washoff on the plant surface and its high acropetal systemicity ensures protection of the new growth, resulting in extended spray intervals and thus, reduced cost of application and strongly reduced amount of chemical per season, as shown in Table 5.

Table 5
Comparison of total amount of product per season, as applied in potatoes against late blight (Phytophthora infestans)

Fungicide	Number of applications	Average dose rate (kg a.i./ha)	Total per season (kg a.i./ha)
mancozeb M 45	8	1.6–2.4	12.8–19.2
metalaxyl	5	0.25	1.25

Beyond this classical form of application against the foliar diseases, as described above, the strong systemicity of metalaxyl opens new approaches for the control of root and crown diseases caused by Peronosporales and systemic downy mildews. Examples of such diseases are given in Table 6.

Root and crown diseases, as shown as types number 2 and 3 in this Table, can now be controlled by granular or drench application of 2 g a.i./m² at intervals of three to six months in perennial crops like avocado (against root rot, caused by *Ph. cinnamoni*) or citrus (against stem canker, caused by *Ph. citrophthora*) and in a single application per season in annual crops, such as tobacco (against black shank, caused by *Ph. nicotianae* var. *nicotianae*). Results of such applications have been published by Urech et al. [17], Morton et al. [9] and Schwinn [12].

The same type of application can successfully be used against systemic downy mildews on perennial crops like hops; for annual crops which suffer from this type of diseases, like sunflower or tropical maize, sorghum and millet, a simple and cheap seed dressing application at dose rates of 175–250 g a.i./100 kg seed gives high protection over the whole infection period, which mostly is no longer than six to ten weeks

Table 6
Major types of diseases caused by Oomycetes

Type No.	Symptoms on ...	Type and site of infection	Site of survival	Typical examples
1	leaves	local; leaves	plant parts	*Phytophthora infestans*/ potato *Plasmopara viticola*/ grape vines
2	shoot and/or leaves	local; roots or shoot	soil	*Phytophthora nicotianae var. nicotianae*/tobacco
		systemic; hypocotyl or roots	soil or seed	*Peronosclerospora maydis*/corn *Plasmopara halstedii*/ sunflower
		systemic; shoot or *leaves*	root-stock	*Pseudoperonospora humuli*/hops
3	root, hypocotyl or crown	local; root, hypocotyl or crown	soil	*Pythium spp.*/vegetables, ornamentals *Phytophthora cinnamomi*/ avocado *Phytophthora cactorum*/ apple tree

(URECH et al. [17], SCHWINN [12]). These diseases can now for the first time be effectivelly controlled, an aspect of high relevance for the production of staple food in the developping countries of Africa and Southeast Asia.

3. Uptake and Transport of Acylalanines

The apoplastic transport pattern and the curative action of metalaxyl and related acylalanines became obvious already during the early screening stages (STAUB et al. [13]). In these tests grape and tomato plants could be protected by drench applications to the soil against *Pl. viticola* and *Ph. infestans*, respectively. Good foliar penetration of metalaxyl was suggested by the excellent curative action which was at least as effective as protective sprays. This was in clear contrast to previously known standard fungicides such as maneb which do not show these two types of activity even at high concentrations.

The curative activity of metalaxyl on grape downy mildew was studied further by LAFON et al. [8] who showed (cf. Table 7) that treatments up to five days after inoculation gave good curative control and prevented sporulation completely. Treatments made at the onset of sporulation, i.e. seven to ten days after inoculation, did not substantially inhibit sporulation from days 11 to 19 after inoculation. Thereafter, however, even these treatments strongly inhibited sporulation. Similar observations on the effectiveness and limitations of curative treatments were made by COHEN et al. [2] with *Ph. infestans* on tomatoes.

Table 7
Curative activity of metalaxyl against Plasmopara viticola. Influence of time lapse between inoculation and treatment (Dose rate 20 g a.i./hl = 200 ppm). Adapted from LAFON *et al.* [8]

Application at days after inoculation (= J)	Treatments				
J	inoculation	inoculation	inoculation	inoculation	inoculation
J + 3	treatment				
J + 5		treatment			
J + 7			treatment		
J + 10				treatment	
Evaluation at					
J + 11	0	0	5	27	51
J + 13	0	0	4	51	51
J + 15	0	0	2	25	50
J + 19	1	1	4	25	70
J + 21	2	1	1	3	31
J + 24	1	3	1	1	27

They showed that curative treatments were fully effective up to 3 days after massive inoculation, but that sprays on macroscopically visible lesions did not completely prevent them from expanding for some time; however, the rate of expansion was strongly reduced (Fig. 3). Since this latter observation covered only days 4 through 7 after inoculation, it does not allow to draw any conclusions on the final fate of such lesions.

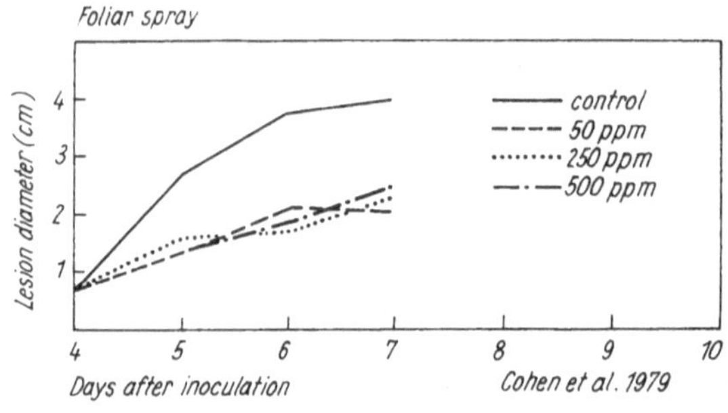

Fig. 3. Effect of metalaxyl on development of lesions (*Phytophthora infestans*/tomato) after curative application (3 days after inoculation) Adapted from COHEN et al. [2]

Even though treated lesions continued to expand, their sporulation was severely reduced as soon as 4h after foliar or 9h after soil drench treatments, respectively (cf. Fig. 4). In field trials it was found that in order to ensure a full knockdown of the pathogen in a curative spray schedule, two applications at an interval of no more than 7 days are necessary.

In addition on the apoplastic transport of metalaxyl from roots into leaves and young fruits, which COHEN et al. [2] showed to occur within one hour in tomato plants, it was also shown in studies on tomato and grape plants to be translocated acropetally after application on green stem parts and on basal leaf parts (STAUB et al. [13]).

In the meantime, the translocation pattern as described above has been confirmed in autoradiographic studies, in which the ^{14}C-labelled chemical was locally applied to the

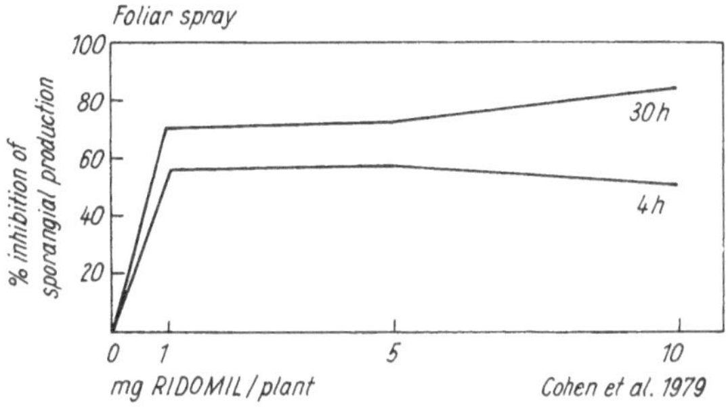

Fig. 4. Effect of metalaxyl on production of sporangia by *Phytophthora infestans* on performed lesions on tomato leaves. Adapted from COHEN et al. [2]

foliage, the green stem or the roots, respectively, of grape cuttings. Under field conditions, especially the uptake by green stems seems to be a significant factor in the excellent performance of the chemical.

LAFON et al. [8] confirmed the acropetal transport of metalaxyl after foliar applications on grape plants in the field by showing protection of the new growth during 30 days after a single application at a dose rate of 50 g a.i./hl, as shown in Table 8.

Table 8
Systemic activity of metalaxyl against Plasmopara viticola on young shoots one month after treatment (after LAFON *et al., [8])*

Treatments (ppm)		percentage of diseased leaf surface
metalaxyl	250	26
metalaxyl	500	9
mancozeb	2800	66
none		85

In closed humidity chambers they could also demonstrate that metalaxyl can be distributed in the vapour phase. This type of activity has also been observed under field conditions. However, it remains to be shown what significance it has for the practical disease control.

Although symplastic transport of acylalanines from treated leaves to new growth seems negligible in most instances, it may be sufficient for disease control under special conditions or on special crops. For example, under light conditions conducive to the assimilate flow from furalaxyl-treated source leaves to non-treated sink leaves of tomato plants, the latter were partially protected from late blight attack (STAUB et al. [13]). On avocado ZENTMYER [19] showed protection against *Ph. cinnamomi* by foliar treatments of potted plants with metalaxyl. The practical significance of the limited symplastic transport remains to be studied.

One of the obvious implications of the systemic nature of acylalanines is their unconventional response to climatic conditions. When tomato plants treated with acylalanines were subjected to artificial precipitation in climate chamber tests, protection against *Ph. infestans* was superior to corresponding treatments without precipitation

(Staub et al. [13]). Experience from field trials confirms these results. Since the development of downy mildews is favored by rainfall, this behaviour of the acylalanines is particulary valuable.

4. Metabolism

In the translocation studies discussed above, biological activity was the measure for transport; thus they did not show whether metalaxyl itself or a conversion product was ultimately responsible for the effects observed. Evidence produced so far from studies directed at this question suggests that the parent molecule itself is largely responsible for the high level of activity in the plant. The results of studies on the metabolism of metalaxyl in plants, animals and soil will be published separately.

5. Site of Action in Fungal Development on and in the Host Plant

For a rational application of systemic fungicides it is important to know at which stage in the disease cycle they inhibit the development of the target fungi.

In studies with *Pl. viticola* on grape leaves by Staub et al. [15] it was shown that contrary to classical protective fungicides zoospore germination, initial penetration processes and the formation of the first haustoria are not affected by metalaxyl, neither after foliar spray nor soil drench treatments. In contrast, the further establishment of the pathogen in the host tissue was practically completely stopped. This was demonstrated by counting the number of haustoria formed in treated and untreated host tissue, respectively. Similarly, on the surface of tomato leaves metalaxyl did not affect the germination process of zoospores of *Ph. infestans*. In this case, however, observations on haustoria were obscured by quick necrotic reactions of the host tissue. Similar necrotic reactions on treated tomato plants were reported by Cohen et al. [2] and on treated lettuce plants by Crute [3, 4]. Crute [4] points out that the necrotic reactions to infections by *Bremia lactucae* on metalaxyl-treated susceptible lettuce plants were indistinguishable from those normally observed on untreated resistant cultivars. This effect may be indirectly caused by metalaxyl by stimulating the host's resistanc mechanisms or by slowing down the fungal growth sufficiently to allow resistance to be expressed even in susceptible host cultivars. Circumstantial evidence from other trials so far tends to support the second of the two possibilities.

The activity of metalaxyl against soilborne fungi was analyzed using *Ph. parasitica* var. *nicotianae* as a test fungus (Staub and Young [16]). In analogy to the results obtained with foliar pathogens it was found that the release of zoospores around tobacco root tips were not inhibited by the chemical. First signs of inhibition during the infection process were again seen only after the fungus had penetrated the host tissue. In addition, sporangia formation from chlamydospores and on infected root tissue was strongly inhibited at about the same concentration as was required for stopping the growth of infections in the root tissue.

6. Risk of resistance

Since systemic fungicides have been introduced into the agricultural practice, acquired resistance has become a problem particulary in relation to benzimidazole compounds (DEKKER [6], GEORGOPOULOS [7]). In view of this development, we studied the potential of target fungi to develop resistance against acylalanines. The results of our model studies (STAUB et al. [14]) can be summarized as follows: Repeated exposure of *Ph. infestans* isolates in agar cultures or on plants to sublethal dosages of metalaxyl failed to produce strains with in vivo resistance to this chemical. Strains which showed a decreasing sensitivity in these tests and either lost their pathogenicity or still were fully sensitive when tested on tomato plants. In an in vivo-selection trial in a climate chamber the continuous use of sublethal dosages of metalaxyl over eleven generations of *Ph. infestans* on potato plants did not result in decreasing sensitivity.
Since then, BRUIN and EDGINGTON [1] reported acquired in vitro-resistance of *Ph. capsici* and of *Pythium* spp. both in vitro and in vivo on peas to metalaxyl and chemically related compounds in model tests.
In addition, DAVIDSE [5] was able to select resistant strains from in vitro-cultures exposed to mutagens, which, in contrast to strains selected by repeated exposure to sublethal dosages of the chemical, still were fully pathogenic.
Recently, the continuous, exclusive, year-round usage of metalaxyl during two years against downy mildew of cucurbits *(Pseudoperonospora cubensis)* by growers in greenhouses in Israel lead to a strong reduction of its effectiveness. Even though there is not yet complete evidence, it is highly probable that this phenomenon is due to acquired resistance of the pathogen population.
In view of the potential risk of resistance of metalaxyl which we have to face, adequate strategies need to be developed and enforced in the market which eliminate or at least reduce or delay the built-up of resistant strains under field conditions and thus safeguard the longevitiy of these biologically interesting and attractive fungicides (SCHWINN [12]). However, in view of the novelty of the resistance phenomenon in phytopathology the theoretical basis for the design of such strategies still is weak and certainly needs improvement.

Diskussion

LYR: Die Substanz hat im Unterschied zu anderen Fungiziden bei *Phytophthora* in Flüssigkultur einen ungewöhnlichen lag-Effekt, d. h. die Wirkung setzt erst nach 6—12 Stunden ein, was auch in Ihren Bildern bezüglich der Haustorienbildung deutlich war. Haben Sie dafür eine Erklärung?

SCHWINN: Da wir den Wirkungsmechanismus noch nicht kennen, es wird seit längerem daran gearbeitet, kann ich das Phänomen nicht erklären.

BURTH: In Versuchen zur Bekämpfung von *Phytophthora nicotianae* an Tomaten in Gewächshäusern zeigte Ridomil im Gegensatz zum Fongarid in fungizid wirksamen Anwendungskonzentrationen beträchtliche Phytotoxizität. Ist diese Phytotoxizität allgemein bei Bodenapplikation in Tomaten zu erwarten?

SCHWINN: Ich kann bestätigen, daß wir nach Bodenapplikation (Gießverfahren) an dieser Pflanze gelegentlich solche Beobachtungen gemacht haben.

GRÜNZEL: Haben Sie eine Erklärung für die hohe spezifische Wirkung des Metalaxyls (erfaßt werden nur Peronosporales aber nicht Saprolegniales oder Zygomycetes)?

SCHWINN: Diese erstaunliche Spezifität können wir bis heute noch nicht erklären. Sie ist überraschender im Fall der Saprolegniales die ja mit den Peronosporales zur gleichen Klasse (Oomyceten) gehören, als im Fall der Zygomyceten, die eine eigene, in vieler Hinsicht von den Oomyceten verschiedene Klasse bilden.

BOCHOW: Ist aus der Translokation des Muttermoleküls von Metalaxyl in der Pflanze eine relativ hohe Stabilität der Substanz abzuleiten und gilt dies auch für den Boden?

SCHWINN: Die Stabilität von Metalaxyl in der Pflanze ist sicherlich geringer als die der Benzimidazole. Es schützt nach Blattapplikation während 14—21 Tagen vor Neubefall. Bei Bodenapplikation kann man davon ausgehen, daß kleine Mengen die für die Bekämpfung der Erreger im Gewebe genügen) während längerer Zeit aus dem Boden (oder vom behandelten Saatgut) aufgenommen werden.

GROSSMANN: Welcher Anteil des applizierten Fungizids dringt unter normalen Bedingungen nach Spritzbehandlungen in die Blätter ein?

SCHWINN: Der Anteil liegt nach unseren bisherigen Erkenntnissen etwas über den von Ihnen genannten Werten anderer systematischer Fungizide. Trotzdem wird auch im Fall von Metalaxyl nur der kleinere Teil vom Blatt aufgenommen. Diese Aufnahme geschieht innerhalb kurzer Zeit.

GROSSMANN: Wie lange hält die Wirkung des Fungizids nach Saatgutbehandlung vor, z.B. bei Mais und Hirsen?

SCHWINN: Die Wirkdauer liegt, abhängig von den Gegebenheiten im einzelnen Fall, in der Größenordnung von 6—10 Wochen. Dies genügt in der Mehrzahl der Fälle, da der Erreger danach die Pflanze ohnehin kaum mehr in größerem Umfang infizieren kann. In Einzelfällen können ergänzende Blattapplikationen nötig werden.

LEROUX: Have you made trials on the activity of metalaxyl on root diseases after foliar treatments?

SCHWINN: As I explained, generally, the downward translocation of metalaxyl is relatively weak. In these cases, a foliar treatment will not protect the roots. However there are a few exceptions. For example, it has been shown by ZENTMYER, that foliar applications on avocado protected the roots from *Phytophthora* root rot.

LEROUX: What is the type and mechanism which explains vapour activity? Codestillation or volatility of the compound alone?

SCHWINN: As far as we know at this point in time, it is the volatility of metalaxyl which causes the remarkable vapor action.

PFLIEGEL: Kann man aus dem Wirkungsmechanismus der analogen Herbizide keine Folgerungen auf den Wirkungsmechanismus des Metalaxyls ziehen?

SCHWINN: Wir sehen bisher keine Zusammenhänge .Wir haben aber auch noch ungenügende Kenntnisse über den Wirkungsmechanismus in Pilzen.

PFLIEGEL: Gibt es außer kommerziellen auch andere Gründe für so viele Namen des Produktes?

SCHWINN: Wir haben versucht, die verschiedenen Anwendungsformen (Blattfungizid, Saatbeizmittel) auch in der Namengebung zum Ausdruck zu bringen (Ridomil, Apron). Im übrigen muß man bei weltweiter Einführung neuer Produkte gelegentlich auch lokale Wünsche berücksichtigen.

Literature

[1] BRUIN, G. C. and L. V. EDGINGTON, Induced resistance to Ridomil of some oomycetes, Amer. Phytopath. Soc., Northeast Regional Meeting (1979), Abstracts, 22
[2] COHEN, Y., M. REUVENI and H. EYAL, The systemic activity of Ridomil against *Phytophthora infestans* on tomato plants, Phytopathology **69** (1979), 645—649
[3] CRUTE, I. R., Studies on new systemic fungicides active against *Bremia lactucae*, 3rd Internat. Congr. Pl. Path., München (1978), Abstracts, 384
[4] CRUTE, I. R., Lettuce mildew-destroyer of quality, ARC Research Review **5** (1979), 9—12
[5] DAVIDSE, L. C., Resistance to acylalanine fungicides in *Phytophthora megasperma* f. sp. *medicaginis*, Neth. J. Path. (in press)
[6] DEKKER, J., Resistance, In MARSH, R. R., (ed.) Systemic fungicides, Wiley, New York 1972, 156—174
[7] GEORGOPOULOS, S. G., Development of fungal resistance to fungicides, In SIEGEL, M. R. and H. SISLER (ed.), Antifungal compounds, Vol. 2, M. DEKKER Inc., New York and Basel, 1977, 439—496
[8] LAFON, R., Y. BUGARET et J. BULIT, Essais d'un nouveau fongicide de la famille des acylalanines (CGA 48988) contre le mildiou de la vigne (*Plasmopara viticola* (B. C.). BERL. et DE TONI), Phytiatrie-Phytopharm. **27**, (1978) 263—274
[9] MORTON, H. V., T. STAUB and T. R. YOUNG, The use of CGA 48988 for the control of *Phytophthora parasitica* var. *nicotianae*, 3rd Internat. Congr. Pl. Path., München (1978), Abstracts, 383
[10] MOSER, H. and C. VOGEL, Preparation and biological activity of the enantiomers of CGA 48988, a new systemic fungicide, 4th IUPAC Congress, Zürich (1978), Abstract Volume II—310
[11] SCHWINN, F. J., T. STAUB and P. A. URECH, A new type of fungicide against diseases caused by Oomycetes, Meded. Fac. Landbouww. Gent, **42** (1977), 1181—1184
[12] SCHWINN, F. J., Control of Phycomycetes — a changing scene, Proc. 11. Brit. Crop. Prot. Conf. Brighton (1979), **3**, 791—802
[13] STAUB, T., H. DAHMEN and F. J. SCHWINN, Biological characterization of uptake and translocation of fungical acylalanines in grape and tomato plants, Z. Pfl. Kr. Pfl-. schutz **85** (1978), 162—168
[14] STAUB, T., H. DAHMEN, P. URECH and F. J. SCHWINN, Failure to select for in vivo resistance in *Phytophtora infestans* to acylalanine fungicides, Pl. Dis. Reptr. **63** (1979), 385—389
[15] STAUB, T., H. DAHMEN and F. J. SCHWINN, Effects of Ridomil® on the development of *Plasmopara viticola* and *Phytophthora infestans* on their host plants, Z. Pfl.kr. Pfl.schutz **87** (1980), 83—91
[16] STAUB, T. H. and T. R. YOUNG, Biological mode of action of CGA 48988 against *Phytophthora parasitica var. nicotianae*, Phytopathology (in press)
[17] URECH, P. A., F. J. SCHWINN, and T. STAUB, CGA 48988, a novel fungicide for the control of late blight, downy mildews and related soil-borne diseases Proc. 9. Brit. Crop. Prot. Conf. Brighton (1977), **2**, 623—31
[18] WIERTSEMA, W. P. and B. H. WISSINK, Fongarid 50 WP, a new fungicide for ornamentals, Meded. Fac. Landbouww. Gent **29** (1977), 1189—1194
[19] ZENTMYER, G. A., New organic fungicides for control of *Phytophthora* root rots, 3rd Internat. Congr. Pl. Path., München (1978), Abstracts, 362

L. C. DAVIDSE and H. G. KRÜSE

Laboratory of Phytopathology, Agricultural University, Wageningen, The Netherlands

Resistance to Acylalanine Fungicides in *Phytophthora megasperma* f. sp. *medicaginis* and *P. megasperma* f. sp., *glycinea*

Introduction

Acylalanines have recently been introduced as systemic fungicides with highly selective activity against fungi of the order Peronosporales (SCHWINN et al. [5, 6]; STAUB et al. [7] and URECH et al. [9]). Their efficacy in the control of numerous diseases caused by these fungi has repeatedly been shown, and in some cases their use might lead to a major breakthrough in chemical control methods.
Although the mechanism of action of the acylalanine fungicides has not yet been elucidated, it might be highly specific in view of the selectivity and systemic behaviour of these compounds. Therefore as has been encountered with other specifically-acting fungicides, development of resistance might be a potential threat for a long-lasting useful life of the acylalanines.
In order to evaluate the potential of target fungi to develop resistance to these fungicides, selection experiments were carried out with *Phytophthora megasperma* f. sp. *medicaginis* and *P. megasperma* f. sp. *glycinea*, the causal organisms of root rot of lucerne and soybean respectively.

Selection of Resistant Mutants

P. megasperma f. sp. *medicaginis*. Resistant isolates were obtained in two ways, according to DAVIDSE [1]: a) through growing the fungus on V8-agar containing metalaxyl (0.5–2 µg/ml) for 2–4 weeks and subculturing mycelium, which showed reduced sensitivity, on media with metalaxyl at higher concentrations and b) through mass selection on V8-agar amended with 2–20 µg metalaxyl/ml from encysted zoospores either non-treated or treated with the mutagen N-methyl-N'-nitro-N-nitrosoguanidine (MNNG).
With the first method 6 mutants were obtained with a relatively low degree of resistance. Although all isolates were able to grow on V8-agar amended with 50 µg metalaxyl/ml, mycelial growth of the most resistant isolate (MET-141) was still inhibited for 68% at 5 µg/ml (Table 1). All these 6 isolates showed a reduced growth rate on medium without metalaxyl and several of them had much less aerial mycelium than the original isolate. Mass selection from non-treated zoospores yielded 18 resistant mutants per 43×10^5 spores. Similar to the isolates obtained by adaptation these mutants also showed a reduced growth rate and/or had less aerial mycelium than the

Table 1
Degree of metalaxyl resistance of the least sensitive mutants of P. megasperma f. sp. medicaginis obtained by different selection procedures

Strain	Number of isolates tested	% inhibition of mycelial growth on V8-agar at different metalaxyl concentrations				
		2	5	10	50	100 μg/ml
7x (wild type)	—	100	nd[1]	nd	nd	nd
MET-141 (adaptation)	6	74	68	74	76	nd
MET-161 (spontaneous)	13	nd	68	nd	nd	nd
MET-5 (MNNG)	6	7	3	0	−4	4

[1] nd = not determined.

original isolate. ED_{50}-values for inhibition of mycelial growth of the spontaneous mutants were all below 5 μg metalaxyl/ml medium (Table 1).

Treatment of encysted zoospores with MNNG and selection on V8-agar containing 2 μg metalaxyl/ml resulted in the rapid development of many resistant colonies. A total of 248 mutants were obtained from 114×10^6 spores. Since so many colonies developed no attempt was made to isolate them all so the actual mutation frequency was higher. About half of the mutants showed a growth rate and colony morphology similar to that of the original isolate, when transferred to fungicide-free media. As distinct from the spontaneous mutants the MNNG-induced mutants attained a much higher level of resistance. ED_{50}-values for inhibition of mycelial growth of 6 randomly chosen mutants were all well above 100 μg metalaxyl/ml (Table 1).

Eight metalaxyl-resistant isolates of the second category were also examined on resistance to furalaxyl, a structurally related fungicide. The ED_{50}-values for inhibition of mycelial growth of all strains were above 100 μg furalaxyl/ml V8-agar, whereas mycelial growth of the original isolate 7 x was already completely inhibited at a concentration of 5 μg furalaxyl/ml.

P. megasperma f. sp. *glycinea*. Mutants of this organism resistant to metalaxyl and furalaxyl were selected from non-treated and MNNG-treated encysted zoospores on

Table 2
Degree of metalaxyl resistance of mutants of P. megasperma f. sp. glycinea

Origin of mutants[1]	Number of isolates	Classification of mutants according to their ED_{50}-value for inhibition of mycelial growth by metalaxyl[2]		
		ED_{50} <50 μg/ml	50 μg/ml < ED_{50} <100 μg/ml	ED_{50} > 100 μg/ml
metalaxyl, spontaneous	2	2	0	0
metalaxyl, MNNG	10	3	3	4
furalaxyl, spontaneous	1	0	0	1
furalaxyl, MNNG	13	2	5	6

[1] Selection procedure
[2] ED_{50}-value of the wild type was c. 0.2 μg metalaxyl/ml V8-agar.

V8-agar containing either 2 µg metalaxyl/ml or 10 µg furalaxyl/ml. The zoospores for these experiments were obtained according to EYE et al. [2]. Out of c. 1.2 × 10^5 non-treated zoospores 2 metalaxyl-resistant mutants were obtained whereas 10 mutants were selected from the same number of MNNG-treated spores. Selection on medium with furalaxyl resulted in 1 spontaneous mutant per 1.9 × 10^5 zoospores. From the same number of MNNG-treated spores 13 furalaxyl-resistant mutants were obtained. All mutants were tested on resistance to metalaxyl. Results are given in Table 2. The degree of resistance was variable but attained a high level, even in one of the spontaneous mutants which was selected on medium containing furalaxyl. All mutants which arose on furalaxyl-amended medium were also resistant to metalaxyl, indicating the presence of cross resistance in these strains.
Seven mutants had a significantly lesser growth rate than the original strain. The growth rate and colony morphology of the other strains were similar to the original one.

Virulence and Resistance in vivo of Metalaxyl-Resistant Mutants

P. megasperma f. sp. *medicaginis*. Virulence and resistance in vivo of the mutants were determined in an assay with lucerne seedlings, cultivar Vernal, developed by IRWIN et al. [3], and modified by DAVIDSE [1]. Data are summarized in Table 3. All 6 isolates obtained by adaptation and all 13 spontaneous mutants selected from zoospores showed a reduced virulence as compared with that of the original isolate or were avirulent. The percentage of diseased seedlings increased with time and after 6 days 2 isolates obtained by adaptation and 3 obtained by mass selection were able to cause a complete damping-off of non-treated plants. One of theses isolates displayed resistance in vivo causing a 90–100% damping-off of the metalaxyl treated plants 6 days inoculation (data not shown). Out of 176 MNNG-induced mutants tested, 81 were as virulent as the original isolate and 33 of them caused a complete damping off of metalaxyl-treated plants within 3 days.

Table 3
Virulence and resistance to metalaxyl of mutants of P. megasperma f. sp. medicaginis in a seedling assay

Origin of mutants	Number of isolated tested	Classification[1] of mutants according to their virulence on control and metalaxyl treated lucerne seedlings					
		Control plants			Treated plants[2]		
		1	2	3	1	2	3
adaptation	6	0[3]	2	4	0	0	6
spontaneous	13	0	5	8	0	0	13
MNNG	176	81	62	33	33	41	102

[1] Classification, according to percentages of plants killed (P), 3 days after inoculation Class 1, 90 < P < 100; class 2, 0 < P < 90; class 3, P = 0.
The original isolate caused a 90–100% kill.
[2] Metalaxyl was applied as a soil drench of Ridomil 25 WP at a final concentration of 20 mg/kg planting mix, 2 days before inoculation. This treatment gave complete control of damping-off cuased by the original isolate.
[3] Numbers of mutants.

The virulence of 7 isolates showing resistance in the seedling assay has also been compared with that of the original isolate in a mature plant assay with Vernal plants (Irwin [3]). Two of the resistant isolates were significantly less virulent than the original isolate. Metalaxyl completely failed to control root rot caused by all 7 isolates, when applied as a soil drench to a final concentration of 20 mg a.i./l of soil. The average disease severity indexes of control and treated plants inoculated with the resistant isolates did not differ significantly, indicating the absence of any effect of the fungicid on the resistant mutants. Disease caused by the original isolate could be completely prevented at fungicide concentrations as low as 2.5 mg a.i. of soil.

P. megasperma f. sp. *glycinea.* The virulence and resistance in vivo of the mutants was determined in the soybean hypocotyl assay (Keen [4]) using 7-days-old seedlings of the cultivar Harosoy, either non-treated or root-treated with metalaxyl. Results are given in Table 4. A considerable number of the isolates was still as virulent in this

Table 4
Virulence and resistance to metalaxyl of mutants of P. megasperma f. sp. glycinea in the soybean hypocotyl assay

Origin of mutants	Number of isolates tested	Classification[1] of mutants according to their virulence on control and metalaxyl treated[2] soybean seedlings					
		Control plants			Treated plants		
		1	2	3	1	2	3
metalaxyl, spontaneous	2	2[3]	0	0	0	0	0
metalaxyl, MNNG	10	7	3	0	4	1	5
furalaxyl, spontaneous	1	1	1	0	1	0	0
furalaxyl, NMNG	6	6	0	0	1	5	0

[1] Classification according to percentages of seedlings killed (P), 4 days after inoculation. Class 1, $90 \leq P \leq 100$; class 2, $0 < P < 90$; class 3, $P = 0$.
10 plants were used per treatment. The original isolate caused a 100% kill.
[2] Twenty four hours before inoculation seedlings were placed in water containing Ridomil 25 WP (5 mg a.i./l). This treatment gave a complete control of disease caused by the original isolate.
[3] Numbers of mutants.

assay as the original isolate. The 3 spontaneous mutants did not show a decrease in virulence, but only one of them was able to cause disease on metalaxyl-treated seedlings. This mutant showed also a high degree of resistance in vitro in contrast with the other 2 (Table 2).

Four of the MNNG-mutants selected on metalaxyl-amended medium were highly virulent on treated plants. These mutants displayed the highest ED_{50}-values for inhibition of mycelial growth (Table 2), which indicates that with virulent strains a high degree of resistance in vitro corresponds with a high degree of resistance in vivo. Both the spontaneous mutant and the MNNG-induced mutant selected on furalaxyl-amended medium, which were highly virulent on treated plants had an ED_{50}-value higher than 100 µg metalaxyl/ml V8-agar (Table 2). Only one of the 5 strains, isolated from furalaxyl-amended medium, which displayed a moderate resistance in vivo, had a comparable high ED_{50}-value, but this isolate showed a reduced growth rate. Three of those 5 strains had an ED_{50}-value between 50 and 100 μg metalaxyl/ml and one had an ED_{50}-value below 50 μg/ml. Thus it seems that also with these mutants the degree of resistance in vivo corresponds with that in vitro.

Stability of Resistance and Competitive Ability of Metalaxyl-Resistant Mutants

The stability of resistance and the competitive ability of 4 metalaxyl-resistant mutants of *P. megasperma* f. sp. *medicaginis* have been determined in the seedling assay. To this end cups with lucerne seedlings were infected with motile zoospores from each of the 4 mutants separatly or with a composite inoculum consisting of equal numbers of zoospores of a mutant and the original strain. Three days after inoculation the seedlings were used as inoculum for a second set of plants. This procedure was repeated 9 more times. After each cycle the presence of the resistant isolate in the population was determined by transferring seedlings to a test set of seedlings which, 1 day before inoculation had been metalaxyl-treated by applying a soil drench of Ridomil 25 WP to a final concentration of 20 mg a.i./kg.

During the course of the experiment no apparent reduction in rate of damping-off of test seedlings occurred with 3 of the mutants, either mixed with the original sensitive strain or separately. The ability of the mixture of the 4th mutants and the sensitive isolate to cause damping-off of the test seedlings started to decline in the 4th cycle but increased again after the 6th cycle and after the 9th cycle again a complete damping-off occurred. This mutant alone did not show any decrease of virulence towards metalaxyl-treated seedlings although the rate at which damping-off of test seedlings as well as of non-treated seedlings occurred was lower than that of the other isolates. The metalaxyl-sensitive isolate was not able to cause disease of the metalaxyl-treated test seedlings in all cases. Since in this experiment a complete infection cycle is completed every 3—4 days the presence of the resistant mutants in the mixed populations after 11 infection cycles indicates that the competitive ability of the resistant isolates is not significantly reduced. It also indicates that the resistance is highly stable in character.

Concluding Remarks

Development of resistance in both *Phytophthora* sp. to the acylalanine fungicide metalaxyl could easily be demonstrated in laboratory experiments. Resistant mutants could be obtained either spontaneously from mycelium and zoospores or after mutagenic treatment of zoospores. With *P. megasperma* f. sp. *medicaginis* the resistance of spontaneous mutants never attained a high level. Moreover, these mutants were less virulent and disease caused by the majority of them could be controlled by the fungicide. A considerable number of the mutagen-induced resistant strains, however, was as virulent as the original isolate and a considerable number of them severely attacked metalaxyl-treated lucerne seedlings and mature plants. Resistance was highly stable and the fitness of the resistant mutants did not appear to be reduced. Cross resistance to the structurally-related compound furalaxyl was present in all the strains tested. With *P. megasperma* f. sp. *glycinea* one of the spontaneous mutants was highly resistant

as virulent as the original isolate and displayed resistance in vivo. After mutagenic treatment of zoospores mutants with similar properties could be selected.
Resistance to acylalanine fungicides has not yet been found under practical conditions. STAUB et al. [8] described isolates of *Phytophthora infestans* with decreased sensitivity to acylalanine fungicides in vitro, but these isolates were usually less virulent and/or fully sensitive to fungicide treatment on potato and tomato plants. These isolates were obtained by adaptation and mass selection without mutagenic treatment and are likely of the same type as those of *P. megasperma* f. sp. *medicaginis* obtained by similar methods.
Since the majority of the mutants which were highly virulent and displayed resistance in vivo, were obtained after mutagenic treatment, one might argue whether this type of mutants will also arise spontaneously. Chemical mutagens, however, affect mutation rates by increasing the frequencies of DNA replication errors and misrepair of genetic damage, which processes are also the basis of spontaneous mutations, so it is unlikely, that MNNG induces a specific type of mutation, which will not occur spontaneously. The fact that one of the spontaneously obtained resistant mutants of *P. megasperma* f. sp. *glycinea* was highly virulent and showed resistance in vivo seems to confirm this.
In view of our results development of resistance to acylalanine fungicides may occur in practice. Therefore measures should be taken to keep the selection pressure as low as possible, especially when acylalanine fungicides are used to control pathogens with a high reproduction rate, a short infection cycle and aerial dispersal.

Diskussion

GEORGOPOULOS: It is unfortunate that in the field of chemical control those discoveries, which create the greatest enthusiasm originally, have quite understandably perhaps, face the problem of resistance.
For the benefit of those who will for the first time use the new compounds that are active against Oomycetes and not acylalanines only, I would like to say that there is strong evidence that resistant strains exist in nature. I would like to suggest that we do not proceed at full speed in the extensive and particularly, in the exclusive uses of these compounds.

SCHWINN: (Kommentar zu GEORGOPOULOS) Because I did not have the time during my lecture I should like to add here a few remarks concerning the risk of resistance. First of all, we have started the studies in this area in an early phase of development of metalaxyl. Our in vitro model studies on *Phytophthora infestans* showed that strains could be adapted to the chemical. However, when we inoculated the resistant strains on tomatoes in the climate chamber, they had either lost pathogenicity, or they were still fully sensitive to the chemical. Thus Dr. DAVIDSE confirmed our findings. We did not use mutagenic agents in our tests, therefore this aspect is new in Dr. DAVIDSE's work. Further, we have run an in vivo-trial with *Phytophthora infestans* on potatoes in which we exposed 10 generations of the pathogen to sublethal dosages of the fungicide. In this test, we did not observe any signs of resistance of the pathogen.
Refering to the remarks of Prof. GEORGOPOULOS, I can inform you that the continuous, exclusive and year-round use of metalaxyl in glasshouses in Israel against the downy mildew of cucumbers *(Pseudoperonospora cubense)* has recently lead — after about

two years of usage — to a breakdown of activity. Even though we have not yet obtained clear evidence, it seems that we are faced with a can of resistance. Thus, we have to be aware that the risk does exist, and we have to take all possible measures to avoid or at least delay it in the future.

LYR: Have you any information on the biochemical background of the resistant mutants?

DAVIDSE: We have just started the research on the biochemical basis of resistance, so we do not have data yet.

LANGCAKE: Most of your NG mutants that are resistant in vitro and fully virulent are nonetheless still sensitive to Ridomil in vivo. Do you have an explanation for this?

DAVIDSE: Because the degree of resistance of the NG mutants is variable and we determined only the resistance in vivo at one concentration, those mutants might be only moderate resistant and are inhibited at the concentration we used in the in vivo experiments.

SCHMIDT: Do you have experiences with metalaxyl-resistant mutants from other *Phytophtora* spp. collected under field conditions?

DAVIDSE: There have not been reports of development of resistance to metalaxyl under practical conditions in The Netherlands. So I do not have any data about resistance in other *Phytophtora* spp.

LEROUX: Is there cross resistance between acylalanine fungicides and milfuram. Do you have results with other antioomycete compounds?

DAVIDSE: We have not looked yet to the effect on these compounds on the metalaxyl-resistant isolates.

References

[1] DAVIDSE, L. C., Resistance to acylalanine fungicides in *Phytophthora megasperma* f. sp. *medicaginis*, Submitted to Neth. J. Plant Path.

[2] EYE, L. L., B. SNEH, and L. J. LOCKWOOD, Factors affecting zoospore production by *Phytophthora megasperma* var. *sojae*, Phytopathology **68** (1978), 1766–1768

[3] IRWIN, J. A. G., S. A. MILLER, and D. P. MAXWELL, Alfalfa seedling resistance to *Phytophthora megasperma*, Phytopathology **69** (1979), 1051–1055

[4] KEEN, N. T., J. J. SIMS, D. C. ERWIN, E. RICE, and J. E. PARTRIDGE, 6a-Hydroxyphaseollin: an antifungal chemical induced in soybean hypocotyls by *Phytophthora megasperma* var. *sojae*, Phytopathology **61** (1971), 1084–1089

[5] SCHWINN, F. J., T. STAUB, and P. A. URECH, Die Bekämpfung falscher Mehltau-Krankheiten mit einem neuen Wirkstoff aus der Gruppe der Acylalanine, Mitt. Biol. Bundesanst. Land.-Forstwirtsch. Berlin-Dahlem **178** (1977), 145–146

[6] SCHWINN, F. J., T. STAUB, and P. A. URECH, A new type of fungicide against diseases caused by Oomycetes, Meded. Fac. Landbouwwet. Rijksuniv. Gent **42** (1977), 1181 to 1188

[7] STAUB, T., H. DAHMEN, and F. J. SCHWINN, Biological characterization of uptake and translocation of fungicidal acylalanines in grape and tomato plants, Z. Pflanzenkr. Pflanzensch. **85** (1978), 162–168

[8] STAUB, T., H. DAHMEN, P. URECH, and F. SCHWINN, Failure to select for in vivo resistance in *Phytophthora infestans* to acylalanine fungicides, Plant Dis. Rep. **63** (1979), 385–389

[9] URECH, P. A., F. J. SCHWINN, and T. STAUB, CGA 48988, a novel fungicide for the control of late blight, downy mildews and related soil-borne diseases, Proc. Br. Crop Prot. Conf. (1977), 623–631

S. G. Georgopoulos, P. Mytilineou and B. N. Ziogas

Laboratory of Phytopathology, Athens College of Agricultural Sciences, Votanikos, Athens 301, Greece

Genetic Activity of a Metalaxyl Preparation in Diploid *Aspergillus nidulans*

Studies of recent years with the use of heterozygous diploid strains of *Aspergillus nidulans* have helped to localise the action of some fungicides inside the fungal nucleus. By contrast to inhibitors of processes unrelated to the hereditary function [3], these chemicals increase the frequency of mitotic recombination in colonies of such heterozygous diploid strains. The agriculturally important fungicides that have been shown genetically active in this test belong either to the benzimidazole [7] or to the aromatic hydrocarbon [3] group with which the new N-(3,5-dichlorophenyl) dicarboximide fungicides have recently been classified [4]. In the former case, the recombinogenic activity is due totally to the ability of the compounds to cause chromosome non-disjunction; in the latter, a breakage-deletion mechanism is at least partly responsible for the increased sectoring [5]. Two antibiotics have also been found active, namely griseofulvin, causing non-disjunction [6] and actinomycin D, causing mitotic crossing-over [5]. This paper will report apparent genetic activity of metalaxyl, one of the acylalanine fungicides which is very important for the control of Oomycetes and will present preliminary data on the mechanism involved.

Materials and Methods

The heterozygous diploid strain of *A. nidulans* that was used has been synthesized by Kappas [5] and has the genotype shown in Figure 1. This strain is prototrophic and produces green conidia but heterozygosity for the conidial color markers and for the several nutritional requirements permits the recognition of yellow and white recombinants and helps in the genetic analysis of such recombinants.
The material obtained from the commercially available formulation Ridomil 25 WP, following two recrystallisations from ethanol will be referred to as metalaxyl [DL-methyl-*N*-(2,6-dimethylphenyl)-N-(2-methoxyacetyl) alaninate] in this paper. This material exhibits the selective toxicity of metalaxyl to Oomycetes described in the literature. A sample of the experimental fungicide Ortho 20615 Technical [2-chloro-*N*- (2,6-dimethyl-phenyl)-N-(tetrahydro-2-oxo-3-furanyl) acetamide], provided by Chevron Chemical Company was also used. The two fungicides are shown in Table 1. Stock solutions were prepared in ethanol for the former chemical and in ethyl acetate for the latter, and small amounts were added to the agar medium to give the appropriate concentrations. Control media contained similar amounts of the organic solvents.
The techniques for measuring growth-inhibitory and recombinogenic activity of the chemicals have been described previously [3, 4]. Because treatments may affect not only the rate of growth but also the final size attained by a colony, the number of color recombinants per given colony area (instead of per given number of colonies, as in previous studies) was recorded. Enough plates were used to provide at least 500 cm² of colony area per treatment. The attempts to elucidate the mechanism(s) of mitotic recombination were

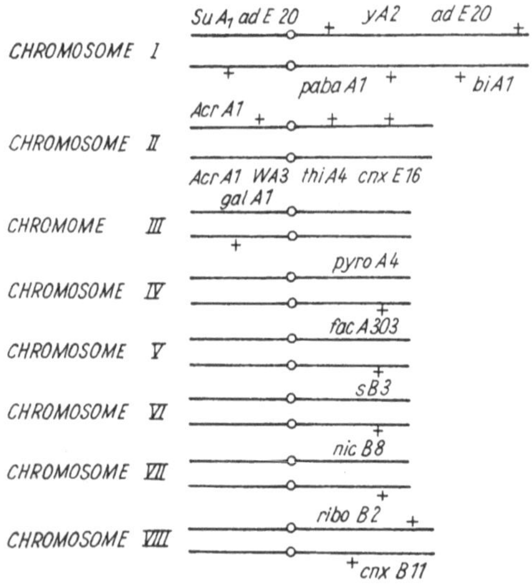

Fig. 1. Genotype of the heterozygous diploid strain of *Aspergillus nidulans* used

Table 1
Fungicides tested for recombinogenic activity

Name	Structural formula	Ref.
Metalaxyl [*DL*-methyl-*N*-(2,6-dimethylphenyl)-*N*-(2-methoxyacetyl)alaninate]	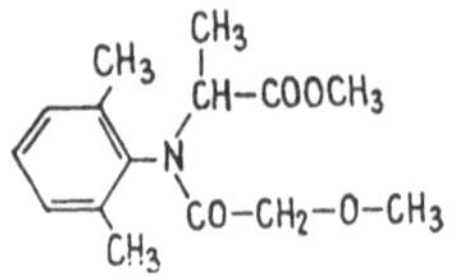	12
Ortho 20615 [2-chloro-*N*-(2,6-dimethylphenyl)-*N*-tetrahydro-2-oxo-3-furanyl)acetamide]	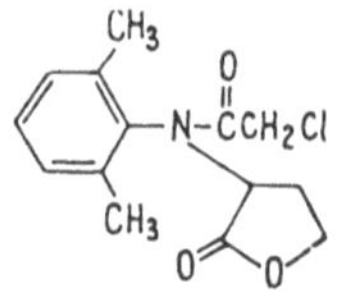	8

based on the analysis of large numbers of recombinants for their nutritional requirements using a 26-pin replicator. In cases of doubt on the ploidy of recombinants, the conidial size was used as a second criterion [Roper, 1952].

Results and Discussion

Although metalaxyl is a fungicide recommended specifically against Oomycetes our preparation inhibits the growth of *Aspergillus nidulans* considerably. In plates with Aspergillus complete medium [9] a concentration of approximately 150.0 µM was required for 50% inhibition of the heterozygous diploid strain. A striking effect of metalaxyl on sectoring was easily recognised in the preliminary tests (Fig. 2). The increase in the number of macroscopically visible sectors was a function of fungicide concentration up to approximately 500 µM (Table 2). Apparently, the number of

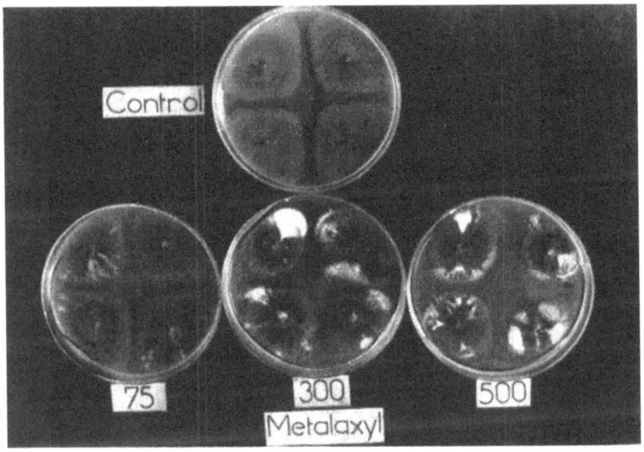

Fig. 2. Sectoring of the heterozygous strain of *Aspergillus nidulans* as affected by metalaxyl concentration (given in μM)

Table 2
Growth inhibition and induction of mitotic recombination by metalaxyl in diploid Aspergillus nidulans

Metalaxyl concentration (μM)	% reduction of colony diameter	Color recombinants per 100 cm²	Induced segregation index[1]
0.0	—	4	—
50.0	10	23	5.75
75.0	27	39	9.75
100.0	39	31	7.75
150.0	46	44	11.00
300.0	58	65	16.25
500.0	71	89	22.25
1000.0	78	73	18.25

[1] Ratio of color recombination in 100 cm² of treated to untreated colonies.

independent events leading to mitotic recombination must increase further, although such further increases are impossible to measure by the number of color sectors in practice. From the fact that even slightly inhibitory concentrations show distinct recombinogenic activity (Table 2) it can be suggested that the mechanism leading to mitotic recombination is at least partly responsible for the toxicity of metalaxyl to this organism.

In sharp contrast with metalaxyl, Ortho 20615 does not induce sectoring in diploid *A. nidulans*. This experimental fungicide was used at concentrations causing up to 40% inhibition of growth but no indication of genetic activity was obtained. The structures of the two compounds have similarities and the antifungal spectra seem to be the same [8, 12] but the mechanisms are apparently different.

The presence of an alanine moiety in the molecule of metalaxyl and its absence in that of Ortho 20615 suggests that metalaxyl may act like parafluorophenylalanine, one of the best known inducers of mitotic recombination in *A. nidulans* [11]. In an attempt to examine this possibility two parafluorophenylalanine-resistent mutants of *A. nidulans* were obtained from Dr. U. Sinha, University of Delhi, India. These mutants were compared to parafluorophenylalanine-sensitive haploids as to metalaxyl sensitivity but no cross-resistance between the two compounds was recognised. A similar test, using appropriate strains of *Penicillium italicum* and *Ustilago maydis* did not show cross-resistance between metalaxyl and benomyl. It must be remembered, of course, that such negative results do not exclude the possibility of similarities in the mechanism of action since the mutations may effect properties (e.g. permeability) other than sensitivity at the site [2].

Increased sectoring may result from ability of a chemical to induce gene mutations and, therefore, homozygosity for the recessive color marker. In the case of metalaxyl this possibility must apparently be excluded because the compound did not increase the number of color sectors in colonies of a haploid green strain.

To determine whether non-disjunction or mitotic crossing-over or chromosome breakage might be involved in the action of metalaxyl analyses of recombinants were performed. Segregant clones were first purified on complete medium and then tested for ploidy and phenotype as already described. Diploid recombinants were classified as chromosomal, showing simultaneous segregation of all the markers on chromosome I or II and subchromosomal showing recombination between markers on the same chromosome. In the latter class no attempt has so far been made to determine whether the segregant region is in the homozygous (following crossing-over) or the hemizygous (following breakage-delection) state.

The genotype of the heterozygous diploid strain used (Fig. 1) permits classification of recombinants as haploids and chromosomal and subchromosomal diploids in the following way:

a. White haploids must require thiamine *(thi)*, reduced nitrogen *(cnx)* and probably one or more of the factors specified by the genes on other chromosomes.
b. Yellow haploids must not require adenine because of the presence of the suppressor gene (*su* A, *ad* E 20) but must require either riboflavine or reduced nitrogen from chromosome VIII and probably one or more of the factors specified by the genes on chromosomes II–VII.
c. White diploid chromosomal segregants must require thiamine and reduced nitrogen.
d. Yellow diploid chromosomal segregants must be prototrophs.
e. White diploid subchromosomal segregants must be prototrophs.
f. Yellow diploid subchromosomal segregants must require adenine.

Obtaining samples of recombinants representative of the various treatments appears to be a main problem in this type of experiment. In some of the earlier studies [1, 7] a maximum of one yellow and one white recombinant per colony were used. In this way sampling two or more times the result of a single segregational event is avoided, but the method may discriminante against a class of recombinants which may form small sectors due to some selective disadvantage on the treated medium. In our first experiments with metalaxyl we picked up a few sectors per colony, trying to avoid discrimination on the basis of sector size. These experiments showed that metalaxyl increases the frequency of both haploidization and formation of diploid segregants but while most of the recombinants obtained on low fungicide concentrations are subchromosomal, mostly haploids are recovered from colonies treated with high concentrations.

Table 3
Classification of mitotic segregants of Aspergillus nidulans as affected by metalaxyl treatment

Metalaxyl concentration (μM)	Number of segregants analysed	Percentage of		
		Haploid Segregants	Diploid Segregants	
			Chromosomal	Subchromosomal
0.0	65	35.0	8.0	57.0
75.0	61	3.5	2.0	94.5
300.0	79	69.6	3.8	26.6
500.0	129	83.7	3.9	12.4

This finding is not easy to explain by a single mechanism of metalaxyl action. For this reason a second experiment was carried out in which colonies were examined under a dissecting microscope and all segregants appearing as groups of more than four yellow or white conidiophores were picked up. The results of this experiment, shown in Table 3, again indicate a difference between low and high metalaxyl concentrations. It is important to note that the increase in the percentage of haploid recombinants is not paralleled by an increase in the percentage of non-disjunctional (chromosomal) diploids. In the case of benomyl, the effect of which is unquestionably due to chromosome non-disjunction, an increase in the percentage of both classes of recombinants is observed [5]. Whether some processes of irregular distribution of chromosomes, leading to haploidization rather than non-disjunction, is involved in the action of metalaxyl remains to be investigated.

The frequency of subchromosomal segregants is highest at the low metalaxyl concentrations (Table 3). In this regard the effect of metalaxyl is reminiscent of the action of formaldehyde [1]. Experiments are in progress to determine whether such recombinants are the result of mitotic crossing-over or chromosome breaks and terminal deletions.

In conclusion, metalaxyl is highly active genetically on diploid strains of *A. nidulans* but from the data obtained so far it is not possible to ascribe its effects on somatic segregation to any one of the know mechanisms alone. From the results presented in this paper no conclusions on genetic activity of metalaxyl on higher organisms can be drawn. It cannot also be assumed that the effects observed in *A. nidulans* are related to the mode of action of metalaxyl in the highly sensitive Oomycetes against which the fungicide is used. We think, however, that both possibilities are worthy of some investigation.

Diskussion

SISLER: Is it possible that the activity in your tests was due to Ridomil contaminants?

GEORGOPOULOS: No, don't believe contamination might have been involved. The compounds were recrystalised twice from an appropriate solvent and the solvent used for the preparation of stock solutions was always added to the control plates as well.

SISLER: Are the pigment precursors known in *A. nidulans*?

Georgopoulos: I don't know if they have been identified chemically, but from the fact that each block is imposed by one gene and w is epistatic over y it is assumed that the pigments are on the same pathway.

Lyr: Have you added any possible precursors to your white and yellow segregants in order to find out what is the background of the action of metalaxyl?

Georgopoulos: No, we have not done this. I suppose it would be possible to overcome the blocks by supplying the appropriate compounds. This, however, can be done with segregants obtained from untreated colonies as well. In this work we concerned ourselves with the ability of the fungicide not to impose as block but to induce new-arrangement of existing genes which have a known effect on pigmentation. In other words, the action of metalaxyl has nothing to do with the biosynthesis of the pigments.

Schwinn: With all due appreciation of the work presented here I have problems to see the relevance of the results, since they have been obtained from a non-target fungus at dose rates much higher than ever used in practice. What can be derived from such findings with regard to the target fungi?

Georgopoulos: I think I made it quite clear that no conclusions can be reached as to the relevance or irrelevance of these observations to the mechanism of action in target fungi, until appropriate experiments with such fungi are conducted. I know of no reports of such experiments. I don't believe that the fact that the concentration at which this effect is observed in *A. nidulans* are much higher than those needed to inhibit the target fungi is evidence against a similarity in the mechanisms involved. As you well know permeability of affinity of the site for the compound may well be the important factors for the difference in concentration.

Note added in proof

In the continuation of this work it was found that not all metalaxyl-containing preparations are capable of induction of sectoring. While genetically active speciments were toxic to both *Pythium* sp. and *Aspergillus nidulans*, non-active samples were inhibitory only to the former fungus. Recrystalisation, however, from either ethanol or acetone or ether or hexane could not separate the two types of activity. Eventually, from the genetically active recrystalised metalaxyl preparations a small fraction has been separated chromatographically. This fraction is not toxic to *Pythium* sp. but is highly active on *A. nidulans* both as an inhibitor of growth and as an inducer of mitotic segregation. Details on these recent experiments and on the nature of the genetically active component will be presented elsewhere.

References

[1] Fratello, B., G. Morpurgo, and G. Sermonti, Induced somatic segregation in *Aspergillus nidulans*, Genetics **45** (1960), 785–800

[2] Georgopoulos, S. G., Development of fungal resistance to fungicides, in Siegel, M. R. and H. D. Sisler (eds.) Antifungal Compounds, Marcel Dekker, Inc., New York, 1977, Vol. 2, 439–495

[3] Georgopoulos, S. G., A. Kappas, and A. C. Hastie, Induced sectoring in diploid *Aspergillus nidulans* as a criterion of fungitoxicity by interference with hereditary processes, Phytopathology **66** (1976), 217–220

[4] GEORGOPOULOS, S. G., M. SARRIS, and B. N. ZIOGAS, Mitotic instability in *Aspergillus nidulans* caused by the fungicides iprodione, procymidone, and vinclozolin, Pestic. Sci. **10** (1979), 389–392

[5] KAPPAS, A., On the mechanisms of induced somatic recombination by certain fungicides in *Aspergillus nidulans*, Mutat. Res. **51** (1978), 189–197

[6] KAPPAS, A. and S. G. GEORGOPOULOS, Interference of griseofulvin with the segregation of chromosomes at mitosis in diploid *Aspergillus nidulans*, J. Bacteriol. **119** (1974), 334–335

[7] KAPPAS, A., S. G., GEORGOPOULOS, and A. C. HASTIE, On the genetic activity of benzimidazole and thiophanate fungicides on diploid *Aspergillus nidulans*, Mutat. Res. **26** (1974), 17–27

[8] LUKENS, R. J., D. CHAN, and G. ETTER, Ortho 20615, a new systemic for the control of plant diseases caused by Oomycetes. Proc. Am. Phytopath. Soc. **5** (1978), 142.

[9] PONTECORVO, G., J. A. ROPER, L. M. HEMMONS, K. D. MACDONALD, and A. W. J. BUFTON, The genetics of *Aspergillus nidulans*, Advan. Genet. **5** (1953), 141–238

[10] ROPER, J. A., Production of heterozygous diploids in filamentous fungi, Experientia **8** (1952), 14–15

[11] SHANFIELD, B. and E. KÄFER, Chemical induction of mitotic recombination in *Aspergillus nidulans*, Genetics **67** (1971), 209–219

[12] URECH, P. A., F. SCHWINN, and T. STAUB, CGA 48988, a novel fungicide for the control of late blight, downy mildews and related soil-borne diseases, Proc. Br. Crop. Conf. (1977), 623–631

H. Buchenauer

Institut für Pflanzenkrankheiten der Universität Bonn, 5300 Bonn, BRD

Zum Wirkungsmechanismus Sterolsynthese-blockierender Fungizide

Einleitung

In den letzten Jahren wurden eine Reihe neuer fungitoxischer Verbindungen mit systemischen Eigenschaften entwickelt, die sich hinsichtlich ihrer Wirkungsweise in Pilzen und ihrer Nebenwirkungen in höheren Pflanzen von den übrigen bisher bekannten systemischen Fungiziden unterscheiden. Als gemeinsames Strukturmerkmal besitzen diese neuen Substanzen einen N-substituierten heterozyklischen Ring, welcher entweder ein Pyrimidin (Triarimol, Fenarimol, Nuarimol), Triazol (Triadimefon, Triadimenol, Biloxazol, Fluotrimazol, Diclobutrazol, CGA 64251, CGA 64250) oder Imidazol (Imazalil, Prochloraz, Phenapronil, Clotrimazol, Miconazol) darstellt. Eine Ausnahme in dieser Hinsicht bildet die Struktur von Triforine (Piperazin-Derivat).

Fungitoxisches Wirkungsspektrum

Vergleichende Untersuchungen zum fungitoxischen Wirkungsspektrum von Triadimefon, Triadimenol, Fenarimol, Nuarimol und Imazalil ergaben, daß sich die Verbindungen auf der Grundlage der ED_{50}-Werte als toxisch bis hoch toxisch gegenüber verschiedenen Pilzarten, die zu den Klassen der Ascomyceten (z.B. *Erysiphe graminis* f. sp. *hordei*, *Sphaerotheca fuliginea*, *Gaeumannomyces graminis*, *Sordaria fimicola*), Deuteromyceten (z.B. *Septoria nodorum*, *Colletotrichum trifolii*, *Botrytis cinerea*, *Cladosporium cucumerinum*, *Pleiochaeta setosa*, *Fusarium culmorum*, *Verticillium albo-atrum*, *V. dahliae*) und der Basidiomyceten (*Rhizoctonia solani*, *Ustilago avenae*) gehören. Gegenüber den Oomyceten und Zygomyceten besaßen die Verbindungen eine schwache bis begrenzte Toxizität (Buchenauer [10], [13]). Das antimykotische Wirkungsspektrum der geprüften Substanzen entsprach weitgehend dem von Triforine (Fuchs und Drandarevski [24]) und dem von Triarimol (Brown and Hall [4]). Es ergab sich keine enge Korrelation zwischen der Toxizität der verschiedenen Verbindungen und der taxonomischen Stellung der in die Untersuchungen einbezogen Pilze. Sogar bei Pilzen, die der gleichen Gattung angehörten, konnten Unterschiede in der Empfindlichkeit gegenüber diesen Wirkstoffen festgestellt werden.

Wirkungsmechanismus in Sporidien von *Ustilago avenae*

Die Substanzen induzierten in Sporidien von *Ustilago avenae* starke morphologische Veränderungen. Die Sporidien, die sich normalerweise durch Sprossung vermehren, wurden unter Fungizideinfluß vielzellig und verzweigt. Auch die Sporenkeimung wurde durch die Substanzen im allgemeinen nur geringfügig beeinflußt, andererseits verursachten die Wirkstoffe deutliche Veränderungen im Keimschlauchwachstum, wie z.B. starke Verkürzung und Verzweigung der Keimschläuche.

Während in Anwesenheit der Substanzen die allgemeinen Stoffwechselprozesse (Atmung, Nuklein- und Proteinsynthese) erst nach relativ langen Einwirkungszeiten (4—6 Stunden) beeinträchtigt wurden, wurde die Ergosterolbiosynthese bereits unmittelbar nach Zugabe dieser Wirkstoffe gehemmt. Demgegenüber wurde der Stoffwechsel der anderen Lipidkomponenten in der Anfangsphase durch die Fungizide nicht nennenswert beeinflußt. Die Hemmung der Ergosterolbiosynthese war begleitet von einer Anreicherung der C-4,4-Dimethyl-, C-4-Methyl- und C-14-Methylsterolfraktion. Da in Gegenwart der Fungizide während der Sterolbiosynthese keine Verbindung jenseits des C-14-Demethylierungsprozesses angetroffen wurden, kann die oxidative C-14-Demethylierung als der primäre Wirkort dieser Substanzgruppe angesehen werden. Darüberhinaus hemmen einige Substanzen (z.B. Triarimol, Fenarimol und Nuarimol) die Einführung der C-22(23)-Doppelbindung und die Reduktion der C-24(28)-Doppelbindung (RAGSDALE und SISLER [36]; SHERALD u. a. [39]; SHERALD und SISLER [40]; RAGSDALE [34]; KATO u. a. [31]; BUCHENAUER [5]; [6]; [7]; [8]; [9]; [10]; [11]; LEROUX und GREDT [32]; SIEGEL und RAGSDALE [41]; VAN DEN BOSSCHE [2]). Die in der Seitenkette vorkommenden Reaktionen finden wahrscheinlich erst nach der C-14-Demethylierung statt (RAGSDALE [35]).

Die Hemmung der C-14-Demethylierung deutet darauf hin, daß von den „mixed function" Oxidasen ein oder mehrere Enzyme, die die C-14-Demethylierung katalysieren, durch die Wirkstoffe besonders gehemmt werden. Es liegt nahe anzunehmen, daß die Fungizide, welche N-substituierte Imidazol-, Pyrimidin- und Triazolderivate darstellen, in ähnlicherweise wie die Arylimidazole mit dem Cytochrom P-450 Bindungen eingehen (BAGGALEY u. a. [1]; WILKINSON u. a. [45]).

Im Sterolstoffwechsel der meisten Pilze wird gewöhnlich die C-14-Methylgruppe vor den C-4-Methylgruppen entfernt. Dadurch, daß in den behandelten Zellen die C-14-Demethylierung blockiert ist, während die C-4-Methylgruppen noch abgespalten werden können, kommt es zur Anreicherung von Sterolzwischenprodukten in den behandelten Zellen, die im ungestörten Sterolstoffwechsel nicht auftreten.

Mit zunehmender Inkubationszeit reicherten sich in den behandelten Sporidien einerseits die freien Fettsäuren stark an, andererseits verminderten sich deutlich die Gehalte an verschiedenen Phospholipiden. Die Anreicherung der freien Fettsäuren in den Sporidien als Folge der Fungizideinwirkung ist möglicherweise z. T. auf einen Abbau bzw. verminderte de novo Synthese der polaren Lipide zurückzuführen. Diese Akkumulation an freien Fettsäuren in den behandelten Sporidien, die als eine Sekundärwirkung der Fungizide aufzufassen ist, verstärkt die fungitoxische Primärwirkung dieser Fungizidgruppe. Von Fettsäuren ist bekannt, daß sie auf Pilzzellen toxisch wirken (RIETH [37]; SUMRELL u. a. [43]). Es kann daher angenommen werden, daß den hohen Gehalten an freien Fettsäuren in den Fungizid-behandelten Zellen möglicherweise eine entscheidende Bedeutung für die Mortalität der Sporidien zukommt, während die primäre Wirkung der Mittel mehr fungistatischer Natur ist.

In diesem Zusammenhang sollte auch auf die Hypothese der Anlagerung der Fungizide an Sterol-Trägerproteine ('sterolcarrier protein') hingewiesen werden. Während der enzymatischen Umwandlung werden die wasserlöslichen Sterolvorstufen an Sterol-

Trägerproteine gebunden und nach Transport an ihren Wirkort wird das Sterol-Trägerprotein wieder freigesetzt (RITTER and DEMPSEY [38]). Möglicherweise lagern sich die Ergosterolbiosynthesehemmer an bevorzugte Stellen des bzw. der Sterol-Trägerproteine an, wodurch entweder die Bindung an die Sterolzwischenprodukte verhindert oder spezifische enzymatische Reaktionen die zur oxidativen Eliminierung der C-14-Methylgruppe führen, gehemmt werden. Die Bildung des Sterol-Träger-protein-Komplexes wird durch die Anwesenheit von Phospholipiden beschleunigt (RITTER and DEMPSEY [38]). Es ist deshalb nicht ausgeschlossen, daß aufgrund des verminderten Phospholipidgehaltes in den behandelten Sporidien die Bildung des Sterol-Trägerprotein-Komplexes gehemmt ist.
Eine Inkubation der Fungizide mit Eilecithin führte nach einer 4-stündigen Inkubationszeit zu einer deutlich höheren Absorption bzw. zu einer Verschiebung der Absorptionsmaxima. Dieser Effekt beruht vermutlich auf einer Interaktion zwischen den Fungiziden und dem Phospholipid (BUCHENAUER [12]; [15]).
Ergosterol ist die Hauptsterolverbindung in vielen Pilzarten und erfüllt offenbar eine wichtige Aufgabe in der Funktion und Struktur der Pilzmembranen (HENDRIX [28]; WEETE [44]). Für die Wirksamkeit der Sterole in den Membranen müssen drei Kriterien erfüllt sein: a) eine freie Hydroxylgruppe an der C-3-Position; b) ein Steroidringsystem mit einer planaren α-Außenseite, die keine Methylgruppen an der C-4- und C-14-Position enthält sowie einer gewölbten β-Außenseite, die durch die beiden Methylgruppen an der C-10- und C-13-Position gebildet werden, und c) eine intakte aliphatische Seitenkette, die sich am C-17 des Steroidkernes befindet (BROCKENHOFF [3]; NES [33]; GRUNWALD [25]; HUANG [29]). Die Interaktionen zwischen Sterolen und Phosphlipiden, die physikalisch zu einer Verdichtung („condensing effect") der Membranen führen, beeinflussen somit entscheidend die Mobilitätscharakteristika (d. h. die inneren Molekularbewegungen) der Membranlipide. Die Sterole regulieren auf diese Weise die Membranpermeabilität und spielen somit eine wesentliche Rolle in den Membranfunktionen (DEMEL und DE KRUYFF [20]; VAN DEENEN [19]; BROCKENHOFF [3]). Die biologische Bedeutung der Sterole in den Pilzen erstreckt sich nicht ausschließlich auf die Funktion der Membranen, sondern sie sind auch zur sexuellen und asexuellen Reproduktion der Arten notwendig, die zur Familie der Pythiaceae gehören (ELLIOTT u. a. [21]; HASKINS u. a. [26]; HENDRIX [27]).
Der durch die Fungizide induzierte Defekt im Membranenaufbau kann die selektive Permeabilität so stark verändern, daß ein Efflux von intrazellulären Bestandteilen eintritt. Diese Erscheinung wurde beispielsweise bei *Candida albicans* nach Behandlung mit Clotrimazol (IWATA u. a. [30]) und Miconazol (SREEDHARA SWAMY u. a. [42]) beobachtet.

Wirkungsweise im Infektionsverlauf von *Erysiphe graminis*

Die verschiedenen Stadien des primären Infektionsablaufes von Gersten- und Weizenmehltau wurden durch die Substanzen unterschiedlich stark beeinflußt. Außer einer verminderten Keimrate, die stark konzentrationsabhängig war, wurden gelegentlich durch Veränderungen an den Appressorien beobachtet. Die auffallendsten strukturellen Veränderungen wurden während der Haustorienentwicklung beobachtet. In den Epidermiszellen behandelter Pflanzen war die Haustorienentwicklung unvollständig, wobei die Schädigung besonders an den fingerartigen Fortsätzen in Erscheinung trat; die Form dieser Fortsätze war beträchtlich verändert und zahlenmäßig reduziert. Die Scheidenmembran, die bei ungestörter Entwicklung eng die Haustorien umgibt, war angeschwollen. Aufgrund dieser starken Schädigung der Haustorien wurde die weitere Mehltauentwicklung, wie sekundäre Myzel- und Haustorienbildung, blockiert. Wenn unterstellt wird, daß diese Wirkstoffgruppe auch bei Mehltaupilzen in den Sterolstoffwechsel eingreift, würde dies zu unvollständigen Membranstrukturen und folglich zu starken Störungen in den Membranfunktionen führen. Die starken Deformationen während der Haustorienentwicklung in den behandelten Pflanzen sind wahrscheinlich auf defekte Membranstrukturen zurückzuführen, wodurch schließlich die innige kompatible Beziehung zwischen Wirt und Pathogen gestört sein würde.

Untersuchungen zur Kreuzresistenz

Die Mehrzahl der geprüften Mutanten von *Cladosporium cucumerinum* und *Botrytis cinerea* wies einen hohen Grad an Kreuzresistenz zwischen Triadimefon, Triadimenol, Fenarimol, Nuarimol, Phenapronil und Imazalil auf. Diese positive Kreuzresistenz zwischen den verschiedenen Verbindungen kann als weiterer Hinweis für einen ähnlichen bzw. identischen Wirkungsmechanismus innerhalb dieser Fungizidgruppe im Pilzstoffwechsel gedeutet werden (BUCHENAUER [12]).

Gasphasenwirkung

Die verschiedenen Substanzen zeigten deutliche Unterschiede hinsichtlich ihrer Wirkung über die Gasphase gegenüber Gerstenmehltau. Die ausgeprägtesten Effekte entfalteten Triadimefon und Triadimenol. Während die beiden Substanzen bereits bei niedrigen Konzentrationen (z.B. 5×10^{-4} M) den Mehltaubefall an Gerstenprimärblättern nahezu vollständig unterdrückten, zeigten die übrigen Substanzen erst bei höheren Konzentrationen (z.B. 10^{-3} M) eine eindeutigere Wirkung über die Gasphase. Bezüglich der Wirksamkeit über die volatile Phase ergab sich bei den Substanzen folgende Reihe: Triadimefon $\geq$ Triadimenol $>$ Nuarimol $\geq$ Fenarimol $>$ Fluotrimazol $>$ Imazalil (BUCHENAUER [10]; [12]). Obwohl die Mehltau-wirksamen Substanzen im allgemeinen eine geringe Flüchtigkeit besitzen, ist eine Rückverteilung der Wirkstoffrückstände auf der Pflanzenoberfläche über die Dampfphase für die Bekämpfung der Echten Mehltaupilze von Bedeutung.

Transformation von Triadimefon in die Diastereomeren von Triadimenol durch Pilze und Pflanzen

Nach Zugabe von Triadimefon zu einer 8 Tage alten Kultur von *Fusarium culmorum* in Nährlösung, wurde diese rasch in die beiden Diastereomeren von Triadimenol (Triadimenol-I und Triadimenol-II) umgewandelt. Nach einer 2-tätigen Einwirkungszeit lagen von der nachweisbaren Fungizidmenge im Filtrat und Myzel nur noch 39,6% als Triadimefon vor. Diese anfangs starke Umwandlungsintensität verlangsamte sich mit fortschreitender Inkubationsdauer. Der Triadimefon-Anteil betrug z. B. nach 3, 6, 10 und 15 Tagen 30, 18, 9 und 3,5%, demgegenüber reicherten sich gleichzeitig die Triadimenol-Diasteromeren an. Nach 15 Tagen erreichte Triadimenol-II 71% und Triadimenol-I 25% des nachweisbaren Fungizidgehaltes. Bei den Diastereomeren von Triadimenol-I handelt es sich um die Threo- und bei den Isomeren von Triadimenol-II um die Erythroform.

Die beiden Isomeren von Triadimenol besitzen eine unterschiedliche fungitoxische wie systemische Wirksamkeit. Um eine entsprechende Hemmung des Myzelwachstums von *Cladosporium cucumerinum* zu erzielen, war im Vergleich zu Triadimenol-I (Threoform) von Triadimenol-II (Erythroform) eine 3—4 mal höhere Konzentration notwendig. Eine deutliche Überlegenheit von Triadimenol-I gegenüber Triadimenol-II kommt in der Bekämpfung des Gerstenmehltaus nach Gießbehandlung zum Ausdruck. Nach Wurzelbehandlung zeigte Triadimenol-I eine 4—5 mal stärkere systemisch-fungitoxische Wirksamkeit als Triadimenol-II. Auch im Vergleich zu Triadimefon und Triadimenol besitzt Triadimenol-I eine deutlichere systemische Wirkung gegenüber Gerstenmehltau. Ähnliche Wirkungsunterschiede wurden nach Spritzbehandlung erzielt. Sowohl nach Wurzel- als auch nach Sproßbehandlung ergab sich in der Wirksamkeit der verschiedenen Verbindungen gegenüber Gerstenmehltau die Reihenfolge: Triadimenol-I > Triadimenol > Triadimefon > Triadimenol-II.

Das Sproßwachstum der Tomatenpflanzen wurde z. B. durch Triadimenol-II stärker gehemmt als durch Triadimenol-I.

In orientierenden Untersuchungen über Aufnahme und Translokation wurden Gerstenkeimlinge (8 Tage alt) im Gießverfahren über die Wurzel mit den verschiedenen Fungiziden behandelt, die Sprosse zu verschiedenen Terminen nach der Fungizidapplikation extrahiert und die fungitoxischen Substanzen in den Extrakten mit Hilfe der Bioautographie-Technik nachgewiesen. Die Bioautogramme der Extrakte von Pflanzen, die mit Nuarimol, Fenarimol und Imazalil behandelt worden waren, zeigten jeweils eine Hemmzone, wobei deren R_f-Werte mit denen der entsprechenden technischen Wirkstoffe übereinstimmten. Andererseits wiesen die Bioautogramme der Sproßextrakte von Triadimefon — behandelten Pflanzen zwei fungitoxische Zonen auf (BUCHENAUER [6]). Unter Verwendung verschiedener Lösungsmittelgemische entsprachen die R_f-Werte der einen Hemmzone denen von Triadimefon, während die der zweiten Hemmzone mit denen von Triadimenol übereinstimmten.

Untersuchungen über die Umwandlung von Triadimefon in Gerstenpflanzen ergaben 3 Tage nach der Behandlung einen Triadimefongehalt von 60,5%, dieser verminderte sich mit zunehmender Inkubationszeit auf 14,6% nach 24 Tagen. Gleichzeitig reicherten sich in den Gerstenpflanzen mit fortschreitender Einwirkungszeit die Transformationsprodukte von Triadimefon, insbesondere die Stereoisomeren von Triadimenol, an. Die Halbwertszeit von Triadimefon in Gerstenpflanzen betrug etwa 4—5 Tage. Während in den ersten Untersuchungsterminen (bis 12 Tage nach Behandlung) ein höherer Anteil an Triadimenol-I als an Triadiemenol-II in den Extrakten der Gerstensprossen nachgewiesen wurde, hatte sich zu Versuchsende (24 Tage) ein Gleichgewicht zwischen beiden Diastereomeren eingestellt (BUCHENAUER [12]; [14]).

Wirkung auf das Wachstum höherer Pflanzen sowie auf den Gibberellin- und Lipidstoffwechsel

Wurzel- bzw. Saatgutbehandlungen mit Triadimefon, Triadimenol, Fenarimol, Nuarimol und Imazalil hemmten das Wachstum der Koleoptilen, Primärblätter und Wurzeln von 7 Tage alten Gersten- und Weizenkeimlingen. Bei vergleichbaren Konzentrationen hemmten die verschiedenen Verbindungen das Wachstum unterschiedlich stark, wobei sich hinsichtlich der Hemmwirkung folgende Reihe ergab: Nuarimol > Fenarimol > Triadimefon > Triadimenol > Imazalil. Im allgemeinen wurde bei den Keimlingen das Wurzelwachstum stärker durch die Fungizide gehemmt als das der Koleoptilen und Primärblätter (BUCHENAUER [12]; BUCHENAUER und GROSSMANN [16]; BUCHENAUER und RÖHNER [17]).

Saatgutbehandlungen mit Triadimefon und Triadimenol verzögerten die Seneszenz der Primärblätter von Gersten- und Weizenkeimlingen. Mit zunehmendem Alter der Keimlinge enthielten die Primärblätter der behandelten Pflanzen im Vergleich zur Kontrolle erhöhte Gehalte an Chlorophyllen, Karotinoiden, Xanthophyllen und Nukleinsäuren. Triadimefon und Triadimenol zeigten im Betacyanin-Test mit *Amaranthus caudatus*-Keimlingen Cytokinin-ähnliche Aktivität (FÖRSTER [22]; [23]).

Behandlungen mit verschiedenen Gibberellinsäuren wirkten der durch die Fungizide induzierten Hemmung des Längenwachstums der Koleoptilen und Primärblätter entgegen. Im Vergleich zu den unbehandelten Sprossen enthielten die Extrakte der behandelten Sprosse verminderte Gehalte an Gibberellin-ähnlichen Substanzen, wobei eine deutliche Beziehung zwischen der Hemmung des Sproßwachstums und den Gibberellingehalten bestand.

Die Fungizide induzierten in Sprossen von Gerstenpflanzen z. T. starke Veränderungen in der C-4-Desmethylsterolfraktion, wobei eine Beziehung zwischen der wachstumsregulatorischen Aktivität der Substanzen und dem Gehalt der Phytosterole wie Campesterol, Stigmaterol und β-Sitosterol festgestellt wurde; mit zunehmender Hemmung des Sproßwachstums verminderten sich die Gehalte an Phytosterolen.

Studien zum Wirkungsmechanismus von Ancymidol, einem Wachstumsregulator bei Pflanzen, der strukturelle Ähnlichkeit mit den Fungiziden Triarimol, Fenarimol und Nuarimol besitzt, deuten darauf hin, daß Ancymidol primär in die Gibberellinsäuresynthese eingreift, indem die Umwandlung von Kauren zum Kaurenol bzw. die Transformation von Kaurenol zum Kaurenal gehemmt werden. Ancymidol hemmt somit die oxidative Demethylierung an der C-19 Position während der Gibberellinbiosynthese in Pflanzen (COOLBAUGH u. a. [18]). Möglicherweise greifen die Fungizide an den gleichen Stellen wie Ancymidol in die Gibberellinbiosynthese ein. In diesem Zusammenhang scheint es erwähnenswert, daß die Fungizide sowohl in Pilzen als auch in Pflanzen einen analogen Wirkungsmechanismus besitzen; sie greifen jeweils in oxidative Demethylierungsreaktionen von verschiedenen Stoffwechselwegen ein. Bei vergleichbaren Wirkstoffkonzentrationen wurde in Pflanzen der Gibberellinsäurestoffwechsel empfindlicher durch die Fungizide gehemmt als die Sterolbiosynthese.

Diese Ergebnisse zeigen, daß die Wirkstoffe analoge Reaktionen im pilzlichen und pflanzlichen Stoffwechsel beeinflussen können, wobei sich die Effekte vermutlich in beiden Fällen auf die Membranfunktion auswirken. Allerdings besteht keine Beziehung zwischen fungitoxischer Wirkung und einer Wachstumshemmung bei höheren Pflanzen.

Diskussion

Banasiak: Die untersuchten Wirkstoffe weisen als gemeinsames Strukturmerkmal den N-substituierten Triazol-Ring auf. Trägt diese Gruppierung zur Wirkungsweise der Verbindungen bei oder erfüllt sie nur eine Trägerfunktion für andere essentielle Strukturelemente?

Buchenauer: Bei den N-substituierten heterozyklischen Verbindungen, wie Pyrimidin-, Triazol- und Imidazolderivaten, befindet sich jeweils ein N-Atom in 3'-Position, welches wahrscheinlich für die Bindung am Angriffsort verantwortlich ist.

Böttcher: Ergosterol und andere Sterole stabilisieren bekanntlich Phospholipidmembranen und somit auch das Plasmalemma. Eine Blockierung der Ergosterol-Bildung muß demnach Membranen schädigen. Wie erfolgen jedoch die beobachteten Zellwandveränderungen, die z.B. bei der Induktion der multipolaren Sproßknospenbildung bei *Ustilago* beobachtet wurden?

Buchenauer: Rasterelektronenmikroskopische Untersuchungen von De Nollin und Borgers zeigten, daß nach Miconazol-Behandlung die normalen Teilungsvorgänge gestört waren; es bildeten sich Aggregationen von miteinander verbundenen Zellen wie sie auch bei *Ustilago avenae*-Sporidien beobachtet werden. Die behandelten *C. albicans*-Zellen hatten ebenfalls ihre bevorzugte polare Position der Knospen und Knospennarben verloren und die Wirkstoffeinwirkung führte sogar zu einer Erhöhung der Knospen und Knospennarben. Die behandelten Zellen tendierten zu einer erhöhten Teilungsrate gegenüber den Kontrollzellen. Diese anomalen Differenzierungs- und Teilungsvorgänge der behandelten Sporidien lassen sich auf eine gestörte Membranzusammensetzung zurückführen. Für die Einleitung der Teilungsprozesse spielt Ergosterol eine Rolle, wie beispielsweise bei *C. albicans* nachgewiesen werden konnte.

Pfliegel: Denken Sie nicht, daß die Wirkung dieser Stoffe vom Redoxpotential des zentralen tertiären (oder quaternären) Kohlenstoffatoms abhängt und die (Hetero)arylsubstituenten (außer ihrer sterischen Wirkung) nur indirekt durch Beeinflussung dieses Redoxpotentials dazu beitragen?

Buchenauer: Während vieles dafür spricht, daß für die gemeinsame Wirkung dieser Substanzgruppe das N-Atom in 3-Position der N-substituierten Ringe von ausschlaggebender Bedeutung ist, stellt Ihre Hypothese über die Wirkung der Verbindungen einen interessanten Aspekt dar.

Schmidt: Gibt es Hinweise zur Wirkungsbeeinflussung von Triadimefon beim gemeinsamen Einsatz mit Halmstabilisatoren, wie z.B. Chlormequat oder Ethephon in Getreide, bzw. werden Halmstabilisatoren in ihrer Wirkung durch Triadimefon beeinflußt?

Buchenauer: Die Halmstabilisatoren bzw. Halmverkürzungsmittel wie Chlormequat oder Ethephon werden routinemäßig zu bestimmten Entwicklungsstadien des Getreides (insbesondere des Weizens und der Gerste)eingesetzt. Bei der Anwendung von Triadimefon liegen allerdings bestimmte Schadschwellen für die einzelnen Krankheitserreger zu Grunde. In den meisten Fällen wird es daher nicht zur gleichzeitigen Anwendung beider Wirkstoffkomponenten kommen. Unter bestimmten Umständen müssen beispielsweise Triadimefon und Chlormequat bzw. Ethephon gleichzeitig appliziert werden. Hier sind bestimmte Interaktionen sowohl auf die Wirtspflanzen als auch auf die Krankheitserreger denkbar. Bisher liegen mir keine Informationen über solche Wechselwirkungen vor.

LYR: Can you explain me, or perhaps Prof. SISLER can do that, why sterol synthesis blocking in mildews should have such a high effect, when the parasite can take up sterols from plant cells? Incorporation of foreign sterols into membranes of several organisms is frequently demonstrated, not only in fungi?

BUCHENAUER: Die Bedeutung von Ergosterol im Vergleich zu Cholesterol für die Zellfunktion wurde beispielsweise an *Saccharomyces cerevisiae* unter anaeroben Bedingungen untersucht. Während sich die Hefezellen in einem ergosterol-haltigen Medium normal vermehrten, trennten sich nach Cholesterolzugabe nach der ersten Vermehrungsrate die Tochterzellen nicht mehr von den Mutterzellen. Die Folge waren Zellaggregationen, ähnliche Effekte wie sie bei *Ustilago avenae*-Sporidien in Anwesenheit der Ergosterol-Biosynthe-Hemmer beobachtet wurden. Auf die „Echten Mehltaupilze" übertragen würde dies bedeuten, daß für die Funktionsfähigkeit der Haustorienmembran vom Pilz synthetisiertes Ergosterol erforderlich ist und, daß die Phytosterole die Funktion von Ergosterol nicht übernehmen können.

SISLER: The action of the group of compounds may not be limited to sterol C-14 demethylation. Oxidative demethylation in other pathways may be concurrently inhibited. Therefore ergosterol does not reverse toxicity. There are also various other explanations. For example, the large quantity of ergosterol precursors present in the treated cells may interfere with the utilization of ergosterol in the membrane. That is, they may occupy a position in the membrane which keeps added ergosterol from entering the membrane.

NAUMANN: Welche Erklärung gibt es für die unterschiedliche Empfindlichkeit nahe verwandter Pilze gegen Triadimefon und die anderen erwähnten Fungizide, wenn man denselben Wirkungsmechanismus bei allen diesen Verbindungen in Betracht zieht?

BUCHENAUER: Für die unterschiedliche Empfindlichkeit taxonomisch nah verwandter Pilze gegenüber dieser Wirkstoffgruppe sind drei verschiedene Mechanismen denkbar:
1. Unterschiede in der Aufnahmerate
2. Strukturunterschiede in den Rezeptormolekülen
3. unterschiedlich starke Inaktivierung der Wirkstoffe

Welcher der in Frage kommenden Mechanismen für die Sensitivitätsunterschiede verantwortlich ist, ist noch weitgehend unbekannt.

GASZTONYI: As you have shown in experiments with *Fusarium culmorum* incubated with triadimefon, this fungus is able not only to reduce the tridimefon to triadimenol, but in this reduction the more active diastereomer of triadimenol (I) is produced in great quantity. And inspite of this fact, the *Fusarium* species (in your experiment *F. culmorum*) are very resistant to tridimefon. Have you any idea for the explanation of this phenomenon?

I think, it may be explained by the differences in the affinity of the acceptor in this fungus or by very low local accumulation of a.i. at the site of action. Have you any experiments in this direction?

The uptake of triadimefon by *Fusarium* spp. is lower than by sensitive *C. cucumerinum*, but it is not so low that it can be an explanation for the resistance.

BUCHENAUER: Ich stimme Ihrer Erklärung über die unterschiedliche Affinität am Angriffsort zu.

Literatur

[1] BAGGALEY, K. H., S. D. ATKIN, P. D. ENGLISH, R. M. HINDLEY, B. MORGAN and J. GREEN, Inhibition of cholesterol biosynthesis by 1-alkylimidazole, Biochem. Pharmacol. **24**, (1975), 1902–1903

[2] BOSSCHE, H. VAN DEN ,G. WILLEMSENS, W. COOLS, W. F. J. LAUWERS and L. LE JEUNE, Biochemical effects of miconazole on fungi. II. Inhibition of ergosterol biosynthesis in *Candida albicans*, Chem.-Biol. Interactions **21** (1978), 59–78

[3] BORCKENHOFF, H., Model of interaction of polar lipids, cholesterol and proteins in biological membranes, Lipids **9** (1974), 645–650

[4] BROWN, I. F. jr. and H. R. HALL, Studies on the activity of triarimol (EL-273) against certain powdery mildew fungi, Phytopathology **61** (1971), 886

[5] BUCHENAUER, H., Systemisch-fungizide Wirkung und Wirkungsmechanismus von Triadimefon (MEB 6447), Mitt. Biol. Bundesanst. Land.-Forstw. Berlin-Dahlem, H. **165** (1975), 154–155

[6] BUCHENAUER, H., Untersuchungen zur systemischen Wirkung und zum Einfluß von ®Bayleton (Triadimefon) auf verschiedene pilzliche Getreidekrankheiten, Pfl. Schutz-Nachrichten (Bayer) **29** (1976), 267–289

[7] BUCHENAUER, H., Mode of action of triadimefon in *Ustilago avenae*, Pestic. Biochem. Physiol. 7 (1977a), 309–320

[8] BUCHENAUER, H., Biochemical effects on the systemic fungicides fenarimol (EL-222) and nuarimol (EL-228) in *Ustilago avenae*, Z. Pfl.Krankh. Pfl. Schutz **84** (1977b), 286–299

[9] BUCHENAUER, H., Studies on the mode of action of imazalil in *Ustilago avenae*, Z. Pfl.Krankh. Pfl.Schutz **84** (1977c), 440–450

[10] BUCHENAUER, H., Mode of action and selectivity of fungicides which interfere with ergosterol biosynthesis, Proc. Br. Crop. Prot. Conf. – Pests and Dis. **3** (1977d), 699–711

[11] BUCHENAUER, H., Inhibition of ergosterol biosynthesis by triadimenol in *Ustilago avenae*, Pestic. Sci. **9** (1978)

[12] BUCHENAUER, H., Untersuchungen zur Wirkungsweise und zum Verhalten verschiedener Fungizide in Pilzen und Kulturpflanzen, Habilitationsschrift (1979a), Bonn

[13] BUCHENAUER, H., Comparative studies on the antifungal activity of triadimefon, triadimenol, fenarimol, nuarimol, imazalil and fluotrimazole in vitro, Z. Pfl.Krankh. Pfl. Schutz **86** (1979b), 341–354

[14] BUCHENAUER, H., Conversion of triadimefon into two diastereomers, tridiamenol-I and triadimenol-II, by fungi and plants, IX. Intern. Congress of Plant Protection, Washington D. C., August 5–11 (1979c), Abstr. No. 939

[15] BUCHENAUER, H., Interaction of different lipid components with various fungicides, Z. Pfl.Krankh. Pfl.Schutz **87** (1980), (im Druck)

[16] BUCHENAUER, H. und F. GROSSMANN, Triadimefon: Mode of action on plants and fungi, Neth. J. Pl. Pathol. **83** (1977), 93–103

[17] BUCHENAUER, H. und E. RÖHNER, Einfluß von Nuarimol, Fenarimol, Triadimefon und Imazalil auf den Gibberellin- und Lipidstoffwechsel in jungen Gersten- und Weizenpflanzen, Mitt. Biol. Bundesanst. Land-Forstw. Berlin-Dahlem, H. 178 (1977), 158

[18] COOLBAUGH, R. C., S. S. HIRANO and C. A. WEST, Specificity of action of ancymidol, a plant growth inhibitor, Plant Physiol. **57** (1976), 77

[19] DEENEN, VAN, L. L. M., Phospholipide: Beziehungen zwischen ihrer chemischen Struktur und Biomembranen, Naturwissenschaften **59** (1972), 485–491

[20] DEMEL, R. A. and B. DE KRUYFF, The function of sterols in membranes, Biochim. Biophys. Acta **457** (1976), 109–132

[21] ELLIOTT, C. G., M. R. HENDRIX, B. A. KNIGHTS and W. PARKER, A steroid growth factor requirement in a fungus Nature **203** (1964), 427–428

[22] FÖRSTER, H., Einfluß von Triadimefon auf den Pigment-, Nukleinsäure- und Proteingehalt junger Gerstenpflanzen, Mitt. Biol. Bundesanstl. Land-Forstw. Berlin-Dahlem, H. **178** (1977), 159

[23] Förster, H., Mechanism of action and side effects of triadimefon and triadimenol in barley plants, 3rd Intern. Congr. Pl. Pathol. München 16–23 August (1978) p. 365
[24] Fuchs, A. and Ch. A. Drandarevski, Wirkungsbreite und Wirkungsgrad von Triforine in vitro und in vivo, Z. Pfl.Krankh. Pfl.Schutz **80** (1973), 403–417
[25] Grunwald, C., Plant sterols, Ann. Rev. Plant Physiol. **26** (1975), 209–236
[26] Haskins, R. H., A. P. Tulloch and R. G. Micetich, Steroids and the stimulation of sexual reproduction of species of *Pythium*, Can. J. Microbio. **10** (1964), 187–195
[27] Hendrix, J. W., Sterol induction of reproduction and stimulation of growth of *Pythium* and *Phytophthora*, Science **144** (1964), 1228–1229
[28] Hendrix, J. W., Sterols in growth and reproduction in fungi, Ann. Rev. Phytopath. 8 (1970), 111–130
[29] Huang, C. H., A structural model for the cholesterol-phosphatidylcholine complexes in bilayer membranes, Lipids **12** (1977), 348–356
[30] Iwata, K., H. Yamaguchi and T. Hiratam, Mode of action of clotrimazole, Sabouraudia **11** (1973), 158–166
[31] Kato, T., S. Tanaka, M. Ueda and Y. Kawase, Inhibition of sterol biosynthesis in *Monilinia fructigena* by the fungicide S-1358, Agr. Biol. Chem. **39** (1975), 169–174
[32] Leroux, P. and M. Gredt, Effects de quelques fongicides systémiques sur la biosynthese de l'ergosterol chez *Botrytis cinerea* Pers., *Penicillium expansum* Link et *Ustilago maydis* (D. C.) Cdn, Ann. Phytopathol. **10** (1978), 45–60
[33] Nes, W. R., Role of sterols in membranes, Lipids **9** (1974), 596–612
[34] Ragsdale, N. N., Specific effects of triarimol on sterol biosynthesis in *Ustilago maydis*, Biochim. Biophys. Acta **380** (1975), 81–96
[35] Ragsdale, N. N., Inhibition of lipid synthesis. In: Antifungal Compounds (Eds. H. Sisler and M. R. Siegel) Marcel Dekker Inc., New York, Vol. II (1977), 333 to 363
[36] Ragsdale, N. N. and H. D. Sisler, Mode of action of triarimol in *Ustilago maydis*, Pestic. Biochem. Physiol. **3** (1973), 20–29
[37] Rieth, H., Die unterschiedliche Bedeutung der Fettsäuren in ihrer fungistatischen Wirkung, Fette-Seifen-Anstrichmittel **79** (1977), 120–121
[38] Ritter, M. C. and M. E. Demsey, Squalene and sterol carrier protein: Structural properties, lipid-binding, and function on cholesterol biosynthesis, Proc. Natn. Acad. Sci. (USA) **70** (1973), 265–269
[39] Sherald, J. L., N. N. Ragsdale and H. D. Sisler, Similarities between the systemic fungicides triforine and triarimol, Pestic. Sci. **4** (1973), 719–727
[40] Sherald, J. L. and H. D. Sisler, Antifungal mode of action of triforine, Pestic. Biochem. Physiol. **5** (1975), 477–488
[41] Siegel, M. R. and N. N. Ragsdale, Antifungal mode of action of imazalil, Pestic. Biochem., Physiol. **9** (1978), 48–56
[42] Sreedhara Swamy, K. H., M. Sirsi and G. Ramananda Rao, Studies on the mechanism of action of miconazole. Effect of miconazole on respiration and cell permeability of *Candida albicans*, Antimicrobi Agents Chemother. **5** (1975), 420–425
[43] Sumwell, G., R. R. Mod and F. C. Maque, Antimicrobial activity of some fatty acid derivatives, J. Americ. Oil Chem. Soc. **55** (1978), 395–397
[44] Weete, J. D., Sterols of the fungi: Distribution and biosynthesis, Phytochemistry **12** (1973), 1843–1846
[45] Wilkinson, C. F., K. Hetnarski and L. J. Hicks, Substituted imidazoles as inhibitors of microsomal oxidation and insecticide synergists, Pestic. Biochem. Physiol. **4** (1974), 299–312

H.-J. PREUSSER

Institut für Mikrobiologie der Technischen Hochschule Darmstadt,
D-6100 Darmstadt/FRG

Zum Wirkungsmechanismus von Imidazolfungiziden

Neben den von Bakterien, Streptomyzeten und Pilzen als Sekundärmetabolite gebildeten Antimykotika Amphotericin B, Nystatin, Pimaricin und Griseofulvin nehmen die synthetischen antifungalen Verbindungen in der modernen Therapie der Mykosen einen immer breiter werdenden Raum ein. Ein prinzipielles Problem bei der Entwicklung neuer Chemotherapeutika gegen humanpathogene Pilze ist die strukturelle und funktionelle Ähnlichkeit von Parasit und Wirt im zellulären Bereich. Beide sind Eukaryonten, in deren Zellen ähnliche stoffwechselphysiologische Prozesse ablaufen, so daß eine Schädigung des Pilzes zugleich eine inhibierende Wirkung auf den Stoffwechsel der Wirtszelle auszuüben vermag. Es sollte das Ziel der pharmazeutischen Forschung sein, solche Antimykotika zu entwickeln, die bei hoher Effektivität gegen den Parasit den Wirtsorganismus so wenig wie möglich schädigen. Neben tierexperimentellen und klinischen Tests kommt den Untersuchungen über den Wirkungsmechanismus neuentwickelter antifungaler Substanzen eine besondere Bedeutung zu. Unsere Arbeitsgruppe hat es sich vorgenommen, aus ausgewählten, medizinisch relevanten Testorganismen den Wirkungsmechanismus von Imidazolfungiziden in vitro zu erforschen und — soweit praktisch durchführbar — mit in vivo Experimenten zu vergleichen. Hierzu bedienen wir uns in erster Linie solcher Verfahren, die Aussagen über feinstrukturelle Veränderungen in den Zellen ermöglichen. Darüber hinaus werden die submorphologischen Ergebnisse durch biochemische Untersuchungen ergänzt, die in einer physiologisch-genetisch arbeitenden Gruppe unseres Instituts zum Teil mit den gleichen Substanzen und Organismen durchgeführt werden.
Über die zeitliche Folge von pathomorphologischen Veränderungen in der Parasitenzelle unter dem Einfluß unterschiedlicher Konzentrationen von Econazol-nitrat, eines Mitte der 60er Jahre entwickelten Antimykotikums aus der Reihe der Imidazolderivate, wurde bereits früher berichtet [5, 6]. Im folgenden sollen die durch Econazol bei humanpathogenen Pilzen ausgelösten Primäreffekte beschrieben und Überlegungen zum Wirkungsmechanismus der Substanz diskutiert werden.

Material und Methoden

Der für die in vitro Versuche verwendete Stamm 101/58 von *Trichophyton rubrum* wurde 1976 von einer Patientin isoliert und wird seitdem auf Malzextrakt-Pepton-Dextrose-Agar kultiviert. Diesen sowie den durch regelmäßige Tierpassagen hochvirulenten Stamm 200/175 von *Candida albicans* stellte uns freundlicherweise Dr. W. DITTMAR, Labor für Mykologie der Firma HOECHST AG in Frankfurt/M., zur Verfügung. Der Wachstums-inhibitor Econazol-nitrat, 1-[2'-(2,4-dichlorphenyl)-2'-(4-chlorbenzyloxy)-äthyl]-imidazol-nitrat, (Hersteller: CILAG-CHEMIE AG, Schaffhausen/Schweiz) wird in Äthanol gelöst. Von der Stammlösung werden entsprechende Verdünnungen mit 0,8% NaCl-Lösung hergestellt. Kultivierungs- und Präparationsmethoden werden an anderer Stelle ausführlich beschrieben [5, 6]. Für die in vivo Versuche wurden einer, an akuter *Candida*-Mykose erkrankten Patientin vor und während einer 3-Tage-Therapie mit 3 × 150 mg Econazol-Suppositorien Vaginalabstriche entnommen und diese wie üblich für elektronenmikroskopische Untersuchungen präpariert [10]. Die Gefrierbruchpräparationen erfolgten nach einem an anderer Stelle beschriebenen Verfahren in einer Hochvakuum-Gefrierätzanlage BAF 301 der Firma BALZERS AG, Balzers/Fürstentum Liechtenstein [9]. Die elektronenmikroskopischen Untersuchungen geschahen in einem Elmiskop 102 der Firma Siemens AG, Berlin, bei 80 kV.

Ergebnisse und Diskussion

Die unbehandelte *Trichophyton rubrum*-Zelle wird von einer, im Querschnitt strukturlos erscheinenden Zellwand umgeben. Der von einer wellenförmig verlaufenden Cytoplasmamembran begrenzte Protoplast ist mit Polysomen dicht gefüllt. Die zahlreich vorhandenen, rundlichen Mitochondrien besitzen lamelläre Cristae (Abb. 1). Fügt man einer wachsenden Flüssigkeitskultur 0,25 μg Econazol pro ml hinzu, so treten nach etwa zweistündiger Inkubationszeit die ersten elektronenoptisch sichtbaren Veränderungen im Zellinneren auf. Die zuvor deutlich nachweisbaren Außenmembranen der Mitochondrien, Kernmembranen und andere intrazelluläre Membransysteme lassen sich nicht mehr darstellen. Lediglich die den Protoplasten umgebende Cytoplasmamembran bleibt sichtbar (Abb. 2). Im Zellwandprofil erscheinen homogen verteilte fein-

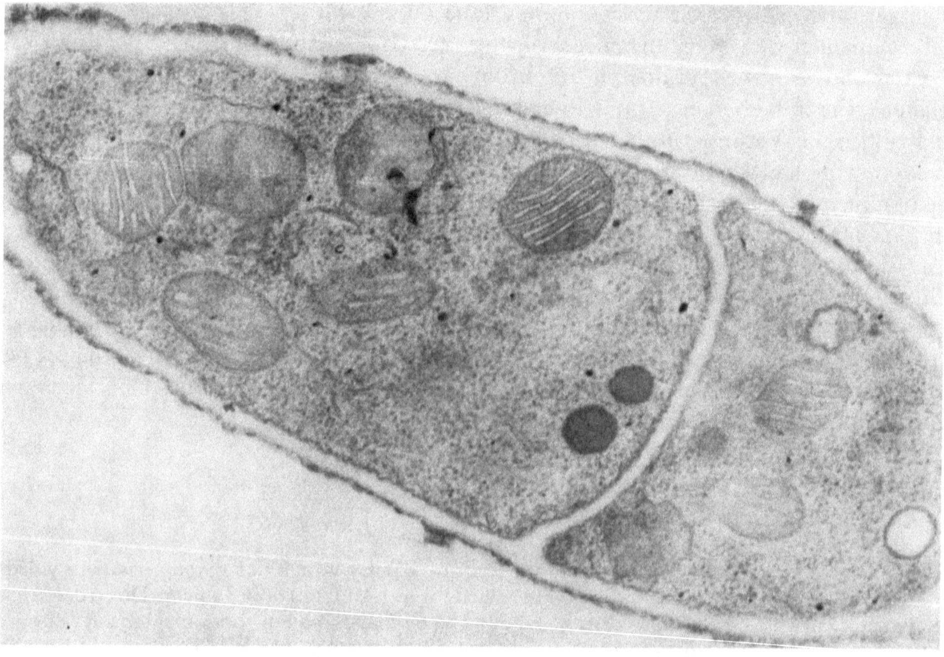

Abb. 1. Ultradünnschicht von *Trichophyton rubrum*; unbehandelte Zelle Vergr. 36000×

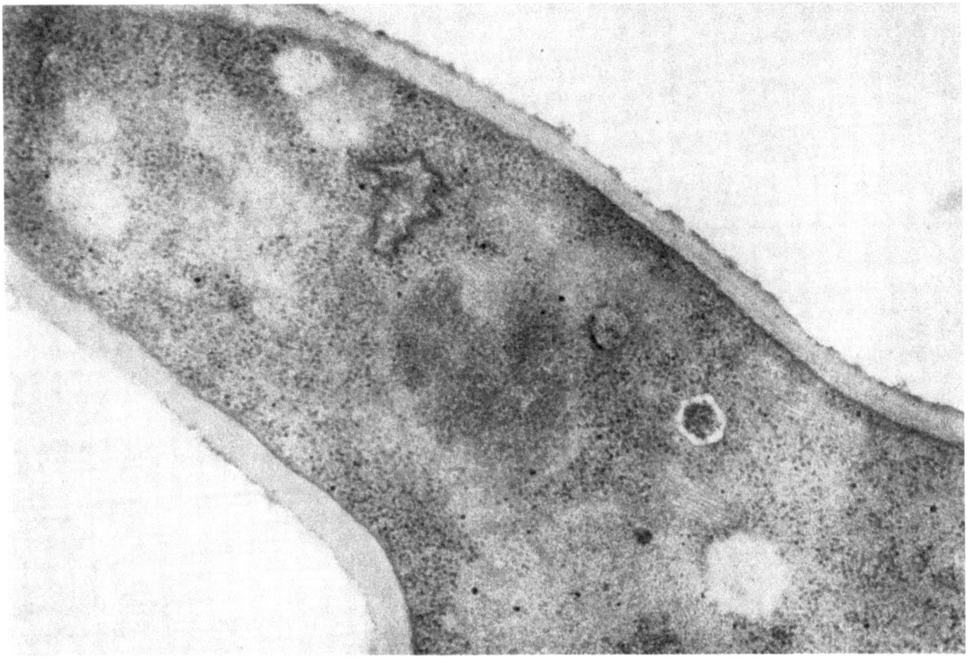

Abb. 2. Ultradünnschnitt von *Trichophyton rubrum* nach zweistündiger Einwirkung von 0,25 µg Econazol pro ml Nährsubstrat; intrazelluläre Membransysteme sind nicht mehr darstellbar, Kontraststeigerung der Zellwand
Vergr. 48000 ×

granuläre Ablagerungen. Ähnliche Strukturveränderungen der Zellwand beobachtet man auch in vivo an *Candida albicans* aus vaginalem Abstrichmaterial während einer 3-Tage-Therapie mit Econazol-Suppositorien: Vor der Behandlung sind die Zellen des *Candida*-Myzels von einer strukturlos erscheinenden Zellwand umgeben (Abb. 3) 4 Std. nach Applikation eines 150 mg Econazol-Suppositoriums nimmt die Zellwand der Sproßzellen eine granuläre Strukturierung an, während die Organellen im Zellinneren noch weitgehend unverändert erscheinen (Abb. 4). 24 Std. nach Behandlungsbeginn sind im Abstrichmaterial nur noch vereinzelt vermehrungsfähige *Candida*-Zellen nachweisbar. Abbildung 5 zeigt ein invasives Myzel innerhalb einer Vaginalepithelzelle. Die die Zellen umgebenden Wände sind stark kontrastiert.

Für die durch Econazol hervorgerufenen primären Strukturveränderungen in den Zellen humanpathogener Pilze bieten sich folgende beiden Deutungsmöglichkeiten an:

1. Das Antimykotikum bewirkt während der Penetration durch die Zellwand eine Freisetzung gebundener Zellwandlipide und -lipoproteine. Bei der nachfolgenden Fixierung bzw. Kontrastierung des Materials für die Elektronenmikroskopie kommt es zu einer Anlagerung von Schwermetallionen an diese Lipide, was zu der beobachteten Kontrasterhöhung führt. Da nach Rattray u. a. [12] der Lipidgehalt der Zellwand von Candida *albicans* mit 0,03–0,6% ihres Trockengewichts nur sehr gering ist, erscheint eine Kontraststeigerung auf diesem Wege recht unwahrscheinlich.

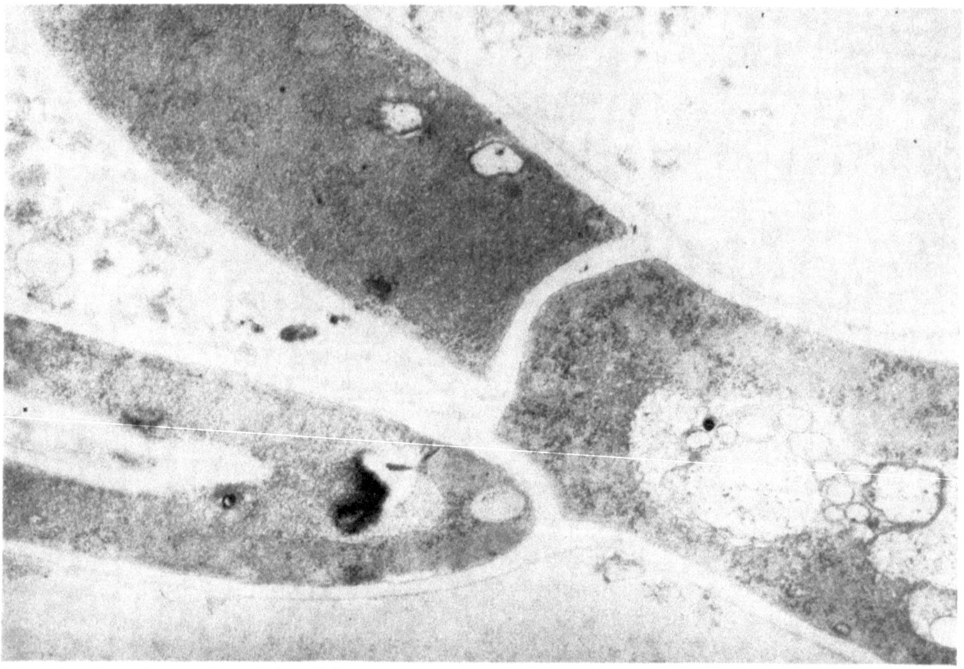

Abb. 3. Myzel von *Candida albicans* im Vaginalabstrich einer an akuter *Candida*-Mykose erkrankten, unbehandelten Patientin Vergr. 24000×

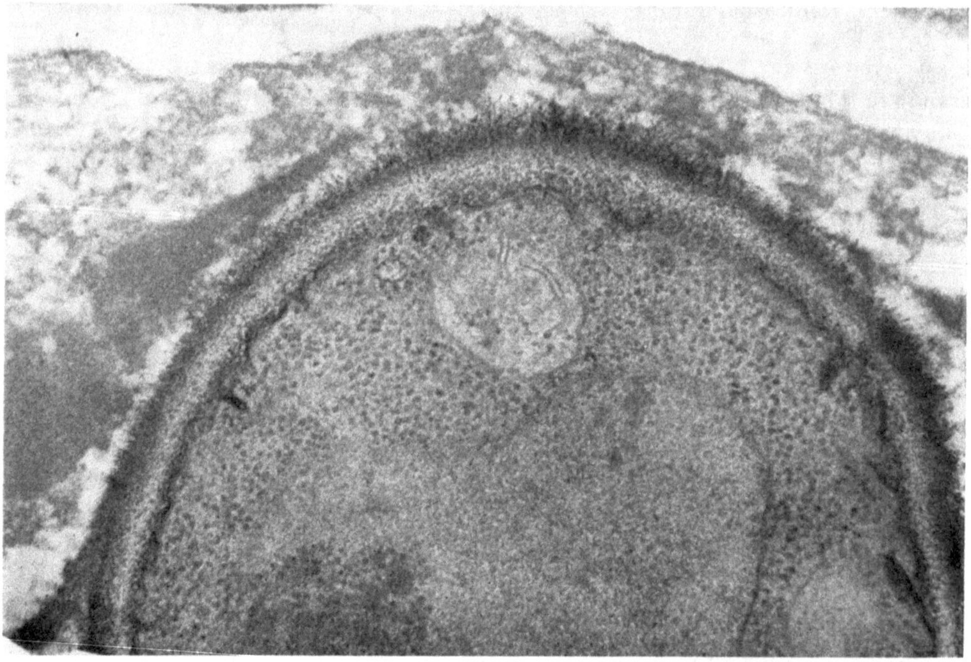

Abb. 4. *Candida albicans*-Zellquerschnitt in Vaginalepithelzelle nach vierstündiger Einwirkung eines 150 mg Econazol-Suppositoriums; erhebliche Kontrasterhöhung des Zellwandprofils Vergr. 72000×

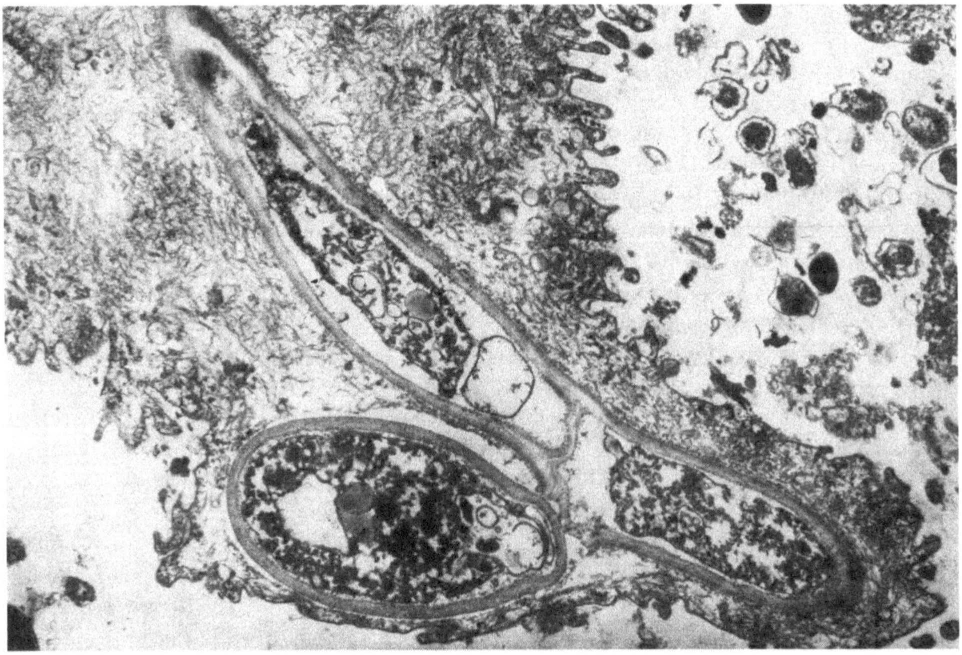

Abb. 5. Invasives Myzel von *Candida albicans* nach 24stündiger Einwirkung eines 150 mg Econazol-Suppositoriums; starke Kontrasterhöhung der Zellwand, Protoplast weitgehend lysiert Vergr. 14 500 ×

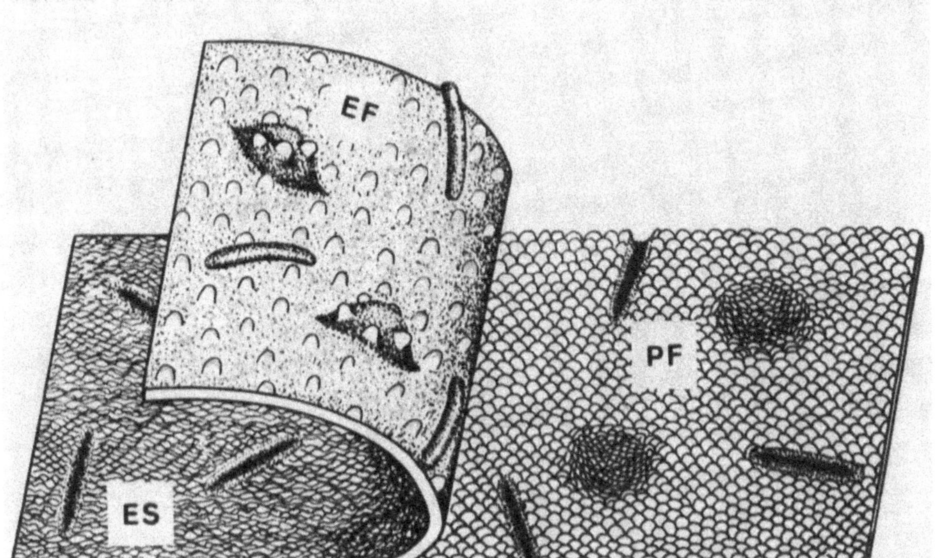

Abb. 6. Schematische Darstellung der Innenstrukturierung der Cytoplasmamembran von *Candida albicans*

2. Das Antimykotikum penetriert durch Zellwand und Cytoplasmamembran und bewirkt eine Extraktion von Lipiden, Phospholipiden und anderen lipidhaltigen Substanzen aus intrazellulären Membransystemen von Zellorganellen. Wie an anderer Stelle berichtet, können in econazolbehandelten Zellen größere Lipidmengen elektronenoptisch und cytochemisch nachgewiesen werden [8]. Darüber hinaus verändert Econazol die Permeabilität der Cytoplasmamembran, so daß die extrahierten Lipide nach Passage der Membran in der Zellwand mittels entsprechender Kontrastierungssubstanzen nachgewiesen werden können. Für diese Deutung sprechen Befunde von KERN u. ZIMMERMANN [2], die in physiologischen Versuchen mit *Candida albicans* durch Econazol eine Permeabilitätserhöhung der Cytoplasmamembran hervorrufen konnten.

Es stellt sich nun die Frage, ob und in welcher Weise Imidazolderivate die Struktur der Cytoplasmamembran verändern. Nach biochemischen in vitro Versuchen an *Candida albicans* mit Miconazol und Clotrimazol scheint die Vermutung zuzutreffen, daß Imidazole die Biosynthese des Membranbausteins Ergosterin inhibieren, wodurch die Stabilität der Membran herabgesetzt wird [1, 16]. Diese Erklärung mag für den Wirkungsmechanismus der Imidazole zwar zutreffen, erscheint jedoch nach Befunden von Hemmtests mit Bakterien unvollständig. Von mehreren Forschergruppen konnte nämlich nachgewiesen werden, daß Imidazolderivate bakterizid auf grampositive Bakterien wirken, also auf Prokaryonten, deren Membranen keine Steroide enthalten [11, 13, 14, 15]. Zur weiteren Aufklärung dieses Problems erschien es daher sinnvoll, das

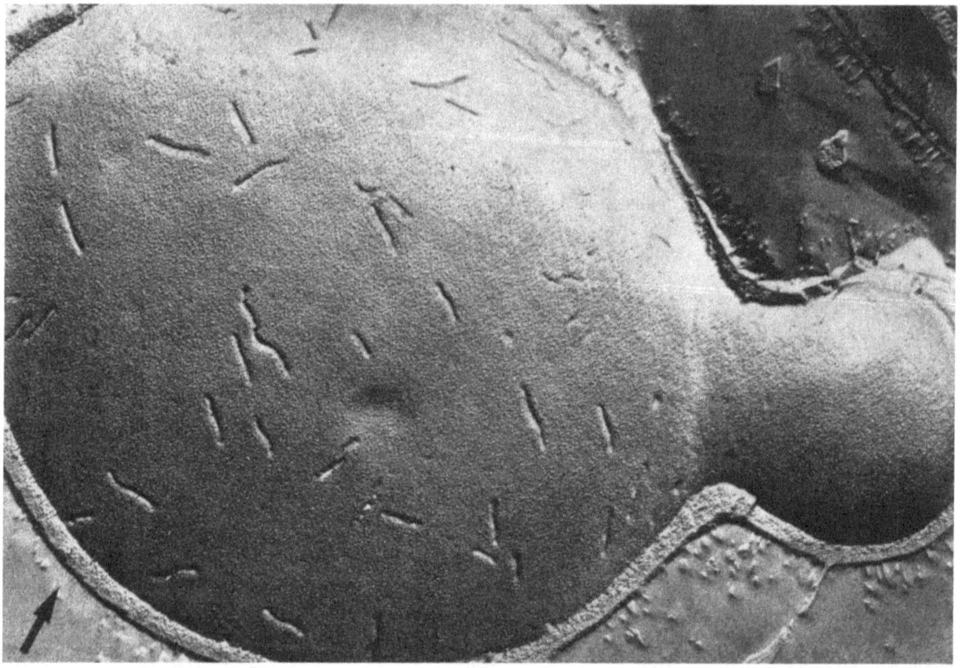

Abb. 7. Gefrierbruchabdruck der inneren Spalthälfte der Cytoplasmamembran einer unbehandelten, sprossenden *Candida albicans*-Zelle; dichter, homogener Besatz mit globulären Proteinen, in das Zellinnere hineinreichende Membraninvaginationen; der Pfeil markiert die Beschattungsrichtung! Vergr. 36000 ×

Abb. 8. Gefrierbruchabdruck der inneren Membranspalthälfte nach 24stündiger Einwirkung von 10 µg Econazol pro ml; Reduzierung der Anzahl globulärer Proteine und gruppenweise Anordnung in hexagonalen Mustern, Abreißen von Teilen der äußeren Membranspalthälfte; der Pfeil markiert die Beschattungsrichtung! Vergr. 100000 ×

innere Gefüge der Cytoplasmamembran von *Candida albicans* mit Hilfe der Gefrierbruchtechnik zu untersuchen. Mit dieser Methode erhält man Ansichten der beiden Spalthälften einer Membran (EF- und PF-Fläche), da die Brüche zumeist im hydrophoben inneren Bereich der Membran verlaufen und so die beiden Lipidlamellen voneinander getrennt werden (Abb. 6).

Die PF-Fläche („plasmatic fracture face") ist dicht und homogen mit etwa 10 nm großen globulären Partikeln besetzt. Sie bestehen nach Kopp [3] aus den hydrophoben Anteilen von Proteinen, die in die protoplasmaseitige Lipidlamelle integriert sind und in das apolare Membraninnere hineinragen. Invaginationen der Cytoplasmamembran senken sich in Form von rinnenförmigen Vertiefungen in das Innere der Zelle (Abb. 7). Nach Linnemans u. a. [4] sollen in diesen Einfaltungen Enzyme der Mannanbiosynthese gebildet werden. Nach sechsstündiger Einwirkung von Econazol verschwinden zunächst die Membraninvaginationen. Unter anhaltendem Econazoleinfluß löst sich dann der dichte Partikelbesatz der PF-Fläche in einzelne Partikelgruppen auf, die oftmals ein hexagonales Muster bilden (Abb. 8). Dazwischen treten immer zahlreicher partikelfreie Bereiche auf. Der Gefrierbruch verläuft bei derartig stark veränderten Membranen nicht mehr exakt zwischen den beiden Lipidlamellen, so daß Teile der äußeren Lamelle abgerissen werden und der PF-Fläche als Bruchstücke anhaften.

Wir nehmen an, daß durch eine verminderte Hydrophobie im Membraninneren die stabilisierenden Wechselwirkungen zwischen beiden Lipidlamellen verringert werden. Dadurch ändert sich der Bruchverlauf während der Präparation und Teile der äußeren Lipidlamelle reißen ab. Sie haften der PF-Fläche als Bruchstücke an. Eine, durch das Antimykotikum verursachte Verringerung des Ergosteringehalts der Membran könnte diese Stabilitätsminderung auslösen. Außerdem ändern sich Anzahl und Anordnung der globulären integralen Membranproteine, denen in erster Linie Transportfunktionen zuzuordnen sind. Durch diese Strukturveränderung im Membrangefüge wird vermutlich auch der kontrollierte Membrantransport erheblich beeinflußt.
Die beobachteten und hier beschriebenen Strukturveränderungen verursachen nicht nur einen Funktionsverlust der Membranen, sondern sie führen direkt oder indirekt auch zum Tod der Zellen. Ausplattierungen derartig stark geschädigter Zellen zeigten nämlich eine Überlebensrate von weniger als 0,5%.

Zusammenfassung

Die zur Therapie von Humanmykosen verwendeten Antimykotika aus der Reihe der Imidazolderivate beeinflussen die Permeabilität der Zellhülle der Pilze. Dieser Effekt beruht vermutlich auf einer Biosynthesehemmung des essentiellen Membranbausteins Ergosterin. Darüber hinaus findet gleichzeitig mit der Permeabilitätserhöhung eine mengenmäßige Abnahme und eine Umorientierung integraler Proteine in der Cytoplasmamembran statt. Dieser Vorgang beeinflußt möglicherweise die Transportfunktion der Zellhülle.
Es scheint, daß unter dem Einfluß der antifungalen Verbindungen lipidhaltige Substanzen aus intrazellulären Membransystemen extrahiert werden und dann durch die defekte Cytoplasmamembran und die Zellwand in das Nährmedium gelangen. Aufgrund der Bindungsfähigkeit von Schwermetallen an Lipide während Fixierung und Kontrastierung können diese Substanzen in der Zellwand elektronenmikroskopisch nachgewiesen werden.
Mittels biochemischer Analysenverfahren sollte es gelingen, die extrahierten Lipide im Nährmedium qualitativ und quantitativ zu bestimmen.

Summary

Imidazole derivatives used as antimycotics in the therapy of human mycoses affect the permeability of the fungal cell envelope. It is suggested that this effect depends on an inhibition of the biosynthesis of the essential membrane component ergosterol. Moreover, the increase of the membrane permeability is accompanied by a decrease in number of integral proteins and by a change of their orientation in the cytoplasmic membrane. This process may affect the transport function of the cell envelope.
It seems that under influence of the antifungal agents lipid containing substances will be extracted from intracellular membrane systems and, then, pass the defect cytoplasmic membrane and the cell wall. Due to the binding power of heavy metals to lipids during fixation and staining these substances can be detected electron microscopically in the cell wall.

Qualitative and quantitative determinations of the extracted lipids should be feasible in the culture medium by means of analytical biochemical methods.

Diskussion

LYR: Welches sind die Ursachen der humantoxischen Wirkung, die Sie erwähnten? Gibt es Anzeichen, daß die Synthese der Steroidhormone blockiert wird?

PREUSSER: Die erwähnten Wirkungen können nicht als toxisch eingeordnet werden.

SCHEWE: Aufgrund der vorgestellten Ergebnisse von Ihnen und Dr. BUCHENAUER besteht die Möglichkeit, daß die hier besprochenen Fungizide eine Peroxidation der Membranphospholipide verursachen. Dafür sprechen folgende Fakten:

- Dr. BUCHENAUER wies eine chemische Modifizierung der Phospholipide nach mit dem Auftreten eines UV-Absorptionsmaximums bei 240 nm; dies spricht für die Bildung konjugierter Diene, wie sie auch bei der Lipidperoxidation eintritt.
- Die Bildung von Lipidperoxiden würde auch die elektronenoptischen Veränderungen bei Anwendung der Gefrierbruchtechnik erklären; die Einführung der hydrophilen -O-O-H Gruppen setzt die Hydrophobizität der unpolaren Fettsäurereste herab und damit die Wechselwirkung mit den integralen Membranproteinen, wodurch deren Solubilisierung bewirkt werden kann.
- Das geschädigte Enzymsystem der Ergosterol-Biosynthese ist das mikrosomale Cytochrom P-450-System. Aus Untersuchungen des entsprechenden Systems der Leber ist bekannt, daß das mikrosomale Elektronentransportsystem durch Lipidperoxidation inaktiviert wird (z. B. KAGAN et al.). Somit könnte die Hemmung der Ergosterolysnthese einen Sekundäreffekt darstellen.

KOCH: Econazol, Miconazol und Clotrimazol haben bei systemischer Anwendung zu einer Induktion von Lebermikrosomen-Enzymen geführt, wodurch mit zunehmender Therapiedauer eine Inaktivierung eintritt.
Welche Beeinflussungen von Proteasen durch Imidazole halten Sie für möglich?

PREUSSER: Die beobachtete quantitative Abnahme integraler Membranproteine unter Imidazoleinfluß kann möglicherweise darauf beruhen, daß durch das Antimykotikum im Plasmalemma lokalisierte membrangebundene Proteasen freigesetzt und/oder aktiviert werden.

Literatur

[1] HALLER, I., Imidazole-antimycotics: experience with clotrimazole, experimental aspects, aims for the future, in: H.-J. PREUSSER (Hsg.), Medical Mycology, Gustav Fischer Verlag, Stuttgart 1980, 241—253

[2] KERN, R. und F. K. ZIMMERMANN, Über den Wirkungsmechanismus des Antimyzetikum Econazol, Mykosen **20** (1977), 113—146

[3] KOPP, F., Electron microscopy of yeasts, in: D. M. PRESCOTT (Hsg.) Methods in Cell Biology, Academic Press, New York 1975, Vol. XI, 23—44

[4] LINNEMANS, W. A. M., P. BOER and P. F. ELBERS, Localization of acid phosphatase in *Saccharomyces cerevisiae*: a clue to cell wall formation, J. Bact. **131** (1977), 638—644

[5] PREUSSER, H.-J., Die Wirkung von Econazol auf die Feinstruktur der Zellen von *Trichophyton rubrum*, Mykosen **18** (1975), 453—465

[6] PREUSSER, H.-J., Effects of in vitro treatment with econazole on the ultrastructure of *Candida albicans*, Mykosen **19** (1976), 304—316

[7] Preusser, H.-J. und R. Bartlett, Die Wirkung von Econazol-nitrat auf *Candida albicans*. I. Der Einfluß auf die Vermehrungsrate und die Ultrastruktur der in vitro-Zellen, Notabene medici **7** (1977), 6–9

[8] Preusser, H.-J. and H. Rostek, Econazole effects on *Trichophyton rubrum* and *Candida albicans* — electron microscopic and cytochemical studies, Mykosen, Suppl. 1 (1978), 314–321

[9] Preusser, H.-J. and H. Rostek, Freeze-fracture studies of the plasmalemma of *Candida albicans* after treatment with econazole-nitrate, Sabouraudia **17** (1979), 389–398

[10] Preusser, H.-J., H. Rostek, M. Klein, H. Becker und W. P. Nass, Elektronenmikroskopische Untersuchungen an *Candida albicans* aus vaginalem Abstrichmaterial während einer 3-Tage-Therapie mit Econazol, Arzneim.-Forsch. (Drug. Res.) **29** (1979) 1432–1437

[11] Raab, W., Mykosebehandlung mit Imidazolderivaten, Springer-Verlag, Berlin 1978

[12] Rattray, J. B. M., A. Schibeci and D. K. Kidby, Lipids of yeasts, Bact. Rev. **39** (1975), 197–231

[13] Schär, G., F. H. Kayser and M. C. Dupont, Antimicrobial activity of econazole and miconazole in vitro and in experimental candidiasis and aspergillosis, Chemotherapy **22** (1976), 211–220

[14] Thienpont, D., J. van Cutsem, J. M. van Nueten, C. J. E. Niemeggers and R. Marsboom, Biological and toxicological properties of econazole, a broadspectrum antimycotic, Arzneim.-Forsch. (Drug Res.) **25** (1975), 224–230

[15] van Cutsem, J. M. and D. Thienpont, Miconazole, a broad-spectrum antimycotic agent with antibacterial activity, Chemotherapy **17** (1972), 392–404

[16] van den Bossche, H., G. Willemsens, W. Cools, W. F. J. Lauwers and L. Lejeune, Biochemical effects of miconazole on fungi — II. Inhibition of ergosterol biosynthesis in *Candida albicans*, Chem.-Biol. Interact. **21** (1978), 59–78

H. D. SISLER, C. P. WOLOSHUK and P. M. WOLKOW

Department of Botany, University of Maryland, College Park, Maryland, U.S.A.

Studies on the Mode of Action of Tricyclazole and Related Compounds

There is considerable interest in compounds which control fungal diseases of plants by modifying the host-parasite relationship through mechanisms other than direct toxicity to fungal growth. It is a reasonable assumption that a diverse variety of modifications of this relationship may lead to the failure of disease development or to a marked diminishment of disease severity.

A recent study by MAITI et al. [5], showed that specific inhibition of cutinase by cutinase antibodies or by diisopropylfluorophosphate protected pea epicotyls from infection by *Fusarium solani pisi*. In neither case did the protecting agent prevent growth of the pathogen on the plant tissue surface.

Several recent studies have been made with compounds which control rice blast disease caused by *Pyricularia oryzae* by mechanisms other than conventional fungitoxicity. CARTWRIGHT et al. [2] showed that 2,2-dichloro-3,3-dimethylcyclopropane carboxylic acid accentuated production of momilactone phytolexins in rice plants challenged by *P. oryzae* and suggested that this may be the basis for rice blast control by the compound.

Probenazole, a compound related to saccharin, is reported by WATANABE [11] to be more effective in preventing penetration of rice plants by *P. oryzae* than as an inhibitor of growth of the pathogen in vitro.

Another compound, tricyclazole, is highly effective in controlling rice blast disease at concentrations below those which inhibit growth of the pathogen in vitro [6]. Even though the compound is essentially devoid of toxicity to *P. oryzae* in the concentration range of 1 to 20 μg/ml [3, 9], it is a potent inhibitor of melanin biosynthesis in this fungus at concentration of 0.1 μg/ml or less [10, 14].

Although tricyclazole is not reported to control diseases caused by *Verticillum dahliae*, it inhibits melanin biosynthesis in this fungus. Since the polyketide-melanin biosynthetic pathway has been partially elucidated in *V. dahliae* [1, 8, 12], the organism was used to study the mode of action of tricyclazole on melanin biosynthesis [10]. These studies showed that tricyclazole primarily blocks the pathway in *V. dahliae* between 1,3,8-trihydroxynaphthalene and melanin. More recently a similar mode of action of tricyclazole has been demonstrated in *Thielaviopsis basicola* [13].

The present paper reports on the effect of tricyclazole and of two compounds with a similar mode of action on growth and melanin biosynthesis in *P. oryzae* and considers some possible relationships between the action of the compounds on this pathogen and disease control.

Results

Fungitoxicity and inhibition of melanin biosynthesis

Tricyclazole (EL 291), 5-methyl-1,2,4-triazolo (3,4-*b*)-benzothiazole [3]; CGA 49104, 1,2,5,6,-tetrahydro-4H-pyrollo-[3,2,1-i,J]-quinolin-4-one [6] and pp-389, 4,5-dihydro-4-methyltetrazolo (1,5*a*) quinazolin-5-one [7], (Fig. 1) specifically control rice blast disease caused by *P. oryzae.* All of the compounds are very weak fungitoxicants but are potent inhibitors of melanin biosyntheses in *P. oryzae* [Tab. 1]. Tricyclazole and CGA 49104, in particular, display outstanding potency as selective inhibitors of melanin biosynthesis. Partial inhibition can be detected at concentrations less than 10^{-8} M and 10^{-7} M for the two compounds respectively [14].

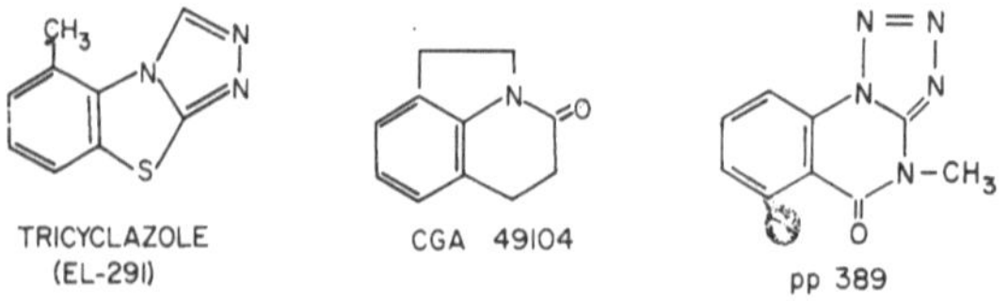

Fig. 1. Structures of tricyclazole, CGA 49104 and pp-389

Table 1
Effect of various concentrations of EL 291 (tricyclazole), CGA 49104 and pp-389 on radial growth (G) and melanization (M) of Pyricularia oryzae on sucrose-nitrate agar medium [10] after 7 days

Compound	µg/ml 0.01		0.1		1.0		10		50	
	G[a]	M[b]	G	M	G	M	G	M	G	M
EL-291	0	2	0	3	0	3	0	3	18	3
CGA	0	1	0	2	0	3	0	3	18	3
pp-389	0	0	0	1	0	2	0	3	0	3

[a] Percent inhibition of growth.
[b] Inhibition of melanization: 0, none; 1, slight; 2, moderate; 3, complete.

Although the three compounds appear dissimilar in several structural features, all share in common the structural configuration shown in Fig. 2. It is believed that the 3 compounds act in a similar or identical manner for the following reasons. (a) All are specific for the control of rice blast disease. (b) All selectively inhibit melanin biosynthesis in *P. oryzae* at non-fungitoxic concentrations. (c) All produce a similar pattern of polyketide pigment excretion in this fungus [14].

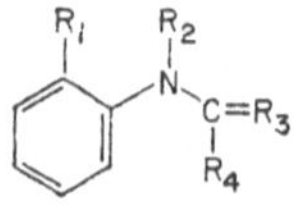

Fig. 2. Configuration common to tricyclazole, CGA 49104 and pp-389 R_1 = H, CH_3 or (CH_2 in ring system): R_2 = N, CH or CH_2 in ring system; R_3 = 0 or (N in ring system); R_4 = S, CH_2 or N in ring system (compare with Fig. 1)

Site of action in melanin biosynthetic pathway of *P. oryzae*

Tricyclazole blocks the pentaketide pathway (Fig. 3) between 1,3,8-trihydroxy-naphthalene and melanin in *V. dahliae* and as a consequence various metabolites such as 2-hydroxyjuglone accumulate. We have investigated the effect of tricyclazole on the pentaketide pathway to melanin in *P. oryzae*. While melanin biosyntheses in this organism is more sensitive to tricyclazole than in *V. dahliae*, the action of the compound is similar in the two organisms. Low concentration (0.01—1.0 μg/ml) block the pathway prior to vermelone in *P. oryzae* (Fig. 3) and this leads primilary to the accumulation of 2-hydroxyjuglone together with some flaviolin and 3,4-dihydro-3,4,8-trihydroxy-1(2H)napthalenone (DTN). Only the latter compound is detectable in untreated cultures (Tab. 2). Higher concentrations (10 μg/ml or greater) inhibit conver-

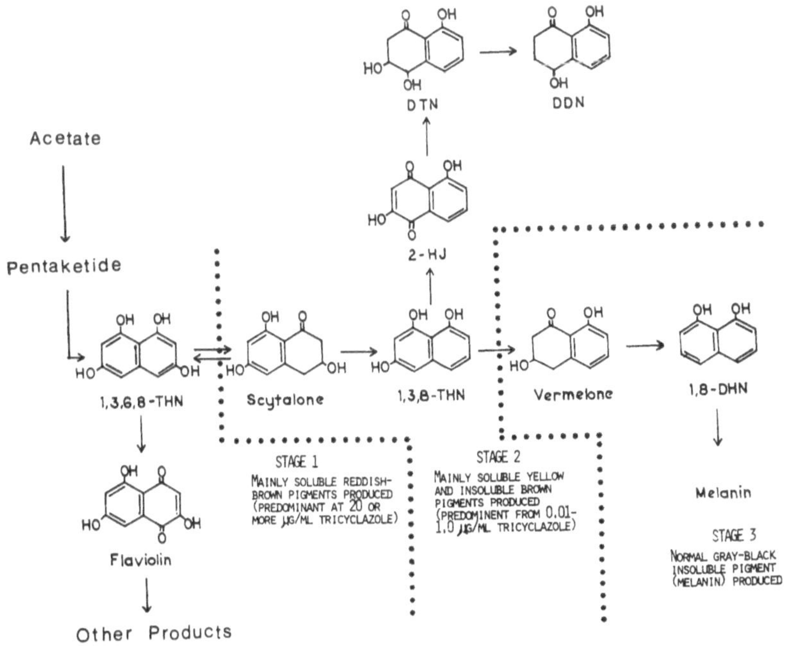

Fig. 3. Scheme showing polyketide pathway to melanin with associated side pathways and indicating pigment production at various stages. Abbreviations: 1,3,8-THN, 1,3,8,-tri-hydroxynaphthalene; 1,3,6,8-THN, 1,3,6,8-tetrahydroxynaphthalene; 2 HJ, 2-hydroxy-juglone; DTN, 3,4 dihydro-3,4,8-trihydroxy-1 (2H) naphthalenone; DDN, 3,4 dihydro-4, 8-dihydroxy-1 (2H) naphthalenone; 1,8-DHN, 1,8-dihydroxynaphthalene

sion of 2-hydroxyjuglone to DTN and apparently partially block the pathway prior to scytalone with the result that accumulation of flaviolin and products derived from this metabolite is accentuated (Fig. 3, Stage I). Products derived from flaviolin include 3,3′-biflaviolin a compound first identified in *T. basicola* [13] and 4-hydroxyscytalone. These metabolites tend to accumulate in older cultures treated with tricyclazole. The metabolite 3,4-dihydro-4,8-dihydroxy-1(2H)-naphthalenone (DDN), derived from DTN (Fig. 3), is found in untreated cultures of both *P. oryzae* and *V. dahliae*. This metabolite accumulates in older cultures of *P. oryzae* treated with 1 μg/ml or less

Table 2
Effect of several concentrations of tricyclazole on accumulation of polyketide metabolites in sucrose-nitrate agar cultures of P. oryzae[a]

Tricyclazole µg/ml	µg/125 ml agar medium		
	2HJ	Flaviolin	DTN
0	0	0	5
0.1	76	11	21
1.0	89	22	9
10.0	84	66	trace

[a] Medium containing inhibitor (0—10 µg/ml) innoculated with conidial suspension and incubated for 5 days extraction and purification of metabolites by procedures of TOKOUSBALIDES and SISLER (10)

of tricyclazole, but its production is strongly suppressed by concentrations of 10 µg/ml or greater.

In the pathway of polyketide-melanin biosynthesis as presently conceived [1, 8, 10, 13], tricyclazole primarily blocks the reduction of 1,3,8-trihydroxynaphthalene to vermelone in the fungi *V. dahliae* [10] *P. oryzae* [14] and *T. basicola* [13]. The precise nature of this inhibition, however, has not been determined.

Since the pattern of pentaketide pigment excretion associated with inhibition of melanin biosynthesis in *P. oryzae* is similar in cultures treated with either tricyclazole CGA 49104 or pp389, the three compounds apparently act at the same site in the melanin biosynthetic pathway of this organism.

Relationship of inhibition of melanin biosynthesis to disease control

Although it is believed that a relationship exists between inhibition of melanin biosynthesis and rice blast control, this relationship has not been established. However, in view of the unusual selectivity and high potency of tricyclazole, CGA 49104 and pp-389 as inhibitors of melanin biosynthesis, it seems likely the type of enyzme inhibited in this pathway is involved in action leading to disease control. The enzyme may not be restricted to melanin biosynthesis in the pathogen and it may also occur in the host plant.

Cultures of tricyclazole treated wild type *P. oryzae* closely resemble buff mutants genetically defective in melanin biosynthesis. Latterell [4] reported that such mutants are not pathogenic. More than 30 buff mutants of *P. oryzae* (p — 2), derived spontaneously or by UV irridiation, have been selected in our laboratory. Thus far, pathogenicity and the biochemical lesion in melanin biosyntheses have been studied in only one of these mutants, p — 2 m_1. This mutant is not pathogenic to rice varieties Nova-66 and Nato which are readily infected by the parent wild-type strain. Accumulation patterns of the pentaketide metabolites 2-hydroxyjuglone, flaviolin and DDN in this mutant are nearly identical with those found in the wild type strain treated with 0.1 µg/ml of tricyclazole.

These results strongly indicate that the observed effects of tricyclazole CGA 49104 and pp-389 are antipathogenic in nature. It remains to be established, however, whether

the genetic suppresion of pathogenicity in p-$2m_1$, or chemical suppression in the wild type are based on the pentaketide-melanin biosynthetic pathway.
Although we have no evidence at present to indicate how these compounds modify the host-parasite relationships one or more of the following are suggested as possible mechanisms.

1. Elicitor release is accentuated and results in a more rapid response of the host defense system such as phytoalexin accumulation or lignification. The elicitor may be a polyketide metabolite, a cell wall component or some other substance. This is regarded as the most attractive hypothesis.
2. The fungicides or accumulating fungal metabolites may inhibit activity or repress production of certain enzymes required for infection.
3. The fungicides may inhibit production of phytotoxins derived from the polyketide pathway (Stage III, Fig. I) or from other pathways.
4. Compounds may act directly on host plant increase rate of mobilization and magnitude of resistance response.

In view of the similarity of the buff mutant and chemically treated wild-type strain of *P. oryzae*, it seems more likely that primary action of the compounds is on the pathogen rather than on the host. Investigations are in progress to clarify this and other questions relating to the mode of action of these compounds.

Acknowledgements

The study was supported by Research Grant A100225 from the National Institute of Allergy and Infectious Diseases, U.S. Public Health Service.
Scientific Publication No. A2758, Contribution No. 5808 of the University of Maryland Agricultural Experiment Station.

Diskussion

LANGCAKE: Are there other plant pathogens in which pathogenicity is linked to melanin biosynthesis ?

SISLER: I am not aware of any.

LANGCAKE: What is the pathogenicity of the albino mutants of *Verticillium albo-atrum* ?

SISLER: I believe the albino strains are pathogenic. Melanin synthesis is localized in the microslerotia in *Verticillium dahliae*. Survival time is reported to be reduced in albino microsclerotia.

SCHWINN: What is your evidence for the loss of pathogenicity ? Have you made corresponding in vivo-tests ?

SISLER: We have tested 3 of the buff mutants on 2 varieties of rice which are highly susceptible to the black wild-type strain. None of the buff mutants were pathogenic in these tests.

GEORGOPOULOS: Although I think you are probably right, I must indicate the frequent loss of virulence following mutagenic treatment. It would be good to show that the block to melanin biosynthesis and the loss of pathogenicity are the result of the same mutation.

References

[1] BELL, A. A., J. E. PUHALLA, W. J. TOLMSOFF and R. D. STIPANOVIC, Use of mutants to establish (+)-scytalone as an intermediate in melanin biosynthesis by *Verticillium dahliae*, Canad. J. Microbiol. **22** (1976), 787–799

[2] CARTWRIGHT, D., P. LANGCAKE, R. J. PRYCE, D. P. LEWORTHY and J. P. RIDE, Chemical Activation of host defence mechanisms as a basis for crop protection, Nature **267** (1977), 511–513

[3] FROYD, J. D., C. J. PAGET, L. R. GUSE, B. A. DREIKORN, and J. L. PAFFORD, Tricyclazole: A new systemic fungicide for control of *Piricularia oryzae* on rice, Phytopathology **66** (1976), 1135–1139

[4] LATTERELL, F. M., Phenotypic stability of pathogenic races of *Pyricularia oryzae* and its implications for breeding of blast resistant rice varieties in proceedings seminar on horizontal resistance to the blast disease of rice. Centro Internacional de Agricultura Tropical, Cali, Columbia (1975), 199–234

[5] MAITI, I. B. and P. E. KOLATTUKUDY, Prevention of fungal infection of plants by specific inhibition of cutinase, Science **205** (1979), 507–508

[6] SCHWINN, F., M. NAKAMURA and G. HANDSCHIN, CGA 49104, a new systemic fungicide against rice blast, IX Int. Congr. Plant Protection (Abstracts) Washington D. C. (1979), 479

[7] SHEPHERD, M. C. and S. R. RAMASWAMY, Control of rice blast with the systemic fungicide pp389, Third Int. Congr. Plant Pathol. (Abstracts) Munich (1978), 358

[8] STIPANOVIC, R. D. and A. A. BELL, Pentaketide metabolites of *Verticillium dahliae*. II. Accumulation of naphthol derivaitves by the aberrant-melanin mutant brm-2, Mycologia **69** (1977), 164–172

[9] TOKOUSBALIDES, M. C. and H. D. SISLER, Effect of tricyclazole on growth and secondary metabolism in *Pyricularia oryzae*, Pestic. Biochem. Physiol. **8** (1978), 76–32

[10] TOKOUSBALIDES, M. C. and H. D. SISLER, Site of inhibition by tricyclazole in the melanin biosynthetic pathway of *Verticillium dahliae*, Pestic. Biochem. Physiol. **11** (1979), 64–73

[11] WATANABE, T., Effects of probenazole (Oryzemate) on each stage of rice blast fungus (*Pyricularia oryzae* CAVARA) in its life cycle, J. Pestic. Sci. **2** (1977), 395–404

[12] WHEELER, M. H., W. J. TOLMSOFF, A. A. BELL and H. H. MOLLENHAUER, Ultra structural and chemical distinction of melanins formed by *Verticillium dahliae* from (+)-scytalone, 1,8-dihydroxynaphthalene catechol and L-3,4-dihydroxyphenylalanine, Canad. J. Microbiol. **24** (1978), 289–297

[13] WHEELER, M. H. and R. D. STIPANOVIC, Melanin biosynthesis in *Thielaviopsis basicola*, Exptl. Mycol. **3** (1979), 340–350

[14] WOLOSHUK, C. P., P. M. WOLKOW and H. D. SISLER, Effect of 3 fungicides specific for control of rice blast disease on growth and melanin biosynthesis by *Pyricularia oryzae*, Pestic. Sci., In press

W. F. POLETAJEVA und S. E. BECKER

Botanisches Institut der Akademie der Wissenschaften der TSSR, Aschchabad, Lehrstuhl für Mikrobiologie der Biologischen Fakultät der Moskauer Staatlichen Lomonossow-Universität, Moskau, UdSSR

Bilanz der Dehydrogenasen und Oxidasen in Beziehung zur Anfälligkeit von zwei Baumwollarten gegenüber verschiedenen Welkeerregern

Es ist bekannt, daß in den mittelasiatischen Republiken zwei Arten von Baumwollwelkekrankheiten vorkommen. Die eine Art wird hervorgerufen durch den Erreger *Fusarium oxysporum* SCHLECHT f. sp. *vasinfectum* (ROBIN) BERKH., der an die feinfaserigen südlichen Sorten von *Gossypium barbadense* L. angepaßt ist. Die andere Art der Baumwollwelke wird hervorgerufen durch *Verticillium dahliae* KLEB., charakteristisch für die mittelfaserigen Sorten der Art *Gossypium hirsutum* L., die in nördlicheren Gebieten kultiviert werden. Hybriden beider *Gossypium*-Arten werden von beiden Erregern nur mäßig befallen.

Da sich die beiden Baumwollarten sowohl im Habitus der Pflanzen als auch in ihrer Reifezeit voneinander unterscheiden, wurde vermutet, daß diese Unterschiede in Beziehung stehen zum Grad der Anpassungsfähigkeit der Pathogene an die Unterschiede im Stoffwechsel, besonders der REDOX-Regime beider Baumwollarten, die sich in der Morphologie und Vegetationsdauer widerspiegeln.

Wie in einer kurzen Mitteilung bereits erwähnt [3], wurden von uns zur Bestätigung der Hypothese über den Zusammenhang zwischen dem Unterschied in der Erregerart und den Unterschieden im Stoffwechsel ihrer Wirtspflanzen Untersuchungen des Zellsaftes von Stengeln junger Baumwollpflanzen durchgeführt. Mit Hilfe einer einfachen Methode wurde die Enzymaktivität auf chromatischem Papier bestimmt, das mit o-Tolydin getränkt war. Zur Ermittlung der Glukoseaerodehydrogenase-Aktivität wurde das Papier mit Glukoselösung getränkt und zur Polyphenoloxidase-Aktivitätsbestimmung mit destilliertem Wasser. Die Enzymaktivität wurde aus dem Durchmesser der Blauzone ermittelt, die sich rings um einen mit Zellsaft gefüllten Zylinder bildet. Durch Interpolation auf die Penicillin-Aktivitätskurve, dargestellt auf halblogarithmischem Papier, kann man die Enzymkonzentration in festgelegten Einheiten bestimmen. Die Ergebnisse dieser Untersuchungen zeigt Abbildung 1, in der die Baumwollsorten und -arten in der Reihenfolge einer zunehmenden Resistenz gegenüber Fusarienwelke und Anfälligkeit gegenüber Verticillose dargestellt sind (Abszisse von links nach rechts). Zur Unterscheidung des Komplexes thermolabiler Komponenten der REDOX-Systeme von den thermostabilen wurde außer dem Nativsaft auch Stengelsaft untersucht, der 30 Minuten bei 60 °C erwärmt wurde.

Das Ergebnis zeigt, daß der Unterschied in der Reihe der Baumwollsorten nach zweiwöchigem Wachstum (punktierte Kurve) noch sehr gering ist. Nach sechs Wachstumswochen (ausgezogene Kurve) ist der Unterschied am größten. Dieser Zeitpunkt fällt zusammen mit der Ausbildung der Resistenzmechanismen in der Pflanze (siehe Hinweise 1, 2, 4, 5, 10), besonders mit einem Ansteigen des Polyphenoloxidase (PPO)-

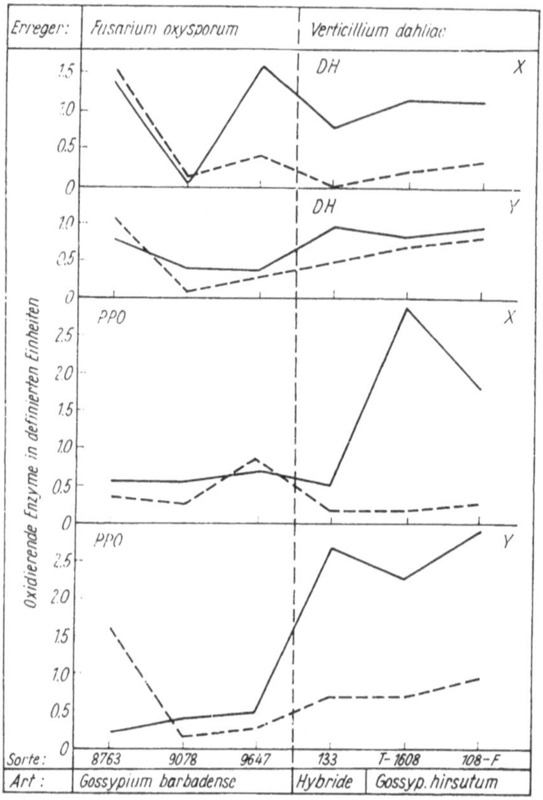

Abb. 1. Konzentration der Dehydrogenase (DH) und der Polyphenoloxydase (PPO) im Zellsaft des Baumwollstengels, ermittelt aus dem Diffusionszonen-Durchmesser und in festgelegten Einheiten ausgedrückt. Die Sorten sind in der Reihenfolge einer zunehmenden Resistenz gegenüber Fusariose und einer Resistenzverminderung gegenüber *Verticillium*-welke angeordnet.
Punktierte Kurve nach 2 Wachstumswochen
Ausgezogene Kurve nach 6 Wachstumswochen

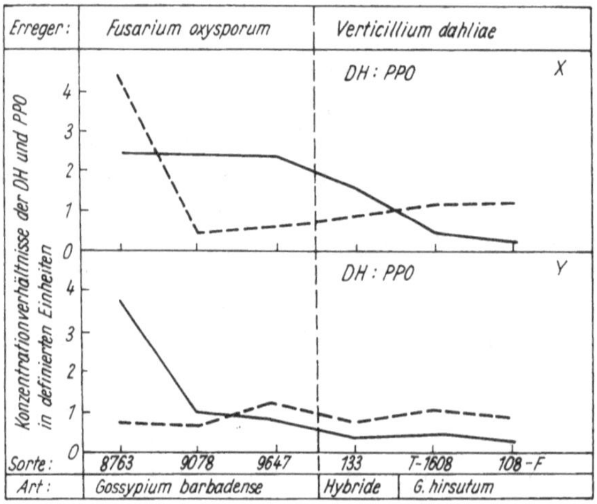

Abb. 2. Verhältnis zwischen den Konzentrationen der Dehydrogenasen (DH) und der Polyphenoloxydasen (PPO) im Stengelsaft von Baumwollpflanzen im Nativzustand und nach Erwärmen für 30 Minuten bei 60 °C. Anordnung der Baumwollsorten wie bei Abb. 1
Punktierte Kurve nach 2 Wachstumswochen
Ausgezogene Kurve nach 6 Wachstumswochen

Spiegels. Die PPO erhöht folgerichtig die Resistenz gegenüber Fusarienwelke und vermindert die Resistenz gegenüber Vertillose.

Bei rechnerischer Auswertung ergibt das Verhältnis der Dehydrogenasen (DH-) Aktivität zur Polyphenoloxidasen (PPO)-Aktivität für jede der untersuchten Sorten (Abb. 2) nach sechs Wachstumswochen eine umso stärker sinkende Kurve je höher die

Resistenz der Sorten gegenüber Fusarienwelke ansteigt. Das zeigt die besondere Anpassungsfähigkeit von *Fusarium oxysporum* an die Art und Sorten mit vorherrschender Dehydrogenasen-Aktivität und von *Verticillium dahliae* an die Art und Sorten mit einem Stoffwechsel, bei dem die Oxidasen überwiegen.

Die morphologischen und Stoffwechselbesonderheiten der Welkeerregerarten stimmen offensichtlich mit dieser Konzeption überein, insofern als das hellgefärbte *Fusarium oxysporum* in seinen Konidien reichlich Karotinoide bildet, die mit Hilfe der NADP-Dehydrogenasen [9] synthetisiert werden. Dieselbe Kategorie von Enzymen (NAD und NADP) synthetisiert auch das Toxin, die Fusarinsäure. Dagegen sind aggressive Stämme von *Verticillium dahliae* aktive Produzenten von Melaninen, die mittels Polyphenoloxidasen gebildet werden [6, 7, 12]. Diese Stämme besitzen schwarze Sklerotien. In die Reihe letzterer gehört auch ein anderer Krankheitserreger an Baumwollpflanzen, *Thielaviopsis basicola* (Berk. ex Br.) Ferraris.

Alle diese drei Erreger von Baumwollkrankheiten, wie auch der die Rhizosphäre besiedelnde Antagonist *Aspergillus terreus* Thom, wurden von uns hinsichtlich ihrer Ausscheidung von Glukoseaerodehydrogenasen und Polyphenoloxidasen in die Umgebung geprüft. Die Untersuchungen wurden in der gleichen Weise durchgeführt wie die Prüfung des Zellsaftes der Baumwollpflanzen, nur mit dem Unterschied, daß die Diffusionszonen auf chromatographischem Papier durch Diffusion aus Agarblöcken entstanden, die aus der Mitte von Pilzkolonien herausgeschnitten und auf das chromatographische Papier aufgelegt wurden.

Die in der Tabelle 1 zusammengestellten Ergebnisse basieren auf Untersuchungen an 10 und mehr Stämmen jeder Art, mit Ausnahme von *Verticillium dahliae*, dessen Stämme nur selten die genannten Enzyme aus dem Myzel in das umgebende Medium abgaben (von mehr als 10 Stämmen nur 2). Aus der Tabelle geht hervor, daß wie erwartet, bei *Fusarium oxysporum* das Verhältnis zwischen der Dehydrogenase und der Oxidase zugunsten ersterer verschoben ist. Die DH-Aktivität liegt im Durchschnitt um das 5-fache höher als die PPO-Aktivität. Dagegen ist bei melaninsynthetisierenden Erregern das Verhältnis zugunsten der PPO verschoben, deren Konzentration bei *Thielaviopsis basicola* um das 2,5-fache höher ist als die DH-Konzentration.

So kann man wahrscheinlich annehmen, daß die Anpassung der Erreger an bestimmte Baumwollarten tatsächlich auf einer Ähnlichkeit der REDOX-Verhältnisse ihres Stoffwechsels mit denjenigen der Baumwollpflanzen beruht. Aus einem Vergleich der Abbildung 2 und der Tabelle 1 ist zu ersehen, daß die Sorten Nr. 8763 und 9078 am stärksten mit *Fusarium*welke befallen werden. Hier liegt das Verhältnis der DH:PPO etwa in der Mitte des Bereiches, innerhalb dessen ein Erreger in der Lage ist, eine Baumwollpflanze zu infizieren. Eine Erklärung dafür ist in der Bildung des Toxins durch den Erreger, die Fusarinsäure, zu suchen, für deren Synthetisierung Dehydrogenasen erforderlich sind.

Dagegen weist der nichtpathogene *Aspergillus terreus*, der die Rhizosphäre beider Baumwollarten besiedelt und dessen Antibiotikum Geodin durch beide Kategorien von Fermenten gebildet wird, nur verhältnismäßig geringe Schwankungen im Verhältnis von DH:PPO auf, das bei 1,0 liegt (Tab. 1). Im Gegensatz dazu variieren sowohl die im Boden lebenden als auch die in den Baumwollpflanzen parasitierenden Stämme von *Fusarium oxysporum* hinsichtlich ihres Verhältnisses DH:PPO und der Menge der gebildeten Enzyme wesentlich stärker. Der Schwankungsbereich erstreckt sich von völligem Fehlen einer Enzymausscheidung bei einigen bodenbürtigen Formen

Tabelle 1

Vergleich des Niveaus und der Aktivitätsverhältnisse von Glukoseaerodehydrogenase (DH) und Polyphenoloxidase (PPO) bei Baumwollpathogenen und ihrem Antagonisten, Aspergillus terreus, der die Rhizosphäre besiedelt

Forschungsobjekt: Arten, Stämme		Anzahl der Untersuchungen	Mittlere Konzentrationen und ihre Grenzwerte für Oxidationsenzyme in festgelegten Einheiten		Verhältnis DH:PPO
			Glukoseaerodehydrogenase (DH)	Polyphenoloxidase (PPO)	
Pathogene	*Fusarium oxysporum* f. sp. *vasinfectum*, infektionsfähig (Aktivität der Fusarinsäure 48 mkg/ml und mehr)	15	2,83 (0–9,5)	0,79 (0–2,15)	3,98 (0–18,8)
	Thielaviopsis basicola (aus Pflz.)	10	2,54 (0–8,20)	6,91 (0,6–16,0)	0,38 (0–2,05)
	Verticillium dahliae (aus Pflz.)	2	0,86 (0,72–1,0)	1,06 (0,73–1,06)	0,90 (1,45–1,88)
Antagonist	*Aspergillus terreus* (aus Rhizosphaere)	19	0,64 (0,23–2,90)	0,67 (0,24–2,55)	1,03 (0,50–1,79)

bis zu einer hohen Aktivität, besonders der DH, bei aggressiven pathogenen Stämmen mit hoher Fusarinsäure-Produktion.

Die Übereinstimmung der Parameter ihrer toxischen Aktivität mit der Dehydrogenasen-Aktivität zeigt Abbildung 3.

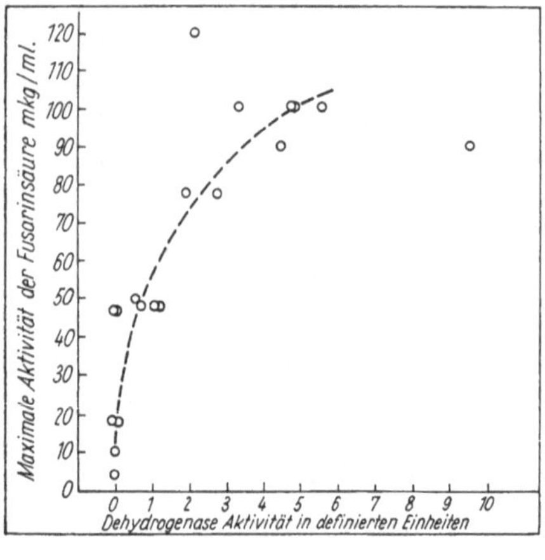

Abb. 3. Abhängigkeit der Bioysnthese-Aktivität der Fusarinsäure bei *Fusarium oxysporum f. sp.* Glukoseaerodehydrogenase im Myzel. Punktierte Kurve mittleres Niveau

Bei einer statistischen Auswertung dieser Ergebnisse wurde der Korrelationskoeffizient zwischen der Fusarinsäure-Aktivität und der Glukoseaerohydrogenase-Aktivität errechnet. Mit S = 0,729 ± 0,124 war die Beziehung des enzymatischen Parameters zur antibiotischen Aktivität genügend hoch gesichert. Eine wesentlich geringere Übereinstimmung ergab bei dieser Berechnung die Toxizität von *Fusarium oxysporum* mit dem Verhältnis DH:PPO (S = 0,504 ± 0,287), sie fehlte völlig für die Aktivität der PPO. Desgleichen gelang es nicht, eine gesicherte Korrelation zwischen den enzymatischen Parametern und der antibiotischen Aktivität bei *Aspergillus terreus* zu erhalten. Hier variieren die enzymatischen Werte bedeutend weniger als die antibiotische Aktivität gegenüber *Fusarium oxysporum* und *Neurospora sitophila*. Das Verhältnis DH:PPO, das im Durchschnitt um 1 lag, schwankte nicht stärker als um das 4-fache (s. Tab. 1). Diese Fakten sind vollkommen logisch, insofern als diese Art nicht pathogen ist und als Bodensaprophyt nicht zu einer Anpassung an den spezifischen Stoffwechsel von Wirtspflanzen gezwungen ist, die für den Prozeß der Pathogenese erforderlich ist.

Leider konnte eine Gegenüberstellung der Parameter der Toxizität für die Pflanzen und der enzymatischen Werte bei den melaninbildenden Pathogenen *Verticillium dahliae* und *Thielaviopsis basicola* bisher nicht vorgenommen werden. Die Ursache liegt einmal darin, daß diese Enzyme von *V. dahliae* nur in geringen Mengen in die Umgebung abgegeben werden, zum anderen fehlen eindeutige Angaben über deren Natur und über quantitative Bestimmungsmethoden der Hauptagenten ihrer Pathogenese, die bisher offensichtlich nur zum Teil untersucht wurden (siehe z.B. 6, 8, 7, 11, 12, 13).

Der Versuch einer Zusammenstellung der Anfälligkeitsstufen von Baumwollsorten gegenüber *Thielaviopsis basicola* mit den für diese Sorten charakteristischen DH- und PPO-Aktivitätsgraden und ihren Verhältnissen zueinander (Tab. 2) führte nicht zu dem erwarteten Ergebnis, da die an Dehydrogenasen reichen Sorten der Art *Gossypium barbadense* stärker befallen werden. Daraus ergibt sich die Vermutung, daß — vorausgesetzt, daß das durch den Erreger produzierte Phytotoxin Hauptagent der Pathogenese ist, — das Toxin in diesem Falle nicht auf dem Wege der Melaninbildung entsteht oder es ist als eine Vorstufe zu betrachten, die noch ohne Beteiligung der Phenoloxidasen synthetisiert wird.

Obgleich der eindeutige Beweis von Beziehungen zwischen der Anpassung der REDOX Bilanz der Enzyme pathogener Pilze an die ihrer Wirtspflanzen und ihrer Aggressivität gegenüber bestimmten Pflanzensorten und -arten bisher noch nicht immer möglich ist, gibt es doch dieses Beispiel für eine spezifische Abhängigkeit zweier Erreger, *Fusarium oxysporum* und *Verticillium dahliae*, von Arten und Sorten der Baumwolle. Eine Beziehung zwischen der Befallsintensität und den jeweils mehr oder weniger stark an Sauerstoff gebundenen Stoffwechseltypen der Baumwollpflanzen — von *G. bardense* bis *G. hirsutum* — spricht für die Bedeutung dieses Parameters. Weiterhin sprechen spezifische Alternativen im Metabolismus pathogener Pilze dafür. Die Kenntnis dieser Prozesse eröffnet Möglichkeiten, mittels synthetischer Inhibitoren oder Immunitätsagenten in diese Prozesse einzugreifen und gezielt die Ausbildung von pathogenen Eigenschaften zu hemmen.

Die praktische Anwendung für den Pflanzenschutz, mit Hilfe systemisch wirkender Verbindungen die Ausbildung eines Stoffwechseltyps bei den Pilzen zu erreichen, der dem der Wirtspflanzen nicht entspricht, oder durch züchterische Maßnahmen den Stoffwechsel der Wirtspflanzen zu beeinflussen mit dem Ziel, eine Adaption des Patho-

Tabelle 2
Zusammenstellung des Befallsgrades der Baumwollsorten durch Thielaviopsis basicola und der Menge der Zellsaftkomponenten mit Dehydrogenase- und Polyphenoloxidaseaktivität

	Arten:	*Gossypium barbadense*			Hybride	G. hirsutum
	Sorten:	8763	9078	9647	133	138–F
Befallshöhe der Baumwollpflanzen durch *Thielaviopsis basicola* in %	Aus vergleichenden Untersuchungen von 3 virulenten Stämmen	43,3	–	28,2	3,6	8,2
	Grenzwerte der Befallsstärke unter Feldbedingungen	82,5–100	20,6–96,2	40,3–100	7,6–68,6	24,4–84,0
Aktivität der Oxidativkomponenten im erwärmten Baumwollzellsaft nach 6 Wachstumswochen in festgelegten Einheiten	Glukoseaerodehydrogenase (DH)	0,80	0,41	0,39	0,95	0,95*
	Polyphenoloxidase (PPO)	0,21	0,43	0,52	2,70	2,90*
	Verhältnis DH:PPO	3,81	0,95	0,75	0,35	0,33*

* verwandt mit der Sorte 138–F der Ferganasorte 108–F.

gens auszuschließen, ist natürlich eine Sache der Zukunft. Aber Schritte in dieser Richtung haben zweifellos eine Perspektive, gerade für das systematische Vorgehen auf dem Gebiet der Phytopathologie.

Diskussion

KOCH: Gibt es Anhalte für die Wirkung von Trichothecenen auf die Baumwolle?

BECKER: Wir haben keine derartigen Ergebnisse. Nach unseren Versuchsergebnissen ist das Hauptwelketoxin die Fusarinsäure.

LYR: Fusarinsäre ist ein hochwirksames und seit längerer Zeit bekanntes Toxin, welches die Zellpermeabilität bei Pflanzen schädigt. Vor kurzem wurde die Verbindung in Japan als starker Hemmstoff der Monoaminooxidase des Nervensystems als Pharmakon patentiert. Das zeigt, daß viele Verbindungen oft unerwartet vielseitige Wirkungen haben.

BECKER: Vom Wirkungsmechanismus der Fusarinsäure war bisher nur bekannt, daß sie sehr starke Erhöhung der Permeabilität der Wirtszellen hervorruft. Ihre Wirksamkeit gegenüber MAO ist sehr interessant und verdient weitere Aufmerksamkeit.

NAUMANN: Worauf beruht nach Ihren Erfahrungen die pathologische Wirkung von *Verticillium dahliae*?

BECKER: Bei *Verticillium dahliae* ist das Toxin-Problem viel komplexer als bei *Fusarium oxysporum*. Hier handelt es sich wahrscheinlich nicht um Fusarinsäure, sondern um andere Toxine wie Verticillin. Über einige Toxine von *V. dahliae* wird Prof. BORODIN noch einiges berichten.

BUCHENAUER: Welche Bedeutung besitzt das Lycomarasmin während der Pathogenese von *Fusarium oxysporum* in Baumwollpflanzen?

BECKER: Möglicherweise ist Lycomarasmin überhaupt nicht am Pathogeneseprozeß beteiligt, da seine Synthese zu spät erfolgt, bei Laborkultur erst nach Monaten. Fusarinsäure wird jedenfalls schneller gebildet und ihre Synthese ist enger mit dem Welkeverlauf verbunden, wie die Übereinstimmung zwischen Welkeintensität und Toxinkonzentration in der Pflanze zeigt.

Literatur

[1] BECKER, S. E., Theoretische Grundlagen der Pathogenese, ökologischer Hintergrund und komplexe Methoden im Kampf gegen die Fusarienwelke. In Sammlung 2: „Ökologisch-physiologische Methoden im Kampf gegen die Fusarienwelke der Baumwollpflanzen" ILIM-Verlag, Aschchabad, 1973, 233–254 (russ.)

[2] BECKER, S. E., Three categories of fungicidal defense substances of plants with systemic action (on the model of cotton diseases). In "Systemic Fungicides", V. International Symposium, Akademie-Verlag, Berlin, 1979, 15–30 (engl.)

[3] BECKER, S. E., W. F. POLETAJEVA, I. D. PUSCHKARJEVA, O. E. MILOVA, The dehydrogenases and polyphenoloxydases balance in connection with the succession of species and sort resistance of cotton to wilt. Abstracts of reports of 3-rd International Congress of Plant Pathology, München, 1978, p. 236 (engl.)

[4] BECKER, S. E. und I. D. PUSCHKARJEVA, Histochemische Untersuchungen des Wirkungsmechanismus des Welketoxins auf Baumwollpflanzen. In der Sammlung 1: „Ökologisch-physiologische Methoden im Kampf gegen die Fusarienwelke der Baumwollpflanzen", ILIM-Verlag, Aschchabad, 1971, 223–231 (russ.)

[5] BECKER, E. S. und I. D. PUSCHKARJEVA, Die Wirkung von Mikroelementen und gewisser organischer Verbindungen auf die Dehydrogenase-Aktivität der Baumwollpflanzen in Beziehung zur Erhöhung ihrer Welkeresistenz. In der Sammlung 3: „Ökologisch-physiologische Methoden im Kampf gegen Fusarienwelke der Baumwollpflanzen", ILIM-Verlag, Aschchabad, 1977, 126–140 (russ.)

[6] BELL, A. A., J. E. PUHALLA, W. J. TOLMSOFF, R. D. STIPANOVIC, Use of mutants to establish (+)-scytalone as an intermediate in melanin biosynthesis by *Verticillium dahliae*. Canad. Journ. Microbiology, **22** (1976a), 787–799

[7] BELL, A. A., R. D. STIPANOVIC, J. E. PUHALLA, Pentacetide metabolites of *Verticillium dahliae*. Identification of (+)-scytalone as natural precurser to melanin, Tetrahedron, **32** (1976b), 1353–1356

[8] BORODIN, G. I. und W. I. RUNOV, Untersuchung über die Natur gewisser toxischer Produkte von *Verticillium dahliae*. In: „Arbeiten der I. Konferenz der Biochemiker der Mittelasiatischen Republiken und Kasachstans". Verlag FAN der Usbekischen SSR, Taschkent, 1967, 431–436 (russ.)

[9] KRITSKY, M. S., Die Probleme „Signal — Antwort" in Untersuchungen über Erscheinungsformen in der individuellen Entwicklung von Pilzen. In der Sammlung: „Mikroorganismen-Ontogenese", Verlag Wissenschaft, Moskau, 1979, 175–185 (russ.)

[10] PUSCHKARJEVA, I. D. und S. E. BECKER, Kernaktivität und Lokalisation der Schutzsysteme gegen Welke in der Baumwollpflanze. In der Sammlung 2: „Ökologisch-physiologische Methoden im Kampf gegen die Fusarienwelke der Baumwollpflanzen", ILIM-Verlag, Aschchabad, 1973, 206–218 (russ.)

[11] RUNOV, W. I., G. I. BORODIN, W. A. GUSEINOV, L. I. TSCHEPENKE, Über die Natur der Virulenz des *Verticillium dahliae*. In: „Arbeiten des 12. Internationalen Botanischen Kongresses", Verlag Wissenschaft, Leningrad, 1975, 501 (russ.)

[12] STIPANOVIC, R. D. and A. A. BELL, Pentacetide metabolites of *Verticillium dahliae*. II. Accumulation of naphthol derivatives by the abberant melanin mutant BRM-2, Mycologia, **69** (1977), 487–488

[13] TSCHEPENKO, L. I., G. I. BORODIN, W. I. RUNOV, B. S. SALICHOVA, Reinigung und Charakterisierung der Verticilline — toxische Substanzen des *Verticillium dahliae*. In der Sammlung: „Immunität und Ruhe der Pflanzen", Verlag Wissenschaft, Moskau, 1972, 151–160 (russ.)

ALENA TOMŠIKOVÁ, JIŘI ŠTĚRBA and DANA NOVÁČKOVÁ (technische Mitarbeit)

Institut für Mikrobiologie und Epidemiologie der Medizinischen Fakultät der Karls Universität, Pilsen, Parasitologisches Institut der Tschechoslowakischen Akademie der Wissenschaften, Prag, ČSSR

Therapeutische Möglichkeiten bei der Adiaspiromykose

Emmonsia crescens EMMONS et JELLISON 1960, dem Erreger der Adiaspiromykose, wurde in den letzten Jahren besondere Aufmerksamkeit gewidmet, anfangs als Ursache einer chronischen Lungenerkrankung der wild lebenden kleinen Nagetiere (CHEVREL et al. [1], DOBY et al. [3], EMMONS et al. [6], OTČENÁŠEK et al. [12], PROKOPIČ et al. [14]), später als Erreger einer schweren chronischen disseminierten Erkrankung des Menschen (CUEVA et al. [2], DOBY et al. [4], KOĎOUSEK et al. [9], LESHCHENKO et al. [10], ŠLAJS et al. [15]).

Wenn auch die Adiapiromykose in der Tschechoslowakei mit Fungizon und Pimafucin (DVOŘÁK et al. [5], OTČENÁŠEK et al. [13]) erfolgreich geheit wurde, bleibt die Therapie dieser Erkrankung stets offen.

Aus diesem Grunde studierten wir bei dieser Erkrankung nicht nur die Wirkung von verschiedenen Chemotherapeutika, sondern auch die Möglichkeit der passiven Immunisierung.

Material und Methoden

Zum Studium benutzten wir 4 *Emmonsia*stämme, die sich voneinander durch Virulenz und antigene Aktivität (TOMŠÍKOVÁ et al. [19]) unterscheiden, weiter *Emmonsia parva* und *Chrysosporium pannorum*.

1. Versuche in vitro

   Zur Feststellung der Empfindlichkeit der saprophytischen Phase aller studierten Stämme gegenüber 22 Antimykotika wurde der Blättchen-Agar-Diffusionstest (TOMŠÍKOVÁ et al. [18]) benutzt.

   Weiter verfolgten wir die Wirkung von Pimafucin, Fonderma und Amphotericin B auf die parasitische Phase von *E. crescens* (TOMŠÍKOVÁ [20]).

2. Versuche in vivo

   Die Wirkung von Amphotericin B, Pimafucin und Fonderma wurde bei experimenteller Adiaspiromykose an 32 6 Wochen alten weißen Mäusen verglichen (TOMŠÍKOVÁ et al. [20]). Die Tiere wurden einmalig intraperitoneal mit 0,5 ml Aleuriensuspension (Dichte $3 \times -10^{5-6}$) infiziert. Der Einsatz von Chemotherapeutika erfolgte gleichzeitig oder 5, 10, 14, 16, 24, 28 und 30 Tage nach der Infektion.

   Die durch die experimentelle Erkrankung verursachte mögliche Spätallergie wurde an Hand des Pfotenschwellungstestes (GRAY et al. [7], MURPHY et al. [11]), festgestellt.

   Die Wirkung des spezifischen Antiserums auf den Verlauf der experimentellen Adiaspiromykose wurde an 7 weißen Mäusen getestet, denen gleichzeitig mit dem, in dem

vorangehenden Versuch erwähnten, Inokulum oder 7, 15, 22, 28, 34 Tage nach der Infektion 0,5 ml des Kaninchen-*Emmonsia*-Antiserums intraperitoneal appliziert wurde.
Die Tiere erhielten die Standard-Labornahrung und Wasser ad libitum und wurden bei 20 – 25 °C gehalten.
Nach der vorher festgesetzten Zeit wurden die Tiere beider Gruppen entblutet und steril seziert. Alle Organe wurden der makroskopischen und mikroskopischen Untersuchung und der Kultivierung auf den mykologischen Nährboden unterzogen.

Ergebnisse

1. Versuche in vitro

Die breitesten Inhibitionszonen im Diffusionstest bildete bei allen untersuchten Stämmen Fonderma. An 2. Stelle war in dieser Hinsicht Mycosynalar, an 3. Stelle Amykon (Abb. 1, 2, 3).
Die Empfindlichkeit der Myzelphase gegenüber den Antimykotika wies eine bestimmte Parallelität mit der Morphologie der Kolonien auf. Im Gegensatz zu Pimafucin und

EMMONSIA – WIRKUNG DER ANTIMYKOTIKA

Inhibitions-zonen in mm	–	–10	–20	–30	–40	–50	–60	–70	–80	–90
Fonderma						●	●		●●●	●
Amykon		●		●	●●●	●				
Fungicidin	●		●●	●	●					
Amphotericin B	●		●	●●	●●					
Candio-Hermal	●			●●	●●●					
Gentiane				●●●	●●●					
Salifungin	●●				●●●	●				
Dequafungan	●	●	●●●		●					
Sporilin	●		●●●	●	●					
Celestoderm	●●		●●●	●						
Etacortin	●●● ●		●●							
Fungiplex	●●● ●			●●						
Visorit	●●	●	●●							
Nitrofungin	●●● ●					●				
Ac. lacticum	●●● ●●				●					
Mycosynalar					●			●●		
Canesten	●		●●							
Pevaryl				●●	●					
Pimafucin			●	●●						
Bromochrom	●●●									
Tonoftal	●●●									
Monistat										

Abb. 1. Die Disken bezeichnen die erreichten Inhibitionszonen

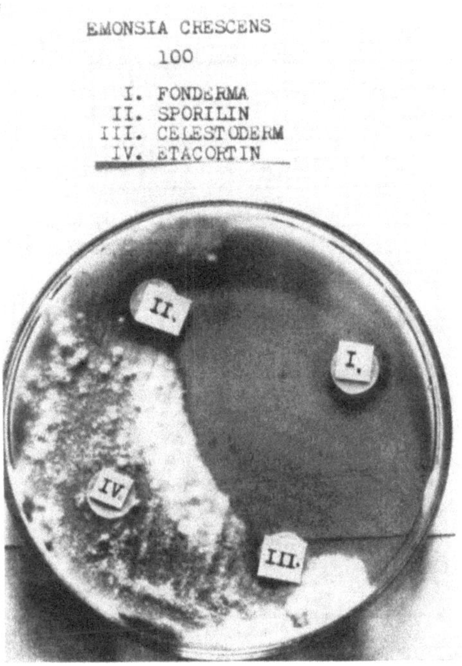

Abb. 2. Hemmzonen beim Blättchen-Test

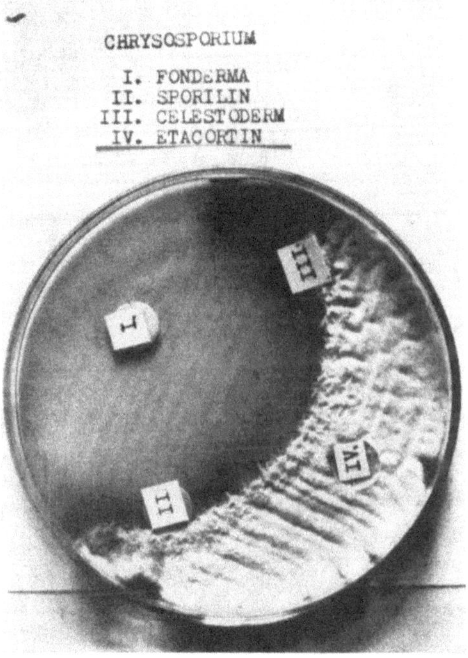

Abb. 3. Hemmzonen beim Agar-Diffusionstest

Amphotericin B, wirkte Fonderma auch auf die parasitische Phase fungizid. Bei 26 °C kam es zu einer Keimung der Adiasporen auf dem SABOURAUDagar, falls Fonderma gleich nach der Entnahme der Adiasporen aus dem Tier appliziert wurde. Ebenfalls wurde die beginnende Keimung unterbrochen.

2. Versuche in vivo

a) Wirkung von Chemotherapeutika:

Bei Tieren, denen das Antigen nur einmal appliziert wurde, entwickelte sich nach 36 Tagen eine disseminierte Adiaspiromykose mit typischem makro- und mikroskopischen Bild der 36 tägigen experimentellen Erkrankung (BOISSEAU-LEBEREUIL, ŠLAJS et al. [16] und ŠTĚRBA et al. [17]). Bei den Versuchstieren war der Befund von der angewandten Therapie, der Zeit der Therapie, dem Therapieeinsatz und den Applikationsgaben abhängig (Abb. 4).

Bei allen Applikationsformen des Pimafucins entwickelte sich eine disseminierte Adiaspiromykose mittleren Grades (++). Nach Verabreichung von Amphotericin B entstand mittelstarke Adiaspiromykose (++) nur bei den Tieren bei denen die Therapie gleichzeitig oder 5 Tage nach der Applikation des Antigens einsetzte. In den anderen Fällen verstärkte sich die Dissemination. Bei der Verabreichung von Fonderma konnten einige Adiasporen um die Infektionsstelle und im Leberhilus nur dann festgestellt werden, wenn die Therapie erst 30 Tage nach der Infektion eingesetzt hatte. Sonst war der histologische Befund negativ, oder es wurden nur einzelne tote Adiasporen angetroffen.

TIEREXPERIMENTE

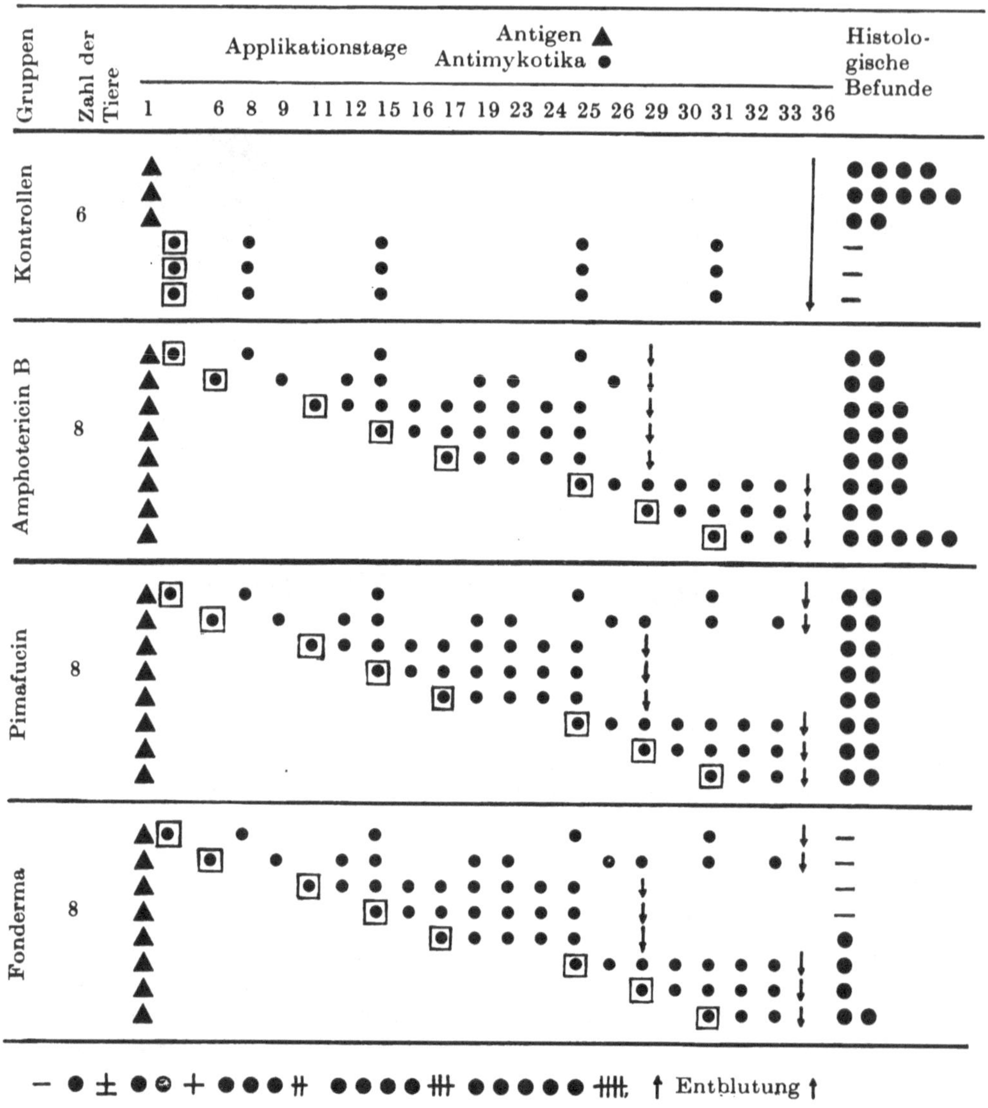

Abb. 4. Die großen Disken bezeichnen die Intensität des pathologischen Prozesses, die kleinen Disken den Applikationstag, die eingerahmten Disken den Applikationsanfang

Bei entwickelter Adiaspiromykose trat auch eine allergische Spätreaktion in Form einer verschieden starken Anschwellung der mit dem Antigen injizierten Pfoten auf.

b) Wirkung von spezifischem Antiserum

Der therapeutische Effekt des Antiserums hing wie bei den Antimykotika von der Zeit der Verabreichung ab. Wurde das Antiserum in den ersten 15 Tagen nach der Infektion appliziert, entwickelte sich keine Erkrankung. Bei späterer Applikation wurden

Tabelle 1
Passive Immunisierung

Gruppen	Applikationswochen < Antigen (A) / Antiserum (AS)									Histologie	Grad der Erkrankung
	1	2	3	4	5	6	7	8	9		
Kontrollen	A									Einstichknoten, Knoten im Fettgewebe, im Leberhilus, im Milzhilus, bei der Magenkonvexität im Peritoneum	++++
I.	A AS 0,5									Geringfügige Einstichknoten, negativ	—
II.	A	AS 0,5				AS 0,5	AS 0,5	AS 0,5	AS 0,5	Geringfügige Einstichknoten, negativ	—
III.	A		AS 0,75			AS 0,5	AS 0,5	AS 0,5	AS 0,2	Geringfügige Einstichknoten, negativ	—
IV.	A			AS 1,0			AS 0,5	AS 0,5	AS 0,5	Einstichknoten (2 × 3 mm), 3 winzige Knoten im Beckenfettgewebe und 1 Knoten auf der Milzkonvexität	+
V.	A				AS 1,0		AS 0,5	AS 0,5	AS 0,5	Einstichknoten, einzelne Adiasporen und geringfügige Aussaat auf der Serose der Bauchwand	++
VI.	A					AS 1,0	AS 0,5	AS 0,5	AS 0,5	Einstichknoten, einzelne Adiasporen und geringfügige Aussaat auf der Serose der Bauchwand	++

entweder einzelne Einstichknoten oder geringfügige Knoten im Beckenfettgewebe und Leberhilus angetroffen. Wie bei den vorangehenen Versuchen zeigte sich bei der mit *Emmonsia* infizierten Kontrollgruppe die kein Antiserum erhalten hatte, das übliche Bild der Organ-Adiaspiromykose (Tab. 1) mit Präzipitinbildung (auch IV.—VI. Gruppe).

Diskussion

Den therapeutischen Effekt eines bestimmten Stoffes auf Mikroorganismen kann man orientierend in vitro und genauer in einem biologischen Versuch beurteilen.
Nach beiden Kriterien bewährte sich Fonderma am besten. Es wirkte sowohl auf die saprophytische als auch auf die parasitische Phase von *Emmonsia crescens* und auch bei verschieden virulenten Stämmen gut fungizid.
Der therapeutische Effekt war vom Stadium der Krankheit und dem Beginn der Therapie abhängig. Die Infektion wurde vollständig unterdrückt, wenn Fonderma gleichzeitig oder einige Tage nach der Infektion verabreicht wurde. Das Antimykotikum reicherte sich im Peritoneum an und schuf hier ungünstige Bedingungen für die Entwicklung der Adiasporen aus den Aleurien.
Die üblicherweise benutzten Antimykotika (Pimafucin und Amphothericin B) haben sich im Tierversuch nicht bewährt. Auch im Plattentest zeigten sie schwächere Wirkung als Fonderma.
Da Fonderma in den Kontrollversuchen keine toxischen Nebenwirkungen aufwies, könnte es als gutes Antimykotikum bei Organ-Mykosen benutzt werden.
Die passive Immunisierung wird bisher bei mykotischen Erkrankungen therapeutisch nicht angewendet. Unsere Versuche zeigen aber gewisse Möglichkeiten auf. Die Ergebnisse der Versuche zur passiven Immunisierung waren mit den biologischen Versuchen unter Einsatz von Fonderma fast identisch. Falls spezifisches Antiserum während der ersten 15 Tage nach der Infektion verabreicht wurde, entwickelte sich keine Krankheit. Auch spätere Applikation von Antiserum beeinflußte die Entwicklung der Erkrankung günstig. Diese zunächst nur an einer kleinen Gruppe von Tieren durchgeführten Versuche zeigen neue Wege in der Therapie schwer heilbarer Organ-Mykosen auf.

References

[1] Chevrel, M. L., R. Souquet, B. Farrend, L. J. Richter, M. Doby-Dubois and L. Louvet, Polyparatisme pulmonaire mycosique. Premier cas humain d'adiaspiromycose associé à une aspergillose, Ann. Anatom. Pathol. (Paris) **9** (1964), 463 to 468
[2] Cueva, J. A. and M. D. Little, Emmonsia crescens infection (adiaspiromycosis) in Honduras, Am. J. Trop. Med. Hyg. **20** (1971), 282—287
[3] Doby, J. M., M. T. Boisseau-Lebreuil and B. Rault, L adiaspiromycose par Emmonsia crescens chez les petites mammiféres sauvages en France, J. Soc. franc. Mycol. méd., Nantes (1969), 23—24
[4] Doby-Dubois, M., M. L. Chevrel, J. M. Doby and M. Louvet, Premier cas humain d adiaspiromycose par *Emmonsia crescens* Emmons et Jellison 1960, Bull. Soc. Pathol. exot. **57** (1964), 240—244
[5] Dvořák, J., M. Otčenášek and B. Rosický, Adiaspiromycosis caused by *Emmonsia crescens* Emmons and Jellison 1960. *E. crescens* and antifungal antibiotics, Studie CASV, Praha-Academia **14** (1973), 86—91

[6] EMMONS, C. W. and W. L. JELLISON, *Emmonsia crescens* species nova and adiaspiromycosis (haplomycosis) in mammals, Ann. NY Acad. Sci. **89** (1960), 91—101
[7] GRAY, D. F. and P. A. JENRRINGS, Allergy in experimental mouse tbc, Am. Rev. Tuberc. Pulm. Dis. **72** (1955), 171—195
[8] JELLISON, W. L., Haplomycosis in Montana rabbits, rodents and carnivores, Publ. Hlth. Rep. **65** (1950), 1057—1063
[9] KOĎOUSEK, R., V. VORTEL, A. FINGERLAND, V. VOJTEK, Z. ŠERÝ, V. HÁJEK and K. KUCERA, Pulmonary adiaspiromycosis in man caused by *Emmonsia crescens*. Report of a unique case, Am. J. Clin. Pathol. **56** (1971), 394—399
[10] LESHCHENKO, V. M. and N. D. SHEKLANOV, Prvý případ adiaspiromykózy v SSSR, Čs, Derm. **50** (1975), 9—12
[11] MURPHY, J. W., J. A. GREGORY, and H. W. LARSH, Skin Testing of Guinea pigs and Footpad Testing of mice with a new Antigen for Detecting delayed hypersensitivity to *Cr. neoformans*, Infection, Immunity **9** (1974), 404—409
[12] OTČENÁŠEK, M., J. DVOŘÁK and J. PROKOPIČ, Isolace *Emmonsia crescens* et JELLISON 1960 na uzemi CSSR, Cs epid. mikrobiol. imunol. **14** (1965), 229
[13] OTČENÁŠEK, M. and J. DVOŘÁK, The effect of Fungizone on experimental adiaspiromycosis of laboratory mice, Mycopath. Mycol. Appl. **33** (1967), 349—352
[14] PROKOPIČ, J., J. DVOŘÁK and M. OTČENÁŠEK, Adiaspiromycosis (haplomycosis) in Czechoslowakia. Preliminary report, Sabouraudia **4** (1965), 35—36
[15] ŠLAJA, J., J. DVOŘÁK and M. OTČENÁŠEK, The morphology of solitary adiaspores of *Emmonsia* from the lung of man, Fol. parasit. **17** (1970), 177—182
[16] ŠLAJS, J., J. ŠTĚRBA and F. HAMÁČEK, Experimentální adiaspiromykóza, Plzen. lék. Sborn. **37** (1972), 79—83
[17] ŠTĚRBA, J., F. HAMÁČEK, A. TOMŠÍKOVÁ and D. NOVÁČKOVÁ, The pattern of experimental, intraperitoneal infection with the fungus *Emmonsia crescens* in mice, Folia Parasitol. (Praha) **20** (1973), 361—366
[18] TOMŠÍKOVÁ, A. and M. OTČENÁŠEK, Orientierende Bewertung der antifungalen Aktivität einiger Antimykotika mittels des Agardiffusionstestes, Castellania 3 (1975), 219—226
[19] TOMŠÍKOVÁ, A., J. DVOŘÁK, M. OTČENÁŠEK and D. NOVÁČKOVÁ, Antigenic activity of *Emmonsia crescens* EMMONS et JELLISON 1960, Acta mikrobiol. palackiense Olomouc **63** (1972), 51
[20] TOMŠÍKOVÁ, A., J. ŠTERBA and J. PROKOPIČ, Beitrag zur Chemotherapie der Adiaspiromykose, Castellania — im Druck

W. WOHLRAB, K.-D. WOZNIAK, K. SCHWABE, B. TSCHIERSCH, R.-P. ZAUMSEIL und K.-M. TAUBE

Hautklinik der Martin-Luther-Universität Halle-Wittenberg
und Zentralinstitut für Krebsforschung der AdW der DDR, Berlin-Buch, DDR

Wirksamkeit und therapeutische Effektivität von 5-Brom-salizyl-4'-chloranilid und Derivaten

Bei der externen Behandlung von Mykosen der Haut und der Hautanhangsgebilde sind dem Einsatz antimykotisch wirksamer Substanzen in verschiedener Hinsicht Grenzen gesetzt. Auch wenn die Kombination von Wirksubstanz und gewähltem Vehikel (Salbe; Emulsion) beim therapeutischen Vorgehen weitgehend dem Charakter der Hautveränderungen sowie der Empfindlichkeit der Haut Rechnung trägt, sind eine Reihe weiterer Faktoren für die Anwendungsmöglichkeiten bestimmend.
Besonders wichtig erscheint die Tatsache, daß Antimykotika entsprechend der Natur der Erreger immer über längere Zeit angewandt werden müssen, so daß neben der fungistatischen und fungiziden Wirkung der Substanz die Reaktion der entsprechenden Kontaktgewebe (z.B. Epidermis) von entscheidender Bedeutung sein können.
Anderseits hängt der therapeutische Effekt jeder externen Therapie u. a. davon ab, inwieweit es gelingt, in der betroffenen Hautschicht eine optimale Wirkstoffkonzentration zu erreichen, wobei die Belastung des Gesamtorganismus möglichst gering gehalten werden sollte. Die Belastungen der Kontaktgewebe sind dabei im Hinblick auf die notwendigerweise angestrebten günstigen Penetrationseigenschaften einer antimykotischen Substanz oft in Kauf zu nehmende Nebenreaktionen, da man in der Regel davon ausgehen muß, daß eine Substanz, die die Barrieren der Hornschicht durchdrungen hat, auch weiter permeiert und schließlich auch resorbiert wird [9; 11].
Einen Beitrag, der Lösung dieser Problematik näher zu kommen, sehen wir darin, den antimykotischen Wirkstoff in eine weniger toxische Form zu überführen, die gegebenenfalls eine Transportform im Sinne von TSCHIERSCH et al. [12] darstellt, wobei der Wirkstoff selbst erst vom Erreger z.B. durch dessen Enzyme freigesetzt wird und so selektiv wirksam werden kann. Entscheidend ist dabei, welcher Grad an Selektivität erreichbar ist.
Eine Möglichkeit der Realisierung dieser Vorstellungen glauben wird durch die Synthese verschiedener *Wirkstoff-Glykoside* gefunden zu haben [10], die einerseits Wirksubstanzen mit wesentlich geringerer Toxizität darstellen könnten, anderseits z.B. bei der Kombination Wirkstoff-Xylopyranosid durch Xylosidasen enzymatisch zur freien Wirkform und dem entsprechenden Zucker gespalten und damit durch eine bevorzugte Schädigung der Pilzzellen eine günstige Erregerselektivität besitzen würde.
Im folgenden sollen am Beispiel des Wirkstoffes 5-Bromsalizyl-4'-chloranilid (BSCA) sowie des 5-Bromsalizyl-4'-chloranilid-0-β-D-xylopyranosid (ZIK 1392) die prinzipiellen Überlegungen am praktischen Beispiel vorgestellt werden (Abb. 1 und 2).
BSCA ist als gut wirksames Antimykotikum bekannt. So beträgt beispielsweise der Hemmwert gegen *Trichophyton mentagrophytes* 1:300000 bis 1:500000. Der gleich-

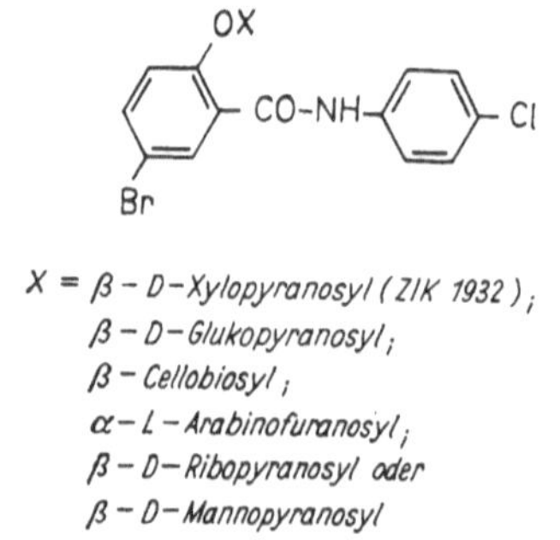

Abb. 1. 5-Bromsalizyl-4'-chloranilid-0-glykoside

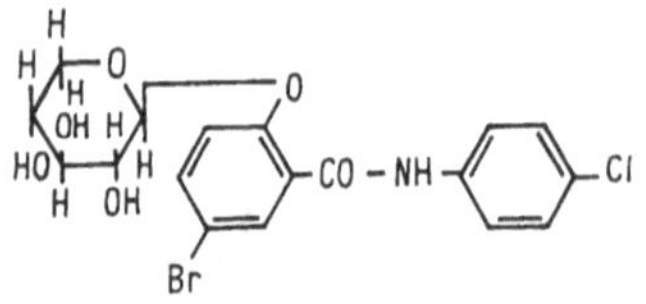

ZIK 1392

Abb. 2. 5-Bromsalizyl-4'-chloranilid-0-β-D-xylopyranosid (ZIK 1392)

zeitige bakteriostatische Effekt, insbesondere gegen Staphylokokken, ist von besonderer Bedeutung [6; 7]. Wichtig ist ferner noch die günstige Penetration in tiefere Hautschichten, wobei z.B. bei 630 µm Hauttiefe noch Konzentrationen mit 2 Zehnerpotenzen über dem Wert zur totalen Wachstumshemmung von *Trichophyton mentagrophytes* nachweisbar sind [6]. Die umfangreichen langjährigen Erfahrungen mit BSCA z.B. im Präparat Multifungin® (KNOLL/AG, BRD) bestätigen die gute antimykotische Wirkung. Als Nebenwirkungen sind photoallergische Reaktionen bzw. lokale Reizerscheinungen beschrieben [1; 4; 8].

Tabelle 1
Biometrische Ergebnisse und autoradiographische Befunde an ^{3}H-Thymidin-markierter Meerschweinchenepidermis nach Kontakt mit 0.06 M 5-Bromsalicyl-4'-chloranilid in O/W-Emulsion (li = *labeling index*; DNS-Si = *DNS-Synthese index*)

	Kontaktzeit in Wochen			
	Kontrollen	1	Kontrollen	4
Biometrie				
N	42	32	36	33
Epidermisbreite [µm]	52,57 ± 7,71	73,27 ± 15,37	60,16 ± 14,13	64,11 ± 15,77
t	–	7,5829	–	1,0974
2 P	–	0,001	–	>0,10
Akanthosefaktor	1,00	1,39	1,00	1,07
Zellgröße [µm²]	167,29 ± 24,53	204,71 ± 42,97	171,73 + 40,35	214,56 ± 52,77
t	–	4,7281	–	3,8059
2 P	–	0,001	–	0,001
Autoradiographie (Str. basale)				
Zellkerne gesamt	3500	7500	4000	10000
Zellkerne markiert	195	551	175	710
li [%]	5,57	7,35	4,38	7,10
DNS-Si	1,00	1,32	1,00	1,62
χ^2	–	11,617	–	35,368
α	–	0,0025	–	0,00025

Biometrische und autoradiographische Untersuchungen [Methodik: 13; 14] über die Reaktion der Epidermiszelle nach Langzeitkontakt mit BSCA am Modell des Meerschweinchenohres zeigten dagegen wesentliche Veränderungen in den Prozessen der epidermalen Proliferation und Differenzierung (Tab. 1).

Zunächst ist nach 1-wöchigem BSCA-Kontakt eine Verbreiterung der Epidermis um ca. 40% nachzuweisen, die sich nach den autoradiographischen Befunden auf eine entsprechend gesteigerte Anzahl markierter Kerne im Stratum basale bezieht. Da aber gleichzeitig auch in der mittleren Zellgröße der Epidermis eine bedeutende Zunahme der Meßwerte abzusichern ist, muß neben der Epidermisverbreiterung auch eine Erhöhung der postmitotischen Generationszeit der Epidermiszellen angenommen werden [2; 3]. Diese Annahme wird durch die Meßwerte nach prolongierter täglicher Applikation von BSCA bis zu 4 Wochen bestätigt. Während hinsichtlich der Epidermisbreite nach 4 Wochen Kontakt analog dem von HARDMEIER und SCHNYDER [5] beschriebenen Adaptationssyndrom ein Rückgang der Veränderungen auf die Ausgangswerte abzuleiten ist, beziehen sich diese „Normalwerte" auf eine geringere Zellanzahl in der Epidermis mit einer wesentlich gesteigerten mittleren Zellgröße. Gleichzeitig zeigt der Vergleich der labeling-Indizes, daß mit einer um ca. 60% gesteigerten Neubildungsrate von Epidermiszellen gerechnet werden muß. Die Generationszeit und Verhornungsrate in der Epidermis muß demnach beträchtlich gesteigert sein.

Im Gegensatz dazu lassen sich nach Langzeitkontakt äquimolarer Mengen von ZIK 1392 unabhängig von der Kontaktzeit weder an der mittleren Epidermisbreite und Größe der Epidermiszellen, noch an der Neubildungsrate im Stratum basale, erkenntlich an den ^3H-Thymidin-markierten Zellkernen, durchgängige Abweichungen von den Normalwerten ableiten (Tab. 2).

Ein Einfluß von ZIK 1392 auf die Vorgänge der epidermalen Proliferation, Regeneration und Differenzierung ist demnach auch bei prolongiertem täglichen Kontakt bis zu 4 Wochen nicht nachzuweisen.

Dem entsprechen auch die an Mäusen vom Stamm C 57 Bl 6/Jena nachgewiesenen Unterschiede bei der Bestimmung der LD_{50} nach i. p. Injektion von BSCA und ZIK 1392 (Tab. 3), wobei für ZIK 1392 6 bis 7-fach höhere Werte als für BSCA ermittelt werden konnten.

Demgegenüber zeigt eine Pilotstudie zur klinischen Wirksamkeit bei Hautpilzerkrankungen im Halbseitenversuch keine Unterschiede in dem bekannten Ansprechen von Interdigitalmykosen der Füße oder oberflächlichen Trichophytien der Leistenregion auf BSCA gegenüber ZIK 1392 (Tab. 4).

Es ist demnach festzustellen, daß am Beispiel des BSCA und dem Xylosid dieser Verbindung (ZIK 1392) nachgewiesen werden konnte, daß durch die Synthese von Wirkstoff-Glykosiden die Wirkungen auf das Kontaktgewebe bzw. die Wirtszellen (z. B. Epidermis) wesentlich reduziert werden und gleichzeitig die für die Wirksubstanz bekannte therapeutische Effektivität nachgewiesen werden kann. Für die verminderte Reaktion der Kontaktgewebe sind dabei verschiedene Eigenschaften der beschriebenen Substanzen verantwortlich, wie geringere allgemeine Toxizität oder die selektive Freisetzung der Wirkform.

Wir halten eine breite klinische Prüfung dieser Verbindungen für aussichtsreich. Anderseits sollte auch bedacht werden, daß dieser Weg bzw. diese Verbindungen auch potentielle Antimykotika für eine systemische Anwendung darstellen könnten.

Tabelle 2
Biometrische Ergebnisse und autoradiographische Befunde an ^{3}H-Thymidin-markierter Meerschweinchenepidermis nach Kontakt mit 0,06 M ZIK 1392 'n Emulsion (li = *labeling index*; DNS-Si = *DNS-Synthese index*)

	Kontaktzeit in Wochen			
	Kontrollen	1	Kontrollen	4
Biometrie				
N	42	22	30	26
Epidermis-breite [µm]	56,31 ± 10,62	61,93 ± 14,29	60,16 ± 17,69	50,19 ± 9,47
t	—	1,7811	—	2,5703
2 P	—	>0,05	—	0,02
Akanthose-faktor	1,00	1,09	1,00	0,84
Zellgröße [µm²]	160,43 ± 30,26	157,23 ± 36,28	161,06 ± 47,36	153,81 ± 29,01
t	—	0,3750	—	0,6777
2 P	—	>0,10	—	>0,10
Autoradiographie (Str. basale)				
Zellkerne gesamt	5500	4000	6000	8000
Zellkerne markiert	296	248	181	250
li [%]	5,38	6,20	3,02	3,13
DNS-Si	1,00	1,15	1,00	1,04
χ^2	—	2,722	—	0,101
α	—	0,05	—	>0,10

Tabelle 3
Bestimmung der LD_{50} von 5-Bromsalizyl-4'-chloranilid (BSCA) und ZIK nach i. p. Injektion bei der Maus (Stamm C 57 Bl 6/Jena)

	LD_{50} [mg/kg]	Grenzen [mg/kg]
BSCA	40,1	34,1 47,1
ZIK	261,7	241,8 283,1

Tabelle 4
Pilotstudie zur klinischen Wirksamkeit bei Hautpilzerkrankungen

ZIK 1392 (2,8% ≙ 0,06 M in O/W-Emulsion) im Vergleich zu
BSCA (2% ≙ 0,06 M in O/W-Emulsion)

7 Patienten: 5 × Interdigitalmykose der Füße;
1 × Mykose der Füße;
1 × Oberflächliche Trichophytie der Leistenregion

Anwendung im Halbseitenversuch über 14 Tage:
jeweils gute Besserung des Befundes ohne Seitenunterschied.

Diskussion

ANDRAS: Haben Sie Angaben über die Bindung des Salicylanilids an Protein in vivo? Ich habe Angaben, daß eine solche vorhanden ist.

WOHLRAB: Die Bindung von 5-Bromsalizyl-4'-chloranilid an Proteine ist bekannt. Für das 5-Bromsalizyl-4'-chloranilid-0-β-D-xylopyranosid ist diese Frage noch nicht geprüft.

ANDRAS: Welche Organismen haben die behandelten interdigitalen Mykosen verursacht, waren es nur Trichophyten oder auch Begleitorganismen wie *Candida*?

WOHLRAB: Bisher wurden mit dieser Substanz nur Mykosen, durch keratinophile Pilze verursacht, behandelt. Weitere klinischen Prüfungen auf andere Erreger sind in Arbeit.

BERTRAM: Werden durch die Glykosidierung des BSCA auch die photoallergischen Nebenwirkungen der Substanz beeinflußt?

WOHLRAB: Diese Frage ist für ZIK 1392 noch nicht geprüft. Wir erwarten allerdings aufgrund des Wirkungsmechanismus dieser Verbindung als Transportform durch selektive Wirkstoff-Freisetzung auch eine Verminderung der photoallergischen Nebenwirkungen.

Literatur

[1] BURRY, I. N., Arch. Dermat. 95 (1967), 287–291

[2] CHRISTOPHERS, E., and E. B. LAURENCE, Curr. Probl. Dermatol. 6 (1976), 87–106

[3] FLAXMAN, B. A., J. Invest. Dermat. 67 (1976), 8–14

[4] HARBER, L. C., and R. L. BAER, J. Invest. Dermat. 58 (1972), 327–342

[5] HARDMEIER, T., U. W. SCHNYDER, R. OETTERLI und D. NEUMANN, Therapeut. Umschau 26 (1969), 97–100

[6] KRAUSHAAR, A., Arzneimittel-Forsch. 4 (1954), 548–551

[7] MEYER-ROHN, J., Dtsch. med. Wschr. 79 (1954), 1297–1299

[8] MISCHER, P., Wien. klin. Wschr. 89 (1977), 188–192

[9] SCHAEFER, H., G. STÜTTGEN, A. ZESCH, W. SCHALLA and I. GAZITH, Curr. Probl. Dermatol. 7 (1979), 80–94

[10] SCHWABE, K., B. TSCHIERSCH, W. WOHLRAB und I. REDSLOB, Verfahren zur Herstellung von 5-Bromsalizyl-4'-chloranilid-O-glykosiden, Patent C 07 H/214 030

[11] STÜTTGEN, G. und H. SCHAEFER, Funktionelle Dermatologie, Springer Verlag, Berlin 1974

[12] TSCHIERSCH, B., K. SCHWABE, E. KLUGE und H. LYR, Verfahren zur Entwicklung von Fungiziden und anderen Schädlingsbekämpfungsmitteln, WP 130 126

[13] WOHLRAB, W., Dermatologica 155 (1977), 97–107

[14] WOHLRAB, W., Dt. Gesundh.-Wesen 34 (1979), 2462–2466

P. LANGCAKE, D. W. CARTWRIGHT and J. P. RIDE

Shell Biosciences Laboratory, Sittingbourne Research Centre, Sittingbourne, Kent ME9 8AG, UK

The Dichlorcyclopropanes and Other Fungicides with Indirect Mode of Action

The main concern of plant pathologists must ultimately be with the control of plant disease. Despite this, much attention is directed towards the killing of fungi. However, there are two other strategies that may be adopted for the chemical control of disease. Both depend on the possibilities of applying compounds that are not in themselves directly fungitoxic.

A. Lethal synthesis. Non-fungitoxic compounds may be applied to plants. They may then be converted to fungitoxic compounds that are responsible for disease control.

B. Action against the host-parasite association. The host-parasite relationship is a unique biochemical entity and may therefore be selectively inhibited by chemicals for disease control. Two aspects of this may be studied. First, the various processes that are involved in pathogenesis may be examined with a view to interfering with them chemically. Secondly, the nature of natural disease resistance mechanisms (that exist in all plants [31]) may also be considered to see how these mechanisms might be made effective in circumstances where they are not normally effective i.e. when the plant is susceptible).

Some selected examples of the different types of plant chemotherapeutants that may act indirectly are given in the following sections. Various aspects of this subject have been reviewed previously [21, 26, 30, 49].

Lethal Synthesis

The idea of masking the active site of a biocidal compound so that it is only unmasked at the sites where its activity is required is an old one. An important group of chemotherapeutants, the sulphanilamide drugs, were discovered in this way. Examination of a large number of dyes that it was thought might selectively stain and kill bacteria in living animal tissue, led to the discovery of the antibacterial properties of prontosil.

Prontosil itself is not active against bacteria. Its activity is due to the release of sulphanilamide moiety which is antibacterial [3].

Pyrazophos is a fungicide active against powdery mildews and several other fungi in vitro. DE WAARD [58, 59] has shown that this compound is rapidly converted to a metabolite, 6-ethoxycarbonyl-2-hydroxy-5-methyl-pyrazolo(1,5-a)pyrimidine (PP), in mycelium of *Piricularia oryzae*, a fungus that is highly sensitive to the fungicide. PP is highly fungitoxic to a range of fungi in vitro. In the insensitive fungi, *Pythium debaryanum* and *Sacharomyces cerevisiae*, this metabolite is not formed and this may explain their lack of sensitivity. Similar results were obtained by GEORGOPOULOS et al. [24].

Recently, GASZTONYI and JOSEPOVITS have claimed that antifungal activity of triadimefon requires an activation step to give triadimenol. This only occurs in sensitive fungi. Insensitive fungi are susceptible to triadimenol [23].

Action against the Host-Parasite Association

Inhibition of fungal virulence factors. In order to be pathogenic, a fungus must have several metabolic capabilities in addition to those required for growth in culture. It must have means of breaching the external defences of the plant. MAITI and KOLATTUKUDY [36] have shown disease control with compounds that inhibit cutinase activity of a fungal pathogen that normally enters the plant by penetration of the cuticle. They used specific antisera prepared against fungal cutinase. These prevented infection of peas by the fungus *Fusarium solani*. Moreover, they suggested that this was of more widespread importance amongst plant diseases since there is evidence that the cuticle extends around the cells of the mesophyll of plant leaves [50]. Cuticular penetration may also be important therefore for pathogens that enter plants through natural openings.

Once inside the plant, the pathogen must avoid or counteract the host defence mechanisms. It must obtain nutrients for its further growth and usually, although not essentially, it produces metabolic disturbances in the host that are evident as the disease symptoms. Two important types of virulence factor involved in producing these metabolic disturbances are phytotoxins and enzymes. Compounds for disease control could affect the production or action of these factors. The success of this approach depends on many things not least of which is that the virulence factor should be the key compound determining virulence.

An apparently specific antitoxic effect concerns the phytotoxin of *Helminthosporium sacchari*, namely helminthosporide (suggested structure 2-hydroxy-cyclopropyl-α-D-galactopyranoside). This toxin is exclusively responsible for the disease symptoms of eyespot in sugar cane. It binds to a protein present in membranes of susceptible clones of sugar cane. The comparable protein in resistant clones of cane does not bind the toxin. The resultant insensitivity of these clones to the toxin explains the resistance to the disease. α-Galactosides such as raffinose, mellibiose and methyl-α-galactoside, which can be considered as analogs of the toxin, compete with the toxin for the binding site. Application of these compounds to leaves inhibits symptom development in response to the toxin [53].

Kinetin has therapeutic activity on the alternariosis disease to tobacco [34]. The fungus produces various chlorosis-inducing toxins and both these, and natural senescence, predispose the tissue to the toxic action of pectic enzymes produced by the fungus. The action of kinetin in delaying this predisposition may be related to its ability to suppress the disease.

The simple phenol catechol is not inhibitory to the growth of *Fusarium oxysporum*. Repeated applications of catechol to tomato and eggplants via the soil maintained plants infected wirh *F. oxysporum* free from disease symptoms for four months. CHET et al. [17] provided evidence that this was due to an effect on virulence of the pathogen. Growth of the fungus in treated plants was not reduced despite the reduction in symptom development. When added to cultures, catechol prevented the fungus from producing wilt inducing materials. It did not inactivate the 'toxins' once formed, nor did it inhibit their toxic activity in the plant.

Several extracellular fungal enzymes are also involved in pathogenesis. From a large number of compounds examined, GROSSMANN and coworkers [25] identified rufianic acid as an inhibitor of fungal pectolytic acid cellulolytic enzymes in vitro. It gave some control of *Rhizoctonia* on bean [37] and considerable effects against the *Fusarium* and *Verticillium* wilts of tomato. The results suggested that the disease protectant effects were due to inhibition of pectic enzymes in vivo but other explanations for disease control cannot be ruled out.

Many naturally occuring phenolic compounds inhibit pectolytic enzyme activity and may contribute to natural disease resistance (4, 12). CARRASCO et al. [13] found a substantial reduction in the colonization of tomato plants by *Fusarium oxysporum* following treatment of the roots with the non-antifungal compounds quinic acid and phenylalanine. They attributed this effect to an enhanced level of the normal endogenous pool of phenolics, but did not provide evidence as to how the phenols promoted resistance. Phenolic compounds, particularly in their oxidised form, may directly inhibit fungal growth as well as extracellular enzyme activity [35].

Alterations of host plant physiology. For many plant diseases a healthy physiological condition of the host is necessary for optimum disease development. Chemicals that cause certain changes in plant physiology can often cause a reduction in the disease that subsequently occurs. Many of the compounds that do this have plant growth regulatory activity, e.g. auxins. Most od them have effects on wilt diseases in particular.

Inhibitors of auxin transport such as the morphactins [10] and related derivatives of fluorene, phenanthrene and xanthene [11] are not antifungal but are effective at controlling *Fusarium* and *Verticillium* wilts of tomato. Several plant growth retardants also control wilt diseases. They include pydanon [9], chlormequat and 1,1-dimethyl piperidium chloride [22], chlorocholine chloride (CCC), N-dimethyl-(α-chloroethyl)-hydrazonium chloride, phosfon D and N-dimethyl aminosuccinamic acid [7]. The mechanism of action of these compounds is not clear, although in some cases, physical obstruction of the vessels, e.g. by tyloses, is implicated [48].

BUCHENAUER and ERWIN [8] found that the growth retardants, N,N-dimethyl pyrrolidinium iodide and N,N-dimethyl piperidinium iodide, reduced the severity of *Verticillium* wilt of cotton and promoted the formation of antifungal compounds within the plant. One of these was probably hemigossypol, a phytoalexin-like compound previously identified in cotton. Its production was promoted by the retardants in the absence of infection.

Rather fewer examples are known of growth regulating compounds that alter resistance to diseases other than wilts. Kinetin is active against several powdery mildews [20]. CCC is known to reduce the severity of eyespot of wheat although susceptibility to other wheat diseases can be increased at the same time [6]. A reduction in brown spot of tobacco has been noted following treatment with maleic hydrazide [42] while VAN ANDEL reported the effects of several compounds against cucumber scab, among them CCC, the activity of which was reversed by the simultaneous application of gibberellin [2]. BUCHENAUER and GROSSMANN [11] found that some of their fluorene derivatives controlled late blight of tomato. Compounds that controlled the wilt diseases were not always active against *Ph. infestans*.

Ethylene has multiple effects on plant physiology and it can both increase and decrease the resistance of plants to disease [39]. Several types of changes in host plant physiology that may affect disease development may be brought about by ethylene [1]. PEGG [40] showed that gassing of tomato plants with ethylene increased the resistance to subsequent infection with *Verticillium albo-atrum* and RETIG [44] found induced resistance to *Fusarium* wilt with the ethylene generator Ethephon.

Stimulation of host resistance mechanisms. The distinction between compounds that control disease by alteration of host plant physiology and those that act through specific resistance mechanisms in the host is an arbitrary one. Two compounds seem particularly appropriate to this section, these being phenylthiourea and the dichlorocyclopropanes.

Phenylthiourea (PTU) gives complete protection of cucumber plants against *Cladosporium cucumerinum* infections of the hypocotyls. For complete inhibition of growth in vitro a concentration of 500 ppm was required although the concentration of PTU in the hypocotyl sap was estimated to be around 10—20 ppm [47]. Since PTU did not give rise to fungitoxic derivatives in the plant, nor did it inhibit pectolytic enzymes of the fungus, an indirect effect through host plant resistance mechanisms was sought. Three effects of PTU were found. First, polyphenoloxidase activity of the plant was eliminated. Secondly, PTU treatment of the plants stimulated peroxidase activity although the compound is without effect on peroxidase in vitrol Thirdly, infection of PTU-treated plants resulted in a strong lignification reaction, detected histochemically, around the sites of attempted penetration by the fungus. This localised lignification response is known to be a resistant response of cucumber cultivars that have genetically determined resistance to scab [27] and it occurs as a resistant reaction in other plants [46]. It may be significant that phytoalexin production as a natural resistant response of cucumber has not been demonstrated as yet and lignification may be an alternative, but equally effective, post-infectional defence mechanism.

SIJPESTEIJN [47] has suggested that inhibition of polyphenoloxidase may take more phenolic compounds available for peroxidase action. These phenols may then be diverted to lignin, a process known to involve peroxidase. Lignin may form a physical barrier around the infection site and it could increase the resistance of the tissue to attack by fungal enzymes [45].

The dichlorocylopropanes (DCPs) are a group of compounds with systemic activity against the blast rice disease. Although a wide range of esters are active, most studies have been conducted with WL 28325 (2,2-dichloro-3,3-dimethyl cyclopropane carboxylic acid). WL 28325 acts as a protectant rather than a therapeutant and is best applied via the roots, rather than to the leaves. Disease protectant effects appear to be highly specific for the rice blast disease. The possibility that WL 28325 has an indirect mode of action was first raised during studies on the effects of the compound on *Piricularia oryzae* in vitro [32]. Although growth of the fungus was inhibited, this was not considered to be relevant to the mode of action of the compound in plants because (a) the effects were weak at the concentrations (ca. 20 ppm) estimated to be present in protected leaves, (b) the effects were not restricted to *P. oryzae*, several other fungi had similar sensitivity despite the specificity of WL 28325 in vivo, and (c) there was no correlation between the antifungal effects against *P. oryzae* of a range of DCPs in vitro and in vivo.

There was no evidence for the conversion of WL 28325 to a fungitoxic derivative in the plant.

On treated plants, *P. oryzae* caused small brown hypersensitive-type lesions reminiscent of those on genetically resistant cultivars of rice [33]. Light and electron microscope studies confirmed this [16]. Thus there were no effects on spore germination, appressorium formation or the penetration of epidermal cells. Subsequently, gross disorganisation of penetrated cells occurred and there was no futher development of the fungus. These degenerative changes (evident macroscopically as browning)in host cells did not occur so rapidly in untreated plants and growth of the fungus was not arrested.

Treatment of rice plants with WL 28325 led to enhancement of peroxidase activity and further increases occurred following infection. Peroxidase activity of untreated plants also increased following infection but levels did not reach those in treated, inoculated plants [14, 33]. Spectroscopic measurements of the accumulation of brown pigments (melanin) associated with the hypersensitive response confirmed that these accumulated more rapidly in treated, infected plants than in non-treated, infected plants [33]. In addition, compounds inhibitory to *P. oryzae* accumulated in infection droplets on treated plants. Infection droplets from untreated plants were considerably less fungitoxic while droplets from comparable non-infected plants, treated or untreated, had no fungitoxicity [15]. These fungitoxic compounds, of which there appeared to be two, had some of the characteristics of phytoalexins and indeed there have been several reports of phytoalexin activity in rice [57].

In order to characterise the phytoalexins, a more convenient source of material was required and it was found that irradiation of dark-grown coleoptiles with a particularly strong source of UV light induced the formation of antifungal compounds with similar chromatographic properties to those induced in infected leaves. The two phytoalexins were isolated from UV-irradiated coleoptiles and characterised chemically by ^{1}H NMR, Mass (EI and FD ionisation), IR and UV spectroscopy as the momilatones A and B [15]. These compounds had been identified previously as plant growth regulants in rice husks [28]. Samples of momilactones A and B prepared from UV-irradiated coleoptiles and authentic momilactones were highly toxic to *P. oryzae*, momilactone B (ED_{50} ca 1 μg/ml) being appreciably more active than A (ED_{50} ca. 5 μg/ml). With the aid of combined GLC-MS it has been established that the same compounds accumulate in infection droplets on treated and untreated leaves.

Identification of the antifungal compounds as the momilactones permitted the development of a GLC technique for their measurement in and around infected tissue. This showed that infection of treated tissue caused a rapid increase in the concentrations of momilactones A and B from the undetectable levels in healthy leaves. These increases were first detected at about 20 h after inoculation. This approximates to the time at which epidermal cells are first penetrated. The compounds accumulated subsequantly with maximum levels of both compounds occuring at about 48 h and this coincided with the appearance of the macroscopically visible hypersensitive browning response. Momilactone A reached 30 μg/g fresh wt. while momilactone B reached

14 µg/g fresh wt. These concentrations are several times in excess of the ED_{50} values against germ tube growth of *P. oryzae*. Accumulation of momilactone B in untreated, infected tissue did not occur during the 60 h time period studied although a small accumulation of momilactone A (ca. 15 µg/g by 60 h) was detected. An interesting point is that in treated infected leaves, both momilactones A and B also accumulated in the apparently healthy tissue immediately surrounding the infection droplets [16]. These antibiotic substances were therefore present in advance of the pathogen, and were well placed to inhibit further fungal growth.

There is evidence that WL 28325 caused changes in the host plant, rather than in the fungus. First, application of picolinic acid to wounds on treated plants causes a hypersensitive-type browning response similar to that seen in infected, treated wounds. Picolinic acid causes a much less intense response in untreated plants [33]. Picolinic acid is reputed to be one of the toxins produced by *P. oryzae* [55]. Secondly, in preliminary experiments, it was found that greater momilactone accumulation occurred in response to mycelial extracts of *P. oryzae* in treated as compared with untreated plants [14]. The third piece of evidence, albeit indirect, is that WL 28325 is most effective when applied via the roots and before infection. Thus, application to the leaves at the time of infection is less effective for disease control. This suggests that some prior interaction of WL 28325 with the plant is necessary for activity.

It is clear then that WL 28325 modifies the plant/fungus interaction such that it behaves in a way similar to that of a genetically resistant cultivar. This resistant response involves a hypersenstive type of melanisation reaction that could be responsible for restricting fungal growth in much the same way as does lignification. It also involves the production of phytoalexin-like compounds (although it has not been shown that these same compounds are produced in plants with natural resistance). The way in which these alterations in plant reaction to the fungus are brought about has not been determined.

Contribution of Host Defence Mechanisms to the Effectiveness of Fungicides

There are some chemicals that have potent antifungal effects in vitro yet which may also cause responses in the host reminiscent of resistant responses. It is difficult to determine the contribution that these indirect mechanisms may make to the overall ability of the compound to control disease. In some case it seems likely that the resistant type response is a consequence, rather than the cause, of the direct inhibition of fungal growth [cf 29].

The oxathiins (e.g. oxycarboxin) and related compounds are known to have potent antifungal activity through effects on fungal succinic acid dehydrogenase [64]. Nevertheles, PRING and RICHMOND [41] found that the infection of oxycarboxin-treated plants by the rust, *Uromyces phaseloi*, was not affected until after haustorial formation. The subsequent ultrastructural changes closely resembled a form of resistance reaction in plants that are not susceptible to rust infection. They implied that host resistance mechanisms may contribute to the disease control afforded by oxycarboxin.

The acylalanines are recently introduced systemic fungicides for control of Oomycete diseases of various crops. Ridomil, an example of this class, inhibits fungal growth in the plant (lettuce, vines or tomato) only after the formation of the first haustoria [18,

19, 51]. Although ridomil has antifungal activity in vitro [52], CRUTE [19] reported that its effects in vivo were more pronounced. The cessation of fungal growth in lettuce was associated with host cell necrosis identical in appearance with the response of non-hosts or resistant cultivars penetrated by incompatible isolates of downy mildew. The observation that strains of *Ph. infestans* resistant to acylalanines in vitro were fully susceptible (or avirulent) in vivo [52] further suggests the involvement of indirect mechanisms in the mode of action of this group.

REILLY and KLARMAN [43] found that several fungicides promoted phytoalexin formation when applied to soybean hypocotyls in the absence of infecting fungi. Benomyl showed good phytoalexin elicitation but interestingly, its fungitoxic breakdown product, MBC, was inactive. The other (non-fungitoxic) breakdown product, butylamine, was an effective phytoalexin elicitor. Benomyl is reported to depress tissue levels of phenylalanine ammonia lyase [54] an enzyme involved in phenolic biosynthesis and therefore of possible importance in disease resistance. It was suggested that this effect of benomyl may account for certain instances of enhanced disease susceptibility in treated plants.

Miscellaneous Compounds with Indirect Action

Probenazole (Oryzemate, 3-allyloxy-1,2-benzisothiazole-1,1-dioxide) is a few fungicide for rice blast control [60]. Benzisothiazole-1,1-dioxide (saccharin) and its N-glucoside are the major compounds found in probenazole-treated rice plants and are believed to be the active principles [56]. Probenazole and its breakdown products were not inhibitory to the growth on agar of several species of bacteria and fungi, among them *P. oryzae* [61]. The fungicide showed different effects on various stages of the infection cycle according to whether or not the host plant was present and the way in which it was applied to the plant (through the roots or directly to conidia). Nevertheless, although probenazole strongly inhibited appressorium formation on and penetration of cellophane membranes by *P. oryzae*, other effects, particularly those an hyphal growth, were much more pronounced in the presence of host tissue. This suggested the mediation of the host plant in the activity. Blast lesions on treated plants were very similar to the small brown hypersensitive lesions seen on genetically resistant cultivars. Treated plants even in the absence of the pathogen had increased peroxidase activity (cf. phenylthiourea and the dichlorocyclopropanes) and contained a fungitoxic factor not present in untreated plants [62]. Elucidation of the chemical nature of this fungitoxic compound will be an important step in the understanding of the mode of action of this fungicide.

The systemic fungicide indar (4-n-butyl-1,2,4-triazole) is among the most specific of fungicides known. Its activity is restricted to control of brown rust of wheat *(Puccinia recondita)*, having no activity even against stem rust of this host [38]. While this high degree of specificity might suggest the involvement of specific resistance mechanisms of the host, an ultrastructural study by WATKINS et al. [63] concluded that it acts through effects on the fungus rather than through host defence mechanisms.

Conclusions

Despite the wide range of compounds that may have indirect action, few, if any, have found extensive practical use so far. It may be questioned therefore whether such compounds have inherent disadvantages compared with those that act directly. The use of fungicides with indirect action should avoid the ecological and toxicological problems of spraying highly active biocidal compounds over large areas of agricultural land. With the exception of those of the lethal synthesis type, such fungicides should not continually expose pathogen populations to the selection pressures that promote the occurrence of resistant strains. In addition, indirect fungicides should not be detrimental to the host as opposed to the pathogen. This is a necessary prerequisite for systemic and therapeutic compounds that must penetrate plant tissue. However, it is probably too early to answer the question. The fungicide industry has not reached maturity as yet, systemic fungicides have been with us for little over ten years and even now the market is still dominated by long-established non-systemic compounds [5]. It may be significant that the agrochemicals industry is discovering compounds that may act indirectly, and others whose activity may be supplemented by indirect mechanisms. The wide use of in vivo screens indicates that few workers are prepared to ignore the contribution of indirect mechanisms to fungicidal effectiveness.

Diskussion

DAVIDSE: Does *Piricularia oryzae* metabolise the momilactones?

LANGCAKE: We have not examined this. It would seem likely that *Piricularia oryzae* is able to degrade the momilactones at least to some extent.

DAVIDSE: What is the significance of your observation that the fungal toxin picolinic acid enhances the browning reaction caused by your compound to the inhibitory effect?

LANGCAKE: We have not carried out any experiments to examine this directly. However, the effect of picolinic acid on treated and untreated plants does indicate that WL 28325 causes a change in the host plant. It may also indicate that picolinic acid plays a role in the normal compatible interaction between *Piricularia* and the rice plant.

SISLER: I believe your results favor an initial action of the fungicide on the fungus Since it does not control other fungal diseases of the rice plant, it seems to me the specificity arises from an initial action on the pathogen *Pyricularia oryzae.*

LANGCAKE: I would certainly not exclude the possibility that WL 28325 has effects on the fungus although as I have shown, there are also effects on the host in the absence of the pathogen. The specificity of WL 28325 could arise from a specific effect on the fungus, as you have implied for tricyclazole. However, the specificity could also arise by an interaction with the normal specificity-determining mechanisms of the plant. We cannot distinguish between these possibilities as yet.

TÜSKE: Es ist bekannt, daß einige Phytoalexine antibakterielle Aktivität besitzen. Dr. LANGCAKE, ist Ihnen bekannt, ob die von Ihnen gefundenen Phytoalexine Momilacton A und B gegen phytopathogene Bakterien wirksam sind?

LANGCAKE: We have not tested the momilactones against the growth of bacteria and I do not know of any other workers who have looked at this. Other terpenoid phytoalexins, such as rishitin from potato, are known to have antibacterial activity in vitro.

SCHMALFUHS: WL 28325 ist chemisch den Säuren der praktisch angewandten Pyrethroide ähnlich. Wurden weitere Pyrethroidsäuren auf antifungale Wirkung untersucht?

LANGCAKE: Pyrethroidsäuren haben eine andere Struktur. Ein Wirkungszusammenhang mit Pyrethroiden besteht nicht.

SCHMALFUHS: Wurden als Metabolite von WL 28325 Produkte einer Ringöffnungsreaktion beobachtet?

LANGCAKE: Im alkalischen Medium erfolgt Ringöffnung leicht.

PFLIEGEL: The WL compound is obviously a new acid intermediate for synthetic pyrethroid candidates. Have perspective insecticides been found on the basis of this intermediate?

LANGCAKE: As far as I can remember, we have not developed any pyrethroid type insecticides based on the cyclopropane carboxylic acid, WL 28325.

LYR: What is the reason that you get in vitro on some media a fungicidal effect and on others not?

LANGCAKE: Antifungal activity is only seen on simple defined growth media, e.g. CZAPEK DOX. Addition of nutrients (amino acids vitamins) partially reverse inhibition. White yeast extract (for example) will completely reverse inhibition of growth.

LYR: Does mechanical wounding induce lactones in rice?

LANGCAKE: No. The gentle wounding that we used for inoculating plants does not induce momilactone formation.

LYR: Is it only a question of screening, that you and others found indirect fungicidal effects just in the combination rice/*Pyricularia*?

LANGCAKE: In my opinion it is only coincidence that these indirect effects have been found on rice. Other compounds are new being recognized as possibly acting indirectly against diseases of other plants.

LYR: Would you expect, that your substance is effective in other host-parasite combinations too?

LANGCAKE: As far as we can tell, WL 28325 is highly specific for the rice blast disease.

LYR: Have you tested a combination of your substance with tricyclazole? Is there any synergistic effect which could decrease the problem of resistance?

LANGCAKE: We have not tested this. There may already be better compounds in the field that could be used in a mixture with tricyclazole.

References

[1] ABELES, F. B., 'Ethylene in Plant Biology', Academic Press, N. York, 1973

[2] VAN ANDEL, O. M., Mode of action L-threo-β-phenylserine as a chemotherapeutant of cucumber scab. Nature **211** (1966), 326

[3] ARIENS, E. J., Modulation of pharmacokinetics by molecular manipulation. in ARIENS, E. J. (ed.) 'Drug Design, Vol II.', Academic Press, N. York and London 1971, p. 77

[4] BACHMANN, O. and R. BLAICH, Occurrence and properties of condensed tannins in Vitaceae, Vitis, **18** (1979), 106–116

[5] BENT, K. J., Fungicides in perspective, Endeavour, N. S., **3** (1979), 7–14

[6] BOCKMANN, H., Phytopathological aspects of chlorocholine chloride application. Euphytica, **17**, Suppl. (1968), 271–274

[7] BUCHENAUER, H., Einfluß einiger Wuchshemmstoffe (CCC, CMH, Amo 1618, Phosfon D, B-995) auf die *Fusarium* und *Verticillium*welke der Tomate, Phytopathol. Z., **72** (1971), 53–66

[8] BUCHENAUER, H. and D. C. ERWIN, Reduction of the severity of *Verticillium* wilt *of* cotton by the growth retardants N,N-dimethyl pyrrolidinium iodide and N,N-dimethyl piperidinium iodide associated with the production of antifungal compounds, Z. Pflkrankh. Pflschutz., **80** (1973), 576–586

[9] BUCHENAUER, H. and D. C. ERWIN, Effect of the plant growth regulator pydanon on *Verticillium* wilt of cotton and tomato, Phytopathol., **66** (1976), 1140–3

[10] BUCHENAUER, H. and F. GROSSMANN, Effect of morphactins on various tomato diseases, Vortr. Gesamtgeb. Bot. No. 3 (1), (1969), 149–159

[11] BUCHENAUER, H. and F. GROSSMANN, Therapeutic effects of fluorene, phenanthrene and xanthene derivatives on fungal diseases of tomato, Nature, **227** (1970), 1267–8

[12] BYRDE, R. J. W., The varietal resistance of fruits to brown rot. II. The nature of resistance in some varieties of cider apple, J. Hort. Sci., **32** (1957), 227–238

[13] CARRASCO, A., A. M. BOUDET, and G. MARIGO, Enhanced resistance of tomato plants to *Fusarium* by controlled stimulation of their natural phenolic production, Physiol. Plant Pathol., **12** (1978), 225–232

[14] CARTWRIGHT, D., Studies on the mode of action of the dichlorocyclopropane fungicides on rice blast disease, PhD thesis, University of Birmingham, UK (1978)

[15] CARTWRIGHT, D., P. LANGCAKE, R. J. PRYCE, D. P. LEWORTHY and J. P. RIDE, Chemical activation of host defence mechanisms as a basis for crop protection, Nature, **267** (1977), 511–513

[16] CARTWRIGHT, D., P. LANGCAKE and J. P. RIDE, Phytoalexin production in rice and its enhancement by dichlorocyclopropane fungicides, Physiol. Plant Pathol., in press

[17] CHET, I., D. HAVKIN and J. KATAN, The role of catechol in inhibition of *Fusarium* wilt, Phytopathol. Z., **91** (1978), 60–66

[18] COHEN, Y., M. REUVENI and H. EYAL, The systemic antifungal activity of Ridomil against *Phytophthora infestans* on tomato, Phytopathol., **69** (1979), 645–9

[19] CRUTE, I. R., Studies on new systemic fungicides active against *Bremia lactucae*, Abstracts of the Third International Congress of Plant Pathology, Munich, 1978

[20] DEKKER, J., Effect of kinetin on powdery mildew, Nature, **197** (1963), 1027–8

[21] DIMOND, A. E. and S. RICH, Effects on physiology of the host and on host/pathogen interactions. in R. W. MARSH (ed.) 'Systemic Fungicides', 2nd edition, Longman, London and N. York, 1977, pp 115–130

[22] ERWIN, D. C., Growth retardant effects on *Verticillium* wilt of cotton, Abstracts of the Third International Congress of Plant Pathology, Munich, 1978

[23] GASTONYI, M. and G. JOSEPOVITS, The activation of triadimefon and its role in the selectivity of fungicide action, Pestic. Sci., **10** (1979), 57–65

[24] GEORGOPOULOS, S. G., J. W. G. GEERLINGS and J. DEKKER, Sensitivity of *Ustilago maydis* to pyrazophos and one of its conversion products and failure to induce resistance with UV treatment, Neth. J. Plant Pathol., **81** (1975), 38–41

[25] GROSSMANN, F., Studies on the therapeutic effects of pectolytic enzyme inhibitors, Neth. J. Plant Pathol., **74** (1968), Suppl. 1, 91–103

[26] GROSSMANN, F., Chemical control of toxigenic plant diseases. in R. K. S. WOOD, A. BALLIO and A. GRANITI (eds.), 'Phytotoxins in Plant Disease', Academic Press, N. York and London, 1972, pp 347–364

[27] HIJWEGEN, T., Lignification a possible mechanism of active resistance against pathogens, Neth. J. Plant Pathol., **69** (1963), 314–317

[28] KATO, T., C. KABUTO, N. SASAKI, M. TSUNAGAWA, H. AIZAWA, K. FUJITA, Y. KATO, Y. KITAHARA and N. TAKAHASHI, Momilactones, growth inhibitors from rice, *Oryza sativa* L. Tetrahedron Letters, **39** (1973), 3861–4

[29] KIRALY, Z., B. BARNA and T. ERSEK, Hypersensitivity as a consequence ,not the cause, of plant resistance to infection, Nature, **239** (1972), 456–8

[30] KUC, J., Biochemical control of disease resistance in plants, World Rev. Pest Control, 7 (1968), 42–55

[31] KUC, J., Plant protection by the activation of latent mechanisms for resistance, Neth. J. Plant Pathol., **83** (1977), Suppl. 1, 463–471

[32] LANGCAKE, P. and S. G. A. WICKINS, Studies on the mode of action of the dichlorocyclopropane fungicides: effects of 2,2-dichloro-3,3-dimethylcyclopropane carboxylic acid on the growth of *Piricularia oryzae* Cav. J. Gen. Microbiol., 88 (1975), 295–306

[33] LANGCAKE, P. and S. G. A. WICKINS, Studies on the action of the dichlorocyclopropanes on the host-parasite relationship in the rice blast disease, Physiol. Plant Pathol., 7 (1975), 113–126

[34] LLOYD, H. L., Therapeutic effect of kinetin on tobacco alternariosis, Nature, New Biol., **240** (1972), 94–96

[35] LYR, H., On the toxicity of oxidised phenols, Phytopathol. Z., **52** (1965), 229–40

[36] MAITI, I. B. and P. E. KOLATTUKUDY, Prevention of fungal infection of plants by specific inhibition of cutinase, Science, **205** (1979), 507–508

[37] MARTIN, J. and F. GROSSMANN, Inhibition of pectic and celluloytic enzymes of *Rhizoctonia solani* and the influence of some inhibitors of the disease process. II Effect of some inhibitors in vivo, Phytopathol. Z., **75** (1972), 97–110

[38] VON MEYER, W. C., S. A. GREENFIELD and M. C. SEIDEL, Wheat leaf rust: control by 4-n-butyl-1,2,4-triazole, a systemic fungicide, Science **169** (1970), 997–998

[39] PEGG, G. F., The involvement of ethylene in plant pathogenesis. in R. HEITEFUSS and P. H. WILLIAMS (eds) 'Physiological Plant Pathology. Encyclopaedia of plant physiology, New Series, Vol 4', Springer-Verlag, Berlin, Heidelberg, N. York, 1976, pp. 582–591

[40] PEGG, G. F., The response of ethylene treated tomato plants to infection by *Verticillium albo-atrum*, Physiol. Plant Pathol., **9** (1976), 215–226

[41] PRING, R. J. and D. V. RICHMOND, An ultrastructural study of the effect of oxycarboxin on *Uromyces phaseoli* infecting leaves of *Phaseolus vulgaris*, Physiol. Plant Pathol., 8 (1976), 155–62

[42] VON RAMM, C., G. B. LUCAS and H. V. MARSHALL, The effects of maleic hydrazide on brown spot incidence of flue-cured tobacco, Tobacco Science, **6** (1962), 100

[43] REILLY, J. J. and W. L. KLARMAN, The soybean phytoalexin, hydroxyphaseollin, induced by fungicides, Phytopathol., **62** (1972), 1113–5

[44] RETIG, N., Changes in peroxidase and polyphenoloxidase associated with natural and induced resistance of tomato to *Fusarium* wilt, Phyiol. Plant Pathol., **4** (1974), 145–150

[45] RIDE, J. P., Lignification in wounded wheat leaves in response to fungi and its possible role in resistance, Physiol. Plant Pathol., **5** (1975), 125–134

[46] RIDE, J. P., The role of cell wall alterations in resistance to fungi, Ann. Appl. Biol., **89** (1978), 302–6

[47] SIJPESTEIJN, A. K., Mode of action of phenylthiourea, a therapeutic agent for cucumber scab. J. Sci. Fd. Agric., **20** (1969), 403–5

[48] SINHA, A. K. and R. K. S. WOOD, The effect of growth substances on *Verticillium wilt* of tomato, Ann. Appl. Biol., **60** (1967), 117–128

[49] SISLER, H. D., Fungicides: problems and prospects. in M. R. SIEGEL and H. D. SISLER (eds) 'Antifungal Compounds, Vol. 1'. MARCEL DEKKER, N. York, 1977, pp 531–547

[50] SITHOLEY, R. V., Observations on the three-dimensional structure of the leaf cuticle in certain plants, Ann. Bot., **35** (1971), 637–9

[51] STAUB, T., H. DAHMEN and F. SCHWINN, Effects of Ridomil on the development of target pathogens on their host plants, Abstracts of the Third International Congress of Plant Pathology, Munich, 1978

[52] STAUB, T., H. DAHMEN, P. A. URECH and F. SCHWINN, Failure to select for in vivo resistance in *Phytophthora infestans* to acylalanine fungicides, Plant Disease Reporter, **63** (1979), 385–9

[53] STROBEL, G. A., The helminthosporoside binding protein of sugar cane, J. Biol. Chem., **248** (1973), 1321–8

[54] SWINBURNE, T. R., The effect of benomyl and other fungicides on phenylalanine ammonia lyase activity in *Hordeum vulgare* and *Phaseolus vulgaris*, Physiol. Plant Pathol., **5** (1975), 81–8

[55] TAMARI, K. ,N. OGASAWARA and J. KAJI, Biochemical products of the metabolism of *Piricularia oryzae*. in 'The Rice Blast Disease', The John Hopkins Press, Baltimore, Maryland, 1965
[56] UCHIYAMA, M., H. ABE, R. SATO, M. SHIMURA and T. WATANABE, Fate of 3-allyloxy-1,2-benzoisothiazole-1,1-dioxide (Oryzemate) in rice plants, Agr. Biol. Chem., **37** (1973), 737–45
[57] UEHARA, K., On the production of phytoalexin by the host plant as a result of the interaction between the rice plant and the blast fungus (*Piricularia oryzae*). Ann. Phytopath. Soc. Japan, **23** (1958), 127–30
[58] DE WAARD, M. A., Mechanisms of action of the organophosphorous fungicide, pyrazophos, Meded. Landb. Hogesch. Wageningen, Nederland, **74** (1974), 14
[59] DE WAARD, M. A., Mode of action of the organophosphorous fungicide pyrazophos. in H. LYR and C. POLTER (eds) 'Systemfungizide', Intern. Symp. Reinhardsbrunn. Akademie-Verlag Berlin 1975, p 197
[60] WATANABE, T., H. IGARASHI, K. MATSUMOTO, S. SEKI, S. MASE and Y. SEKIZAWA, The characteristics of probenazole (Oryzemate) for the control of rice blast, J. Pest. Sci., **2** (1977), 291–6
[61] WATANABE, T., Effects of probenazole (Oryzemate) on each stage of the rice blast fungus (*Piricularia oryzae* CAV) in its life cycle, J. Pest. Sci., **2** (1977), 395–404
[62] WATANABE, T., Y. SEKIZAWA, M. SHIMURA, Y. SUZUKI, K. MATSUMOTO, M. IWATA and S. MASE, Effects of probenazole (Oryzemate) on rice plants with reference to controlling rice blast, J. Pest. Sci., **4** (1979), 53–9
[63] WATKINS, J. E., L. J. LITTLEFIELD and G. D. STATLER, Effect of the systemic fungicide 4-n-butyl-1,2,4-triazole on the development of *Puccinia recondita* f. sp. *tritici* on wheat. Phytopathol., **67** (1977), 985–9
[64] WHITE, G. A. and G. D. THORN, Structure-activity relationships of carboxamide fungicides and the succinic acid dehydrogenase complex of *Cryptococcus laurentii* and *Ustilago maydis*, Pest. Biochem. Physiol., **5** (1975), 380–95

Dr. P. LANGCAKE
Shell Biosciences Laboratory,
Sittingbourne Research Centre,
Sittingbourne, Kent ME9 8AG,
England.

Dr. D. W. CARTWRIGHT
Dept. Applied Biology,
Chelsea College,
Hortensia Road,
London, United Kingdon.

Dr. J. P. RIDE
Dept. of Microbiology,
The University,
Birmingham B15 2TT,
United Kingdon.

Helmut Junge, Helmut Bochow und Marthe Jacob

Sektion Gartenbau der Humboldt-Universität zu Berlin, DDR

Wirkungen von Kombinationen systemisch aktiver Substanzen auf die Pathogenese von Pflanzenerkrankungen

Neben dem Interesse, das den Wirkungsmöglichkeiten und -mechanismen einzelner systemisch aktiver Pflanzenschutzmittel zukommen muß, verlangt auch zunehmend die Frage an Aufmerksamkeit, inwieweit sich durch eine kombinierte Anwendung systemisch aktiver Substanzen neue Situationen zur Einflußnahme auf eine Phytopathogenese ergeben.

Erstens erfordert die Praxis des Pflanzenschutzes aufgrund parallelen Auftretens von Schaderregern, in Anbetracht von Komplexkrankheiten und durch verschiedene Einflußnahmen und biologische Prozesse, wie der Wachstumsregulierung und pathologischer Vorgänge, die Notwendigkeit des parallelen oder gemeinsamen Einsatzes von Wirkstoffen. Zweitens läßt sich gerade bei Substanzen, die über das System der Pflanzen zur Wirkung kommen, theoretisch an eine Reihe von Wechselwirkungen bei kombinierter Anwendung denken, die es im positiven Falle für Erweiterungen der Wirkungsphäre und in negativen Fällen zur Fixierung neuer Grenzen, zu studieren gilt.

Ausgehend von Erkenntnissen aus den Wirkungsmechanismus gegenwärtig bereits näher bekannter systemisch aktiver Verbindungen ergeben sich dabei folgende Möglichkeiten für wissenswerte Interaktionen bei kombiniertem Einsatz hinsichtlich Auswirkungen auf die Pathogenese von Pflanzenkrankheitserregern (Junge [3]).

- Eine gegenseitige Beeinflussung bei Aufnahme und Transport der Wirkstoffe in der Pflanze, insbesondere von Veränderungen der Transportgeschwindigkeit.
- Veränderungen des Resistenzverhaltens der Pflanzen durch variierte Wirkungen auf Stoffwechselprozesse der Pflanze bzw. Wirkungen nicht toxischer Komponenten auf die Pathogene und
- Veränderungen in der Aktivität der Wirkstoffe durch deren chemostrukturellen Umbau in der Pflanze oder die Beeinflussung der Affinität in den Rezeptorregionen der Organismen.

Wir wandten uns dem Studium dieser Fragen zu (Junge [3]) unter Verwendung zunächst bekannter systemischer Pflanzenschutzmittel, vornehmlich Fungizide, und Wachstumsregulatoren, in kombiniertem Einsatz. Als Untersuchungsobjekt wurde die Pathogenese systemisch eine Pflanze befallender Erreger, als hauptsächlichste Zielobjekte innertherapeutischer Maßnahmen, herangezogen.

Aus dem Kreis der Untersuchungen sollen im folgenden einige Erfahrungen über das Auftreten von Wechselwirkungen beim kombiniertem Einsatz von systemischen Fungiziden und Wachstumsregulatoren an *Antirrhinum*-Pflanzen bei einem Befall durch

Verticillium albo-atrum RKE. et BERTH. dargestellt werden, wobei gleichzeitig den Mechanismen solcher Wechselwirkungen nachgegangen wurde.

Im Ergebnis mehrerer Versuche ergab sich zunächst nach Wundinfektion ca. 62 Tage alter *Antirrhinum*-Pflanzen ein etwa 95%iger *Verticillium*-Befall, der in Abhängigkeit von der Jahreszeit zu Welkeerscheinungen führte, die nach etwa 30 Tagen durch einen Welkeindex zwischen 50 im Frühjahr und 80 im Spätsommer gekennzeichnet waren.

Die Anwendung von Benomyl (500 ppm) 5 Tage prä- oder 7 Tage postinfektionell im Gießverfahren erbrachte eine fast vollständige Unterdrückung von Welke und Befall der Pflanzen. Dagegen erwies sich Thiophynat-Methyl mit 715 ppm nur bei präinfektioneller Applikation als wirksam, nicht aber bei postinfektioneller Behandlung, wo Welke und Befall nur gering unter der künstlich infizierten Kontrolle blieben. Auch Triforin, das mit 285 ppm postinfektionell durch Spritzen der Pflanzen appliziert wurde, konnte nur einen geringen Bekämpfungserfolg erreichen.

Gießbehandlungen mit Fenaminosulf (210 ppm) oder Prothiocarb (1000 ppm) verminderten den Befall der Pflanzen nicht, die Welkesymptome wurden jedoch teilweise erheblich verstärkt.

Die postinfektionelle Anwendung (1. Tag) der Wachstumsregulatoren im Spritzverfahren hatte neben der Beeinflußung des Pflanzenwachstums auch eine Veränderung des Krankheitsbefalles zur Folge. So wurde durch Daminozid in sehr hoher Dosierung (42500 ppm) vorwiegend bei Versuchen im Spätsommer und Herbst die Welke der Pflanzen deutlich reduziert, während beim Erregernachweis in den Pflanzen gegenüber

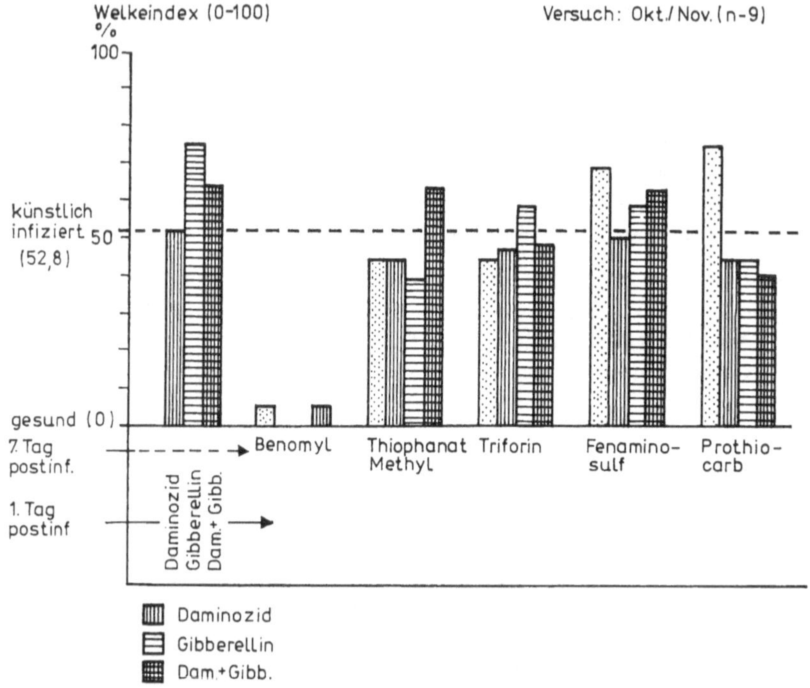

Abb. 1. Wirkung alleiniger und kombinierter Anwendung von Systemfungiziden und Wachstumsregulatoren auf die Ausprägung der Welkesymptome an *Antirrhinum*-Pflanzen nach künstlicher Infektion durch *Verticillium albo-atrum*

der Kontrolle nur geringe Verminderungen zu finden waren. Die Einschränkung der Welkesymptome war dabei im Frühjahr weniger ausgeprägt. Gibberellin (3,5 ppm) verstärkte dagegen in den Frühjahrs- und Winterversuchen das Krankheitsbild auffällig und hatte in Spätsommerversuchen nur gering welkemindernde Tendenzen. Auch hierbei blieb der alternative Befallsnachweis unbeeinflußt. Die Kombination der Wachstumsregulatoren Daminozid und Gibberellin hatte im allgemeinen additive Wirkungen.

Die kombinierte Anwendung der Wachstumsregulatoren vor bzw. nach den Systemfungiziden wurde hinsichtlich auftretender Wirkungen durch Anwendung eines hier nicht näher zu beschreibenden mathematischen Modells mit additivem Charakter zu Erwartungswerten in Beziehung gesetzt, um echte Wechselwirkungen zu kennzeichnen (Junge [3]).

Wurde Benomyl nach den Wachstumsregulatoren appliziert, so blieben die auftretenden Wechselwirkungen meist gering und ausschließlich in den Spätsommerversuchen waren sichtbare Minderungen der fungiziden Wirkung zu beobachten. In den Kombinationsbehandlungen mit Thiophanat-Methyl und Triforin konnte dagegen eine teils beträchtliche Einschränkung der Fungizidwirkung festgestellt werden, die vornehmlich bei Anwendung des Systemfungizides in Kombination mit Gibberellin auftrat. Hier waren stets die Krankheitsausprägung und oft noch stärker der Befallsnachweis verändert (Abb. 1, Abb. 2).

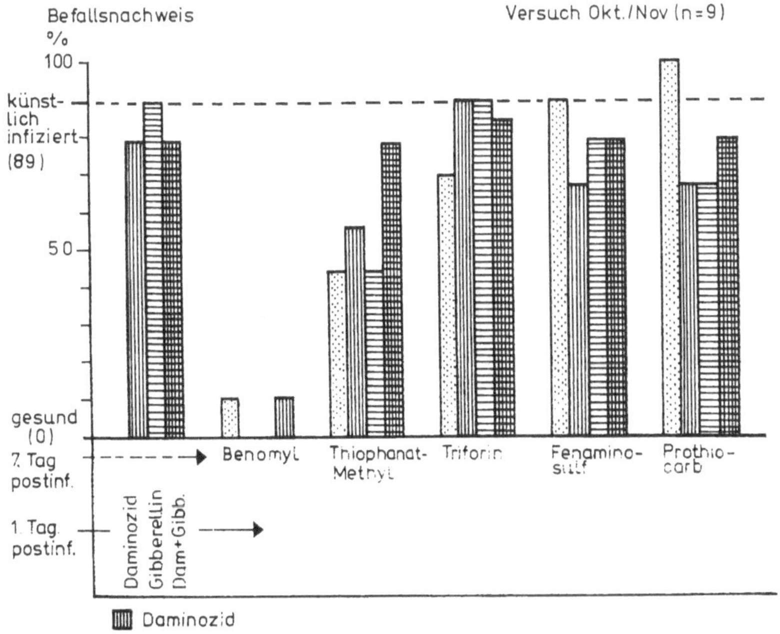

Abb. 2. Wirkung alleiniger und kombinierter Anwendung von Systemfungiziden und Wachstumsregulatoren auf den Befall von *Antirrhinum*-Pflanzen nach künstlicher Infektion durch *Verticillium albo-atrum*

Kamen die alleine unwirksamen Systemfungizide Fenaminosulf oder Prothiocarb in der Folge nach den Wachstumsregulatoren zum Einsatz, so waren häufig beträchtliche welke- und befallsmindernde Wirkungen zu bemerken, die in starken Wechselwirkungen Niederschlag fanden.

Um die Ursachen der beobachteten Wechselwirkungen einzugrenzen, wurden Laborversuche mit abgeschnittenen Sprossen von *Antirrhinum*-Pflanzen angesetzt. Der Nachweis des akropetalen Transportes des Fungizides in der Pflanze erfolgte dabei durch einen Agar-Bio-Test mit dem Erreger *Verticillium albo-atrum*. Es konnte gezeigt werden, daß durch Einstellen der Sprosse in eine Benomyl-Suspension (500 ppm) um die auf Agar ausgelegten Stengelstücke von verschiedenen Sproßhöhen eine deutlich von unteren zum oberen Sproßteil abgestufte Hemmhofbildung induziert wurde, während die Wirkung von Chlormequat (1500 ppm) nicht von der Wasserkontrolle abwich.

Ein Einstellen für 24 h in Chlormequat-Lösung vor dem Einstellen der Sprosse für 24 h in Benomyl-Suspension erhöhte dagegen den Fungizidnachweis in den 3 untersuchten Sproßhöhen deutlich (Abb. 3).

Auch ein Einstellen in ein Gemisch beider Substanzen konnte eine geringe Erhöhung in der Hemmhofbildung hervorrufen, die aber nur knapp über den Erwartungswerten aus dem additiven mathematischen Modell lag. Dagegen rief die Anwendungsfolge von

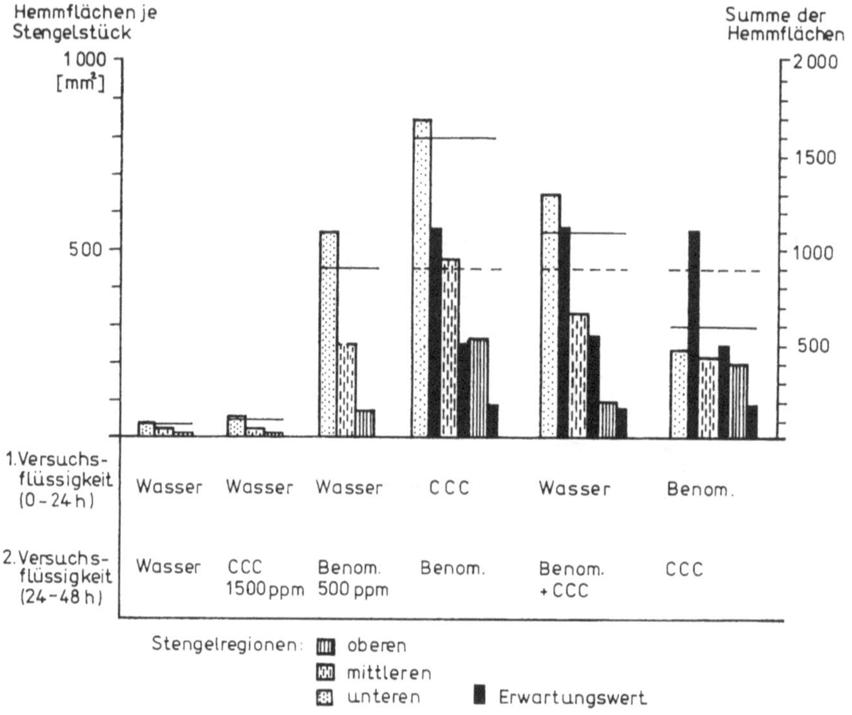

Abb. 3. Darstellung der Hemmflächen durch Auslegen 1 cm langer Teile aus unteren, mittleren und oberen Stengelregionen auf mit *Verticillium albo-atrum* homogen beimpften Biomalz-Agar-Platten, nach Einstellen der *Antirrhinum*-Sprosse in Versuchsflüssigkeit im Vergleich zum Erwartungswert

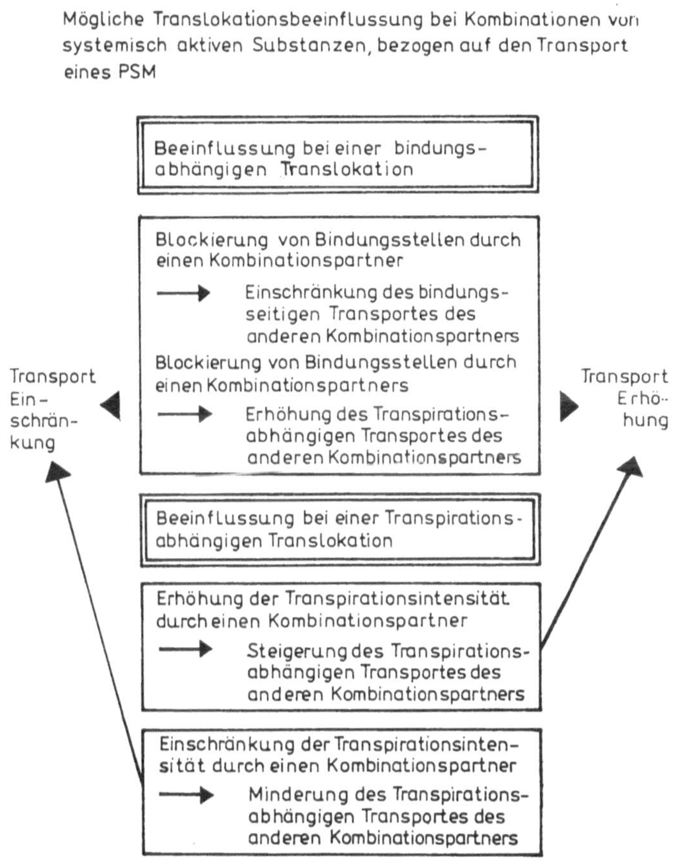

Abb. 4. Mögliche Translokationsbeeinflußung bei Kombinationen von systemisch aktiven Substanzen, bezogen auf den Transport eines Pflanzenschutzmittels

Chlormequat nach Benomyl eine fast vollständige Identität des Fungizidnachweises in allen Stengelhöhen hervor, der in der Summe erheblich unter dem nach alleiniger Anwendung von Benomyl blieb.

Da beide Substanzen im schwach sauren Xylemsaft der Pflanzen als Kationen angesprochen werden können, führen wir die Transportbeeinflußung bei der kombinierten Anwendung beider Mittel auf eine Transport-Konkurrenz der Wirkstoffionen zurück. Der im Verhältnis zum Transpirationsstrom langsamere Transport der Substanzen könnte dabei mit einer ständig wechselnden Sorption und Desorption der Wirkstoffionen an Carboxylgruppen der Zellwände verdeutlicht werden. Käme hier dem Chlormequat eine höhere Bindungsaffinität zu, so würde die von Erwin u. a. [1] postulierte Bindung der aktiven Substanz von Benzimidazolen an Pflanzenbestandteile in den Stengeln durch vorhergehende oder gleichzeitige Applikation von Chlormequat vermindert bzw. durch nachfolgend verwendetes Chlormequat wieder aufgehoben worden sein. Parallel zur Verdrängung aus dem Gefäßsystem wäre dann ein schnellerer akropetaler Transport des fungiziden Wirkstoffes bedingt. Eine langfristige Nachlieferung durch sukkzessive Desorption von diesen Bindungen in Stengeln und ähnlich auch im Wurzelbereich und damit Disponibilität des Wirkstoffes in basisnahen Stengelbereichen zur effektiven Bekämpfung beginnender Tracheomykosen würde damit vermindert werden.

Ähnliche Effekte, die für diese Deutungsweise der Wechselwirkungen sprechen, konnten im Labortest auch für beobachtete Interaktionen beim gemeinsamen Einsatz von systemischen Nematiziden, die gegen einen Pflanzenbefall durch *Meloidogyne hapla* CHITWOOD eingesetzt wurden, für Benomyl und Wachstumsregulatoren, die den Befall von Pflanzen durch Wurzelnematoden beeinflußten, gefunden werden.
Die postulierten Möglichkeiten der Transportbeeinflußung infolge des gemeinsamen Einsatzes systemisch aktiver Substanzen wurden schematisiert (Abb. 4), woraus in Abhängigkeit der Potenz und des Umfanges von transpirationsseitig bzw. bindungsseitig realisiertem Transport, sowohl eine Hemmung als auch eine Förderung der Bewegung der einzelnen Wirkstoffe in der Pflanze abgeleitet werden kann.

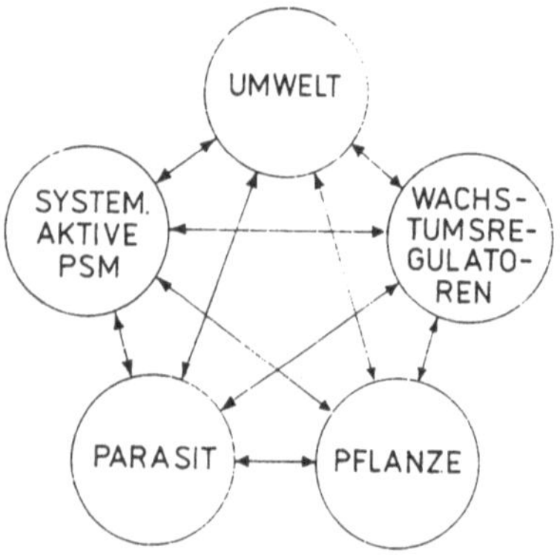

Abb. 5. Wechselbeziehungen zwischen den Elementen der ,Wirt-Parasit-Umwelt-Wirkstoff'-Systeme, die für die Erklärung von Wechselwirkungen, wie sie beim kombinierten Einsatz systemisch wirksamer Substanzen entstehen können, herangezogen werden müssen

Für die Diskussion der bei kombiniertem Einsatz systemisch aktiver Verbindungen auftretenden Wechselwirkungen vornehmlich auf die Pathogenese von Tracheomykosen werden damit von uns hauptsächlich Beeinflußungen im Transportgeschehen der Wirksubstanz in der Pflanze gesehen. Es können jedoch in Anbetracht des gesamten Beziehungsgefüges (Abb. 5), Wirt—Parasit—Umwelt—Wirkstoff' auch noch weitere Ursachen an der Entstehung der beobachteten Wechselwirkungen beteiligt sein, wie dies durch Untersuchungen von ELAMAYEN [2], TEUTSCH [4] u. a. Autoren angedeutet wird. Vornehmlich werden dabei nach unseren Erfahrungen die Wirt—Parasit-Beziehungen — z.B. lokalisierter oder systemischer Pflanzenbefall — d. h. von der Pathogenese abhängige Verschiedenheiten eine Schlüsselrolle bei der Auswirkung von Wirkstoffinteraktionen spielen. Die weitere Erfassung erscheint uns zur gezielten Effektivitätssteigerung des Einsatzes systemisch aktiver Verbindungen im Pflanzenschutz bedeutungsvoll.

Diskussion

GROSSMANN: Kann der hemmende Einfluß der Wachstumsregulatoren auf den therapeutischen Effekt von Thiophanat-Methyl darauf beruhen, daß die Umwandlung dieser Verbindung in MBC verlangsamt wird?

BOCHOW: Bei kurzen Versuchszeiten ist für diese Möglichkeit der Ergebnisinterpretation keine Grundlage gegeben.

GROSSMANN: Worauf ist es zurückzuführen, daß in den Versuchen mit abgeschnittenen Sprossen von Löwenmäulchen das Benomyl so langsam in die oberen Sproßteile transloziert wurde? War die Luftfeuchtigkeit sehr hoch?

Bochow: Es ist richtig. Die Untersuchungen liefen bei Anwendung relativ hoher Luftfeuchte (entsprechend geringer Transpiration). Wir waren an einem schnellen Transport in unseren Versuchen nicht interessiert.

Becker: Ihre Untersuchungen mit Gemischen von Fungiziden und Wuchsstoffen sind sehr interessant und zwar negative Ergebnisse nicht weniger als positive. Hatten Sie auch solche? Interessant ist vor allem inwieweit Wechselwirkungen zwischen Pathogen und Wirtspflanze betroffen werden und die Substanzen den Charakter von Schutzsubstanzen annehmen. Wir hatten bei einigen Mischungen (z.B. Griseofulvin mit Dioxyphenylalanin) unerwartet negative Ergebnisse, da diese beiden Substanzen, die beide das Pilzwachstum hemmen, einen entgegengesetzten Wirkungsmechanismus haben. Bei Behandlung mit der Mischung entwickelt sich das Pathogen fast so gut wie in der unbehandelten Kontrolle.
Es ist zu vermuten, daß Mischungen aus Griseofulvin und Benomyl oder Colchizin, wobei eine Komponente auf die RNS-Synthese und das Wachstum und die andere auf den Kernmetabolismus und die Kernteilung einwirkt, auch negative Ergebnisse geben werden. Es scheint, daß man derartig negative Ergebnisse mancher Gemische voraussagen kann und dies ist auch eine Hauptaufgabe der Wissenschaft.

Bochow: Die Ausführungen von Frau Prof. Becher betreffen einen wichtigen Aspekt der bei der Wechselwirkung von systemisch aktiven Verbindungen in der Pflanze zu erwarten ist. Wir verwendeten hier Substanzen, bei denen vom Wirkungsmechanismus allein schon her, eine negative Blockierung nicht zu erwarten war und konzentrierten uns auf wechselseitige Translokationsbeeinflussungen.

Buchenauer: Welche funktionellen Wechselwirkungen zwischen Wachstumsregulatoren und Halmstabilisatoren wurden analysiert?

Bochow: Im Falle der Kombinationswirkung von Systemfungiziden und Wachstumsregulatoren bei Getreide und der Bekämpfung der Halmbruchkrankheit sowie bei anderen Möglichkeiten der Wechselwirkung zwischen entsprechenden Verbindungen interessieren uns nicht nur Erklärungen bestimmter Effekte, sondern vor allem die Nutzung der Einwirkung von Wachstumsregulatoren auf das Transportsystem der Pflanze und damit gezielte Beeinflussungen der Translokation von systemischen Pflanzenschutzmitteln.

Literatur

[1] Erwin, D. C., Wang, M. C., Sims, J. J., Translocation of 2-(4-thiazolyl)benzimidazole in cotton, Phytopathology **60**, 1970, 1291 (Abstr.)

[2] Elamayen, M. M., Abu, Untersuchungen über Anwendungsmöglichkeiten systemischer Wirkstoffe zur Bekämpfung von Wurzelgallenälchen und anderen Schaderregern, Dissertation, Humboldt-Universität Berlin, 1973

[3] Junge, H., Wirkungen der gemeinsamen Anwendung verschiedenartiger, systemisch aktiver Substanzen auf das Wachstum der Pflanzen und deren Befall durch bodenbürtige Krankheitserreger, Dissertation, Humboldt-Universität Berlin, 1980

[4] Teutsch, H., Wechselwirkungen zwischen Wachstumsregulatoren und systemischen Fungiziden im Hinblick auf Entwicklung und Krankheitsbefall von Tomatenpflanzen, Dissertation, Universität Hohenheim, Stuttgart-Hohenheim, 1977

R. Kälberer, F. Müller und F. Grossmann

Institut für Phytomedizin der Universität Hohenheim, 7000 Stuttgart 70, Bundesrepublik Deutschland

Anreicherung systemischer Fungizide an Infektionsstellen obligater und nichtobligater Parasiten

Über die Akkumulation systemischer Fungizide an Infektionsstellen ist schon mehrfach berichtet worden (Bates und Tweedy [1], Upham und Delp [5], Houseworth [2] u. a.). Diese Angaben beschränkten sich aber auf Infektionen durch obligate Parasiten, überwiegend Rostpilze. In unsere eigenen Untersuchungen bezogen wir dagegen auch nichtobligate Erreger ein. Ferner versuchten wir, einen Beitrag zur Klärung der Ursachen der Akkumulation zu leisten. Die folgenden Ausführungen stützen sich im wesentlichen auf die Dissertation von Kälberer [3].

Zur Methodik

Für die Untersuchungen standen die systemischen Fungizide Carbendazim und Pyracarbolid zur Verfügung, die beide mit ^{14}C markiert waren. Sie wurden den Pflanzen in den meisten Fällen in einer Nährlösung über die Wurzeln verabreicht. Teilweise erfolgte die Applikation auch auf die Blätter.
Bohnen *(Phaseolus vulgaris)*, Sorte „Saxa", wurden an den Primärblättern mit *Uromyces appendiculatus* bzw. *Colletotrichum lindemuthianum* inokuliert; Gerste *(Hordeum distichon)*, Sorte „Firlbecks Union", mit *Erysiphe graminis* f. sp. *hordei* oder mit *Rhynchosporium secalis*. Die Versuche wurden in einer Klimakammer durchgeführt.
Die Auswertung wurde entweder autoradiographisch vorgenommen oder, nach Verbrennung der zu untersuchenden Proben, mit Hilfe eines Flüssigszintillationsspektrometers der Fa. Packard (Tricarb 3380). Hinsichtlich von Einzelheiten muß auf Kälberer [3] verwiesen werden.

Infektion von Bohnen durch Rost *(Uromyces appendiculatus)*

Nach Verabreichung der Fungizide über die Wurzeln war die zu erwartende Anreicherung im Bereich der Pusteln deutlich zu erkennen (Abb. 1). Wie ebenfalls aus der Abbildung hervorgeht, befindet sich um die Pusteln herum eine Zone geringerer Radioaktivität. Die Akkumulation der Fungizide in den Pusteln selbst ist offenbar mit einem Entzug aus dem unmittelbar angrenzenden Gewebe verbunden.
Zur genaueren Erfassung der quantitativen Verhältnisse wurden einzelne Pusteln und die sie umgebenden Ringe mit Injektionsnadeln bzw. dünnen Röhrchen aus den Blättern ausgestanzt, wie dies die folgende Skizze veranschaulicht:

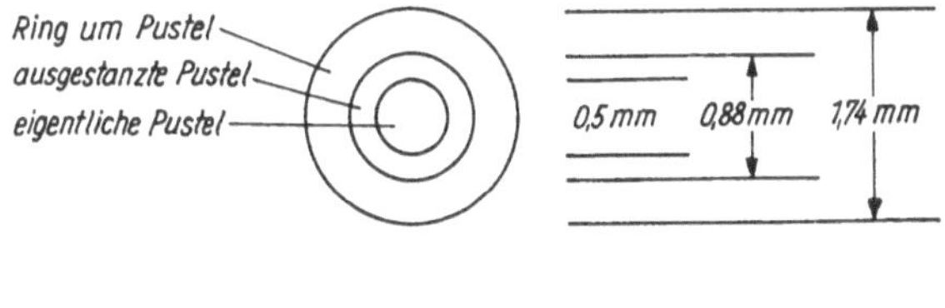

Abb. 1. ^{14}C-Akkumulation in Rostpusteln auf Bohnen-Primärblättern nach Applikation von ^{14}C-Carbendazim (links) bzw. ^{14}C-Pyracarbolid (rechts); Applikation über die Wurzeln 5 Tage nach der Inokulation; Aufnahmedauer 7 Tage. Autoradiographie

Die außerhalb der Ringe liegende Blattfläche wurde als „pustelfreies Gewebe" bezeichnet. Die Bestimmung der Radioaktivität in den verschiedenen Bereichen führte zu den in Abb. 2 wiedergegebenen Resultaten. Die Radioaktivität ist im Bereich der Pusteln eindeutig am höchsten. Dagegen ist sie in den Ringen um die Pusteln, im Vergleich zum pustelfreien Gewebe, eher herabgesetzt, bes. nach ^{14}C-Carbendazim-Appli-

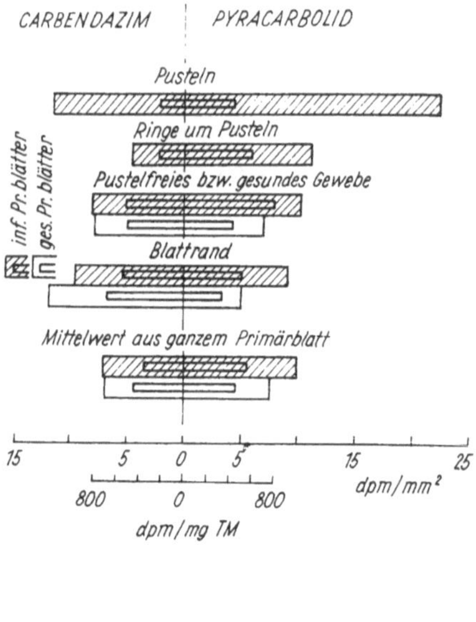

Abb. 2. Verteilung der ^{14}C-markierten Substanz in rostinfizierten und gesunden Bohnen-Primärblättern. Die breiten Säulen geben die auf die Blattfläche bezogenen, die schmalen Säulen die auf die Trockenmasse bezogenen ^{14}C-Gehalte an. Übrige Daten wie in Abb. 1

kation. Da die Anreicherung in den Pusteln mit einem Entzug aus den unmittelbar benachbarten Bereichen einhergeht, ist die Gesamtradioaktivität injizierter Primärblätter im Vergleich zur Aktivität gesunder, nicht infizierter Primärblätter nicht (Carbendazim) oder nur wenig (Pyracarbolid) erhöht.
Auch nach Spritzung der ^{14}C-markierten Fungizide auf die Blätter, unter praxisnahen Applikationsbedingungen, ließ sich eine Akkumulation im Bereich der Pusteln erkennen. Eine exakte quantitative Auswertung war aber dadurch erschwert, daß ein erheblicher Teil der Fungizide auf der Blattoberfläche zurückblieb.

Infektion von Bohnen durch Brennflecken *(Colletotrichum lindemuthianum)*

Unter den Versuchsbedingungen bildeten sich Läsionen durch *C. lindemuthianum* fast ausschließlich an den Blattadern. Wie Abbildung 3 zeigt, kommt es an diesen Stellen zu einer deutlichen Anhäufung ^{14}C-markierter Substanz nach Anwendung von ^{14}C-Carbendazim. Die Infektion durch einen nichtobligaten Erreger bewirkt also eine ähnliche Fungizidakkumulation wie die durch Rostpilze. Nähere Untersuchungen ergaben daß die Radioaktivität infizierter Blattrippen etwa doppelt so hoch war wie die gesunder. Andererseits wurde durch eine starke Infektion der Blattadern der akropetale Transport des Fungizids weitgehend unterbunden: in Abbildung 3 rechts bleibt die Blattspitze hell.

Abb. 3. Einfluß von Infektionen durch *Colletotrichum lindemuthianum* auf die Verteilung ^{14}C-markierter Substanz in Bohnen-Primärblättern (links: nicht infiziert; rechts: infiziert). Applikation von ^{14}C-Carbendazim über die Wurzeln, 7 Tage nach der Inokulation. Aufnahmedauer 3 Tage

Infektion von Gerste durch Mehltau *(Erysiphe graminis* f. sp. *hordei)*

Nach Applikation von ^{14}C-Carbendazim über die Wurzeln reichert sich die radioaktive Substanz rasch in der Spitze gesunder Primärblätter der Gerste an (Abb. 4, links). Dies ist prinzipiell auch bei mehltauinfizierten Primärblättern der Fall, doch ist hier zusätzlich eine Akkumulation an den Infektionsstellen erkennbar. Der Einfluß des Mehltaubefalls wird besonders deutlich, wenn die Infektion auf Querstreifen erfolgt, die durch nichtinfizierte Streifen voneinander getrennt sind (Abb. 4, rechts). Die infizier-

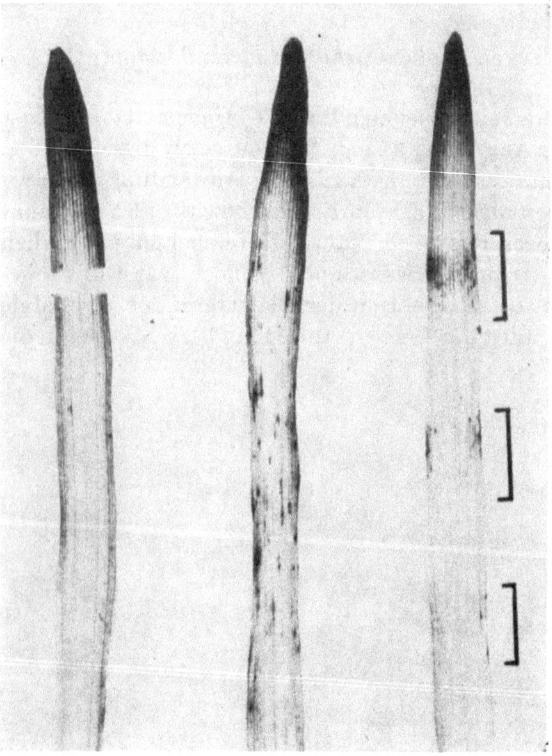

Abb. 4. Verteilung der ^{14}C-markierten Substanz in gesunden und mehltauinfizierten Primärblättern von Gerste. Applikation von ^{14}C-Carbendazim über die Wurzeln, 2 Tage nach der Inokulation. Aufnahmedauer 4 Tage

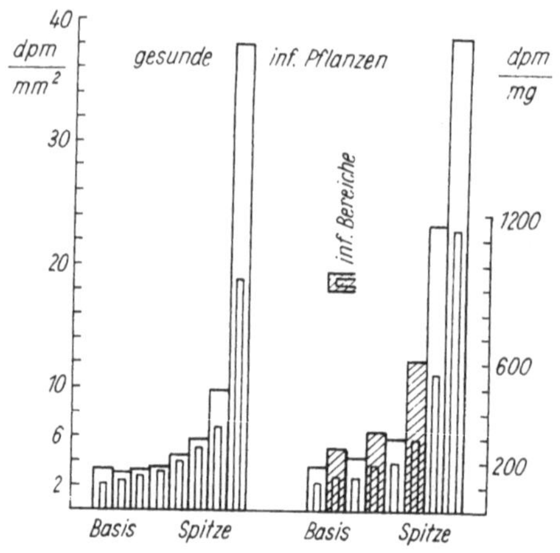

Abb. 5. ^{14}C-Verteilung in gesunden (links) und streifenweise mit Mehltau infizierten (rechts) Gerstenprimärblättern. Die breiten Säulen geben die flächenbezogenen, die schmalen Säulen die trockenmasse-bezogenen ^{14}C-Gehalte an. Übrige Daten wie in Abb. 4

ten Streifen zeichnen sich zur Blattspitze hin immer deutlicher ab, weil dort das Gesamtniveau der Radioaktivität ansteigt.

Die genauen quantitativen Verhältnisse sind aus Abbildung 5 zu ersehen. Jeweils 1 cm lange Blattabschnitte, von der Basis zur Spitze hin, wurden auf ihre Radioaktivität untersucht. Beim gesunden Blatt ergibt sich ein deutlicher Gradient, mit einer besonders starken Anreicherung in der Spitzenregion. Beim infizierten Blatt wird dieser Gradient immer wieder unterbrochen durch eine gewisse Akkumulation in den vom Mehltau befallenen Abschnitten. In ihnen ist die Radioaktivität höher als in den benachbarten nichtinfizierten Bereichen.

Infektion von Gerste durch *Rhynchosporium secalis*

Auch die von diesem Erreger auf Gerstenblättern hervorgerufenen Läsionen bewirken eine Anhäufung radioaktiver Substanz nach Applikation von ^{14}C-Carbendazim über die Wurzeln. Im Autoradiogramm ist die Akkumulation besonders markant, wenn die Infektionen punktförmig gesetzt werden (Abb. 6). In den infizierten Bereichen ist die Radioaktivität mehr als doppelt so hoch wie im umgebenden Gewebe.

Durch die Versuche mit *Rhynchosporium secalis* wird noch einmal bestätigt, daß auch Infektionen durch nichtobligate Parasiten zu einer Fungizidanreicherung führen.

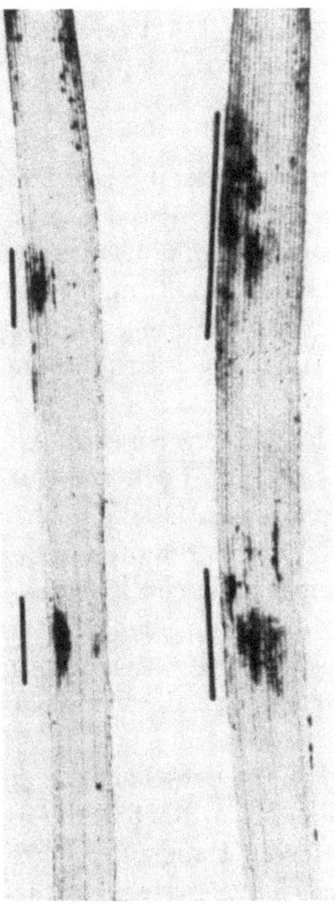

Abb. 6. Verteilung der radioaktiven Substanz in punktförmig mit *Rhynchosporium secalis* infizierten Gersten-Primärblättern. Die Striche markieren die infizierten Bereiche. Applikation von ^{14}C-Carbendazim über die Wurzeln, 13 Tage nach der Inokulation. Aufnahmedauer 4 Tage

Ursachen der Fungizidakkumulation

Die Untersuchungen zu dieser Frage wurden vor allem am Beispiel des Bohnenrostes durchgeführt. Dabei ergab sich zunächst, daß die Akkumulation umso intensiver wird, je weiter der Krankheitsprozeß fortschreitet (Abb. 7). Dies bedeutet darauf hin, daß die Transpiration, die nach Aufbruch der Pusteln (unter den Versuchsbedingungen etwa 7—8 Tage nach der Inokulation) stark erhöht ist, eine wesentliche Rolle spielt.

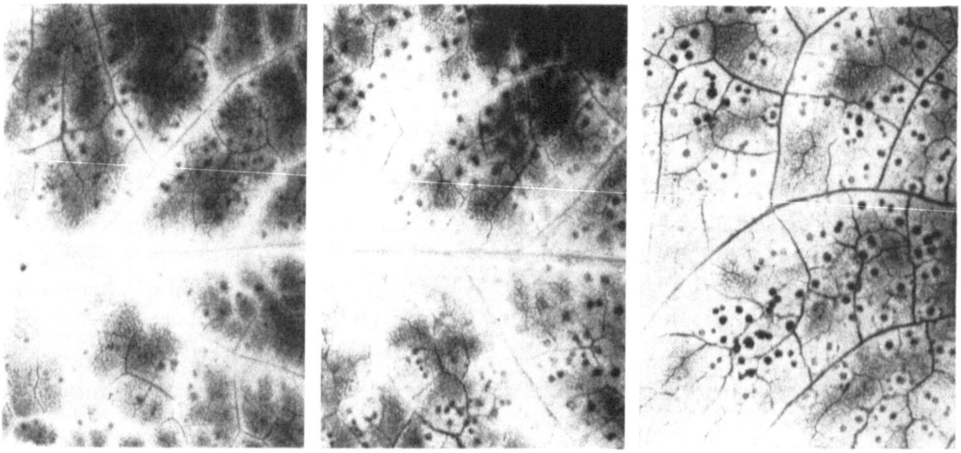

Abb. 7. Einfluß des Applikationstermins auf die Fungizidanreicherung in Pusteln des Bohnenrostes. Applikation von ^{14}C-Carbendazim über die Wurzeln 1, 5 und 10 Tage nach der Inokulation (von links nach rechts). Aufnahmedauer jeweils 7 Tage

Auch andere Autoren (z.B. PRUSKY u. a. [4]) messen der Transpiration eine entscheidende Bedeutung für die Fungizidanreicherung bei. Im Unterschied zu den genannten Autoren konnten wir jedoch eine gewisse, allerdings nur schwache Anhäufung auch schon vor Pustelaufbruch feststellen (vgl. hierzu Abb. 7, links). Daß außer der (zweifellos dominierenden) Transpiration auch andere Faktoren an der Fungizidakkumulation beteiligt sein können, zeigt sich ferner daran, daß auch in wasserdampfgesättigter Atmosphäre eine Anreicherung zu beobachten ist.
Um Hinweise auf die Bedeutung der Atmung für die Fungizidanhäufung zu erhalten, wurden rostinfizierte Bohnenblätter 5 Tage nach der Inokulation, also noch vor Aufbruch der Pusteln, unter anaeroben Bedingungen (N_2-Atmosphäre) im Dunkeln über die Blattstiele mit den ^{14}C-markierten Fungiziden versorgt. Auch unter diesen Bedingungen kommt es nach 4-stündiger Behandlungsdauer zu einer gewissen Fungizidakkumulation an den Infektionsstellen. Ähnliche Befunde wurden mit mehltauinfizierten Gurkenpflanzen (3 Tage nach der Inokulation) bei verlängertem Aufenthalt in einer N_2-Atmosphäre erhalten. Diese Ergebnisse deuten darauf hin, daß die Fungizidanreicherung nicht von der Atmung abhängig ist.
Auch eine „sink"-Wirkung der Infektionsorte scheint hierfür nicht entscheidend zu sein. Örtliches Auftragen von Benzyladenin führte weder bei Bohnen- noch bei Gerstenblättern zu einer Attraktion der über die Wurzeln applizierten Fungizide. Übrigens blieb auch lokales Auftragen normal konzentrierter und eingeengter Kultur-

filtrate von *Colletotrichum lindemuthianum* (bei Bohnen) und *Rhynchosporium secalis* (bei Gerste) in dieser Hinsicht erfolglos.

Schließlich ist die Fungizidakkumulation offenbar auch nicht auf eine bevorzugte Aufnahme durch die Pilze selbst zurückzuführen. Jedenfalls weisen Uredosporen des Bohnenrostes sowie Konidien und ektoparasites Myzel des Gerstenmehltaus nach Wurzelapplikation der ^{14}C-markierten Fungizide eine signifikant geringere Radioaktivität auf als das befallene Blattgewebe.

Andererseits kommt es beim Bohnenrost — solange die Pusteln noch nicht aufgebrochen sind — nur dann zu einer Fungizidanhäufung, wenn der Pilz aktiv ist. Inaktiviert man den Erreger im Fleckenstadium (4 Tage nach Inokulation) mit einem nicht markierten Fungizid und verabreicht anschließend den markierten Wirkstoff über die Wurzeln, so läßt sich keine Akkumulation mehr feststellen; die Infektionsherde heben sich auf dem Autoradiogramm sogar als helle Stellen ab (Abb. 8).

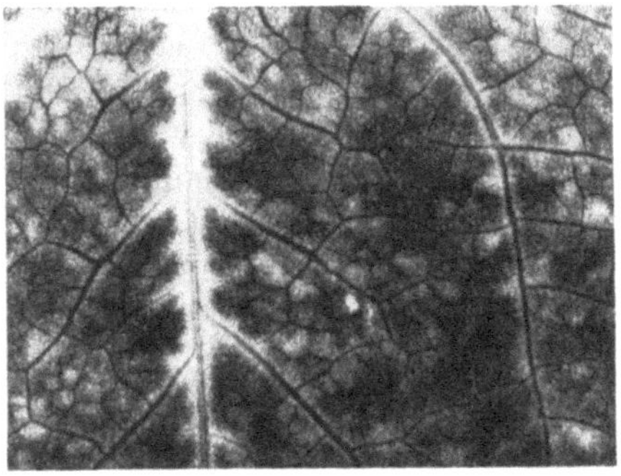

Abb. 8. Autoradiogramm eines rostinfizierten Bohnen-Primärblatts, das 4 Tage nach der Inokulation zunächst mit nicht markiertem Pyracarbolid (zur Inaktivierung des Erregers) und anschließend mit ^{14}C-Pyracarbolid über die Wurzeln behandelt wurde

Wahrscheinlich hängt die Fungizidakkumulation vor Aufbruch der Pusteln mit einer lokalen Hypertrophierung des Gewebes an den Infektionsstellen zusammen, die nur durch einen aktiven Erreger aufrechterhalten wird. Der Grad der Anreicherung ist hierbei aber in der Regel wesentlich geringer als nach Pustelaufbruch, wenn infolge der gesteigerten Transpiration ein erheblich stärkerer Wassertransport zu den Infektionsstellen hin einsetzt.

Diskussion

Müller: Ist es denkbar, daß der Ort der Infektion ein Ort besonders starken Stofftransportes ist, der im Sinne des „Kinetin-Effektes" eine bevorzugte Stoffversorgung erfährt, wodurch das Fungizid mitgerissen wird?

Grossmann: Dieses „Mitreißen" wäre vor allem dann zu erwarten, wenn es sich um einen symplastischen Transport handeln würde. Die untersuchten Fungizide werden aber fast ausschließlich apoplastisch transloziert.

Müller: Ist als Alternative zu den genannten Hemmungen auch eine Blockierung des Fungizids an den Infektionsorten denkbar? Die Radioautogramme bei Monokotylen deuten darauf hin.

Grossmann: Es ist denkbar, daß es vor allem bei nichtobligaten, pertotrophen Parasiten zu einer Schädigung der Leitbahnen kommt, die dann zu einer „Stauung" der im Xylem translozierten Fungizide führen könnte.

Lyr: Die Frage der Akkumulation vor Aufbruch der Pusteln ist hoch interessant. Leider war Ihre Applikation von Benzyladenin ohne Effekt, aber obligate Parasiten entwickeln eine Attraktionskraft und eine Juvenilität des Umgebungsareals, wo Substanzen aus der Umgebung einströmen könnten. Das muß nicht nur symplastisch, sondern kann auch apoplastisch erfolgen. Vielleicht ist außer Cytokinin auch ein anderes Hormon wie Gibberellinsäure wirksam.

Grossmann: Wir haben keine Untersuchungen mit Gibberellinsäure durchgeführt. Die Tatsache, daß vor Pustelaufbruch die Fungizidanreicherung nicht wesentlich stärker ist als die Trockenmassezunahme im Infektionsbereich, deutet aber darauf hin, daß in dieser Phase im wesentlichen die örtliche Verdickung (Hypertrophierung) entscheidend ist.

Lyr: Wenn Rostpusteln sehr viel Fungizid akkumulieren aus dem umgebenden Gewebe, könnten dann nicht andere Pilze diese Räume leicht infizieren?

Grossmann: Sicher ist es möglich, daß an diesen Stellen neue Infektionen eintreten. Dies geschieht aber auch ohne Akkumulation durch Infektionsstellen, wenn bei leicht beweglichen Fungiziden nach kurzer Zeit eine Anreicherung an den Blatträndern bzw. -spitzen erfolgt, während die basalen und inneren Teile der Blattfläche vom Fungizid „entblößt" werden.

Tüske: Werden die Fungizide in sterilen, mechanischen Blattschädigungen auch angereichert oder ist dies spezifisch für Infektionsstellen von Pilzen?

Grossmann: Mechanische Schädigung des Gewebes durch Nadelstiche oder Verätzung durch Chemikalien führt nur zu einer schwachen, vorübergehenden Akkumulation der Insektizide an den Wundrändern.

Davidse: Do you think that the accumulation of the fungicides at the infection area's in a standard atmosphere can be explained by a higher temperature of these areas? A slightly higher temperature will enhance the transpiration causing an influx of the fungicide into the area.

Grossmann: We did not measure the temperature in the infected areas. So far as I know, however, the temperature differences between infected and non-infected areas in case of fungal pathogens are in the range of 1 °C. I have some doubts whether this would have a pronounced effect on transpiration.

Josepovits: Können die beobachteten Anreicherungen an den Infektionsstellen nicht als Sekundäreffekte erklärt werden? Nach den Angaben von Prof. Bochow kann man damit rechnen, daß bei der Adsorption im Blatt eine Wechselwirkung zwischen Systemfungiziden und Naturstoffen vorhanden ist. Wenn an der Infektionsstelle Adsorptionszentren durch Abbau und Naturstoffen frei werden und in der Umgebung der Infektionsstelle durch Kinetin oder ähnliche Naturstoffe die Sättigung von Adsorptionsstellen erhöht wird, ist mit einer sekundären Strömung der Fremdstoffe in entgegengesetzter Richtung zu rechnen.

Grossmann: Ich möchte solche Effekte nicht ausschließen, doch sind mir keine Daten über das Adsorptionsverhalten der Befallstellen bekannt.

Bochow: Ist bei der Anreicherung der Fungizide im Infektionsbereich wirklich eine Bindung durch den infizierenden Pilz auszuschließen? Es scheint doch methodisch schwer zu erfassen zu sein? Reicht dabei die alleinige Untersuchung der Uredosporen aus?

Grossmann: Es ist nicht auszuschließen, daß das intramatrikale Myzel der Rostpilze höhere Radioaktivitäten aufweist als die Uredosporen. Andererseits erhielten wir vergleichbare Ergebnisse auch bei der Feststellung der Radioaktivität in Konidien und ektoparasitischem Myzel des Gerstenmehltaus, so daß dort eine bevorzugte Bindung allenfalls in den im Blatt verbleibenden Haustorien stattfinden könnte, was jedoch nicht sehr wahrscheinlich erscheint.

Literatur

[1] Bates, J. J., and B. G. Tweedy, The effect of oxycarboxin on the bean rust disease caused by *Uromyces phaseoli*, Phytopathology **61** (1971), 884

[2] Houseworth, L., Accumulation of systemic fungicides and naturally occurring compounds at sites of *Uromyces phaseoli* and *Puccinia recondita* infections, Diss. University of Missouri, Columbia 1973

[3] Kälberer, R., Einfluß pilzlicher Infektionen auf die Verteilung systemischer Fungizide in höheren Pflanzen, Diss. Univ. Hohenheim, Stuttgart 1979

[4] Prusky, D., B. Jacoby and A. Dinoor, Is the accumulation of a systemic fungicide at the infection site related to its eradicative action ?, Pestic Biochem. Physiol. **12** (1979), 75–78

[5] Upham, P. M. and C. J. Delp, Role of benomyl in the systemic control of fungi and mites on herbaceous plants, Phytopathology **63** (1973), 814–820

Г. И. Бородин, С. М. Ходжибаева и В. И. Рунов

Институт микробиологии АН УзССР, Ташкент, СССР

Биохимия Взаимоотношений Патоген-Токсин-Растение-фитоалексин при Вертициллезном Вилте Хлопчатника

Механизм взаимоотношения между патогеном и растением является одной из наиболее важных проблем биологии. Решение этой проблемы позволит дать биохимическое толкование таким вопросам, как механизм устойчивости сортов к возбудителям, механизм специализации патогена, наследование и формирование признаков иммунитета и вирулентности.

О наличии токсических метаболитов у грибов *Verticillium* известно из работ Green [13], Kiessig et al. [15], Stoddart et al. [16], Красильникова и др. [6] и многих других. Однако изучение их химического состава начато недавно. Нами (Бородин и др. [1]) было показано, что возбудитель болезни содержит в мицелии и накапливает в культуральной жидкости до концентраций выше 1 мг/мл так называемый токсичный белково-пигментный комплекс (БПК), вызывающий в концентрациях 10—20 мг/мл симптомы вилта. Аналогичное вещество, получившее название белково-липидно-полисахаридный комплекс (БЛПК), было изолировано и изучено у *V. dahliae* Малышевой и др. [7], а у *V. albo-atrum* Keen et al. [14]. Этими исследователями было показано, что БПК и БЛПК состоят из полисахарида, протеина, липида и фосфорсодержащих компонентов. Нами в составе БПК установлено наличие аминокислот (8,4—12,8%), глюкозамина и галактозамина (1,8—3,3%), глюкозы (12—16%), галактозы (5,7—8,1%), маннозы (4,2—5,5%), липида (20—25%) и фосфора (1,2—1,6%), показано присутствие железопротеиновых комплексов, уроновых кислот, зольных элементов и других веществ.

Аминокислотный состав БПК *V. dahliae* и БЛПК *V. albo-atrum* приведен в табл. I. По-видимому, грибы рода *Verticillium* продуцируют близкие по своей природе токсические вещества, на что указывает аналогия составов БПК у различных специализированных (к сортам 108-ф и Ташкент-I) форм гриба и наличии подобных токсических метаболитов у непатогенного для хлопчатника *V. lateritium* (Чепенко и др. [18]). Имеющиеся сведения позволяют представить БПК как набор сильно различающихся по молекулярному весу биополимеров, построенных из однотипных, регулярно повторяющихся фрагментов.

Методом бумажного электрофореза в 0,75 М муравьиной кислоте из водных экстрактов мицелия *V. dahliae* были получены вещества, по химическому составу аналогичные фрагментам БПК. Из них наиболее интересен токсичный полисахарид-протеиновый комплекс, разрушающийся на полисахарид, фрагменты протеиновой цепи, продукты неполного распада комплекса, получившие название вертициллинов А и В (Чепенко и др. [10]) и железотригидроксаматциклопептид — вертициллин, состоящий из серина, глицина и оксиорнитина в соотношении 1:2:3, и близкий по физикохимическим свойствам феррикроцину, описанному Zähner et al. [17] (Рис. 1). В составе вертициллина А обнаружено производное уридина, что позволяет предполагать участие этого метаболита в биосинтезе клеточной оболочки.

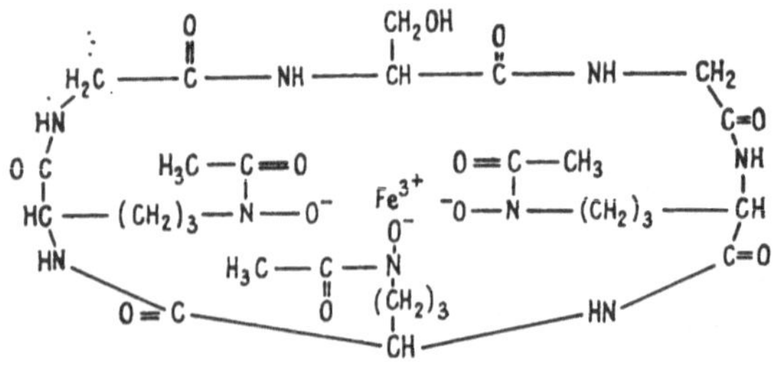

Рис. 1. Структурная формула аналога вертициллина — феррикроцина по Zähner et al. (1963)

На определенную связь токсических метаболитов *V. dahliae* с клеточной оболочкой указывает также присутствие в мицелии и в составе БПК (мицелий) арабитолпроизводных близких тейхоевой кислоте (ТК) и состоящие из P — 6,8%; C — 37,8%; N — 1,95% и H — 7,5% (Бородин и др. [4]). Выделялись они вместе с остатками протеиновой цепи (~ 3,5%) и полисахарида (Рис. 2).

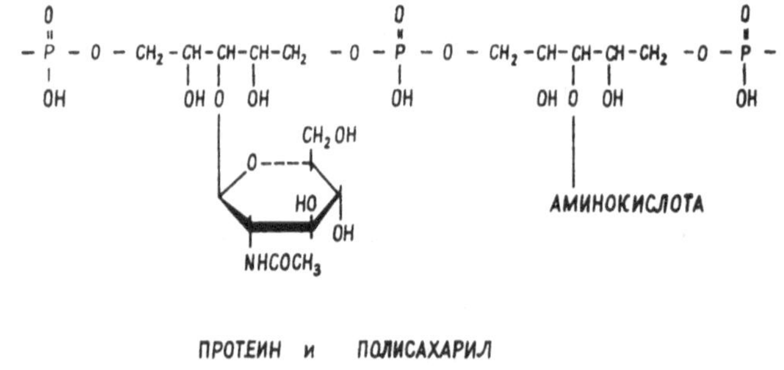

Рис. 2. Предполагаемая структура арабитолсодержащего фрагмента БПК (ТК) из мицелия *Verticillium dahliae*

В основе механизма действия токсических веществ *V. dahliae* на растение лежит нарушение функции клеточных мембран, видимо связанное с образованием каналов проводимости для ионов калия. Было установлено, что различные фракции БПК в концентрации 5—20 мкг/мл и вышерассмотренный арабитолсодержащий фрагмент ТК в концентрации 1,25 мкг/мл вызывают падение сопротивления у искусственных бимолекулярных фосфоролипидных мембран (БФМ) от $1{,}5 \cdot 10^{-10}$ ом до $6 \cdot 10^{-9}$ и $7 \cdot 10^{-9}$ ом (соответственно). При замене в калийнатриевом буфере (KCl — 10 мМ, NaCl — 10 мМ, pH — 6,5) K^+ на Na^+ способность препаратов понижать сопротивление мембран исчезала.

Таблица 1
Аминокислотный состав БЛПК, БПК V. albo-atrum и V. dahliae и фосфорсодержащего фрагмента (ТК) V. dahliae (микромоль %)

Аминокислоты	*V. albo-atrum* БЛПК по Keen et al. 1972	*Verticillium dahliae* БПК	фрагмент ТК
Орнитин	сведения остутст.	8,1	3,5
Лизин	4,15	4,1	5,5
Гистидин	3,02	2,4	1,9
Аргинин	1,47	1,1	1,8
Аспарагиновая кислота	9,23	6,6	8,9
Треонин	11,00	9,3	5,8
Серин	7,41	9,8	6,6
Глютаминовая кислота	8,55	7,5	12,8
Пролин	10,37	8,2	3,9
Глицин	8,42	11,0	20,6
Аланин	9,63	10,6	11,5
Цистин	2,46	2,4	следы
Метионин	0,34	следы	отсут.
Валин	8,42	7,5	5,6
Изолейцин	4,48	3,4	3,3
Лейцин	4,26	3,3	3,4
Тирозин	4,14	2,7	2,2
Фенилаланин	2,57	2,2	2,1

Избирательность действия токсических веществ *V. dahliae* на транспорт калия через мембран проверялась также в серии опытов на проростках хлопчатника. Показано, что у растений, выращенных на средах содержащих 0,005% калия и перенесенных в раствор не содержащий этого элемента, убыль калия из растения существенно усиливалась при воздействии на них препаратами БПК. В этих опытах использованы 3 фракции БПК, из лиофилизатов культуральной жидкости *Verticillium dahliae* (штамма ХЛ-1-3 специализированный к хлопчатнику сорта 108-ф) и его авирулентного γ-мутанта (Р-177). Фракции получали хроматографически на колонке с Дауэкс 1 (2,5 × 50 см). Выявлена корреляция между содержанием сахара и фосфора в препаратах и интенсивностью их действия на скорость вывода ионов калия из растения (Табл. 2). Активность фракций из БПК авирулентного γ-мутанта существенно не отличалась от аналогичных фракций исходной культуры. С помощью метода электронной микроскопии выявлены существенные нарушения в структуре мембран хлоропластов и других внутриклеточных структур (Бородин и др. [3]). Деструктивные изменения в хлоропластах больных вилтом растений отмечены также Воронковым [5].

Известно, что фиксация солнечной энергии и последующая ассимиляция углекислоты всязаны с мембрано-энергетическими процессами, протекающими в хлоропластах и нарушение функции мембран должно существенно понижать уровень используемой растением углекислоты. Изучение

Таблица 2
Влияние состава фрагментов БПД Verticillium dahliae (штамма ХЛ-1-3 и его γ-мутанта П-177) на скорость вывода ионов калия из растения хлопчатника

пп	% вещества в препарате				% от контроля
	сахара	амино-сахара	фосфора	липида	потерянного растением калия
			ХЛ-1-3		
1	1,5	0,2	0,2	8,3	33
2	11,9	1,9	0,3	26,6	64
3	15,5	1,8	10,0	16,0	82
			Р-177		
1	1,3	0,6	0,2	8,9	55
2	11,3	2,6	0,6	28,0	78
3	29,5	1,9	10,3	35,6	84

газообмена растений с помощью газоанализатора ,,Инфралит,, в комбинации с блоком автоматической продувки газовой смеси МН-5126 и 4[ех] канальным распределительным блоком РБ-3 (время работы каждого канала 3 мин, скорость продувки — 0,7 л в мин) подтвердило данное представление. Токсические метаболиты гриба в первую очередь нарушали фотосинтетический аппарат растения, что вело к резкому ингибированию процесса фотолиза воды и при высоких концентрациях препарата (500 мг/кг растения) уже через 15 часов на свету вместо поглощения углекислоты наблюдалось ее выделение (дыхание, фотодыхание)(Рис. 3, А). При понижении концентрации токсического вещества в 2—5 раз время проявления признаков отравления растягивалось до 2—5 суток и возникали побочные явления (отторжение листьев и др.), по-видимому, развивающиеся под действием ферментативного аппарата реакции ,,сверхчувствительности клетки''.

Нарушение фотосинтетического процесса сопровождалоь подавлением фотофосфорилирования. Показано, что при введении АДФ в черешок листа здоровых растений возникает 2,0—2,3 кратное увеличение скорости потребления CO_2 (Салихова, Бородин и др. [9]). У больных и обработанных препаратами растений оно резко снижается, что можно расценивать как нарушение процесса фотофосфорилирования (Рис. 3, Б). Подавление фотофосфорилирующей активности отмечено так же Воронковым [5] при изучении хлоропластов больных вилтом растений.

Нарушение обмена веществ при заражении растений патогеном или под действием токсических метаболитов *V. dahliae* ведет к 1,5—2,0 — кратному увеличению активности каталазы, пероксидазы и полифенолоксидазы (Рунов и др. [8]), образованию фитоалексинов госсиполового ряда и антифунгальных липидсодержащих соедеинений (Рис. 4). В основе этого процесса, по нашему мнению, лежит появление в клетке биологически активных конкурентноспособных веществ в концентрациях, превышающих необходимые для нормального протекания метаболитических процессов. Включая резервные пути окисления и активируя обмен ве-

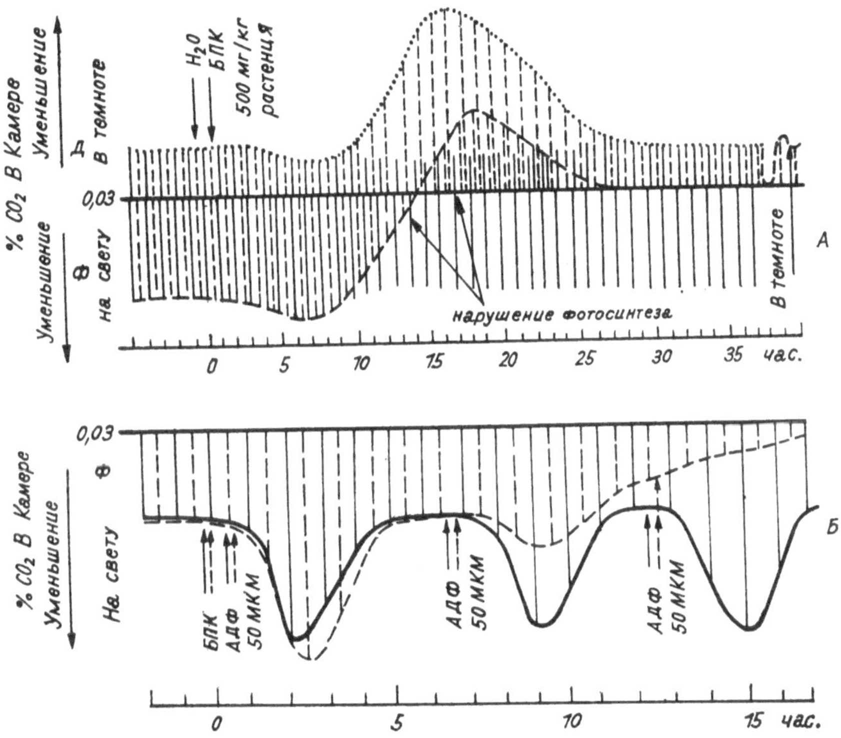

Рис. 3. Действие БПК и его фграрментов на газообмен листьев хлопчатника (А) и фотофосфорилирование (Б) (Д-дыхание, Ф — фотосинтез)
——— контроль на свету и в темноте;
.... действие препаратов на выделение CO_2 (дыхание) через 12 мин после выключения света;
— — — действие препаратов на интенсивность потребления CO_2 (фотосинтез)

ществ, саморегулируемая биологическая система клетки, с одной стороны, искусственно повышает уровень необходимых продутков метаболизма, поддерживая нормальное протекание биологических реакций, а с другой, — способствует быстрой утилизации ингибитора. Мобилизация резервных путей окисления ведет и увеличению в клеточном соке промежуточных продуктов биосинтеза и частично окисленных веществ клетки хозяина, среди которых встречаются вещества токсичные для фитопатогенов.

Поэтому образование антифунгальных веществ при заражении *V. dahliae* или под действием его метаболитов не является специфичным процессом и может вызываться целым рядом биологически активных соединений в слабоингибирующих концентрациях (микроэлементы, продукты биосинтеза других организмов и др.). Так, 20-часовая замочка семян в 0,1% растворе $CuSO_4$ повышает количество фунгицидных веществ у проростков хлопчатника на 5 сутки роста в 3, а на 10- в 1,8 раза, в 0,25% растворе БПК — на 5 сутки в 1,8, а на 10- 1,1 раза и в 0,5% лиофилизате куль-

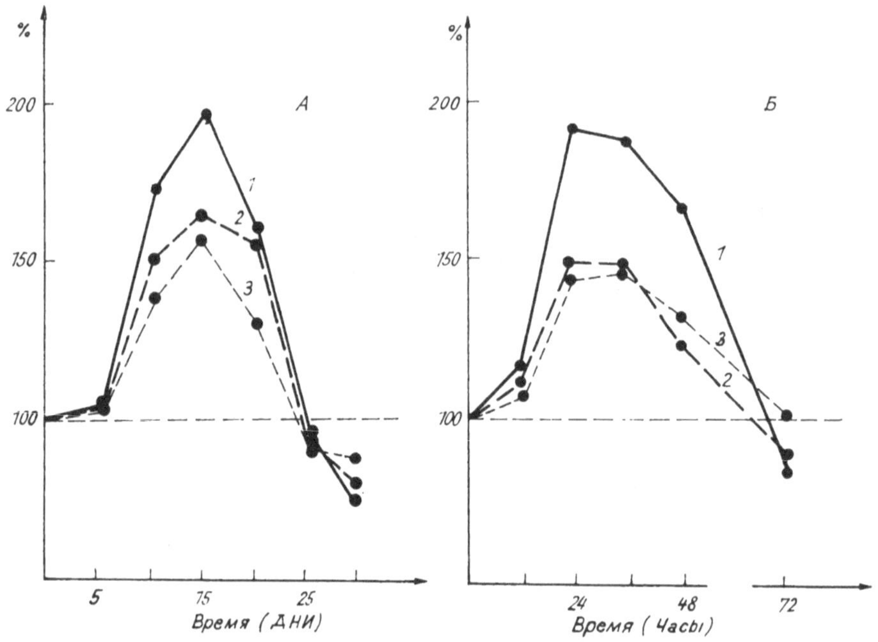

Рис. 4. Влияние заражения грибом *Verticillium dahliae* (А) и действия препаратов БПК (Б) на активность оксидоредуктаз в листьях растения хлопчатника (1 — пероксидаза, 2 — полифенолоксидаза, 3 — каталаза) (в % к контролю)

туральной жидкости *Nostoc muscorum* — на 5 сутки в 1,8, а на 10- 1,7 раза. Обработка препаратами семян хлопчатника повышает их всхожесть (~ на 10%) и на 1—2 сутки развития увеличивает в 2—3 раза (по сравнению с контролем) интерсивность их роста, что обеспечивает более энергичное прохождение наиболее подверженного заболеванию начального этапа развития растений. Однако уже на 4 сутки контрольные растения (вначале развивающиеся более медленно) начинают догонять опытные и к 10 суткам их длина и интересивности роста выравниваются. Интенсификация роста растений под действием препаратов сопровождается повышением в их клетках содержания низкомолекулярных пептидов. У обработанных препаратами растений количество низкомолекулярных пептидов на 2—7 сутки возрастало относительно контроля в 2—4 раза, а затем, снижалось (Рис. 5).

Полученные данные позволяют рассмотреть круг взаимоотношений ,,патоген-токсин-растение-фитоалексин-патоген,, (Рис. 6) и наметить терапевтический путь повышения устойчивости растения к болезням. Мы полагаем, что устойчивость определяется интенсивностью и скоростью ответной реакции растения на проникновение возбудителя — чувьствителностью растения к индуктивным веществам патогена (БПК и др.). Формирование признаков вирулентности, видимо, направлено на преодоление тканевоспецифичных свойств вида растения и связано с генетическим аппаратом патогена, так и растения.

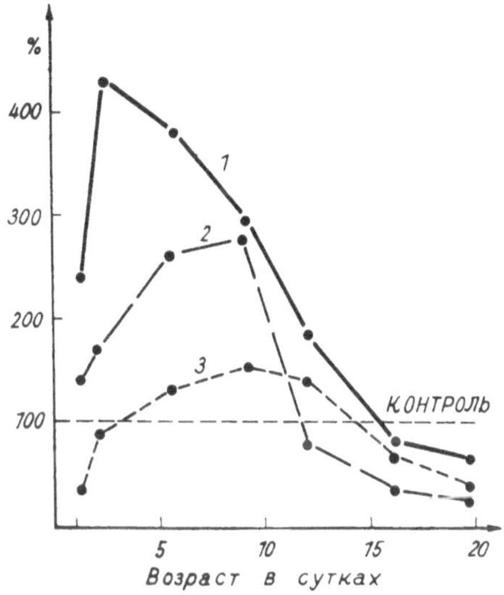

Рис. 5. Влияние препаратов (1—0,5% лиофилизата культуральной жидкости *Nostoc muscorum*, 2—0,1% раствора $CuSO_4$ и 3 — 0,25% раствора БПК) на интенсивность образования низкомолекулярных белков в листьях хлопчатника (в % от контроля)

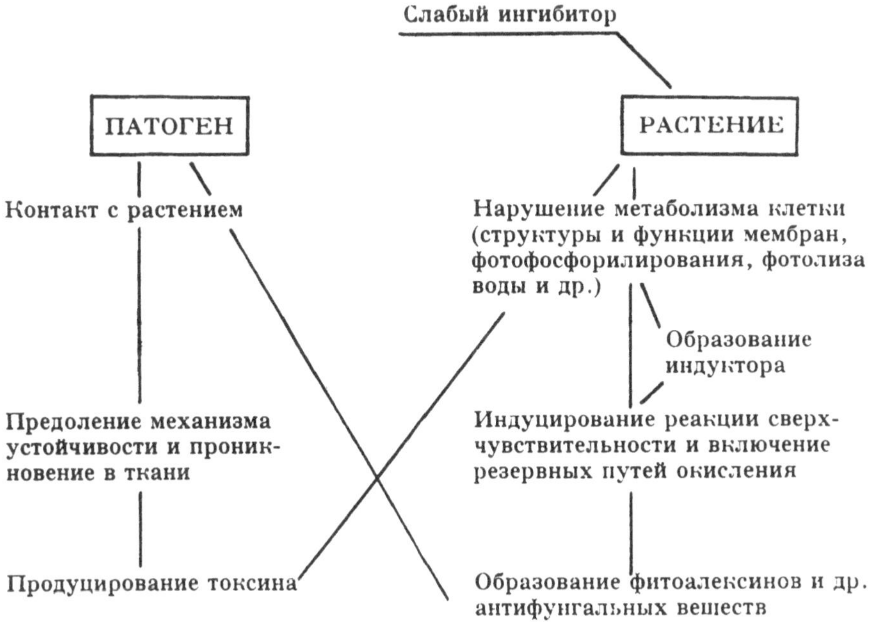

Рис. 6. Схема взаимоотношения патоген-токсин-растение-фитоалексин-патоген при вертициллезном вилте хлопчатника

Наиболее важным моментом патогенеза является прониковенце возбудителя внутрь растения. На этом этапе, как и на последующих, важную роль выполняют токсические вещества возбудителя, находящиеся на поверхности клеточной оболочки патогена. Избирательно нарушая клеточные мембран и тем самым активируя репаративные системы растения, они облегчают процесс проникновения возбудителя. При этом, видимо, если индуцирование реакции сверхчувствительности протекает медленно и под воздействием значительных количеств токсических метаболитов, патоген получает возможность какойто период беспрепятственно развиватсья и своевременно перемещается из зоны, подвергнутой антифунгальной блокаде в области неповрежденных тканей. При достаточной чувствительности растения к метаболитам патогена растение к метаболитам патогена растение локализует его по месту проникновения и ликвидирует. Используя химические препараты, нарушающие обмен веществ, можно воздействовать на биохимические процессы, участвующие в сфере взаимоотношений патоген-растения и искусственно повысить иммунитет. Однако следует учитывать, что процесс наведенной устойчивости не идентичен процессу, развивающемуся при контакте растения с патогеном; он одномоментен, затрагивает только часть комплекса реакций взаимоотношения и функционирует короткое время. Исходя из этих соображений, искусственное повышение устойчивости растений логичнее использовать в народном хозяйстве в качестве профилактического средства (допосевная обработка семян) или в сочетании с антибиотикотерапией в периоды угрозы массового поражения растений возбудителями.

В заключении авторы выражают искреннюю благадорность зав. отделом биофизики Института биохимии АН УзССР член.-корр. АН УзССР Ташмухамедову Б. А. за помощь в изучении действия препаратов на БФМ и зав. лабораторией иммунитета растений ИНЭБР АН УзССР Авазходжаеву М. за помощь в изучении действия токсических веществ *V. dahliae* на скорость образования фитоалексинов госсиполового ряда.

Diskussion

Schmalfuhs: Welche fungiziden Wirkstoffe werden zur Bekämpfung von *Verticillium* an Baumwolle in der UdSSR eingesetzt?

Borodin: Es werden zahlreiche in der UdSSR produzierte Fungizide eingesetzt, die sowohl für den Einsatz in Baumwolle als auch zur Bekämpfung von *Verticillium* geeignet sind.

Schmalfuhs: Haben die angewandten Fungizide auf die Zusammensetzung und Struktur der besprochenen Eiweißkomplexe einen Einfluß?

Borodin: Dies hängt von der Konzentration der Eiweißkomplexe ab.

Литература

[1] Бородин Г. И. и В. И. Рунов, Изучение природы некоторых токсических продуктов *Verticillium dahliae* Kleb., Труды ,,I конференции биохимиков республик Средней Азии и Казахстана. ,,ФАН“ УзССР, Ташкент, 1967, 434—436

[2] Бородин Г. И., Б. С. Салихова, С. М. Ходжибаева, Л. И. Чепенко, О. Ф. Карпачева, В. И. Рунов, О. С. Отрощенко и А. С. Садыков. О характере взаимоотношения патоген-токсин-растение при вертициллезном вилте хлопчатника. Доклады ,,II конференции биохимиков республик Средней Азии и Казахстана“, ,,ЫЛЫМ“, Фрунзе, 1976, сер. ,,биохимия растений“, 37—39

[3] Бородин Г. И., Ж. Сафиязов, С .М. Ходжибаева, О. Ф. Филатова и В. И. Рунов. Действие токсических веществ *Verticillium dahliae* Kleb. на клетки листьев хлопчатника. Узб. биол. журнал № 4 (1978), 15—18
[4] Бородин Г. И., Л. И. Чепенко, С. М. Ходжибаева, В. И. Рунов, О. С. Отрощенко, А. С. Садыков и С. З. Мухамеджанов. Токсические вещества у различных биотипов, рас, видов *Verticillium dahliae* Kleb. В сб. ,,Фитотоксические свойства почвенных микроорганизмов'', ,,ВНИИСХМ'', Ленинград 1978, 235—243
[5] Воронцов Л. А., Влияние вилта на фотосинтетический аппарат и окислительный метаболизм хлопчатника. В сб. ,,Физиология хлопчатника'', ,,Колос,,, Москва 1977, 178—220
[6] Красильников Н. А., С. М. Ходжибаева и Т. Г. Мирчик, Образование токсинов грибом *Verticillium dahliae* Kleb. возбудителем вилта хлопчатника. ,,Аргохимия'' № 10 (1965) 120—134
[7] Малышева К. М. и С. Щ. Зельцер, Белково-липидно-полисахаридный комплекх культуральной жидкости и мицелия гриба *Verticillium dahliae* Kleb. возбудителя вертициллезного вилта хлопчатника. Док. АН СССР, 179 (1968), 231—234.
[8] Рунов В. И. и Г. И. Бородин, Физиология и биохимия возбудителей вилта хлопчатника, ,,ФАН'' УзССР, Ташкент, 1970
[9] Салихова Б. С., Г. И. Бородин, В. И. Рунов и Л .И. Чепенко, Действие токсических соединений мицелия *Verticillium dahliae* Kleb. на газообмен листьев хлопчатника. Узб. биол. журнал № 5, (1970), 28—31
[10] Чепенко Л. И., Г. И. Бородин, В. И. Рунов и Б. С. Салихова, Выделение и характеристика вертициллинов — токсических веществ *Verticillium dahliae* Kleb. В сб. ,,Иммунитет и покой растений''. ,,Наука'', Москва 1972, 151—160
[11] Чепенко Л. И., Г. И. Бородин, Л. Н. Тен, В. И. Рунов, О. С. Отрощенко и А. С. Садыков, Методы выделения и физико-химическая характеристика вертициллина *Verticillium dahliae* Klel. В сб. ,,Экология и физиология микроорганизмов, водорослей и водных растений''. ,,ФАН'', АН УзССР, Ташкент 1973, 17—25
[12] Чепенко Л. И., Е. А. Васильева, Г. И. Бородин и В. И. Рунов, О природе некоторых пигментов *Verticillium lateritium* Узб. биол. журнал № 6 (1974), 8—10
[13] Green, R. J., Jr., A preliminary investigation of toxin produced in vitro by *Verticillium albo-atrum*, Phytopathology **44** (1954), 433—437
[14] Keen, N. T., M. Long and D. C. Erwin, Possible involvement of a pathogen-produced protein lipopolysaccharide complex in *Verticillium* wilt of cotton, Physiological Plant Pathology **2** (1972), 317—331
[15] Kiessig, R., and R. Haller-Kiessig, Beiträge zur Kenntnis einer infektiösen Welkekrankheit der Luzerne (*Verticillium albo-atrum* R and B), Phytopathologische Zeitschrift, **31** (1957), 185—222
[16] Stoddart, J. L., and A. H. J. Carr, Properties of wilt-toxins produced by *Verticillium albo-atrum* Reinke Birth, Annals of Appied Biology **58** (1966), 8192
[17] Zähner, H., W. Keller-Schierlein, R. Hütter, K. Hess-Leisinger und A. Deer, Stoffwechselprodukte von Mikroorganismen, 40. Mitt. Sideramine aus Aspergillaceen, Arch. Mikrobiol., **45** (1963), 119—135

М. Н. Талиева, В. В. Мазин, Л. В. Рункова

Главный ботанический сад АН СССР, Москва, СССР

Об Участии Фитогормонов в Защитных Реакциях Растения к Фитопатогенным Грибам

Поверхность листьев часто служит входными воротами инфекции, и покровной ткани принадлежит роль первой линии обороны растения, препятствующей прорастанию спор патогена и его проникновению в субэпидермальные ткани. Изучение барьерной функции покровой ткани луков было направлено на выявление физиолого-биохимических параметров активных защитных реакций в фазе первичных контактов растения-хозяина и патогена. Изучали особенности распределения фенольных, индольных соединений и веществ, обладающих цитокининовой активностью, в эпидермальной и субэпидермальной тканях листьев видов *Allium*, а также изменение уровня этих веществ под влиянием инфекции *Botrytis allii* и *Peronspora destructor* Определение анализируемых веществ проводили раздельно в энидермисе и мезофиле листьев луков здоровых и инокулированных конидиями фитопатогенов.

Инокулировали средние части одновозрастных листьев в константных оптимальных для заражения условиях влажности и температуры по ранее описанной методике (1). Полифенолы измеряли спектрофотометрически. Сумму хлорогеновых кислот определяли по Цуккеру и Аренсу (2) сумму флавоноидов — по реакции с $AlCl_3$ после смывания колонок Al_2O_3. Индольные вещества исследовали методом двух биотестов (3). Определение цитокининов проводили модифицированным метедом амарантусового биотеста с использованием проростков *Amaranthus caudatus* L. (4). Ферменты — фенилаланинаммиаклиазу (ФАЛ) и пероксидазу анализировали одновременно в параллельных навесках интактной и инокулированной ткани листьев, ФАЛ определяли методом Цуккера (5), пероксидазу спектрофотометрически по реакции с гваяколом.

Было установлено, что общий уровень фенольных веществ — флавоноидов (ФЛ) и хлорогеновых кислот (ХК) в эпидермальной ткани луков значительно выше, чем в мезофилле, что подтверждает ее барьерные функции в отношении патогенных организмов (табл. 1). Из сопоставления анатомического строения листьев луков с полученными данными по полифенолам видно, что они сосредоточены в поверхностом слое толщиной 2,8—4,3 мкм (высота эпидермальных клеток у разных видов луков). На долю эпидермиса, вес которого составляет в 1 г листа от 4,7 до 13,6%, приходится основная часть фенольных соединений — от 69 до 91,2% ФЛ и от 53,3 до 90% ХК (6). Мезофилл, составляющий основную часть навески, содержит незначительное количество фенолов. Подобное распределение фенольных веществ еще раз подчеркивает ту важную роль, которую эти соедин-

Таблица 1
Содержание фенольных соединений в эпидермисе и мезофилле листьев лука (мкг на 1 г веса сырого вещества).

Вид лука	Сумма ХК		Сумма ФЛ	
	эпидермис	мезофилл	эпидермис	мезофилл
A. nutans	467	51	5076	781
A. altissimum	40	17	1203	540
A. aflatunense	32	28	1268	152
A. pskemense	311	24	1878	189
A. galatum	128	21	1424	109
A. oschanini	315	78	1036	98

ения играют в защитных реакциях растительного организма в ответ на инфекцию. Общебиологическое значение наблюдаемого распределения полифенолов в листьях луков, по-видимому, выступает за узкие рамки отдельных защитных реакций. Сравнительный анализ индольных веществ в эпидермисе и мезофилле листьев луков показал, что качественный и количественный состав их существенно различается. Содержание индольных соединений в эпидермальной ткани варьирует у исследованных видов, луков но, как правило, значительно уступает их содержанию в мезофилле листьев. Налицо и различия качественного порядка: в эпидермисе преобладают вещества с Rf 0,6—0,9, ингибирующие рост колеоптильного теста. Под влиянием инокуляции спорами фитопатогенных грибов в листьях лука неблюдалось резкое изменение уровня фенольных и индольных веществ. Так, через 16—20 часов под влиянием метаболитов, выделяемых прорастающими спорами, уровень индольных веществ как в эпидермисе, так и мезофилле повышается. Имеет место изменение количественного содержания веществ индольной природы, так и их качественного состава. Особенно резике изменения наблюдаются в мезофилле листьев всех видов луков (7). Содержание полифенолов — ФЛ и ХК также неизменно повышалось и достигало максимума з эпидермисе (1). Это свидетельствует о гистогенной специфике реакции растения на инфекцию. Итак, начальные этапы заражения стимулируют в растении биосинтез фенольных и индольных веществ. Полученные данные позволяют заключить о сопряженном характере изменений этих 2-х групп соединений, иллюстрирующем их тесную функциональную взаимосвязь на примере больного растения. Изучение уровня цитокининов в листьях луков показало, что минимальной цитокининовой активностью обладала водная фракция незараженных листьев (120% от контроля) со значением Rf 0,0—0,13 на бумажной хроматограмме. Эфирная фракция экстрактов таких листьев содержала лишь вещества — ингибиторы с Rf 0,0—0,13; Rf 0,13—0,25; Rf 0,25—0,38; Rf 0,75—0,88; Rf 0,88—1,00, которые подавляли образование пигмента соответственно на 35%, 30%, 30%, 26%, 22%, после инкубирования отрезков листьев во влажной камере было отмечено значительное увеличение цитокининовой активности зон хроматограммы с Rf 0,0—0,13 и Rf 0,13—0,25, достигавшей 180% и 150% соответственно. Небольшое ингибирование (18%, 21%) было отмечено в зоне хроматограммы с Rf 0,63—0,75 и Rf 0,75—0,88. Инокуляция листьев лука конидиями *Botrytis allii* вызвала значительное увеличение цитокининовой активности в зоне хроматограммы с Rf 0,0—0,13 до 220%, т. е. увеличивается количество вещества, находящегося в водной фракции, проявляющего цитокининовую активность и обнаруженного в интактных, инкубированных во влажной камере и зараженных листьях. Высокая активность биотеста (220%) обнаружена также в зоне хроматограммы с Rf 0,63—0,75. Указанное вещество не проявило активности ни в интактных ,ни в инкубированных во влажной камере листьях, а только в ответ на заражение *Botrytis allii*. Остальные зоны хроматограммы экстрактов зараженных листьев лука проявили достоверное стимулирующее биотест действие в пределах от 130% до 150%. Таким образом, под влиянием прорастающих конидий гриба наблюдается как количественное увеличенис цитокининовой активности в зоне хромагограммы с Rf 0,0—0,13, так и появление нового вещества, обладающего такой активностью (Rf 0,63 до о,75). Полученные данные всидетельствуют о том, что изменение цитокининовой активности является одной из мобильных реакций растения на заражение и подтверждают физиологическую сопряженность двух систем физиологически активных соединений — ауксинов и цитокининов. Изучение активности ФАЛ и пероксидазы, ферментов на разных условиях контролирующих процесс лигнификации струкутуных элементов ткани, параллельно с определением суммы полифенолов показало тесную сопряженность изменения этих показателей (табл. 2). Более высокий уровень активности ФАЛ и пероксидазы, аккумуляция фенольных веществ в ин-

Таблица 2
Содержание полифенолов и активность ФАЛ в листьях под влиянием инокуляции конидиями Botrytis allii.

Полифенолы (в мкг/1 г)				Активность ФАЛ (в усл. ед./1 г)			
$К_{16}$		О		$К_{16}$		О	
эпидермис	мезофилл	эпидермис	мезофилл	эпидермис	мезофилл	эпидермис	мезофилл
			A. v ctorialis				
263	65	464	87	20	10	70	30
389	92	660	80	30	20	85	25
			A. altissimum				
34,4	10.2	46.4	13.4	40	22	76	14
130	29	200	27	30	20	65	20
			A. pskemense				
410	11	923	90	130	97	360	78
665	40	892	37	132	10	180	18
284	25	455	82	168	34	1104	46

Ппримечание. $К_{16}$ — листья через 16 час. инкубации в темноте, О — ионкулированные листья через 16 час.

фицированной ткани на начальном этапе взаимодействия растения и патогена создают основу неспецифической защитной реакции и являются ее необходимыми физиолого-биохимическими параметрами. Эти показатели — необходимое, но не всегда достаточное условие активной защитной реакции, конечным этапом и результатом которой в ряде случаев является синтез лигнина.

Литература

[1] Талиева М. Н., Л. В. Рункова, Изменение содержания фенольных соединений покровной ткани луков под влиянием инокуляции конидиями *Botrytis allii* Munn Микология и футопатология 10 (1976), 108—110

[2] Zucker, M., J. Ahrens, Quantitative assay of chlorogenic acid, Plant Physiol. **33** (1958), 246—249

[3] Рункова Л. В., Исследование ауксинов методом биотестов В кн. Рост ратений и природные регуляторы Москва Наука 1977, 52—59

[4] Мазин В. В., Л. С. Шашкова, Л Н. Андреев, Е. И. Комизерко, И. М. Жлоба, В. И. Кефели, Специфичность влияния кинетина на образование амарантина у щирицы (*Amaranthus caudatus* L.) и на рост каллуса семядоли сои (*Glycine soja* L.) Доклады АН СССР **23** (1976), 506—509

[5] Zucker, M., Sequential induction of phenylalanin ammonia — lyase and a lyase-inactivating system in potato tuber disks, Plant Physiol. **43** (1968), 365—374

[6] Рункова Л. В., М. Н. Талиева, О распределении фенольных соединений в листьях некоторых видов рода *Allium* Доклады АН СССР **225** (1975), 232—234

[7] Рункова Л. В., М. Н. Талиева, О распределение индольных соединений в листьях луков и их изменении под влиянием грибной инфекции. В сб. ,,фитогормоны — регуляторы роста растений'' Москва Наука 190, 64—73

J. AUGUSTIN, Ľ. DROBNICA and P. KRISTIÁN

Department of Microbiology and Biochemistry, Slovak Polytechnical University, Bratislava, Czechoslovakia

Antimicrobial Activity of Dithiocarbamates and Other Sulfur Containing Compounds

Some synthetic organic compounds decompose to isothiocyanates spontaneously in aqueous solutions at laboratory temperature. Thus they resemble aglucones which are formed by enzymatic hydrolysis of glucosinolates. On the basis of this analogy with natural substances, MARTIN [3] introduced the German name „Synthetische Senfölbildner“ for synthetic compounds, which spontaneously decompse to isothiocyanates. The name appeared in English literature as "Mustard oil formers". Presently, all these compounds belong to the group of antimicrobially active sulfur compounds. A more detailed review dealing with physico-chemical properties of mustard oil formers can be found in our previous papers (AUGUSTIN [1], DROBNICA et al. [2]). In this paper physico-chemical constants of decomposition of the compounds are given.
The rate of decomposition of monothiocarbamates, dithiocarbamates and their esters and 3-substituted rhodanines in acidic, neutral and alkaline solutions was calculated from spectrophotometric measurements in absorption maxima in ultraviolet region. The results, expressed as observed rate constants k_{obs}, dependent on pH value of the reaction mixture and half life of decomposition are given in Tables 1–6.
The only primary product of decomposition of monothiocarbamates in acicid, neutral and mild alkaline conditions at laboratory temperature is an amine and carbonyl sulfide. Isothiocyanate, once reacted with OH^- ions cannot be regenerated by reverse reaction.

Table 1
Rate constants k_{obs} of decomposition n-monosubstituted monothiocarbamates in neutral and acid conditions at 25 °C 0,1 M phosphate buffer (equation 1)

$R-NH-CS-O^-$ R =	$\frac{k_{obs}}{[H^+]} \cdot 10^6$ § ($l \cdot mol^{-1} \cdot min^{-1}$)	$t_{1/2}$ at pH 7.4 (min)
n-butyl-	2.09	11.5
benzyl-	1.96	12.3
4-bromobenzyl-	1.23	19.7
4-methoxyphenyl-	52.5	0.38
phenyl-	29.7	0.67
4-bromophenyl-	14.4	1.38
4-acetylphenyl-	6.25	3.84

0,1 M phosphate buffer pH 6,0–7,6
0,1 M borate bu/fer pH 7,2–9,1

Table 2
Rate constants k_{obs} of decomposition of n-aryl- and n-aralkyl dithiocarbamates in acidic solutions at 25 °C (equation 2)

$R-NH-CS-S^-$ R =	$\frac{k_{obs}}{[H^+]} \cdot 10^4$ $(l \cdot mol^{-1} \cdot min^{-1})$	$t_{1/2}$ at pH 5 (min)
benzyl-	0.00645	1070
4-methoxyphenyl-	1.81	3.82
4-etoxyphenyl-	1.89	3.67
phenyl-	2.91	2.38
4-bromophenyl-	3.00	2.31
4-acetylphenyl-	3.40	2.00

Table 3
Rate constants k_{obs} of decomposition of n-aryl- and n-aralkyl-dithiocarbamates at pH 7.4 and 37 °C, 0.1 M phosphate buffer (equation 3)

$R-NH-CS-S^-X^+$ R =	X =	$k_{obs} \cdot 10^{-2}$ (min^{-1})	$t_{1/2}$ (min)
4-acetylphenyl-	Na^+	1.30	53.3
4-bromophenyl-	Na^+	1.29	53.8
phenyl-	NH_4^+	1.96	35.4
4-methoxyphenyl-	Na^+	3.00	22.8
benzyl-	NH_4^+	920	0.075

Table 4
Rate constants k_{obs} of decomposition of monothiourethans in mild alkaline solutions. 0.1 M phosphate buffer, pH 8.0 25 °C (equation 4)

$R-NH-CS-O-C_6H_5$ R =	$k_{obs} \cdot 10^{-2}$ (min^{-1})	$t_{1/2}$ (min)
n-butyl-	1.05	66.0
benzyl-	0.86	80.7
4-bromobenzyl-	0.72	96.3
4-methoxyphenyl-	29	2.40
4-methylphenyl-	21	3.30
4-bromophenyl-	71	0.97
4-acetylphenyl-	87	0.80

Alkali metal and ammonium salts of N-arylsubstituted dithiocarbamic acids decompose to antimicrobially inactive amines even in mild acidic solutions. At pH 5,0, what is typical for the cultivation of yeasts, half life of decomposition is within the range of 2—4 minutes. N-alkylsubstituted dithiocarbamates are much more stable under the same conditions, with half life about 24 hours. Under neutral and mild alkaline conditions such as pH 7,4, what is typical for bacterial cultivations, aryl derivatives decompose to corresponding isothiocyanates with half ranging from 20 to 60 minutes.

Table 5
Rate constants k_{obs} of decomposition of n-monosubstituted dithiocarbamic acid esters in mild alkaline solutions. 0.1 M phosphate buffer pH = 7.4 at 25 °C (equation 5)

$R_1-NH-CS-S-R_2$ R_1 =	R_2 =	$k_{obs} \cdot 10^{-2}$ (min^{-1})	$t_{1/2}$ (min)
benzyl-	propyl-	stable	a
benzyl-	benzyl-	stable	a
benzyl-	phenyl-	25	2.77
4-methoxyphenyl-	propyl-	0.88	78.8
phenyl-	propyl-	0.79	88.4
4-bromophenyl-	propyl-	0.90	77.0
phenyl-	benzyl-	3.4	20.4

a, Half life at 37 °C at given pH ~ 42 hours

Table 6
Half life of decomposition of n-substituted rhodanines in mild alkaline solution. 0.1 M Mc Ilvaine buffer pH 7.4 25 °C (equation 6a, 6b)

R–	$t_{1/2}$ (min)
phenyl-	160
4-methoxyphenyl-	180
4-etoxyphenyl-	239
4-acetylphenyl-	74.3
4-bromophenyl-	98.3
4-dimethylaminophenyl-	257
4-nitrophenyl-	59
4-methylphenyl-	204
4-carbethoxyphenyl-	74.3
3-nitrophenyl-	66.2
3-acetaminophenyl-	192
methyl-	670
n-butyl-	670
benzyl-	264
4-bromobenzyl-	210

O-Phenyl esters of N-arylmonosubstituted monothiocarbamic acids decompose to isothiocyanates quantitatively in neutral and mild alkaline solutions and in organic solvents such as dimethyl sulfoxide. N-alkyl- and N-aralkyl- derivatives are at least by two orders more stable under the same conditions.

O-Alkyl esters of N-monoaryl- and N-monoalkyl-monothiocarbamic acids are stable in acidic, neutral and strong alkaline solutions at laboratory temperature and do not form isothiocyanates.

Alkyl esters of N-alkyl and N-aralkyl substituted dithiocarbamic acids are stable in acidic neutral and mild alkaline solutions at laboratory temperature. Alkyl esters of N-arylsubstituted dithiocarbamic acids decompose to isothiocyanates in neutral and mild alkaline solutions.

N-arylsubstitued rhodanines decompose to isothiocaynates in neutral and mild alkaline solutions, however, N-alkylsubstituted rhodanines decyclise without formation of isothiocyanates.

All the above mentioned reactions are pH dependent. It means that ionization equilibrium itself plays a decisive role in their stability in solutions and in their antimicrobial activity.

$$\text{R-NH-CSO}^- \xrightarrow{+H^+} \text{R-NH}_2 + \text{COS} \quad (1)$$

$$\text{R-NH-CS-S}^-\text{X}^+ \xrightarrow{+H^+} \text{R-NH}_2 + \text{CS}_2 \quad (2)$$

$$\text{R-NH-CS-S}^-\text{X}^+ \rightleftharpoons \text{R-NCS} + \text{SH}^- \quad (3)$$

$$\text{R-NH-CS-O-C}_6\text{H}_5 \rightarrow \text{R-NCS} + \text{HO-C}_5\text{H}_6 \quad (4)$$

$$\text{R-NH-CS-S-R}_2 \rightleftharpoons \text{R}_1\text{-NCS} + \text{HS-R}_2 \quad (5)$$

$$\text{Ar-N-SC-S-CH}_2\text{-CO} \xrightarrow{+H_2O} \text{Ar-NCS} + \text{HOOC-CH}_2\text{-SH} \quad (6a)$$

$$\text{Alk-N-CS-S-CH}_2\text{-CO} \xrightarrow{+H_2O} \text{Alk-NH-CS-S-CH}_2\text{-COOH} \quad (6b)$$

References

[1] Augustín, J., Mustard oil formers. Physico-chemical, biochemical and biological aspects, Assoc. Prof. Thesis, 237 pp. Slovak Polytechnical University, Bratislava, 1970

[2] Drobnica, Ľ., P. Kristián, and J. Augustín, The chemistry of the -NCS group. In: Patai S. (Ed.). The chemistry of cyanates and their thio derivatives, Interscience, New York 1977, Vol. 2, 1003–1221

[3] Martin, D., Synthetische Senfölbildner mit bakterizider, fungizider, nutritiver und das Wachstum von Viren hemmender Wirkung, Akademie-Verlag, Berlin, 1962

H. Bochow und M. S. H. Moustafa

Sektion Gartenbau der Humboldt-Universität zu Berlin, DDR

Untersuchungen zur Fungizidtoleranz von *Botrytis cinerea* Pers.

Zur Auffindung weiterer Hinweise über die praktische Bedeutung der Entwicklung einer Fungizidresistenz von Pilzen nach Anwenden von Benzimidazol-Fungiziden, im besonderen bei der Nacherntebehandlung von Kopfkohl (*Brassica oleracea* conv. *capitata* (L.) Alef var. *capitata*) zur Bekämpfung des Fäuleerregers *Botrytis cinerea* Pers. während der Kopfkohllagerung, wurde das Verhalten einiger *B.cinerea*-Isolate von Benomyl-behandeltem und unbehandeltem Kopfkohl untersucht. Zusammenstellungen über bisherige Erfahrungen einer Resistenz bei *Botrytis cinerea* gegenüber Benzimidazolen liegen bei Fehrmann [2] vor. Die Nacherntebehandlung von Kopfkohl mit Benzimidazolen ist bei Bochow und Hentschel [1] beschrieben.

Tabelle 1
Fungizidempfindlichkeit von 3 Botrytis cinerea-Isolaten, R 2 U, R 2 B und R 2 U'

Fungizide	ppm	Isolate R 2U		R 2B		R 2U'	
		m	v	m	v	m	v
Benomyl	0,1	32	99	90	0	90	0
	1,0	0		13	96	0	
	10,0	0		5	171	0	
Thiophanat-	0,5	0		15	61	0	
Methyl	5,0	0		0		0	
	50,0	0		0		0	
Carbendazim	0,1	53	34	35	38	35	47
	1,0	0		0		23	8
	10,0	0		0		0	
Glycophen	0,1	45	1	48	4	47	0
	1,0	0		24	19	20	13
	10,0	0		18	68	0	
Thiram	5,0	80	3	90	0	90	0
	50,0	0		35	23	43	10
	500,0	0		25	13	23	22

m = Mittlerer Durchmesser der Pilzkolonie auf Agar nach 10 Tagen Wachstum
v = Variationskoeffizient (s %)

Gemessen am Myzelwachstum auf Agar zeigt Tabelle 1 für drei *B. cinerea*-Isolate (Populationsisolierungen) die Reaktion gegenüber verschiedenen Konzentrationen der Fungizide Benomyl, Thiophanat-Methyl, Carbendazim, Glycophen und Thiram an. Es handelt sich dabei um folgende Populationsisolierungen:

R 2 U: Isoliert von 4 Monate lagerndem Weißkohl, unbehandelt; geprüft kurze Zeit nach Inkulturnahme des Pilzes.

R 2 B: Isoliert ebenfalls von 4 Monate lagerndem Weißkohl der gleichen Partie wie bei R 2 U, jedoch nach voraufgegangener einmaliger Benomyl-Behandlung des Pflanzenmaterials vor der Einlagerung.

R 2 U': Eine Subkultur von R 2 U, die nach Inkulturnahme des Isolates Wachstum auf 0,1 ppm Benomyl-Agar zeigte.

Die Verhaltensweise der getesteten Isolate belegt, daß sowohl die einmalige Benomyl-Behandlung des Lagerkohls bei der nachfolgenden Besiedlung durch *B. cinerea* als auch die einmalige Subkultur der Isolate auf Benomyl-Agar (R 2 U') bereits zur Herausbildung fungizidresistenter Formen des Pilzes führte (R 2 B zu R 2 U, R 2 U' zu R 2 U). Die Resistenz des Isolates R 2 B erwies sich als mäßig gegenüber Benomyl (bis zu 10 ppm), Thiophanat-Methyl (0,5 ppm) und in einer unerwarteten Kreuzresistenz gegenüber Glycophen (10 ppm) und Thiram (500 ppm). Beim Isolat R 2 U' ergab sich bei mäßiger Resistenz gegenüber Benomyl gleichfalls Kreuzresistenz zu Glycophen und Thiram. Bisher war bei *B. cinerea* lediglich Kreuzresistenz zwischen Benzimidazolen und Glycophen beobachtet worden (HOLZ [3], LEROUX u. a. [5]).
Eine Überprüfung der genannten Isolate auf ihre Pathogenität durch Inokulation auf Scheiben von Kohlrabiknollen (*Brassica oleracea* var. *gongyloides* L.) ergab, gemessen an der Induktion von Fäulnis, keinerlei Unterschiede in Bezug zur Fungizidempfindlichkeit der *Botrytis*-Isolierungen. Den schwach fungizidresistenten Isolaten R 2 B und R 2 U' war die gleiche Pathogenität zu eigen, wie dem fungizidsensiblen Isolat R 2 U (Tab. 2).

Tabelle 2
Pathogenität der Botrytis cinerea-Isolate R 2 U, R 2 B und R 2 U' auf Kohlrabiknollenscheiben. Fäulniszone in mm um das Inokulum (Mittel aus 5 Wiederholungen), 7 Tage Inkubation

Isolate	Fäulniszone mm nach 5 Tagen	nach 7 Tagen
R 2 U	19	30
R 2 B	21	28
R 2 U'	23	29

$GD_{0,05} = 4{,}9$

Zur Induktion einer höheren Fungizidresistenz wurden Transferationen der *Botrytis*-Isolate auf verschieden konzentriertem Benomyl- und Thiophanat-Methyl-Agar vorgenommen. 7-malige Übertragungen erfolgten dabei unter Laborbedingungen sowohl jeweils geradlinig auf Benomyl- bzw. Thiophanat-Methyl-Agar als auch jeweils kreuzweise von Benomyl- auf Thiophanat-Methyl-Agar und umgekehrt. 3 von 8 subkultivierten Isolaten (R 2 U, R 2 B', ein R 2 B-Isolat, das einmal auf 10 ppm Benomyl wuchs, und K 2, ein Isolat von unbehandeltem Weißkohl vom Freiland aus Kaha, Ägypten) zeigten nach den 7-maligen Fungizid-Agar-Passagen eine hohe Resistenz gegenüber Benomyl bzw. Thiophanat-Methyl (Tab. 3): R 2 $U_{rb, rt}$; R 2 $B'_{rb, rt}$; K 2_{rt}. Die Resistenz blieb auch nach 3-maliger Kultur auf fungizidfreiem Agar konstant und schien damit genetisch fixiert. Sie belief sich nach Testung bei allen 3 For-

Tabelle 3
Fungizidempfindlichkeit von 3 Botrytis cinerea-Isolaten nach 7-maliger Übertragung auf Benomyl- und Thiophanat-Methyl-Agar

Ausgangs-isolat	Induzierte Form	Übertragen von Fungizid-Agar ppm		Wachstum (+) bzw. (−) nach 7 Transferationen auf Benomyl- u. Thiophanat-M.-Agar (ppm)		
R 2U	$R\,2U_{rb}$	Benomyl		Benomyl	10,0	(+)
				Thioph.	50,0	(+)
	$R\,2U_{rt}$	Thioph.	0,25	Benomyl	10,0	(+)
				Thioph.	50,0	(+)
R 2B′	$R\,2B'_{rb}$	Benomyl	0,1	Benomyl	10,0	(+)
				Thioph.	50,0	(+)
	$R\,2B'_{rt}$	Thioph.	0,5	Benomyl	10,0	(+)
				Thioph.	50,0	(+)
K 2		Benomyl	0,1	Benomyl	10,0	(−)
				Thioph.	50,0	(−)
	$K\,2_{rt}$	Thioph.	0,5	Benomyl	10,0	(+)
				Thioph.	50,0	(+)

men auf Verträglichkeiten von 100 ppm Benomyl, 5.000 ppm Thiophanat-Methyl und 1.000 ppm Glycophen, und ergab damit wiederum Kreuzresistenz zwischen Benomyl und Glycophen. Die Resistenzentwicklung gegenüber Thiophanat-Methyl erschien dabei in diesem Trainingstest schneller als an Benomyl vorsich zu gehen (z. B. K 2_{rt}, ausschließlich tolerant gegenüber Thiophanat-Methyl).

Im Trainingstest veränderten die Isolate im Verlaufe der Subkultur auf dem Fungizid-Agar von der 4. bis 5. Transferation ab, ihr morphologisches Aussehen. Im Vergleich zur Ausgangskultur (Stammkultur) verlor sich die Sporulation des Pilzes, die Myzelien waren deformiert, das Wachstum überwiegend resupinat und die Sklerotienbildung unterschiedlich.

Ein Vergleich des Myzelwachstums der nach den Passagen fungizidresistenten Isolate R 2 B'_{rt}; R 2 U_{rt} und K 2_{rt} mit den entsprechenden Ausgangs- oder Stammkulturen auf fungizidfreiem Agar, ergab eine deutlich verringerte Fitness der resistenten *Botrytis*-Formen (Abb. 1).

In Pathogenitätsversuchen auf Knollenscheiben von Kohlrabi erwiesen sich die fungizidresistenten Formen im Gegensatz zu den Ausgangskulturen als apathogen.

Es wurde der Frage nachgegangen, ob die vorliegende Verknüpfung von hoher Benzimidazol-Fungizidresistenz und geringer Fitness bzw. Apathogenität bei den resistenten *Botrytis*-Isolaten bei Rekultivierung der Pilze auf Kohlblattmaterial stabil bleibt. Zugleich wurde geprüft, wie sich eine Mischung von fungizid-resistentem und -sensiblem Pilzmaterial bei Inokulationen auf Kohlblättern verhält. Hierzu wurde eine Myzelmischpopulation des fungizidresistenten Isolates R 2 B'_{rb} mit der entsprechenden fungizidsensiblen Ausgangskultur des gleichen Isolates (R 2 B′) durch Zusammenwuchs auf Agar hergestellt. Diese Mischpopulation und die jeweiligen Partner allein — fungizidresistent, apapthogen R 2 B'_{rb} und fungizidsensibel, pathogen R 2 B′ — wurden 3mal nacheinander auf Weißkohlblätter inokuliert, eine Serie mit unbehandelten Blättern, eine zweite auf Blätter, die zuvor mit einer 0,1 %igen Benlate-Suspension besprüht wurden. Jede Inokulation der Kohlblätter wurde unterbrochen durch eine

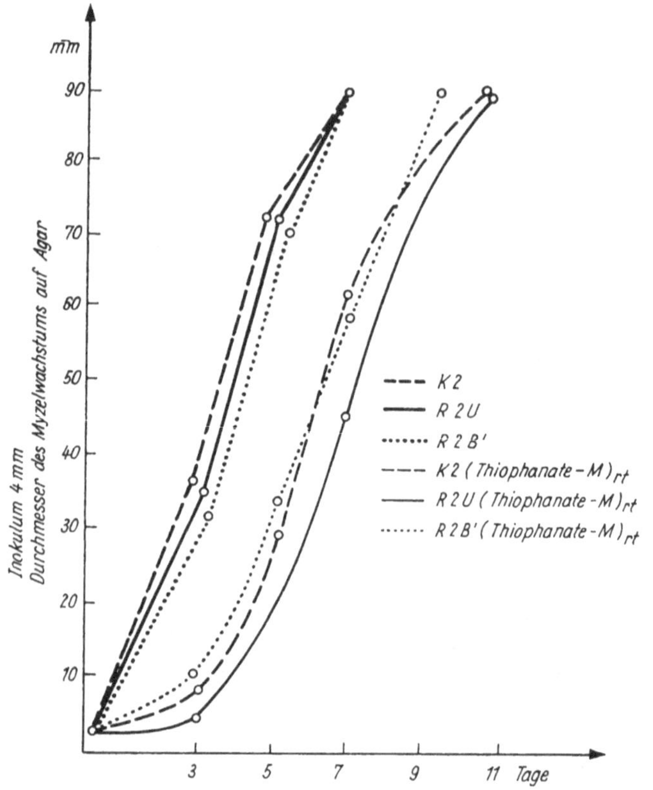

Abb. 1. Myzelwachstum MBC-resistenter (K 2_{rt}; R 2 U_{rt}; R 2 B'_{rt}) und sensibler *Botrytis cinerea*-Isolate (Ausgangsformen der resistenten Isolate: K 2; R 2 U; R 2 B') auf fungizidfreiem Agar

Tabelle 4

Pathogenitätstest mit dem apathogenen, Benomyl- und Thiophanat-M.-resistenten B. cinerea Isolat R 2 B'_{rb} auf Weißkohlblättern, unbehandelt und mit 0,1%iger Benlate-Suspension besprüht, vergleichsweise zur pathogenen, fungizidsensiblen Ausgangsform R 2 B' sowie einer Mischpopulation beider Formen (1:1). Inkubationszeit 14 Tage. 3-malige Folgetestung als Passage

Isolate	Fungizidbehandlung der Blätter	Fäulniszone (mm) — relativ Test 1		Test 2		Test 3	
R 2B'	unbehandelt	(39)	100	(41)	100	(41)	100
	Benomyl	(10)	26	(15)	37	(20)	49
R 2 B'_{rb}	unbehandelt	(10)	100	(13)	100	(21)	100
	Benomyl	(11)	110	(12)	92	(20)	95
R 2B' +	unbehandelt	(34)	100	(36)	100	(40)	100
R 2 B'_{rb}	Benomyl	(14)	41	(16)	44	(23)	57

15-tätige Rekultivierung der Pilzisolate auf fungizidfreiem Agar, um zur Fortsetzung der Blattinfektionen ein ausreichendes Inokulumpotential zu gewinnen. Tabelle 4 zeigt die Fäulnisinduktion der gewählten Isolate bei den 3 sukzessiven Tests auf Kohlblättern sowie den Grad der Einschränkung des Befalles durch die Benomyl-Vorbehandlung der Blätter. Im 3. Test waren nicht nur die Ausgangskultur und Mischpopulation sondern auch die vormals apathogene, fungizidresistente Form pathogen. Be-

Tabelle 5
Fungizidempfindlichkeit der Botrytis cinerea-Isolate R 2 B', R 2 B'_{rb} und der Mischpopulation R 2 B' + R 2 B'_{rh} nach den Wirtspassagen (= Tab. 4)

Isolate	Wachstum auf Benomyl-Agar (ppm) in mm nach 10 Tagen					
	0	0,1	1,0	2,5	5,0	10,0
R 2 B'	8,7	7,5	5,1	–	–	–
R 2 B'_{rb}	5,5	5,3	5,4	5,3	5,0	4,6
R 2 B' + R 2 B'_{rb}	7,7	7,3	6,5	5,3	5,2	4,2

merkenswert ist dabei die Beeinflussung der 3 verschiedenen *Botrytis*-Formen durch die Blattbehandlung mit Benomyl. Während die fungizidsensible Form in vivo erwartungsgemäß durch die Benomyl-Behandlung der Blätter reduziert wurde, war dies bei der Mischpopulation entsprechend der Partnerkombination nur begrenzt der Fall und bei der fungizidresistenten Form waren Fäulnisinduktion auf behandelten Blättern nahezu gleich.

Eine Überprüfung der Fungizidempfindlichkeit der Formen nach den Kohlblattinokulationen in vitro auf Benomyl-Agar (Tab. 5) bewies, daß die Differenzierungen der Benomyl-Resistenz bei den 3 Formen im Verlaufe der Tests konstant geblieben waren. Die Untersuchungen belegen für die vorliegenden *Botrytis cinerea*-Isolate, daß die festgestellte Benzimidazolresistenz und die Fitness und Pathogenität dieser Formen nicht unmittelbar in Verbindung stehen. In vivo vermochte sich die Pathogenität und Fitness zu restaurieren bzw. (Mischpopulationsform) zu erhalten bei Verbleiben der Fungizidresistenz, so daß für beide Merkmale eine getrennte genetische Fixierung anzunehmen ist, wie sie auch von Sozzi und Gessler [6] vermutet wird.

Unsere Beobachtungen über das Verhalten fungizidresistenter *Botrytis cinerea*-Formen legten die Vermutung nahe, daß die resistenten Formen gegenüber den sensibleren nicht nur reaktionsunempfindlicher gegen die geprüften Benzimidazole waren, sondern eventuell auch über eine Detoxifikation der Fungizide sich selbst ständig günstigere Wuchsbedingungen auf behandelten Substraten schafften, sofern sie im Wachstum zunächst ‚Fuß gefaßt' hatten. Wir überprüften deshalb Möglichkeiten einer Detoxifikationswirkung von wachsendem Myzel der Pilzisolate auf Benomyl, wie sie z.B. von Kataria und Grover [4] als eine mögliche Mechanismus-Form für die Fungizidresistenz angesehen wird.

Das Isolat R 2 B'_{rb} und seine Ausgangskultur (fungizidsensibel) R 2 B' wurden parallel auf fungizidfreier Biomalznährlösung kultiviert. Nach 3-wöchiger Wachstumszeit wurde von Wiederholungen mit gleichem Myzelgewicht bei den Kulturen jeder Form die Kulturflüssigkeit entnommen und steril filtriert. Diese sterile Kulturflüssigkeit wurde anschließend mit 1,0 ppm Benomyl versetzt und runde Filterpapierscheiben damit getränkt. Die Scheiben wurden auf fungizidfreiem Agar ausgelegt, der mit der fungizidsensiblen Ausgangskultur R 2 U inokuliert wurde. Abbildung 2 zeigt nach 10 Tagen Wachstum dieses Pilzes bei 1 unter dem Einfluß der von der fungizidresistenten Form herrührenden Benomyl-versetzten Kulturfiltrate das Wuchsbild und bei 2 dasjenige unter Einwirkung der von der sensiblen Form stammenden Filtrate. Wäh-

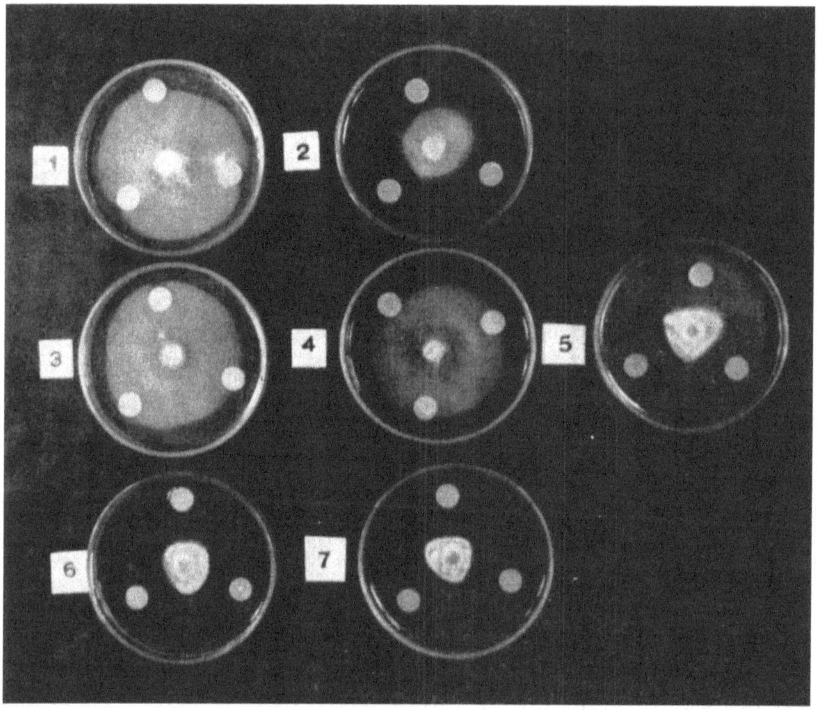

Abb. 2. Wachstum eines fungizidsensiblen *Botrytis cinerea*-Isolates (R 2U) unter dem Einfluß von Kulturfiltraten MBC-resistenter und -sensibler *Botrytis*-Formen, mit Benomyl-Zusatz.

1 = Kulturfiltrat von Isolat R 2 B'_{rb} MBC-resistent + Benomyl
2 = Kulturfiltrat von Isolat R 2 B′ MBC-sensibel + Benomyl
3 = Kulturfiltrat von Isolat R 2 B'_{rb} nach Wachstum bei Benomyl-Zusatz
4 = Kulturfiltrat einer schwach MBC-resistenten Form des Isolates R 2 B′ nach Wachstum bei Benomyl-Zusatz
5 = Kulturfiltrat von Isolat R 2 B′ nach Wachstum bei Benomyl-Zusatz
6 = Benomyl-Suspension pH 5,8
7 = Benomyl-Suspension pH 4,5

rend im ersten Falle das hinzugesetzte Benomyl keinerlei Wirkung ausübte — aufgrund einer vermuteten Detoxifikation durch das Kulturfiltrat — ließ sich im zweiten Falle eine Hemmwirkung des Fungizides allerdings auch etwas eingegrenzt, beobachten.
In einem weiteren Versuch wurden wiederum die beiden *Botrytis*-Formen auf Biomalzflüssigkeitsmedium kultiviert, jedoch unter Zusatz von 1,0 ppm Benomyl bereits zu Beginn der Kultur. Die fungizidresistente Form wuchs, während die sensible im Wachstum stagnierte. Die Verwendung der steril filtrierten Kulturlösungen im Papierscheibentest gegen die fungizidsensible Form auf Agar zeigt in der Abbildung 2 bei Nr. 3 das Wachstum bei getränkten Filterpapierscheiben mit Benomyl-Kulturfiltrat von der fungizidsensiblen Form. Bei einer der Wiederholungen trat bei der Vorkultur der fungizidsensiblen Form in der 1,0 ppm Benomyl enthaltenen Nährlösung Wachstum auf. Die Verwendung des Kulturfiltrates von dieser Variante im Filterscheibentest gegen den fungizidsensiblen Typ des *Botrytis*-Pilzes ergab das in Abbildung 2 un-

ter Nr. 4 dargestellte Wuchsbild der Testform. Nr. 5 zeigt dagegen die Verhältnisse unter Verwendung von Kulturfiltrat der resistenten Form, die mehr als bei Nr. 4 auf eine weitgehende Detoxifikation des Benomyls verweisen.
Ein direkte Tränkung von Filterpapierscheiben mit einer 1,0 ppm Benomyl-Suspension, die jedoch auf verschiedene pH-Werte eingestellt wurde, führte zur Wachstumsbeeinflußung des fungizidsensiblen Testpilzes wie unter Nr. 6 und Nr. 7 in der Abbildung 2 wiedergegeben. Die Suspension bei Nr. 6 und Nr. 7 hatte einen pH-Wert von 5, 8 und Nr. 7 von 4,5. Innerhalb dieses Bereiches bewegten sich auch die pH-Werte der verwendeten Kultufriltrate in den dargestellten Versuchsvarianten: Nr. 1 = 4,59; Nr. 2 = 4,35; Nr. 3 = 4,52; Nr. 4 = 4,82; Nr. 5 = 5,58. Die eingetretenen Beeinflussungen der Benomyl-Wirkung durch Kulturfiltrate von *Botrytis cinerea* können deshalb ihre Ursache nicht in pH-Veränderungen der Medien haben. Auch hier nicht dargestellte Kontrollen von Kulturfiltraten der Isolate ohne Benomyl-Zusätze übten keinen unterschiedlichen Einfluß aus. Sie verhielten sich wie Aqua dest.
Es wird angenommen, daß die untersuchten Formen durch oder aus dem wachsenden Myzel Detoxifikationswirkungen auf Benomyl hervorrufen, wobei die fungizidresistenten Formen zu diesem weit mehr befähigt sind als die entsprechend sensibleren Formen. Wachstum des Pilzes und Detoxifikation des Benomyls stehen dabei in positiver Wechselwirkung.

Zusammenfassung

Es konnte bestätigt werden, daß *Botrytis cinerea* einen hohen Grad an Anpassungsfähigkeit besitzt und leicht zur Herausbildung von resistenten Formen gegenüber Spezialfungiziden, wie Benzimidazolen, neigt. Bei lang anhaltender Einflußdauer der Fungizde, wie bei der Behandlung lagernden Pflanzenmaterials, gilt dies besonders. Auffällig ist die leichte Entwicklung von Kreuzresistenz, die zwischen MBC-Fungiziden Glycophen und Thiram beobachtet wurde.
Es zeigte sich ferner, daß die Eigenschaften Fungizidresistenz und Fitness bzw. Pathogenität bei *Botrytis cinerea* unabhängig voneinander gesteuert werden können. Eine Kopplung hoher Fungizidresistenz mit geringer Fitness ist in einer Population von *B. cinerea* veränderlich, indem auf natürlichen Substraten bei Beibehaltung der Resistenz die Fitness sich wieder entwickeln kann.
Bezüglich des Resistenzmechanismus von *B. cinerea* gegenüber Benzimidazolen (Benomyl) scheinen unter anderem auch von wachsenden Myzel ausgehende, über die Kulturflüssigkeit zum Tragen kommende Detoxifikationserschcinungen bedeutsam zu sein. Sie wurden durch entsprechende Experimente ermittelt.

Literatur

[1] Bochow, H. und K. D. Hentschel, Phytosanitäre Maßnahmen zur Bekämpfung pilzlicher Fäuleerreger bei der Gemüselagerung, Nachrbl. Pflanzenschutz DDR **33** (1979), 97–101

[2] Fehrmann, H., Systemische Fungizide — ein Überblick. II. Fungizidresistenz phytopathogener Pilze, Phytopath. Z. **86** (1976), 144–185

[3] Holz, B., About one case of resistance of *Botrytis cinerea* on grapes against the new contact fungicides Ronilan, Rovral and Sumisclex for control of *Botrytis cinerea*, Biol. Bundesanst. f. Land- u. Forstw., Berlin (West) u. Braunschweig, Jahresbericht 1978, H. 44

[4] Kataria, H. R. and R. K. Grover, Adaptation of *Rhizoctonia solani* to systemic and non-systemic fungitoxicants, Z. Pflanzenkrankh. **81** (1974), 472–478

[5] Leroux, P. Fritz, R. und M. Gredt, Laboratory studies on strains of *Botrytis cinerea* Pers. resistant to dichlozoline, dichloran, quintozene, vinchlozoline and 26019 R (glycophene), Phytopath. Z. **89** (1977), 347–358

[6] Sozzi, D. and C. Gessler, Fungicide (MBC) resistant mutants of *Fusarium oxysporum* f. sp. *lycopersici* and *Botrytis cinerea*, pathogenicity and fitness, Phytopath. Z. **97** (1980), 19–24

F. BÖTTCHER und I. A. SAMSONOVA

Ernst-Moritz-Arndt-Universität, Sektion Biologie, Greifswald, DDR

Zur fungiziden und genetischen Wirkung von Benomyl

Als Ursache der fungiziden Wirkung von Benzimidazolderivaten wird eine Blockierung der Mikrotubuli-Assemblierung angenommen [3]. KAPPAS und Mitarbeiter konnten zeigen, daß bei *Aspergillus nidulans* auch mitotisches Nondisjunction von Chromosomen durch Benomyl induziert werden kann [4, 5]. Inzwischen wurde Benomyl ebenfalls bei Hybriden zwischen *A. nidulans* und *A. rugulosus* [6], *Penicillium citrinum* und *P. cyaneo-fulvum* [1], *Saccharomyces cerevisiae* [7] und *Saccharomycopsis lipolytica* [9] erfolgreich zur Erzeugung mitotischer Segreganten eingesetzt. Diese Befunde waren nicht unerwartet, da Mikrotubuli offensichtlich eine essentielle Komponente des mitotischen Apparats darstellen. Bei Untersuchungen an der ascomycetalen Hefe *Pichia guilliermondii* und dem Heterobasidiomyceten *Rhodosporidium toruloides* fanden wir jedoch, daß fungizide Wirkung und mitotische Rekombination nicht miteinander korreliert sind.

Von uns wurden prototrophe Hefestämme untersucht, die für chromosomale Gene heterozygot sind. Die an *P. guilliermondii* G27 × G65 und *R. toruloides* Sp23-1 erhaltenen Ergebnisse sind repräsentativ für beide Arten. Die Hefen wurden auf Vollmediumagar (MA) kultiviert, der in 1 l Aqua dest. 40 g Biomalz, 5 g Glucose, 10 g Pepton, 5 g Hefeextrakt und 20 g Agar enthielt. Benomyl wurde in Dimethylformamid gelöst und dem VMA zugesetzt. Die Konzentration des Lösungsmittels im VMA betrug maximal 1%, da bei Konzentrationen über 2% toxische Effekte möglich sind. Die Zellen von *P. guilliermondii* kultivierten wir mindestens 7 Tage, die langsamer wachsenden Zellen von *R. toruloides* mindestens 10 Tage bei 30° oder 22 °C. Die sich auf dem VMA entwickelnden Kolonien wurden zum Erfassen auxotropher Segreganten auf Minimalmediumagar [2] überstempelt und danach einer Auxanographie unterworfen.

Fungizider Effekt

Die Beziehungen zwischen der Benomyl-Konzentration im VMA (D) und der Überlebendenfrequenz (N/N_0) manifestieren sich bei beiden Arten als „Vieltreffer"-Kurven, die sich durch die Gleichung $N/N_0 = 1 - (1 - e^{-kD})^n$ beschreiben lassen, in der k eine die Sensibilität der Zellen charakterisierende Konstante und n für die die Inaktivierung (den fungiziden Effekt) notwendige Trefferzahl darstellt (Abb. 1). Bei allen untersuchten Stämmen ist n größer als 10^5. Dieser Befund bestätigt die Annahme, daß die Tubulin-Untereinheiten der Mikrotubuli die Targets sind, deren Schädigung die Ursache der fungiziden Wirkung von Benomyl ist. Wahrscheinlich kann sich eine Zelle nicht mehr reproduzieren, wenn 10^5 oder mehr Tubulin-Untereinheiten nicht mehr assemblierungsfähig sind (Abb. 2).

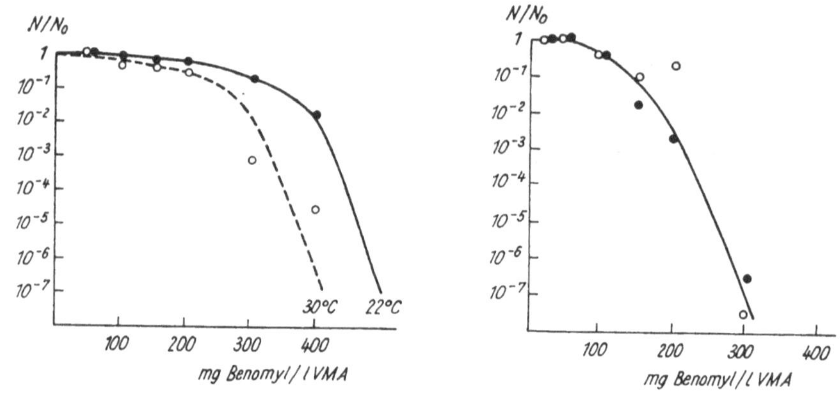

Abb. 1. Abhängigkeit der Überlebensraten (N/N_0) von der Benomyl-Konzentration im Anzuchtmedium (VMA) bei *P. guilliermondii* (a) und *R. toruloides* (b). Bei Benomyl-Konzentrationen über 400 mg/l war N/N_0 niedriger als 10^{-8}. Inkubationstemperatur 22 °C (●) oder 30 °C (○)

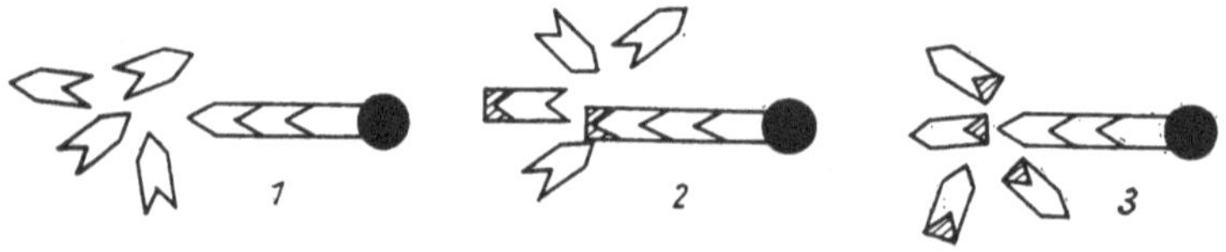

Abb. 2. Modellvorstellungen zur fungiziden Benomyl-Wirkung. 1 = Mikrotubuli-Assemblierung ohne Benomyl, 2 = Blockierung der Mikrotubuli-Enden durch Benomyl (Modell A), 3 = Tubulin-Untereinheiten sind durch Bindung von Benomyl nicht mehr assemblierungsfähig (Modell B). Die experimentellen Befunde sprechen für Modell B

Mitotische Segregation

Benomyl induziert bei den von uns untersuchten Pilzen keine Mutationen. Mitotische Segregationen können jedoch bei *P. guilliermondii* mit großer Frequenz entstehen, obwohl andere Agenzien, die bei einigen Pilzen mit hoher Effizienz mitotischen Chromosomenverlust bewirken, nicht oder nur wenig rekombinogen sind (Tab. 1). Bei *R. toruloides* führt Benomyl nicht zur Entstehung mitotischer Segregationen, obwohl die fungizide Wirkung stärker ist als bei *P. guilliermondii* (vergl. Abb. 1a und 1b) und auch das Chromosomenverlust in der Mitose verursachende Acriflavin [8] weitaus rekombinogener ist als bei dieser ascomycetalen Hefe (Tab. 2).

In den Tabellen sind nur mitotische Segreganten berücksichtigt worden, die sich als auxotrophe Kolonien auf dem VMA entwickelten. Die Konzentration des Benomyls im VMA betrug 100—200 mg/l. In diesem Bereich ist die Segregantenfrequenz nicht von der Benomylkonzentration abhängig und die inaktivierende Wirkung gering (vergl. Abb. 1). Bei der Entwicklung der Kolonien von *P. guilliermondii* auf dem Benomyl-VMA erfolgen weitere mitotische Segregationen, so daß auch die entstehenden prototrophen Kolonien viele auxotrophe Zellen enthalten. Die zum Vergleich angeführten Segregantenfrequenzen bei Einwirkung von Griseofulvin (400—800 mg/l), p-Fluor-

Tabelle 1

Mitotische Segregation bei P. guilliermondii G25 × G67: $\frac{\text{ade}}{+}\frac{\text{his}}{+}\frac{+}{\text{met}}\frac{+}{\text{arg}}$

	Kon-trolle	Benomyl 22°	30 °C	Griseo-fulvin	p-Fluor-Phe	Acri-flavin
Anzahl geprüfter Kolonien	3685	2870	2413	1400	400	2198
Anzahl auxotropher Segreganten	2	273	0	0	0	11
Segregantenfrequenz in %	0,05	9,5	0,1	<0,07	<0,25	0,5
Anzahl von Segreganten mit dem Phänotyp						
ade	1	80	2	0	0	0
his	0	46	0	0	0	0
met	1	78	0	0	0	7
ade his	0	5	0	0	0	0
met arg	0	25	1	0	0	3
ade met	0	1	0	0	0	0
his met	0	11	0	0	0	0
ade met arg	0	2	0	0	0	0
his met arg	0	20	0	0	0	0
ade his met arg	0	5	0	0	0	1

Tabelle 2

Mitotische Segregation bei R. toruloides Sp23-1: $\frac{\text{phe}}{+}\frac{\text{arg}}{+}\frac{\text{met}}{+}\frac{\text{nic}}{+}\frac{+}{\text{ino}}$

	Kon-trolle	Benomyl 22 °C	30 °C	Griseo-fulvin	p-Fluor-Phe	Acri-flavin
Anzahl geprüfter Kolonien	12972	3870	5060	2685	2803	5468
Anzahl auxotropher Segreganten	19	0	3	24	16	189
Segregantenfrequenz in %	0,2	<0,03	0,06	0,9	0,6	3,5
Anzahl von Segreganten mit dem Phänotyp						
phe	12	0	3	13	0	7
arg	3	0	0	0	12	126
met	0	0	0	1	2	15
nic	4	0	0	3	2	32
ino	0	0	0	6	0	3
phe nic	0	0	0	0	0	1
phe ino	0	0	0	1	0	0
arg met	0	0	0	0	0	1
arg nic	0	0	0	0	0	3

phenylalanin (400 mg/l) und Acriflavin (400 mg/l) wurden unter gleichen Versuchsbedingungen wie bei Benomyl festgestellt.

Offensichtlich kommen die inaktivierende und rekombinogene Wirkung von Benomyl nicht über den gleichen Mechanismus zustande: Bei *P. guilliermondii* ist die Frequenz

komplett auxotropher Klone bei 22 °C, nicht aber bei 30 °C erhöht, während die fungizide Wirkung bei 22 °C geringer ist als bei 30 °C (Abb. 1a); bei *R. toruloides* ist Benomyl ein effektives Fungizid, erzeugt jedoch keine mitotischen Segregationen.
Die Entstehung auxotropher Kolonien auf Benomyl-VMA beruht wahrscheinlich auf einem prämitotischen Chromosomenverlust, der vor der Chromatidentrennung erfolgt, denn bei einem Chromosomenverlust nach der Chromatidentrennung, einem Nondisjunction- oder Crossingover-Ereignis entsteht neben dem auxotrophen auch ein prototrophes Rekombinationsprodukt, das eliminiert werden muß, wenn eine auxotrophe Kolonie entstehen soll. Da die auf dem Benomyl-MVA überlebenden Zellklone nicht benomylresistenter sind als der heterozygote Ausgangsstamm und außerdem die rekombinogene Wirkung bei Benomyl-Konzentrationen auftritt, die nur eine geringe inaktivierende Wirksamkeit haben, ist diese Möglichkeit auszuschließen.
Ultrastrukturelle Beobachtungen belegen, daß die Evolution bei Pilzen zu unterschiedlichen Mitosemechanismen geführt hat. Zwischen einigen Basidiomyceten und Ascomyceten bestehen z.B. Differenzen in der interphasischen Lokalisation der dem Nukleus assoziierten Organelle (NAO = SPB), die auch für die unterschiedliche Reaktion von *R. toruloides* und *P. guilliermondii* auf die rekombinogene Wirkung von Benomyl verantowrtlich sein könnten.

Literatur

[1] Anne, J. and H. Eyssen, Isolation of inter-species hybrids of *Penicillium citrinum* and *Penicillium cyaneo-fulvum* following protoplast fusion, FEMS Microbiol. Letters **4** (1978), 87–90

[2] Böttcher, F. und I. A. Samsonova, *Rhodosporidium Banno*: Dosiseffektbeziehungen, Mutageneffektivität und Mutantenspektrum bei der Induktion Auxotrophie verursachender Mutationen durch ultraviolettes Licht und N-Methyl-N'-nitro-N-nitrosoguanidin, Z. Allgem. Mikrobiol. **18** (1978), 637–646

[3] Davidse, L. C., Mode of action, selectivity and mutagenicity of benzimidazole compounds, Neth. J. Pl. Path. **83** (1977), Suppl. 1, 135–144

[4] Hastie, A. C., Benlate induced instability of *Aspergillus* diploids, Nature **226** (1971), 771

[5] Kappas, A., S. G. Georgopoulos and A. C. Hastie, On the genetic activity of benzimidazole and thiophanate fungicides on diploid *Aspergillus nidulans*, Mutation Res. **26** (1974), 17–27

[6] Kevei, F. and J. F. Peberdy, Induced segregation in interspecific hybrids of *Aspergillus nidulans* and *Aspergillus rugulosus* obtained by protoplast fusion, Molec. Gen. Genet. **170** (1979), 213–218

[7] Maraz, A., M. Kiss and L. Ferenczy, Protoplast fusion in *Saccharomyces cerevisiae* strains of identical and opposite mating types, FEMS Microbiol. Letters **3** (1978), 319–322

[8] Samsonova, I. A. and F. Böttcher, Acriflavine-induced mitotic segregation in *Rhodosporidium toruloides* and *Candida guilliermondii*, Abstr. 14th Internat. Congress of Genetics, Moscow 1978, 89

[9] Stahl, U., Zygote formation and recombination between like mating types in the yeast *Saccharomycopsis lipolytica* by protoplast fusion, Molec. Gen. Genet. **160** (1978), 111–114

G. CASPERSON und H. LYR

Institut für Pflanzenschutzforschung der Akademie der Landwirtschaftswissenschaften, Kleinmachnow, DDR

Elektronenmikroskopische Untersuchungen zum Wirkungsmechanismus von Pentachlornitrobenzol (PCNB) bei *Mucor mucedo* (L.) FRES. und *Phytophthora cactorum* (LEB. et COHN) SCHROET.

PCNB wird im Pflanzenschutz als Fungizid besonders zur Bekämpfung von Bodenpilzen eingesetzt. Es wirkt relativ selektiv. So reagiert *Mucor mucedo* im Agar-Plattentest sehr empfindlich auf PCNB ($ED_{50} = 5$ ppm), während *Phytophthora cactorum* mit einer $ED_{50} > 200$ ppm den unempfindlichen Pilzen zuzuordnen ist. Bisher sind der Wirkungsmechanismus und die Ursachen der Selektivität weitgehend ungeklärt, was für die gesamte Gruppe der sogenannten „aromatischen Hydrocarbon-Fungizide" zutrifft, zu denen Wirkstoffe, wie Chloroneb, Diphenyl, Dichloran und Tetrachlornitrobenzol und neuerdings die Carboxamid-Fungizide gezählt werden. Trotz der chemischen Verschiedenheit dieser Verbindungen konnte mehrfach die Ausbildung einer Kreuzresistenz bei einigen Pilzarten festgestellt werden (TILLMAN u. SISLER [6], GEORGOPOULOS [2], BISCHOFF u. LYR [1]), so daß ein gemeinsamer Wirkungsmechanismus zu vermuten ist.

Veränderungen in der Ultrastruktur bei *Mucor* treten bereits nach 2stündiger Einwirkzeit und Konzentrationen unter 10 ppm auf, die bei 10, 20 und 50 ppm immer ausgeprägter werden (Abb. 1—4). Bei 10 ppm beginnt in einigen Mitochondrien Destrukturierung der Cristae und der inneren Mitochondrienmembran, der perinukleäre Raum ist stellenweise erweitert und die Zellwand ist stark verdickt (Abb. 2). Nach einer 2stündigen Behandlung mit 20 ppm PCNB sind in den meist leicht angeschwollenen Mitochondrien nur wenige Cristae noch deutlich zu erkennen, die innere Mitochondrienmembran beginnt sich an einigen Stellen aufzulösen. Der perinukleäre Raum ist stark erweitert, ohne daß an den Membranen Veränderungen zu sehen sind (Abb. 3). 50 ppm PCNB ruft eine starke Vakuolisierung der gesamten Zelle hervor, das Plasmalemma hebt sich von der stark verdickten Zellwand ab und ist stellenweise nach innen eingestülpt. In den großen Vakuolen sind häufig Membranreste oder Vesikeln anzutreffen. Der Zellkern und die Mitochondrien weisen schwere Schäden auf (Abb. 4). Die Verdickung der Zellwand ist in einem bestimmten Bereich annähernd proportional zur aufgewendeten Wirkstoffmenge (Abb. 5).

Bei *Phytophthora* sind das Endoplasmatische Retikulum und der Golgi-Apparat normalerweise ausgeprägter ausgebildet als bei *Mucor* (Abb. 6). Nach einer PCNB-Behandlung mit 1 ppm erweitert sich das ER stellenweise zu Zisternen, die aber nicht, wie bei einer typischen Vakuolenbildung, abgeschnürt werden (Abb. 7). Bei 20 ppm PCNB sind die Mitochondrien geschwollen und die tubulären Cristae werden weitgehend zerstört (Abb. 8). Die Zellwand zeigt im Gegensatz zu Mucor keine im Elektronenmikroskop auffallenden Veränderungen.

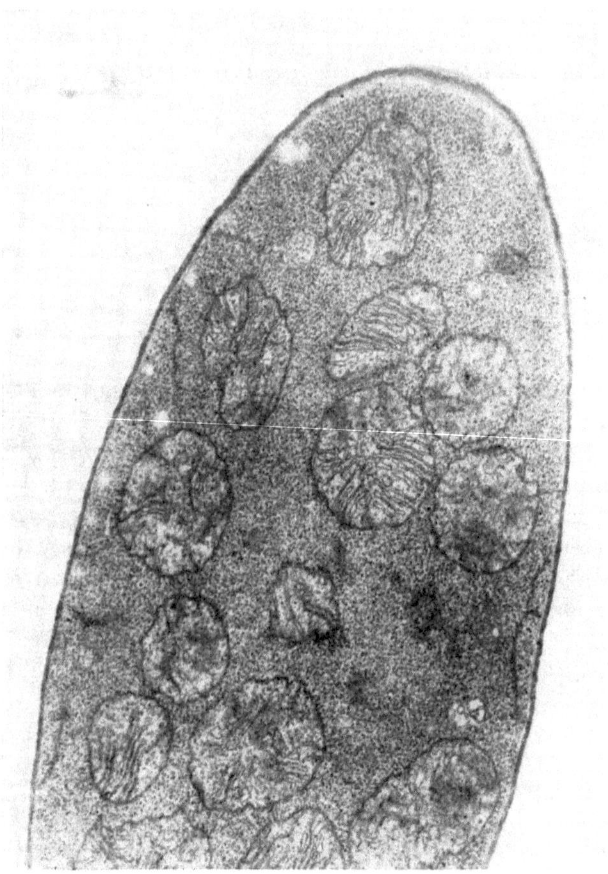

Abb. 1. Hyphenspitze von *Mucor* ohne Wirkstoff, Mitochondrien mit gut erkennbaren Cristae, Zellkern mit doppelter Kernmembran, keine Vakuole, multivesikulärer Körper, ER schwach ausgeprägt

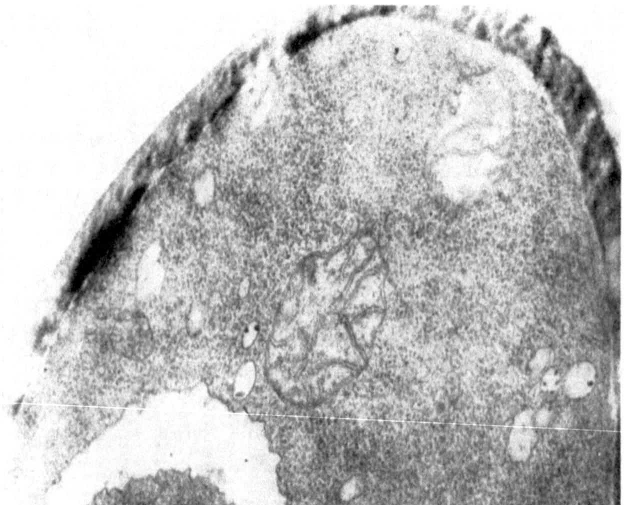

Abb. 2. Mucor nach 2 Stunden Behandlung mit 10 ppm PCNB, beginnende Destrukturierung in den Mitochondrien, Erweiterung des perinukleären Raumes, Verdickung der Zellwand
a) Zellwand der unbehandelten Zelle

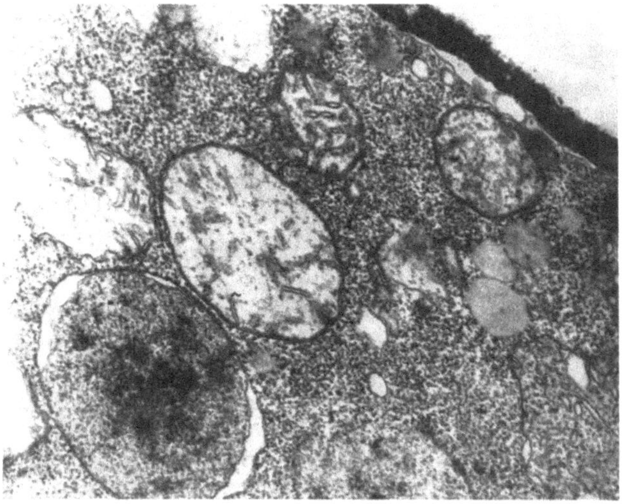

Abb. 3. Mucor nach 2 Stunden Behandlung mit 20 ppm PCNB, partielle Lysis der inneren Mitochondrienmembran (→) und der Cristae, Erweiterung des perinukleären Raumes

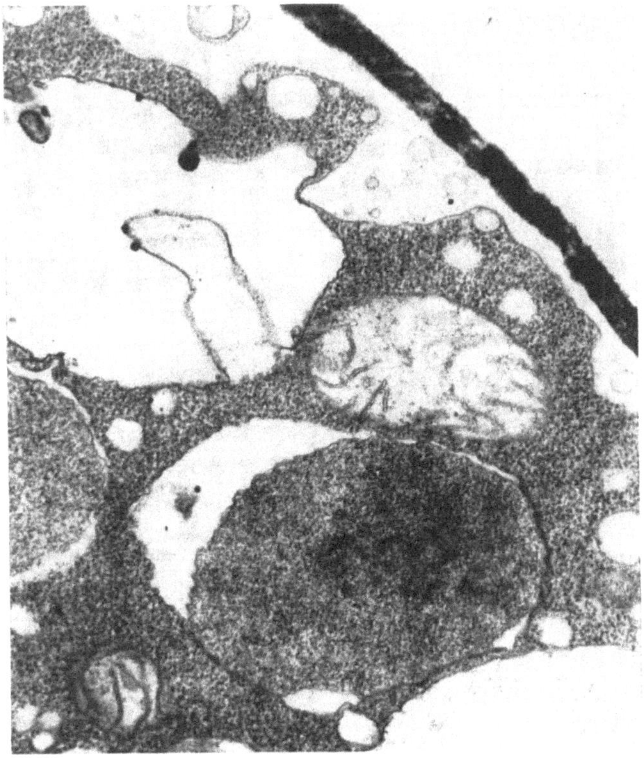

Abb. 4. *Mucor* nach 2 Stunden mit 50 ppm PCNB, starke Vakuolisierung der Zelle, das Plasmalemma hebt sich von der stark verdickten Zellwand ab, Zellkern und Mitochondrien schwer geschädigt

Bei *Mucor* und bei *Phytophthora* sind die Veränderungen an den Mitochondrien weitgehend gleichartig, obwohl die Mitochondrien bei diesen beiden Pilzarten sich im Strukturtyp prinzipiell unterscheiden. Wir halten es für wahrscheinlich, daß der primäre oder zumindest ein primärer Angriffspunkt von PCNB in der inneren Mitochon-

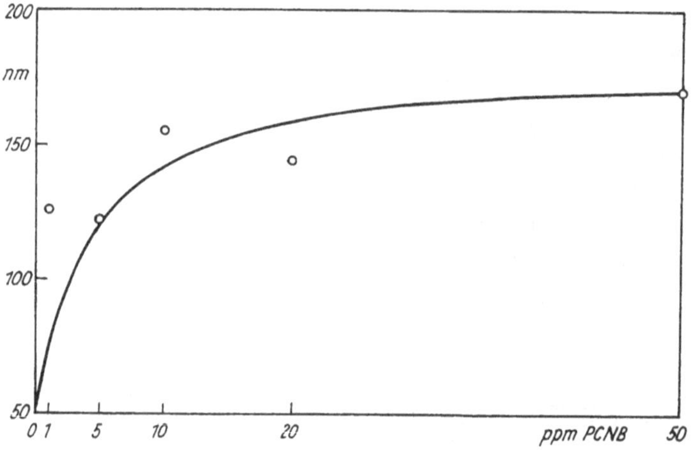

Abb. 5. Zellwandverdickung bei Mucor in Abhängigkeit von der PCNB-Konzentration

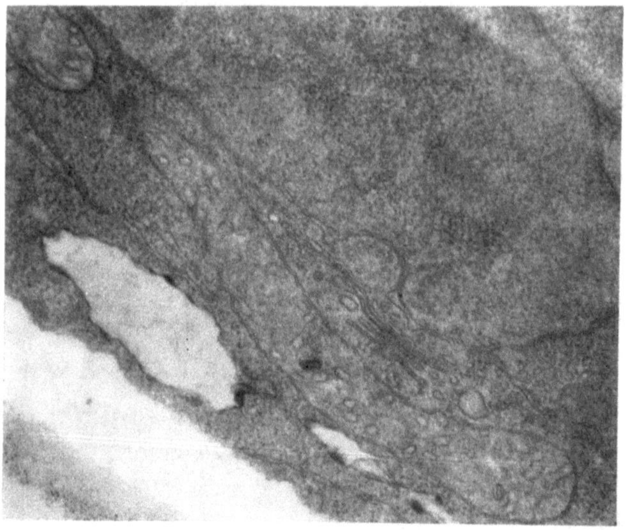

Abb. 6. *Phytophthora* ohne Wirkstoff, Mitochodorien mit tubulären Cristae, Zellkern mit Centriol, ER und Golgi-Apparat

drienmembran liegt. Die Veränderungen an den Mitochondrien stimmen weitgehend mit denen überein, die wir nach Behandlung mit Chloroneb erhielten (WERNER u. a. [7, 8]). Es wurde hier eine Bindung des Wirkstoffs vermutlich an spezifische Proteine der Mitochondrienmembran, nachgewiesen (WERNER u. LYR [9]). Die gesteigerte Proteaseaktivität nach Chloronebeinsatz könnte die Ursache für die Membranlysis sein. Über die Zellwandsynthese nach Wirkstoffeinfluß wird hier an anderer Stelle (LYR u. CASPERSON [3]) ausführlich berichtet. Die beobachteten Veränderungen an der Kernmembran könnten mit der von GEORGOPOULOS [2] beschriebenen Störung der Kernteilung in Zusammenhang stehen.

Nach Untersuchungen von NAKANISHI und OKU [4, 5] beruht die Selektivität gegenüber PCNB einmal auf einer Aufnahmebarriere unempfindlicher Pilze und zum anderen in der Möglichkeit zur Detoxifikation von PCNB zu Pentachloranilin und Pentachlorthioanisol, die bei den resistenten Stämmen besonders ausgeprägt ist, so daß ein

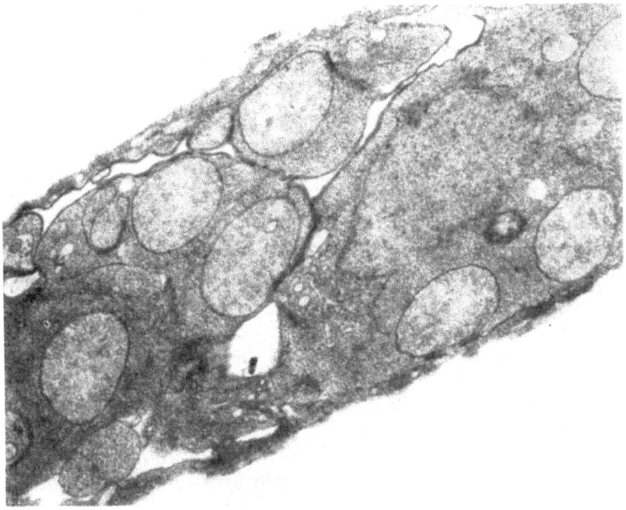

Abb. 7. *Phytophthora* nach 2 Stunden Behandlung mit 1 ppm PCNB, Erweiterung des ER zu Zisternen

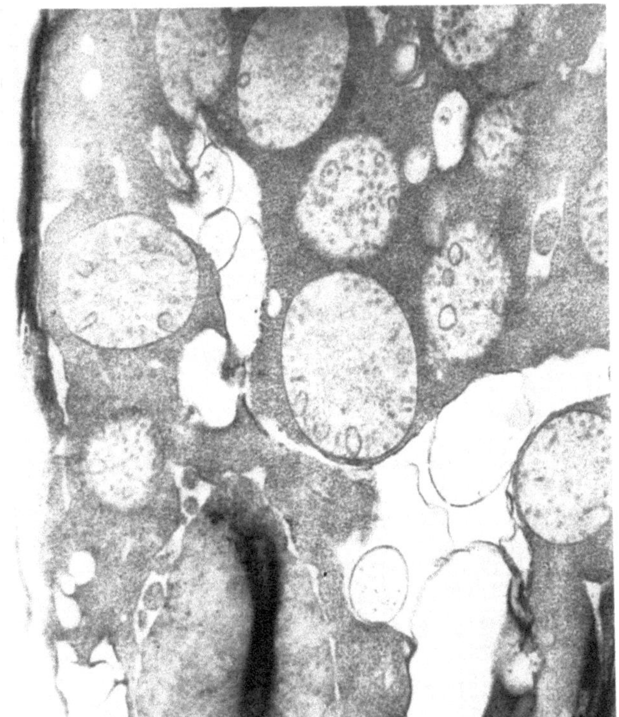

Abb. 8. *Phytophthora* nach 2 Stunden Behandlung mit 20 ppm PCNB, Schwellung der Mitochondrien und Lysis der tubulären Cristae, ER zu Zisternen erweitert

Gleichgewicht zwischen Aufnahmegeschwindigkeit und Abbaugeschwindigkeit hergestellt werden kann.

Literatur

[1] Bischoff, G. und H. Lyr, Das Resistenzverhalten von *Mucor mucedo* gegenüber Chloroneb und anderen Fungiziden, Arch. Phytopathol. **16** (1980), 111–118
[2] Georgopoulos, S. G., Development of fungal resistance to fungicides. In: M. R. Siegel and H. D. Sisler (Hsg) Antifungal Compounds Vol. 2 Marcel Dekker Inc, New York, Basel 1977, 339–495
[3] Lyr, H. und G. Casperson, Zellwandbildung bei *Mucor mucedo* unter dem Einfluß von Fungiziden und anderen Wirkstoffen, Symposium Reinhardsbrunn 1980
[4] Nakanishi, T. and H. Oku, Metabolism and accumulation of pentachloronitrobenzene by phytopathogenic fungi in relation to selective toxicity, Phytopathol. **59** (1969), 1761–1762
[5] Nakanishi, T. and H. Oku, Mechanism of selective toxicity of fungicide: Absorption, metabolism and accumulation of pentachloronitrobenzene by phytopathogenic fungi, Annals phytopathol. Soc. Japan **36** (1970), 67–73
[6] Tillman, R. W. and H. D. Sisler, Effect of Chloroneb on the growth and metabolism of *Ustilago maydis*, Phytopathol. **63** (1973), 219–225
[7] Werner, P., H. Lyr und G. Casperson, Die Wirkung von Chloroneb, seinen Abbauprodukten, sowie von chlorierten Phenolen auf das Wachstum und die Ultrastruktur verschiedener Pilzarten, Arch. Phytopathol. u. Pflanzenschutz **14** (1978), 301–312
[8] Werner, P., H. Lyr und G. Casperson, Die Wirkung von Chloroneb, seinen Abbauprodukten und chlorierten Phenolen auf das Wachstum und die Ultrastruktur einiger Pilzarten, In: H. Lyr und C. Polter (Hsg) Systemfungizide, V. Intern. Symposium Reinhardsbrunn, Abh. Akad. Wiss. DDR Nr. 2 (1979), 299–310
[9] Werner, P. und H. Lyr, Der Wirkungsmechanismus von Chloroneb und die möglichen Ursachen der Resistenz gegenüber dieser Verbindung, Symposium Reinhardsbrunn 1980

L. Ferenczy, A. Franko*, E. Hegedüs*, A. Gajáry*, T. Pfliegel*

Department of Microbiology, Attila József University, Szeged, Hungary
CHINOIN Pharmaceutical and Chemical Works Ltd., Budapest, Hungary

Structure-Activity Studies on Salicylanilides

In the early sixties, production of two salicylanilides was started in the CHINOIN Pharmaceutical and Chemical Works. The possibility to manufacture these compounds — namely Devermin® and Trinoin® — is based on Hungarian patent No. 156.582. With this patent a wide-ranging development program was started. The primary aim was the synthesis of new compounds with high antimicrobial activity. Well-known molecules were also synthesized in order to get more information on structure-activity relationships.

Synthesis of Salicylanilides

Theoretically there are various methods of manufacturing salicylanilides.
The compounds tested by us for their antimicrobial activities were synthesized according to the above-mentioned patent. This method — called „thionilating" — is simple, rather cheap and gives good yields.

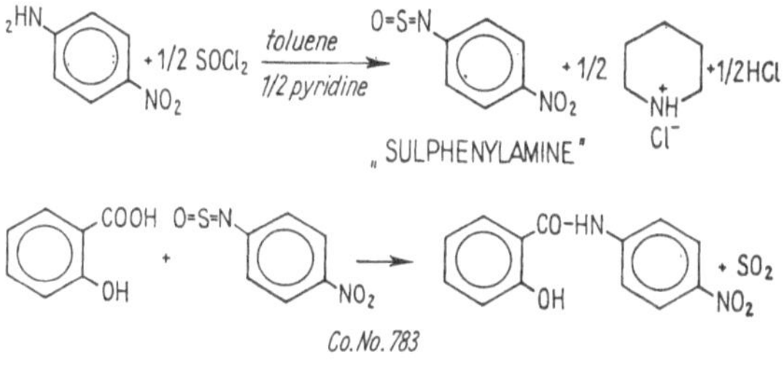

Fig. 1

Figure 1 shows „thionilating" in a shortened form for the molecule coded 783 where the so-called „sulphenylamine" is synthetized from an aromatic amine with thionylchloride and later the sulphenylamine is brought into reaction with salicyclic acid. About 70 compounds were prepared by using this method — mainly new ones.

Antimicrobial Activity

Antimicrobial activities were determined by using serial dilution method. Media contained 1% glucose and 0.5% yeast extract, pH was 5.5 and 7.2, respectively. These media were solidified with 2% agar. The salicylanilide derivates were dissolved in dimethylsulfoxide, and were added to the warm nutrient media. The concentration of dimethylsulfoxide did not exceed 1 percent by volume of agar. Agar plates containing varying concentrations of test compounds were inoculated with the suspension of the organisms and incubated at the appropriate temperature. The minimum inhibitory concentration (in μg/ml) was determined and recorded.

The first screen was made in the CHINOIN Works with a few common phytopathogenic fungi. The second screen was made at ATTILA JÓZSEF University, Department of Microbiology. Activities were determined in vitro against Gram-positive and Gram-negative bacteria, yeast, and filamentous fungi, including phytopathogenic ones, and Dermatophytes. This widening of the spectrum gave an opportunity to find new effects which may be important from industrial, human and veterinary points of view.

Structure and Activity

According to their ligands six groups of tested compounds were formed.

Group 1. (Fig. 2)

This group contains the simplest derivates. As early as in 1960, BAICHWAL et al. [1] established that the hydroxy group of the salicyl moiety is necessary for activity, but according to our experiments it is not sufficient. Forming salts does not improve activity.

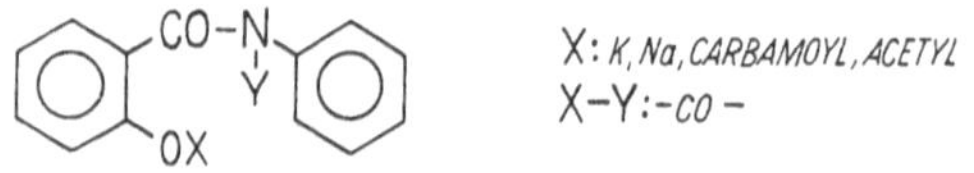

Fig. 2

Group 2. (Fig. 3.)

Substitutions in the salicyl ring only were carried out: nitro derivates were synthesized. The nitro substitution in the 5-position increases activity, but not to an acceptable degree: only Gram-positive bacteria were inhibited.

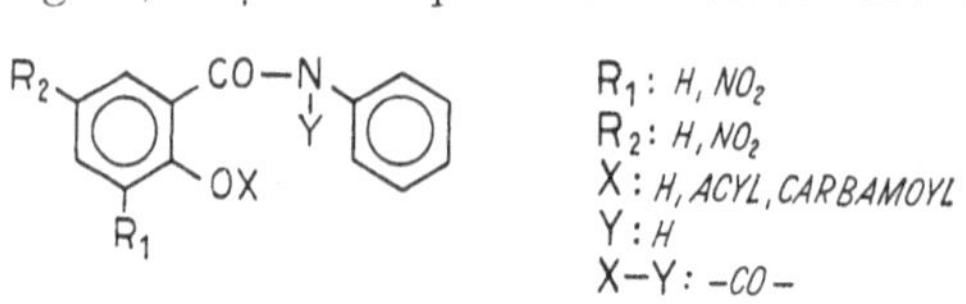

Fig. 3

Group 3. (Fig. 4.)

In this group the number of compounds is great, so we were able to compare the activity of chloro and nitro derivates. The activity of chloro derivates is well-known from the reports of BAICHWAL et al. [1]. SMITH and READ [5] SCHÖNBERGER et al. [3], WILLIAMSON and METCALF [6]. The aim of examination in this group was to clarify the effect of the relative position of the chloro and nitro substituents.

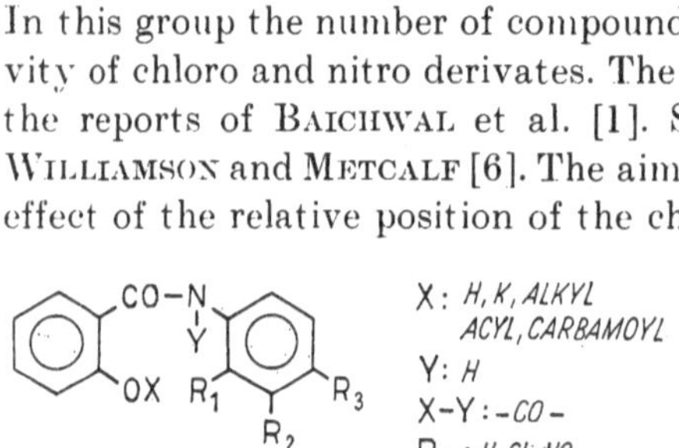

Fig. 4

Table 1 shows that chloro substitution in 2′-position is less active than in 4′-position. According to the investigations of SCHÖNBERGER et al. [3] and SCHULTZ [4] the latter

Table 1

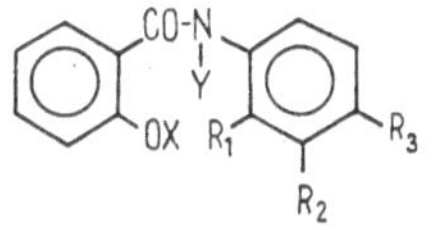

Co. No.	X	Y	R_1	R_2	R_3	Minimum Inhibitory Concentration µg/ml		
						Phytopathogenic fungi	Dermatophytes	Gram-positive bacteria
1293	—H	—H	—H	—H	—Cl	3–12	1,5–6	3–12
997	—H	—H	—Cl	—H	—H	50	3–6	3–12
801	$—COCH_3$	—H	$—NO_2$	—H	—H	>50	>50	>50
806	$—COCH_3$	—H	—H	$—NO_2$	—H	50	5	>50
532	$—COCH_3$	—H	—H	—H	$—NO_2$	0,7–6	1,5–3	0,7–3
783	—H	—H	—H	—H	$—NO_2$	0,7–6	0,3–1,5	0,7–3
1293	—H	—H	—H	—H	—Cl	3–12	1,5–6	3–12

is equal to that of 3′-position in activity. Our data correspond well to the data of Williamson and Metcalf [6] who reported, that 2′—5′-dichloro derivates are less active than 3′-4′-dichloro derivates. The main factor in maintaining high activities is the position of the nitro group. Williamson and Metcalf [6] supposed that a strong electron-with-drawing center is essential for activity. Our experiments have shown that the activity of 2′-nitro derivates is practically zero, of 3′-nitro derivates is medium, and only the 4′-nitro derivates have high activity. Similar results are reported by Clark and Hans [2] in the case of nitroanilines. In this group some 4′-nitro derivates were synthetized, which differed in the ortho position of the salicyl moiety. In Table 2 we compared the activity of alkyl, acyl, carbamoyl derivates with the activity of

Table 2

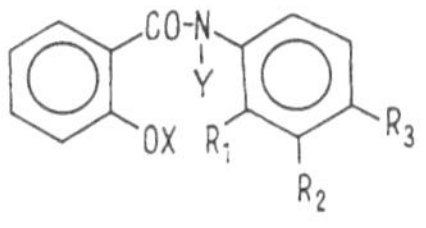

Co. No.	X	Y	R_1	R_2	R_3	Minimum Inhibitory Concentration µg/ml		
						Phytopathogenic fungi	Dermatophytes	Gram-positive bacteria
798	CH_3H_7	—H	—H	—H	$—NO_2$	>50	>50	>50
719	—Co—		—H	—H	$—NO_2$	5 [50]	5	5
532	$COCH_3$	—H	—H	—H	$—NO_2$	0,7–6	1,5–3	0,7–3
769	COC_2H_5	—H	—H	—H	$—NO_2$	5	5	5
534	$CONHCH_3$	—H	—H	—H	$—NO_2$	3–12	1,5–6	1,5–6
783	—H	—H	—H	—H	$—NO_2$	0,7–6	0,3–1,5	0,7–3

compound Code No. 783 which had an hydroxy group in the ortho position. When the hydroxy group is blocked with an alkyl group, activity is reduced considerably .A similar effect, though to a lesser degree, is caused by acyl and carbamoyl groups. A similar effect of these ligands may be detected in Group 6. in the case of 2′-5-dichloro-4′-nitro derivates.

Group 4. (Fig. 5.)

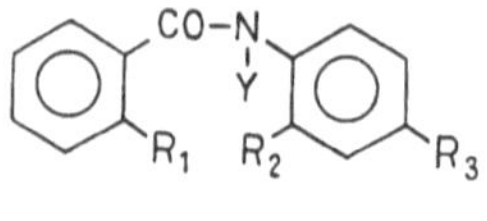

R_1: Cl, NH_2
Y: H, CH_3
R_2: OH, Cl
R_3: H, Cl, NO_2

Fig. 5

In this group not a single active compound could be found. The advantageous effect of Cl and NO_2 groups in the anilide moiety were unable to balance the disappearance of activity caused by the lack of the hydroxy in the salicyl moiety. These results confirm the statements made in connection with Groups 1. and 3.

Group 5. (Fig. 6.)

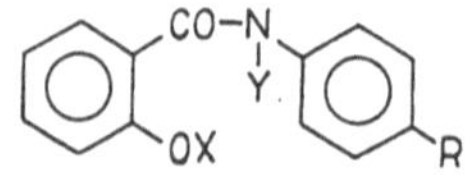

X: H, K, CARBAMOYL
Y: CH_3
R: Cl, NO_2

Fig. 6

Realizing the role of the ligands in both rings, we decided to examine the activity of some N-methyl derivates of the carbonylamide chain. These compounds are inactive in vitro. Similar data can be found in the literature (BAICHWAL et al. [1] and CLARK and HANS [2]).

Group 6. (Fig. 7.)

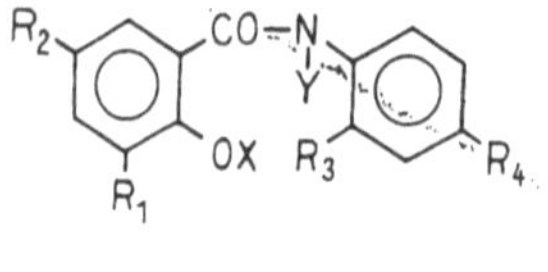

R_1: H, Br
R_2: Br, Cl
X : H, K, ACYL, CARBAMOYL
R_3: H, Cl
R_4: H, Br, NO_2
X–Y : –CO–

Fig. 7

The compounds of this group were synthesized by utilization of the results obtained in the former five groups, so eight out of fifteen were highly active.

All members of this group are substituted with halides in the salicyl moiety at least. The titromoderivates were active but they were not able to surpass any of the chloronitro derivates. Comparing the 4′-nitro derivative (Group 3.) with the 5-chloro-4′-nitro and 2′-5-dichloro-4′-nitro derivates we can observe that the chloro substitution in position 5 is more or less indifferent, but in position 2′ slightly reduces activity. In other words, the 2′-5-dichloro-4′-nitro derivates are not so active as the 5-chloro-4′-nitro derivates of Group 3. These data can easily be explained: chloro substitution weakens the electron-withdrawing power of the nitro group. It seems that the mode of action of salicylanilides is short-circuiting the potential gradient of oxidative phosphorylation by their electron-withdrawing nitro group which can be potentiated by halo substitution if the —I-effect only can exert its action or weakened by the same if the +E-effect of the halogen comes into the foreground.

References

[1] BAICHWAL, R. S., BAXTER, R. M., KANDEL, S. I. and WALKER, G. C., Antifungal action of salicylanilide II., Can. J. Biochem. Physiol. **38** (1960), 245—251

[2] CLARK, N. G. and HANS, A. F., Antifungal activity of substituted nitroanilines and related compounds, J. Sci. Food Agr. **12** (1961), 751—757

[3] SCHÖNENBERGER, H., HOLZEHEU-ECKARDT, J., BAMANN, E., and MEHNERT, B., Zusammenhänge zwischen chemischer Konstitution und antimykotischer Wirksamkeit bei isomeren Oxybenzamiden, Arzneimittel-Forsch. **12** (1962) 1162—1166

[4] SCHULTZ, H. W., Synthesis and antifungal activity of anilides of salicylic acid and o-coumaric acid, J. Pharm. Sci. **52** (1963), 503—505

[5] SMITH, R. J. and READ, W. H., Investigation on fungitoxic derivates of salicylaldehyde I. The 3-, 5 and 3,5-chlorinated derivates of salicylanilide and salicyl-parachloroanilide, Ann. Appl. Biol. **49** (1961), 102—109

[6] WILLIAMSON, R. L. and METCALF, R. L., Salicylanilides: A new group of active uncouplers of oxidative phosphorilation, Science **158** (1967), 1694—1695

J. GORSKA-POCZOPKO and S. WITEK

Institute of Industrial Organic Chemistry, Warszawa, Poland and Institute of Organic and Polymer Technology Technical University of Wroclaw, Poland

Fungicidal Properties of Some New Quaternary Ammonium Salts

Some quaternary ammonium salt are well known as antimicrobial agents, used mostly as protectant of organic materials such as wood, timber, plastics and s.a. against biocorrosion.

In this paper some new quaternary ammonium salts are described and possibility of their use as agricultural fungicides or bactericides are considerated.

These new compounds were synthesized in the Institute of Organic and Polymer Technology of Wroclaw Technical University, by quaternisation of tertiary amines with substituted benzyl chlorides (GROBELNY, WITEK [2])or chloromethyl alkyl ethers.

Structure and purity (above 97%) of synthesized ammonium salts were confirmed by elemental analysis, TLC and H^1-N.M.R and UV spectrums. The details of syntheses will be published separately (WITEK, GROBELNY [6], WITEK, OŚWIĘCINSKA [7]).

Some of the alkoxymethylene derivatives showed surface — activity and bactericidal-properties (KOMKOV, PANKRATOV [3], VASHKOV et al. [5]), formyl — and nitrovinyl benzyl salts were not proved till now for above mentioned activity.

The fungicidal properties of new quaternary ammonium salts were tested in the Institute of Industrial Organic Chemistry in Warszawa (FULDE et al. [1], PTASZKOWSKA et al. [4]).

Biological Activity of New Quaternary Ammonium Salts

First observations were made accordingly to the procedure of I^{st} fungicidal screening using six test fungi. The inhibition of spores germination was tested in vitro on *Alternaria tenuis* and *Botrytis cinerea* and the inhibition of mycelium growth on *Rhizoctonia solani*, *Fusarium culmorum* and *Aspergillus niger*. The mildew control was tested in vivo on *Erysiphe graminis* on wheat. Particulary positive results especially against *Alternaria tenuis* and *Botrytis cinerea* were obtained with three groups of compounds, namely:

- alkoxymethylene morpholinium and piperidinum chlorides: (5 compounds from 37 investigated were assigned to further screenings)
- (β-methyl-β-nitrovinyl) benzylammonium clhorides: (23 compounds assigned to further screenings from 44 investigated)
- (methoxy-formyl) benzylammonium chlorides (4 compounds from 18 investigated were assigned to further screenings).

Compounds from other allied groups were characterized with lower fungicidal activity — from about 150 tested only few passed to further screenings.

During the II_{nd} screening the assigned samples of quaternary ammonium salts were tested more in detail in laboratory and under greenhouse conditions on such test objects as: *Alternaria tenuis*, *A. brassicae*, *Botrytis cinerea*, *Fusarium oxysporum*, *Phoma* sp., *Helminthosporium* sp., *Phytophthora cactorum*, *Ph. nicotianae*, *Tilletia tritici* *Uromyces dianthi*, *Puccinia triticina*, *Septoria apii*. All samples from alkoxy methylene piperidinum and morpholinium chlorides passed to IIIrd screening, as well as 11 samples from (β-methyl-β-nitrovinyl) benzylammonium chlorides and 1 sample from (methoxy-formyl) benzylammonium chlorides.

IIIrd screening is performed in orchard and field on microplots using various test objects and is carried out during minimum 3 years.

Not only fungicidal but also bactericidal activity is investigated.

Structure — Activity Relationships

Till now best investigated are the fungicidal properties of alkoxymethylene piperidinium and morpholinium chlorides. The results of first investigations in slide — germination test are shown on Table 1.

Table 1

Antifungal activity of alkoxymethylene piperidinium and morpholinium chlorides in slide germination test

R	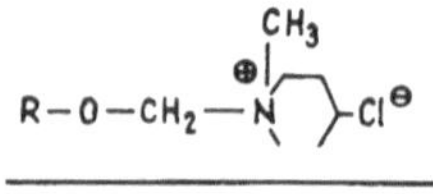			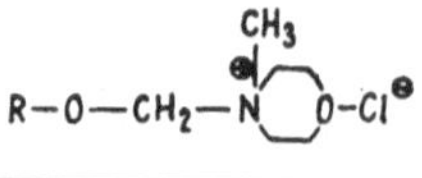		
	Min. inhibitory concentration in μg/ml (ED_{100})					
	Comp. No.	*Alternaria tenius*	*Botrytis cinerea*	Comp. No.	*Alternaria tenuis*	*Botrytis cinerea*
C_8H_{17}	I	1000	±1000		—	—
$C_{10}H_{21}$	II	1000	1000		—	—
$C_{12}H_{25}$	III	10	10	VII	100	100
$C_{14}H_{29}$	IV	100	100	VIII	±10	<10
$C_{16}H_{33}$	V	100	10	IX	±10	10
$C_{18}H_{37}$	VI	±10	±10	X	±10	±10

The activity increased with the number of carbon atomes in the alkyl chain and was best by substituted alkyl groups ranging from $C_{12}H_{25}$ to $C_{18}H_{37}$. This relation was more determinated in respect to morpholinium than to piperidinium salts. Compounds: III, V, VIII, IX, X were also good in laboratory/greenhouse investigations on *Phytophthora cactorum*, *Ph. nicotianae*, *Septoria apii*, some Uredinales a.o. and showed systemic properties. In field examination compound VIII was very active against apple scab *(Venturia inaequalis)* and celery large leaf spot *(Septoria apii)*. Compounds V and IX controlled well *Phytophthora infestans* on potatoes, *Peronospora destructor* on

Table 2

Antifungal activity of some β-methyl-B-nitrovinyl-benzylammonium chlorides in slide germination test

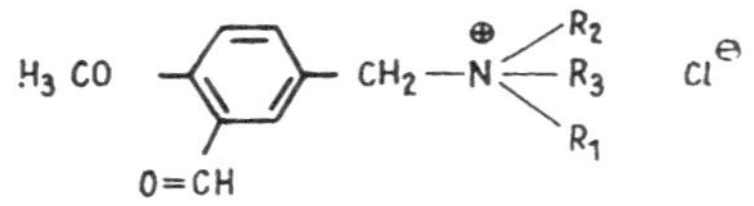

R_1	R_2	R_3	Min. inhibitory concentration µg/ml (ED_{100})								
			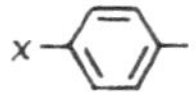			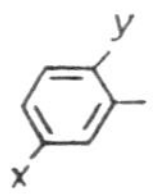			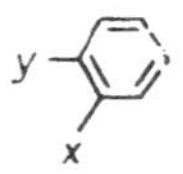		
			Comp. No.	*Alternaria tenuis*	*Botrytis cinerea*	Comp. No.	*Alternaria tenuis*	*Botrytis cinerea*	Comp. No.	*Alternaria tenuis*	*Botrytis cinerea*
$—CH_3$	$—CH_3$	$—C_{10}H_{21}$	XI	1000	100	XIX	±100	100	XXVI	±1000	±1000
$—CH_3$	$—CH_3$	$—C_{12}H_{25}$	XII	100	<10>1	XX	±10	<10	XXVII	100	10
$—CH_3$	$—CH_2—CH_2—OH$	$—C_{10}H_{21}$	XIII	100	100	XXI	±1000	±10	XXVIII	±10	10
$—CH_3$	$—CH_2—CH_2—OH$	$—C_{12}H_{25}$	XIV	100	<10>1	XXII	±100	<10>1	XXIX	100	10
$—CH_2—CH_2—O—$	$CH_2CH_2—$	$—C_{10}H_{21}$	XV	100	100	XXIII	100	<10>1	XXX	100	10
$—CH_2CH_2—O—$	$CH_2CH_2—$	$—C_{12}H_{25}$	XVI	100	±10	XXIV	100	<10>1	XXXI	10	10
$—CH_2CH_2—O—$	$CH_2CH_2—$	$—C_{14}H_{29}$	XVII	10	<10	—	—	—	XXXII	10	±10
$—CH_2CH_2—O—$ $—CH_3$	$CH_2CH_2—$ $—CH_3$	$—C_{12}H_{25}$	XVIII	100	<10	XXV	±10	<10	—	—	—

onion and *Pseudomonas lachrymans* on cucumber. Compounds IX and III were active against *Drepanopeziza ribis* on currants, remarkably inhibiting the defoliation.

A large group of (β-methyl-β-nitrovinyl) benzylammonium chlorides is demonstrated on example of some representatives containing in the alkyl chain the C atoms number ranging 10, 12 and 14 respectively (Table 2).

Generally the dodecyl derivatives were more active than decyl ones. Remarkable activity showed all morpholinium compounds, including also trimethyl derivatives (compounds XVIII, XXV). Activity against *Alternaria tenuis* increased in compounds substituted with the nitrovinyl group in position 3, and group-OCH_3 in position 4, for example compounds XXVIII, XXXI, XXXII.

Some compounds from (methoxy-formyl) benzylammonium chlorides are demonstrated on Table 3. For comparison only the compounds with substituents similar to those showed on Table 2 were selected, independent of their lower fungicidal activity. The general trends in structure — activity relationships, observed by methoxy-formylo derivatives were respectively similar to those of β-methyl-β-nitrovinyl ones.

Table 3
Antifungal activity of N-(3-formyl-4-methoxy)benzyl-ammonium chlorides in slide germination test

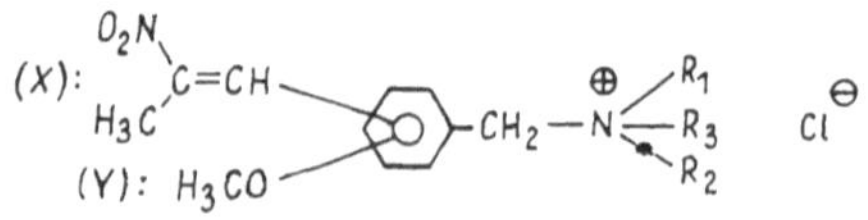

Compounds				Min. inhibitory concentration µg/ml (ED-100)	
No.	R_1	R_2	R_3	*Alternaria tenuis*	*Botrytis cinerea*
XXXII	$-CH_3$	$-CH_3$	$C_{10}H_{21}$	100	1000
XXXIII	$-CH_3$	$-CH_3$	$C_{12}H_{25}$	100	100
XXXIV	$-CH_3$	$-CH_2CH_2OH$	$C_{10}H_{21}$	100	± 10
XXXV	$-CH_3$	$-CH_2CH_2OH$	$C_{12}H_{25}$	$\gg 10$	1000
XXXVI	$-CH_2CH_2-C-CH_2CH_2$		C_8H_{17}	1000	1000
XXXVI	$-CH_2CH_2-O-CH_2CH_2$		$C_{10}H_{21}$	100	100
XXXV	$-CH_2CH_2-O-CH_2CH_2$		$C_{12}H_{25}$	100	100

Conclusions

Described above new quaternary ammonium salts seems to posses interesting fungicidal properties. In search for new agricultural chemicals they can be taken in account both as fungicides and as bactericides. They require fundamental research especially in area of their mechanism of action, metabolic pathways and application possibilities as well as toxicity to mammals and beneficial organisms.

Literature

[1] FULDE, S., BAKUNIAK, E., GORSKA-POCZOPKO, J., GWIAZDA, M., Archievments of Inst. Organic Industrial Chemistry in the field of synthesis and investigations of new pesticides, Materiały z XX Sesji Naukowej I. O. R., (1980), in press

[2] GROBELNY, D., WITEK, S., Transformation of chloromethyl benzaldehydes into corresponding (2-nitro alkenyl-1) benzyl chlorides or acetates, J. pract. Chem. (1980), in press

[3] KOMKOV, I. P., PANKRATOV, V. A., Surface — active salts of quaternary ammonium bases, Zh. Prikl. Khim. 43 (1970), 1367—1371 (Russ.)

[4] PTAZCKOWSKA, J., GORSKA-POCZOPKO, J., OŚWIĘCIMSKA, M., Badania własnosci grzybobójczych niektórych substancji z grupy czwartorzędowych soli amoniowach,Organika — Prace Nauk. IPO 2 (1967), 203—209

[5] VASHKOV, V. I., KOMKOV, J. P., ODINTES, E. E., PANKRATOV, V. A., New desinfecting agents from a group of surface-active salts of quaternary ammonium bases, Tr. Tsentr. Nauch. Issled. Dezinfek. Inst. 18 (1967), 100—106 (Russ.)

[6] WITEK, S., GROBELNY, D., Synthesis of (2-nitrovinyl) benzyl ammonium salts, Polish J. Chem., (1980) in press

[7] WITEK, S., OŚWIĘCIMSKA, M., Synthesis of some alkoxymethylene ammonium chlorides, Polish J. Chem. (1980) in press

J. GÖTZ

Institut für Kartoffelforschung der Akademie der Landwirtschaftswissenschaften, Groß Lüsewitz, DDR

Resistenz von kartoffelpathogenen *Fusarium* spp. gegenüber Benzimidazolen

1. Einleitung

Fusarium spp. zählen gegenwärtig zu den wichtigsten Lagerfäuleerregern an Kartoffelknollen. Nach BURTH [5] sind auf dem Gebiet der DDR an ca. 90% der auftretenden Lagerfäulen Pilze der Gattung *Fusarium* beteiligt. Dabei sind die Species *Fusarium sulphureum* SCHLECHT. *(F. s.)* und *F. solani* var. *coeruleum* (SACC.) BOOTH *(F. s. c.)* als dominierende Erreger der Trockenfäule an Kartoffelknollen anzusehen (JAHNKE [14], GÖTZ [10]). Zur Bekämpfung dieser und anderer Mykosen werden in zunehmendem Maße Präparate eingesetzt, die als Wirkstoff ein Fungizid der Benzimidazolgruppe enthalten (BRAZDA u. PETT [4]). In der Pflanzenkartoffelproduktion konnte durch Beizung der Knollen mit solchen Mitteln eine bedeutende Verminderung der Verluste durch Lagerfäulen und eine Erhöhung der Pflanzgutqualität erreicht werden. Jedoch besteht bei langjährigem Einsatz dieser Präparate die Gefahr, daß sich fungizidresistente Erregerpopulationen herausbilden können.

Die Entstehung resistenter Pilzarten bei längerer Anwendung von Fungiziden ist bereits seit einiger Zeit bekannt. In der Praxis zeigte sich das Auftreten von Resistenz meist dadurch, daß der Wirkungsgrad der angewendeten Präparate ständig nachließ. So traten z.B. nach dreimaliger Anwendung von Benomyl in einem holländischen Gartenbaubetrieb starke *Botrytis*-Infektionen an Cyclamen auf, die durch weiteren Einsatz dieses Wirkstoffes nicht bekämpft werden konnten (BOLLEN u. SCHOLTEN [3]). Zahlreiche weitere Beispiele ließen sich hinzufügen. In umfassenden Übersichten geben FEHRMANN [9] sowie HOFFMANN u. KIEBACHER [11] eine große Zahl pilzlicher Krankheitserreger an, die gegenüber verschiedenen Fungizidgruppen Resistenz aufweisen. Unter diesen Erregern befinden sich auch einige *Fusarium*-Arten, jedoch fehlen Angaben über die eingangs erwähnten kartoffelpathogenen Species. Untersuchungen von JAHN [12] sowie BURTH u.a. [6] belegen, daß es relativ leicht möglich ist, über steigende Fungizidkonzentrationen kartoffelpathogene *Fusarium* spp. zu gewinnen, die Resistenz gegenüber Benzimidazolen aufweisen.

Vorliegende Untersuchungen in vitro hatten zum Ziel, Resistenz bei *F. s.* und *F. s. c.* gegenüber den Benzimidazolen Benomyl, Thiabendazol und Carbendazim zu induzieren. Darüberhinaus sollte geklärt werden, ob bei resistenten *Fusarium*-Linien Kreuzresistenz gegenüber den genannten Mitteln auftritt. Von besonderem Interesse war die Prüfung der „fitness" resistenter Stämme, wobei Myzelwachstum, Sporulation, Konidienkeimung und Pathogenität geprüft wurden. Nicht zuletzt sollte untersucht werden, ob die erworbene Fungizidresistenz stabil ist.

2. Material und Methoden

Als Pilzmaterial kamen jeweils zwei aus trockenfaulen Knollen isolierte Stämme von *F. s.* und *F. s. c.* zur Anwendung. Als Nährmedium diente Biomalzagar mit 300 ppm Chloramphenicolzusatz zur Ausschaltung unerwünschter Kontaminationen mit Bakterien. Da die Prüfmittel Benomyl, Thiabendazol und Carbendazim schwer wasserlöslich sind, erfolgte ihre Zugabe zum Nährmedium in gelöster Form. Als Lösungsmittel fand Aceton (20—40 ml/l Nährmedium) Verwendung. Die Kontrollvariante wurde mit der entsprechenden Menge Aceton versetzt. In Anlehnung an die Methodik von Valášková [18, 19] wurde von bebrüteten Platten (Nährmedium ohne Fungizid) mit einem Korkbohrer (d = 5 mm) Pilzmyzel ausgestanzt und auf wirkstoffhaltigen Agar in vorgestanzte Löcher übertragen. Nach der Inkubation wurde das Wachstum des Pilzmyzels gemessen (Koloniedurchmesser minus 5 mm). Danach erfolgte je nach Wachstum die nächste Passage auf Nährboden mit gleicher oder höherer Wirkstoffkonzentration. Die Kontrolle wurde stets auf wirkstofffreies Medium übertragen. Zur Ermittlung der Kreuzresistenz wurden resistente Linien auf Nährmedien mit den drei verwendeten Benzimidazolen überimpft. Die Prüfung der Stabilität der erworbenen Resistenz erfolgte durch laufende Passagen der resistenten Stämme auf wirkstofffreiem Medium und anschließende Überimpfung auf Substrat mit letalen Fungizidkonzentrationen. Die Zahl der Wiederholungen betrug bei allen genannten Prüfvarianten jeweils 30.
Zur Bestimmung der Keimfähigkeit von Konidien wurden Objektträger mit dünnen Biomalzagarfilmen beschichtet und Pilzsporen ausgesät. Nach 7 bis 10stündiger Inkubation in feuchten Kammern erfolgte das mikroskopische Auszählen der gekeimten Sporen unter Berücksichtigung von insgesamt 200 Konidien je Variante.

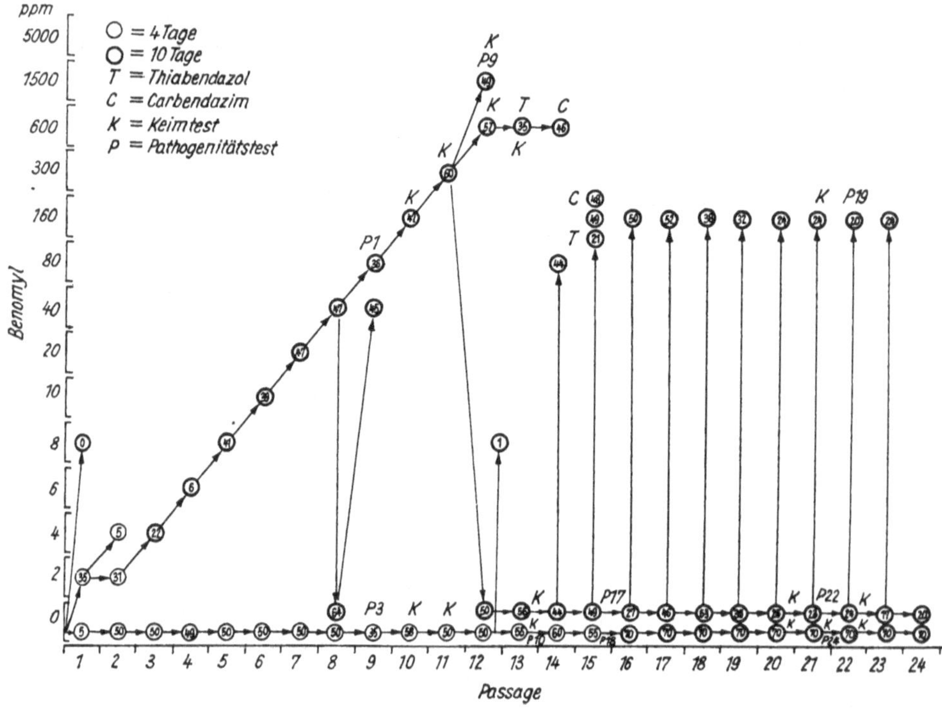

Abb. 1. Einfluß von Benomyl auf das Wachstum von *Fusarium sulphureum* (Stamm 8/1) in vitro (Zahlen im Kreis = Koloniedurchmesser in mm)

Die Pathogenität der Pilzstämme wurde auf ganzen Knollen getestet. Als Kartoffelsorten kamen ‚Mariella', ‚Adretta', ‚Astilla' und ‚Sitta' zur Anwendung. Nach Waschen und Desinfektion (3% Formalin, 20 Min.) wurden die Knollen 5mal an ihrem größten Umfang mit einem Stahlstift von 5 mm Durchmesser und 10 mm Länge verletzt. Die Inokulation erfolgte durch Übertragung eines Myzelflöckchens in jede Verletzung von jeweils 2 Knollen je Prüfvyriante. Nach 14tätiger Inkubation in feuchten Kammern bei 20 °C erfolgte die Auswertung anhand folgender Werte: 0 — keine Fäulnis, 1 — schwache Fäulnis, 2 — mittlere Fäulnis, 3 — starke Fäulnis.

3. Ergebnisse

Da sich beide Pilzstämme von *F. s.* und *F. s. c.* ähnlich verhielten, wurde bei der graphischen Darstellung der Ergebnisse (Abb. 1—5) jeweils nur ein Stamm berücksichtigt. Die Abbildungen 1—3 zeigen das Myzelwachstum von *F. s.* (Stamm 8/1) unter Einfluß von Benomyl, Thiabendazol bzw. Carbendazim. Aus den Abbildungen geht hervor, daß die letale Dosis bei den verwendeten Mitteln immer < 8 ppm war (1. Passage). Carbendazim hat dabei die beste Wirkung. Bereits bei 2 ppm wurde das Myzelwachstum deutlich gehemmt (Abb. 3), während es bei 2 ppm Benomyl und Thiabendazol im Vergleich zur Kontrolle noch recht gut war (Abb. 1 u. 2). Abb. 1—3 verdeutlichen, daß es relativ schnell möglich war, resistente *F. s.*-Linien bei allmählicher Er-

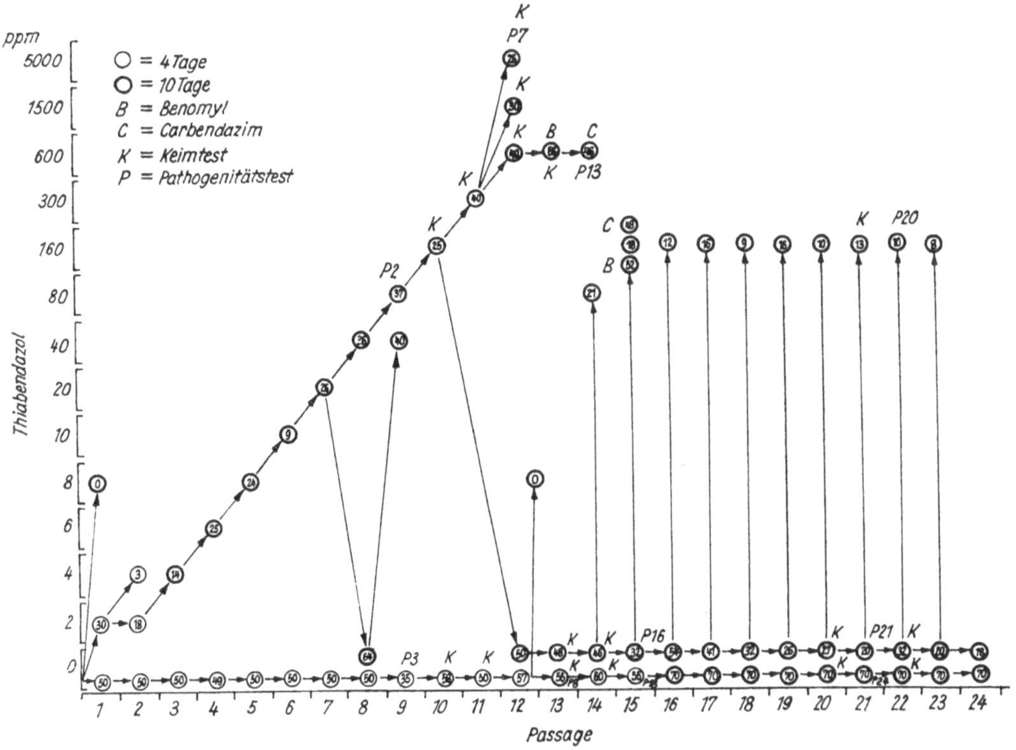

Abb. 2. Einfluß von Thiabendazol auf das Wachstum von *Fusarium sulphureum* (Stamm 8/1) in vitro (Zahlen im Kreis = Koloniedurchmesser in mm)

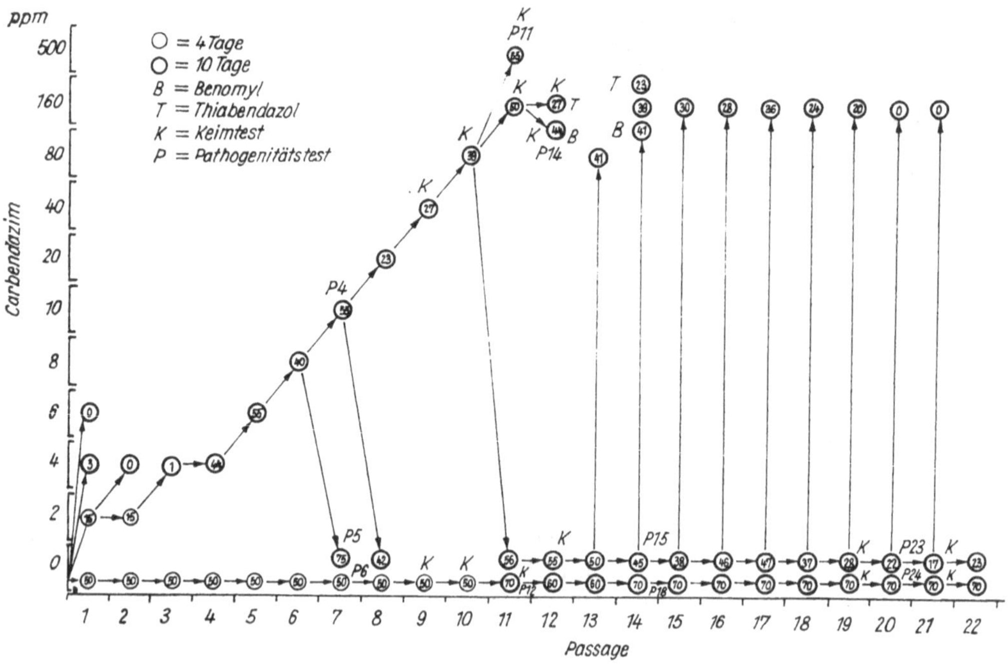

Abb. 3. Einfluß von Carbendazim auf das Wachstum von *Fusarium sulphureum* (Stamm 8/1) in vitro (Zahlen im Kreis = Koloniedurchmesser in mm)

höhung der Wirkstoffkonzentration zu erhalten. Bereits nach der 4. Passage (4—6 ppm konnte durch ständige Steigerung der Fungizidkonzentration das Myzelwachstum bei allen drei Mitteln kaum noch verringert werden. Bei dem zweiten verwendeten *F. s.*-Stamm (12/2) war dies nach der 5. Passage zu beobachten. Ein Rückgang des Koloniedurchmessers trat nur sporadisch auf (Passage 6, Abb. 2; Passage 8 und 9, Abb. 3). Selbst bei sehr hohen Wirkstoffkonzentrationen (5000 ppm Benomyl und Thiabendazol sowie 500 ppm Carbendazim) war noch ein gutes Myzelwachstum zu verzeichnen. Es muß jedoch auch darauf hingewiesen werden, daß das resistente Myzel im Vergleich zur Kontrolle in allen Prüfvarianten ein langsameres Wachstum zeigte.

Die Überimpfung der gegen 600 ppm Benomyl resistenten Kolonien auf Biomalzagar mit gleicher Konzentration von Thiabendazol bzw. Carbendazim ergab bei beiden *F. s.*-Stämmen nur geringfügige Unterschiede im Wachstum (Abb. 1), was auf eine Kreuzresistenz gegen die 3 Fungizide schließen läßt. Fast gleiche Werte ergaben sich für die Passagen der gegen 600 ppm Thiabendazol resistenten Kolonien auf Agar mit 600 ppm Benomyl bzw. Carbendazim (Abb. 2). Ebenso fielen die Ergebnisse bei der Übertragung des gegen 160 ppm Carbendazim resistenten Myzels auf Substrat mit 160 ppm Thiabendazol bzw. Benomyl aus (Abb. 3).

Zur Ermittlung der Stabilität der erworbenen Fungizidresistenz wurden das gegenüber 300 ppm Benomyl und 160 ppm Thiabendazol sowie das gegen 80 ppm Carbendazim resistente Myzel auf wirkstoff-freies Medium überimpft. Bei ständiger Weiterführung des Myzels auf diesem Medium erfolgten parallel Rückimpfungen auf Substrat mit 160 ppm Fungizid. Aus Abbildung 1 geht hervor, daß nach vier Passagen auf wirkstofffreiem Medium noch Resistenz gegenüber den drei verwendeten Benzimidazolen vorlag (Passage 15). Während insgesamt 12 Passagen auf reinem Biomalzagar (Passage 23) war eine allmähliche Abnahme des Koloniedurchmessers zu verzeichnen Die Parallelbeimpfungen von Substrat mit 160 ppm Benomyl zeigten einen ähnlichen Rückgang des Myzelwachstums. Fast gleiche Werte ergaben die Überimpfungen des Myzels mit Thiabendazolresistenz (Abb. 2). Hier war der Rückgang des Myzelwachstums bei den Parallelbeimpfungen von thiabendazolhaltigem Medium (160 ppm) gegenüber dem wirkstofffreien Medium deutlich geringer. Auch carbendazimresistentes Myzel von *F. s.* ergab nach 4 Passagen auf reinem Biomalzagar noch Resistenz gegenüber allen drei Mitteln (Passage 14, Abb. 3). Während weitere Passagen auf Medium ohne Carbendazim kaum zu Veränderungen des Myzelwachstums führten, war nach der 10. Passage und Rückimpfung auf Agar mit 160 ppm Carbendazim kein Myzelwachstum mehr zu verzeichnen, was auf den Verlust der Carbendazimresistenz hindeutet. Bemerkenswert ist, daß in allen Versuchsvarianten das fungizidresistente Myzel bei Übertragungen auf fungizidfreies Medium immer ein geringeres Myzelwachstum zeigte, als die Kontrolle. Die Ergebnisse stimmen bei beiden verwendeten *F. s.*-Stämmen nahezu überein.

In den Abbildungen 4 und 5 ist das Myzelwachstum von *F. s. c.* (Stamm 34/3) unter dem Einfluß von Benomyl bzw. Thiabendazol dargestellt. Die letale Dosis liegt bei diesen Mitteln für *F. s. c.* bei etwa 8 ppm. Durch Passagen auf Medium mit steigenden Wirkstoffkonzentrationen wurde eine Resistenz gegenüber 20 ppm Benomyl und Thiabendazol erzielt. Eine weitere Erhöhung der Fungizidkonzentration steht noch aus. Die gegen 20 ppm resistenten *F. s. c.*-Stämme wurden für 16 Monate auf sterilisierter Erde konserviert und anschließend auf Substrat mit 20 ppm Benomyl bzw. Thiabendazol ausgesät. Beide *F. s. c.*-Stämme waren nach der Konservierung fungizidresistent geblieben. Der Durchmesser der Pilzkolonien war gegenüber der fungizidempfindlichen Kontrolle verringert. Kreuzresistenz sowie Stabilität der erworbenen Resistenz wurden für *F. s. c.* noch nicht geprüft. In den bisher durchgeführten Unter-

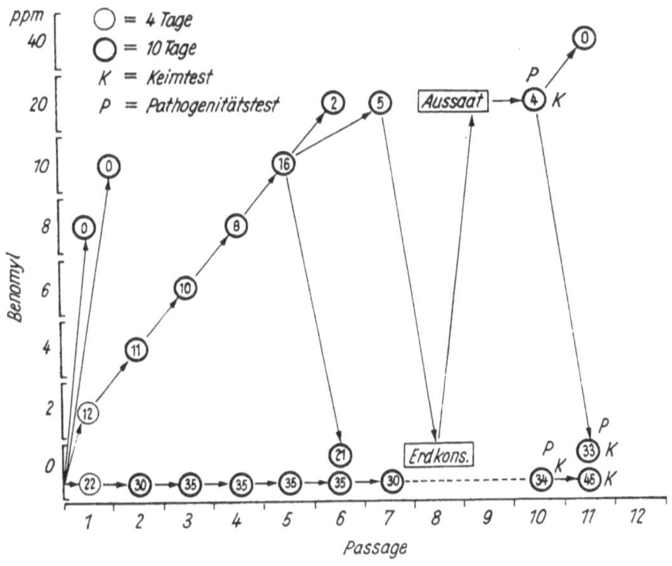

Abb. 4. Einfluß von Benomyl auf das Wachstum von *Fusarium solani var. coeruleum* (Stamm 34/3) in vitro (Zahlen im Kreis = Koloniedurchmesser in mm)

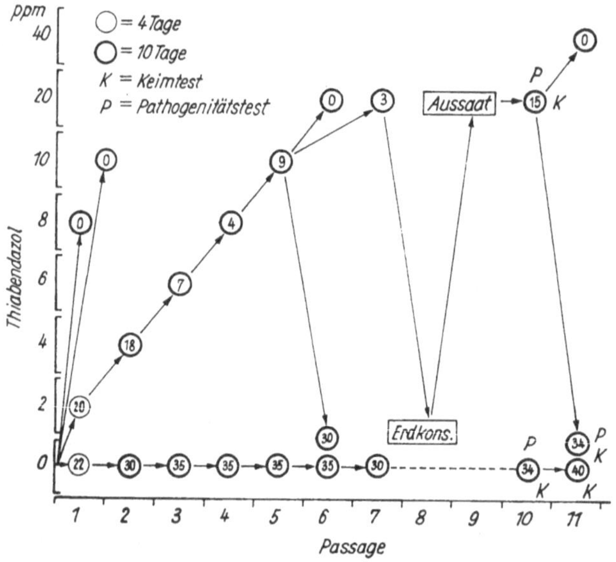

Abb. 5. Einfluß von Thiabendazol auf das Wachstum von Fusarium *solani var. coeruleum* (Stamm 34/3) in vitro (Zahlen im Kreis = Koloniedurchmesser in mm)

suchungen zur Carbendazimresistenz von *F. s. c.* wurden lediglich 8–10 ppm Carbendazim toleriert.
Die Sporenkeimung wurde bei den Passagen getestet, die in den Abb. 1–5 mit „K" gekennzeichnet sind. In fast allen Fällen betrug die Zahl keimfähiger Sporen wie bei der Kontrolle 90–100%. Lediglich in 2 Fällen lag sie unwesentlich darunter (85 und 88%).
Die Sporulation des fungizidresistenten Myzels ist bei beiden *Fusarium*-Arten als gut einzuschätzen. So bereitete es keine Schwierigkeiten, von diesem Myzel ausreichend Sporen für die Keimteste zu gewinnen. Zum anderen zeugte die gute Ausfärbung des resistenten Myzels, die sich kaum von der Kontrolle unterschied, von einer guten Sporulation.
In den Abbildungen 1–5 ist bei den mit „P" gekennzeichneten Passagen ein Pathogenitätstest durchgeführt worden. Die Ergebnisse dieser Teste sind für *F. s.* in Tabelle 1 dargestellt. Daraus geht hervor, daß bis auf eine Ausnahme (P 19) alle fungizidresistenten *F. s.*-Linien pathogen waren. Hinzu kommt, daß die Befallswerte im Vergleich zur Kontrolle nicht oder nur wesentlich geringer waren. Über die Pathogenität der fungizidresistenten Species *F. s. c.* können keine Aussagen getroffen werden, da sowohl das fungizidresistente Myzel, als auch die unbehandelte Kontrolle apathogen waren.

4. Diskussion

Vorliegende Versuchsergebnisse zeigen, daß die letale Dosis von Benomyl, Thiabendazol und Carbendazim für die kartoffelpathogenen *Fusarium*-Arten *F. s.* und *F. s. c.* < 10 ppm. Zu dem gleichen Ergebnis kam Jahn [13]. Auch bei anderen *Fusarium*-Arten wirken Benzimidazole in vitro bereits bei einer Konzentration unter 10 ppm (Bollen u. Fuchs [2], Fehrmann [9], Lashin u. Henriksen [15]). Unterschiede in der Wirksamkeit von Benomyl, Thiabendazol und Carbendazim gegenüber *F. s.* sowie in der Wirksamkeit von Benomyl gegen *F. s.* und *F. s. c.* konnten von Jahn [13] in vitro

Tabelle 1
Pathogenität von fungizidresistenten F.s.-Stämmen (Be = Benomyl, Ca = Carbendazim, Th = *Thiabendazol)*

Test-Nr. (vgl. Abb. 1–3)	Resistenz Mittel	gegen ppm	Kartoffel-sorte	Befallswert	
				Stamm 8/1	Stamm 12/2
P 1	Be	80	‚Mariella'	2,7	2,9
P 2	Th	80	‚Mariella'	2,8	2,9
P 3	–	0	‚Mariella'	3,0	3,0
P 4	Ca	10	‚Mariella'	3,0	3,0
P 5	Ca	8	‚Mariella'	3,0	3,0
P 6	–	0	‚Mariella'	3,0	3,0
P 7	Th	5000	‚Adretta'	0,9	1,0
P 8	–	0	‚Adretta'	1,7	1,5
P 9	Be	1500	‚Astilla'	1,1	1,1
P 10	–	0	‚Astilla'	1,1	1,5
P 11	Ca	500	‚Adretta'	1,1	1,0
P 12	–	0	‚Adretta'	1,7	1,9
P 13	Th, Be, Ca	600	‚Mariella'	2,9	2,4
P 14	Ca, Be	160	‚Mariella'	2,6	2,2
P 15	Ca	80	‚Mariella'	2,4	2,9
P 16	Th	160	‚Mariella'	2,8	2,9
P 17	Be	300	‚Mariella'	2,0	2,4
P 18	–	0	‚Mariella'	3,0	3,0
P 19	Be	160	‚Sitta'	2,3	0
P 20	Th	160	‚Sitta'	2,0	2,2
P 21	Th	160	‚Sitta'	2,4	2,4
P 22	Be	300	‚Sitta'	2,4	2,2
P 23	Ca	80	‚Sitta'	2,8	3,0
P 24	–	0	‚Sitta'	3,0	2,9

nicht festgestellt werden. Die langsamere Entstehung fungizidresistenten *F. s. c.*-Myzels spricht jedoch dafür, daß *F. s. c.* empfindlicher gegenüber den verwendeten Benzimidazolen reagiert als *F. s.*. Für *F. s.* konnte der Nachweis erbracht werden, daß selbst sehr hohe Fungizidkonzentrationen (bis zu 5000 ppm) toleriert werden. Gleichzeitig konnte für diese Art Gruppenresistenz gegenüber den verwendeten Benzimidazolen festgestellt werden. Kreuzresistenz von Pilzen (u. a. auch *Fusarium* spp.) gegenüber Benzimidazolen wurde bereits mehrfach nachgewiesen (Dekker [7], Bartels-Schooley u. McNeill [1], Valášková [18]). Die vorliegenden Versuchsergebnisse belegen, daß die erworbene Fungizidresistenz bei *F. s.* relativ stabil ist. So war nach 13 Passagen auf Agar ohne Wirkstoff noch Resistenz gegenüber Benomyl und Thiabendazol vorhanden. Gegen Carbendazim war die Resistenz erst nach 10 Passagen auf wirkstoff-freiem Medium verlorengegangen. Zu ähnlichen Ergebnissen kam Valášková [19] bei *F. oxysporum,* wobei in Abhängigkeit von der tolerierten Fungizidkonzentration 9–13 Passagen auf fungizidfreien Medium benötigt wurden, um die erworbene Benomylresistenz zu brechen. Auch für *F. s. c.* deutet sich eine relativ hohe Stabilität der erworbenen Fungizidresistenz an. Dafür spricht die längere Konservierung dieser Species auf sterilisierter Erde ohne Verlust der Resistenz gegen Benomyl und Thiabendazol. Das stets geringere Myzelwachstum aller fungizidresistenten *Fusarium*-Linien im Vergleich zur Kontrolle steht wahrscheinlich in Zusammenhang mit der erworbenen Resistenz. Trotz des reduzierten Myzelwachstums waren Sporulation und Keimfähigkeit der Sporen bei den fungizidresistenten Stämmen beider Arten im Vergleich zur Kontrolle kaum verändert. Darüberhinaus konnte für *F. s.* nachgewiesen werden, daß die Fungizidresistenz keinen Einfluß auf die Pathogenität des Erregers hat. Auch Magie u./Wilfret [16] sowie Valášková [17] berichten über unveränderte Pathogenität von fungizidresistenten *Fusarium*-Arten.

Die vorliegenden Versuchsergebnisse weisen für die praktische Kartoffellagerung auf die Gefahr einer Resistenzbildung hin. Obwohl in ALV-Anlagen, die bereits über vier Jahre Beizung von Pflanzkartoffeln mit benzimidazolhaltigen Präparaten durchführen, bisher kein Wirkungsabfall der eingesetzten Mittel beobachtet wurde (Brazda u. Pett [4]), ist künftig bei stärkerem Umfang der Pflanzkartoffelbeizung eine Resistenzbildung nicht auszuschließen. Durch eine unterschiedliche Mittelverteilung auf der Knollenoberfläche ist eine Selektionsgrundlage der die Knollen kontaminierenden Erregerflora denkbar. So besteht bei *Fusarium* spp. die Möglichkeit, daß sich in Lagerhäusern mit Beiztechnik fungizidresistente Pilzpopulationen im Staub anreichern können (Götz [10]). Eine zielgerichtete Bekämpfung (beispielsweise durch Desinfektion) wäre dann aufgrund der nachgewiesenen Stabilität der Resistenz sowie der „fitness" der resistenten *Fusarium* spp. notwendig. Ackerboden als Selektionsmedium von fungizidresistenten Populationen wird aufgrund des raschen Abbaus von Benzimidazolen im Boden (Dunsing [8]) sowie in Verbindung mit einer geregelten Fruchtfolge für wenig wahrscheinlich gehalten (Jahn [13], Burth u. a. [6]).

Die vorliegenden Laborbefunde lassen es erforderlich erscheinen, die Erregerflora im Staub von Lagerhäusern mit Beiztechnik auf den Gehalt an fungizidresistenten pilzlichen Erregern (insbesondere *Fusarium* spp.) zu untersuchen, um gegebenenfalls rechtzeitig Bekämpfungsmaßnahmen einleiten zu können. Darüberhinaus empfiehlt es sich, das Erregerspektrum an solchen Knollen zu untersuchen, die trotz chemischer Behandlung Fäulniserscheinungen zeigen.

Literatur

[1] Bartels-Schooley, J., B. H. MacNeill, Comparison of the mode of action of three benzimidazoles, Phytopathology **61** (1971), 816–819

[2] Bollen, G. J., A. Fuchs, On the specifity of the in vitro and in vivo antifungal activity of benomyl, Neth. J. Pl. Path. **76** (1970), 299–312

[3] Bollen, G. J., G. Scholten, Acquired resistance to benomyl and some other systemic fungicides in a strain of *Botrytis cinerea* in cyclamen, Neth. J. Pl. Path. 77 (1971), 83–90

[4] Brazda, G., B. Pett, Stand und Entwicklung der chemischen Bekämpfung von Lagerfäulen und Auflaufkrankheiten der Kartoffel, Fortschrittsberichte f. d. Landwirtschaft und Nahrungsgüterwirtschaft Akad. Landw.-Wiss. DDR **18** (1980), 2, 40 S.

[5] Burth, U., Chemische Bekämpfung von Lagerfäulen und Auflaufkrankheiten der Kartoffel unter besonderer Berücksichtigung der *Fusarium*-Trockenfäule, Tag.-Ber., Akad. Landw.-Wiss. DDR, Berlin (1978), 157, 229–241

[6] Burth, U., M. Jahn, B. Pett, J. Götz, G. Brazda, Möglichkeiten der Herausbildung resistenter Erreger bei der Kartoffelbeizung und ihre Bedeutung für das Beizverfahren, Arch. Phytopathol. u. Pflanzenschutz, Berlin **15** (1979) 4, 255–264

[7] DEKKER, J., Acquired resistance to fungicides, Annual Review of Phytopathology, Palo Alto, Calif. **14** (1976), 405–428

[8] DUNSING, M., Rückstandstoxikologische Untersuchungen von Präparaten zur Nacherntebehandlung von Pflanzkartoffeln, Diss. Akad. Landw.-Wiss. DDR (1977), 154 S.

[9] FEHRMANN, H., Systemische Fungizide — ein Überblick. II Fungizidresistenz phytopathogener Pilze, Phytopath. Z. **86** (1976), 144–185

[10] GÖTZ, J., Vorkommen und Verbreitung kartoffelpathogener Fusarien an der Knolle und im Lagerhaus, Diss. Akad. Landw.-Wiss. DDR (1977), 110 S.

[11] HOFFMANN, G. M., H. KIEBACHER, Resistenzbildung bei pflanzenpathogenen Pilzen gegen systemische Fungizide, Z. Pflanzenkrankheiten u. Pflanzenschutz **83** (1976), 368–382

[12] JAHN, M., Untersuchungen zur Entwicklung und Charakterisierung von Beizmitteln zur Bekämpfung von Lagerfäulen bei Pflanzkartoffeln, Diss. Akad. Landw.-Wiss. DDR (1977), 93 S.

[13] JAHN, M., Wirkung von Benzimidazol-Fungiziden auf kartoffelpathogene *Fusarium*-Arten, Abhandlungen Akad. Wiss. DDR (1979) 2, 141–150

[14] JAHNKE, C., Untersuchungen über die in Lagerhäusern der DDR Trockenfäule verursachenden *Fusarium*-Arten, Arch. Phytopathol. u. Pflanzenschutz, Berlin **12** (1976), 379–391

[15] LASHIN, S. M., J. B. HENRIKSON, Control of gangrene and *Fusarium* dry rot on potato tubers with thiabendazol, Tidskr. Planteavl **81** (1977) 3, 310–314

[16] MAGIE, R. O., G. J. WILFRET, Tolerance of *Fusarium oxysporum* f. sp. *gladioli* to benzimidazole fungicides, Pl. Dis. Reptr. **58** (1974), 256–259

[17] VALÁŠKOVÁ, E., Resistance to benzimidazole fungizides in *Fusarium* spp. and its relation to pathogenicity, Abstr. Third Intern. Congr. Pest. Chem. IUPAC Helsinki, 1974/75

[18] VALÁŠKOVÁ, E., Differential sensitivity of *Fusarium* spp. to benzimidazoles and the development of cross-resistance, Sbor. ÚVTIZ-Ochr.rostl. **12** (1976), 273–282

[19] VALÁŠKOVÁ, E., The stability of acquired benomyl resistance in *Fusarium oxysporum*, Abhandlungen Akad. Wiss. DDR (1979), 2 103–110

J. Kołodynski, S. Ułaszewski, R. Witkowska, S. Witek* and T. M. Lachowicz

Department of Genetics of the Microbiological Institute Wrocław University, Poland
*Institute of Organic and Polymer Technology, Technical University of Wrocław, Poland**

Antifungal Activity of Some Enolophosphate Insecticides

Three enolophosphate compounds: mebromvinphos, bromphenvinphos and chlorphenvinphos (Fig. 1) proposed as insecticides have been tested in aspect of antifungal activity in different species of yeast. All examined insecticides showed strong antifungal activity (at final concentration 100 ppm) after 24 hours action in complete liquid

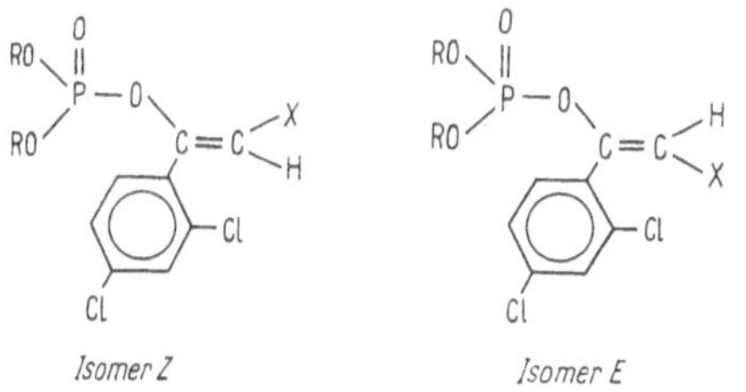

Fig. 1. Insecticides under investigation

Chlorphenvinphos : $R = C_2H_5$, $X = Cl$
Bromphenvinphos : $R = C_2H_5$, $X = Br$
Mebromvinphos : $R = CH_3$, $X = Br$

Table 1
The fungicidal action of insecticides on different species of yeast

species	lethal effect
1. *Saccharomyces cerevisiae*	+++
2. *Candida albicans*	+
3. *Hansenula canadensis*	++
4. *Zygosaccharomyces bailli*	+++
5. *Zygosaccharomyces rouxii*	+++
6. *Torulaspora delbrueckii*	+++
7. *Torulopsis delbrueckii*	+++
8. *Kluyveromyces lactis*	+++
9. *Kluyveromyces thermotolerans*	+++
10. *Schizosaccharomyces pombe*	++

Percent of viability:
+ 10.0%
++ 1.0%
+++ 0.1%
determined after 24 hours action.
Final concentration of insecticide 100 ppm.

medium containing 2% of glucose as a carbon source. Percent of viability was determined from colony counts on the same medium solidified with Difco agar (Table 1).
The action mechanism of these chemicals was studied more exactly in *Saccharomyces cerevisiae*, known as a convenient model for testing potential mutagens (SIEBERT [4]). Obtained data show, that the kinetics of cell viability depends on concentration and kind of insecticides applied. It depends also on strains (respiratory sufficient rho$^+$ or respiratory deficient rho$^-$), which were used in these experiments. The most efficient compound appeared to be mebromvinphos, which causes 80% of lethality within 10 minutes in case of respiratory sufficient strain (rho$^+$). Bromphenvinphos at the same conditions as chlorphenvinphos cause about 50% of lethality. As shown in Figure 2 the respiratory deficient (rho$^-$) mutant is more resistant to all tested insecticides. It is not quite clear why rho$^-$ strain is more resistant than the rho$^+$. May be the resistance is connected with the rate of metabolism, which is much slower in rho$^-$ cells and the changed structure of their membranes.
The data in Table 2 show, that both, respiration and fermentation decrease rapidly after 15 minutes of insecticide action. Both values are significantly higher than the loss of cell viability, which is still about 90%. The direct destroing action of these chemicals on protoplasts can be also observed. Protoplasts sensitivity suggests, that the cell surface is apparently directly involved in the loss of cell viability.
The effects of insecticides on amino acid and sugar transport is shown in Figures 3 and 4. As can be generally seen, amino acids transport is immediately inhibited after insecticides action. Moreover, progressive outflow of amino acids from the cell can be also detected. The decreasing rate of sugar transport during insecticide treatment parallels with the loss of cell viability. The cell death seems to be a result of direct insecti-

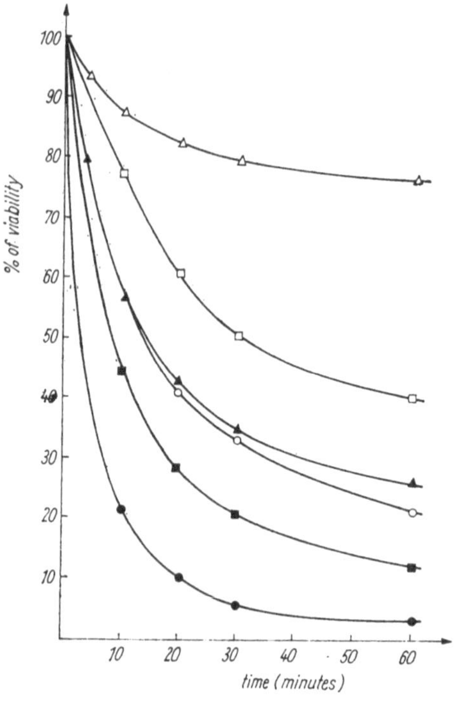

Fig. 2. Effect of insecticides [1000 ppm] on cell viability of respiratory sufficient [rho$^+$] and respiratory deficient [rho$^-$] strain D225-5A

Mebromvinphos for rho$^+$ [●] and for rho$^-$ [○] strain
Bromphenvinphos for rho$^+$ [▲] and for rho$^-$ [△] strain
Chlorphenvinphos for rho$^+$ [■] and for rho$^-$ [□] strain
Percent of viability was determined by methylene blue staining procedure

Table 2
*The effect of insecticides action on cell viability, respiration and fermentation. Percent of initial activity**

cell activity	membrom vinphos	bromphen-vinphos	chlorphen-vinphos	solvent (DMSO)
viability of intact cells	90	94	91	100
viability of protoplasts	57	6	50	100
respiration	10	32	35	100
fermentation	45	52	52	100

The initial activities were:
viability: 100%
respiration: 80 µl O_2/mg cells (1 hour and fermentation: 296 µl CO_2/mg cells) 1 hour (measured coventional Umbreit method)
final concentration of chemicals: 100 ppm
time of action: 15 minutes

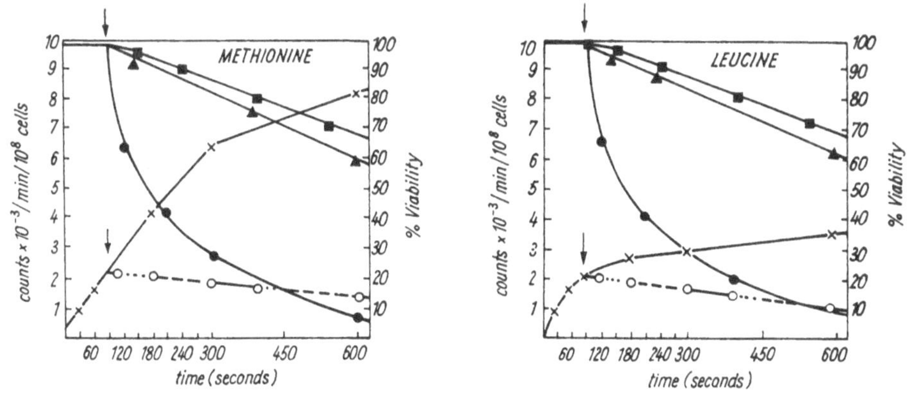

Fig. 3. Effect of insecticides action on cell viability and amino acid uptake activity [●] mebromvinphos, [▲] bromphenvinphos [■] chlorphenvinphos. [○] amino acid uptake activity in the presence of three insecticides [×] control uptake activity.
Arrows indicate time of insecticide addition.
Percent of viability was determined from colony count on complete medium

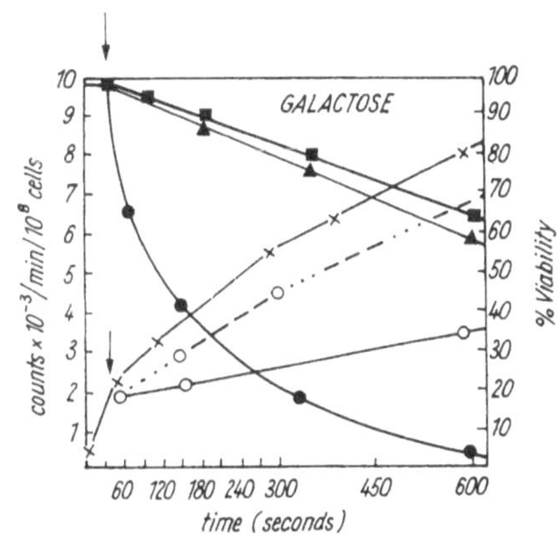

Fig. 4. Effect of insecticides [1000 ppm] on cell viability and sugar uptake activity

[●] mebromvinphos, [▲] bromphenvinphos, [■] chlorphenvinphos

[○—○] sugar uptake activity in the presence of mebromvinphos

○····○] bromphenvinphos and [○– –○] chlorphenvinphos.

[×] control uptake activity.

Arrows indicate time of insecticide addition.

Percent of viability was determined from colony count on complete medium

cides action on a membrane structure as well as on energy producing system. Similar effects of membrane damage have been observed by Cirillo [1] and Ułaszewski [5] in the case of polyene antibiotics action and inositol-less death in *Saccharomyces cerevisiae*. Polyene antibotics (nystatin and filipin) interact only with membranes containing sterols and cause an unusual increase of membrane leakiness (Lampen [3]). Phospholipids-sterol ratio in the membrane structure play an important role in the sensitivity of cells and organelles against various fungicides (Ferenczy [2]).
It may be assumed that the lethal effect of insecticides action, close associated with the loss of cell membrane and organelles integrity is a consequence of enolophosphate-sterols combining.

Acknowledgements

This work was sponsored by the inter-ministry program MR-11-15 Nature-Oriented Enviroment Management Fundamental Research.

References

[1] Cirillo, V. P., Harsch, M., Lampen, J. O., Action of polyene antibiotics filipin and nystatin and n-acetylcandidin on the yeast cell membrane, J. Gen. Microbiol. **35** (1964), 249–259
[2] Ferenczy, L., Kevei, F., Berek, J., and Teren J., Some new data on the antifungal effects of steroid glucosides. Symp. on Mechanism of Action of Fungicides (Eds.: Lyr, H. and Rawald, W.), Akad. Verlag, Berlin 1970 pp 191–199
[3] Lampen, J. O., Arnow, P. M., and Safferman, R. S., Mechanism of protection by sterols against polyene antibiotics, J. Bacteriol. **80** (1960), 200–205
[4] Siebert, D., Lamperle, E., Genetic effect of herbicides: induction of mitotic gene conversion in *Saccharomyces cerevisiae*, Mutation Res. **22** (1974), 111–116
[5] Ulaszewski, S., Woodward, J. R., and Cirillo, V. P., Membrane damage associated with inositolless death in *Saccharomyces cerevisiae*, J. Bacteriol. **136** (1978), 49–54

M. E. LADYGINA

Plant Physiology Chair, Moscow State University, USSR

Antiviral Activity of Histones

The basic nuclear proteins — histones — are typical for all the eucariots and enter into the composition of the nucleus chromatin, taking part in the formation of different chromatin levels, i.e. fullfilling a structural function. Another function of histones consists in nonspecific regulation of the gene activity by blocking definite parts of the genes (gene expression). The eucariot nuclei contain several fractions of histones: H1, H2A, H2B, H3, H4 etc., which differ in the primary structure and in the length of the polypeptide chain (Table 1 and Fig. 1). Histone molecules contain many basic, i.e.

Table 1
Characteristic of thymus histones called typical

Class	Fractions	Number of sub-fractions	Number of amino acids	Mol. wt. abt.	N-term. amino acids	C-term.
Highly lysine-rich	H1	3–6	215	21,500	Ac-Ser	Lys
Moderately lysine-rich	H2A	1	129	14,005	Ac-Ser	Lys
	H2B	1	125	13,775	Pro	Lys
Arginine-rich	H3	3	135	15,324	Ala	Ala
	H4	2	102	11,282	Ac-Ser	Gly

positively charged, amino acids of arginine and lysine. It should be emphasized that the histone molecule is asymmetric: nonpolar (uncharged) amino acids are collected at one end of the chain, and charged ones — arginine and lysine — at the other. The uncharged end of the chain winds into a tangle, while the other, charged, end hangs out as a "tail" (Fig. 2).

In 1965, MÜLLER discovered that histones released from the nucleus with the aid of DNA-ase possessed bactericidal properties [1]. A number of other authors also revealed antimicrobial and antiviral properties of histones [2–24]. These investigations were carried out with histones isolated from animal tissues with respect to animal and human viruses. The antiviral activity of histones isolated from plant tissues was discovered by us in comparative studies of the antiviral properties of histones of animal (the histones of the calf thymus) and plant (histones isolated from the nuclei of kidney bean and tobacco leaves) origin on plant tissues. The experiments were conducted with indicator bean and tobacco plants having a necrotic response to infection. Leaves

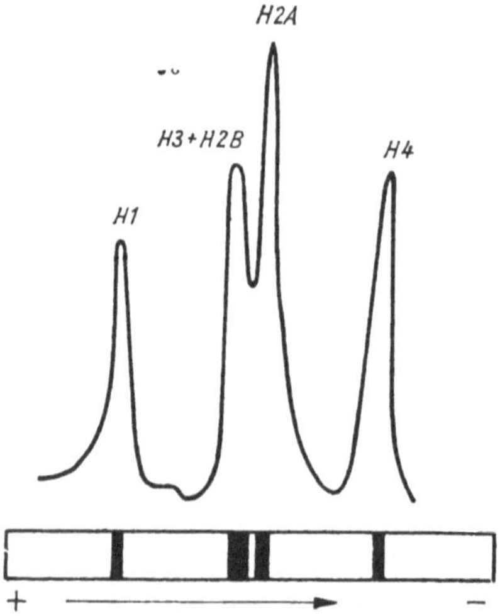

Fig. 1. Densitogram and electropherogram of thymus histones

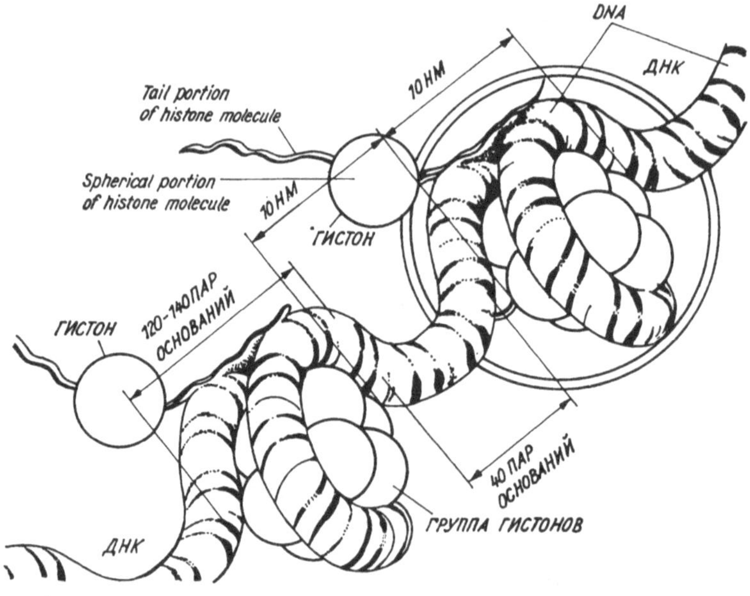

Fig. 2. Scheme of nucleohistone structure

of TMV (common strain) were inoculated. The results of the experiments are presented in Table 2.

It is seen that the thymus histones, as well as those of the bean leaves, inhibit TMV reproduction. But the bean histones isolated from leaves after TMV inoculation did not show any antiviral activity. To get a better insight into the changes in the antiviral properties of histones, we studied the electrophoretic characteristics of histones isolated from TMV-infected leaves (Fig. 3).

Table 2
Effect of thymus and bean leaf histones on the formation of necroses in bean leaves on TMV inoculation

Object	Thymus	Bean	Tobacco
Histone concentration, μg/ml	500	450	450
% of inhibition of necrose formations	34.6 ± 5.6	75.2 ± 7.2	No inhibition effect

Quantitative and qualitative changes were revealed in the electrophoregrams of histone from infected leaves as compared to those of healthy ones. The lysine fraction H1 and the arginine fraction H4 decreased, but the fraction H3 and the amount of fraction H5, typical for plant histones, increased. The reduction in fraction H4 was accompanied by the appearance of a minor band. Thus, a relationship between the quantitative and the qualitative characteristic of histones, on the one hand, and their antiviral activities, on the other, is observed. The data of Table 2 show that thymus histones are capable of inhibiting TMV, i.e. we observe the absence of specificity of histones in the manifestation of their antiviral properties. It was surprising to see that a histone isolated from tobacco leaves in our experiment did not exhibit any antiviral activity. Knowing from the literature that different histone fractions are inequivalent as regards their antiviral activity, we decided to carry out electrophoretic control of the total histones of bean and tobacco leaves (Fig. 3). The electropherograms showed that the bean and tobacco histones had basic fractions inherent in plant histones. But tobacco histones also contain minor fractions which are not found in bean leaves. We believe that the histones could have been subjected to proteolytic degradation in the course of isolation; when degraded, they lose their antifungal properties [5]. Determination of the proteolytic activity of tobacco and bean leaves showed that tobacco leaves display a considerable protease activity, whereas none was observed in bean leaves. These data do not indicate that bean leaves contain no protease at all: it is evidently inhibited, what is noted for pathogen-resistant plants. Our beans show a "hypersensitive" response to TMV incorporation, forming necroses. As for Trapesond tobacco, with which we worked, it belongs to the TMV-susceptible varieties. One more assumption can be made as to why bean and tobacco histones differ in their antiviral properties. It is possible that the histones of the resistant and susceptible varieties are different in quality or change differently under the effect of virus infection, but this assumption can only be checked by excluding the effect of proteases on the histones during isolation. The nature of interaction between histones possessing antiviral activity and the virus particles in the cell in vivo is not yet clear, however, by analogy with the literature concerning animal histones one can assume three possible mechanisms of antiviral action of histones in plants:

I. *Complexing of free histones with virus RNA and DNA* [3, 4]. The authors believe that the following complex is formed: histon — — — nucleic acid of the virus and the nucleic acid is excluded from the process of virus reproduction. It has been established that the synthesis of virus nucleic acid is mainly repressed by the lysine fraction H1. This viewpoint is confirmed by the data obtained by us with regard to the effect of

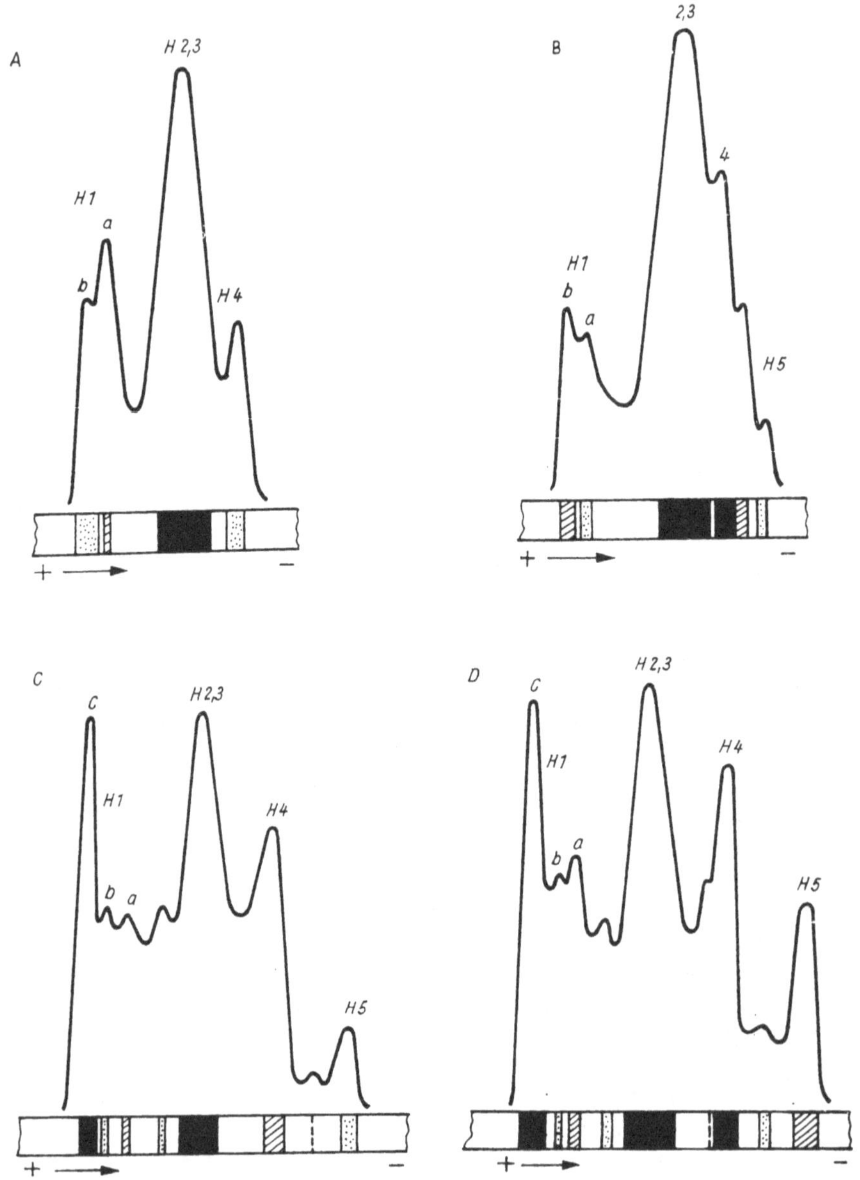

Fig. 3. Densitograms and electropherograms of histones isolated from leaves of healthy (A) and TMV-infected (B) bean plants

different methods for introducing histones on the manifestation of their antiviral activity*) (Table 3):

1. Rubbing the upper surface of the leaf with a histone solution and subsequent TMV inoculation after a four-hour exposure to the histones.
2. Simultaneous rubbing of the upper surface of the leaf with a histone solution and a virus suspension (direct contact of the histone with the virus).

Table 3
Effect of different methods for introducing thymus histones on the manifestation of their antiviral activity

Introduction method*	1	2	3	4
Efficiency of histone action (d)**	22 ± 4	7 ± 6	16 ± 6	10 ± 10

* see text
** Index *d* is the calculated difference of the average paired data. The larger is d, the greater is the difference between the number of necroses in the control and in the experiment, i.e. the stronger is the inhibiting effect of the histones.

3. Rubbing the lower surface of the leaf with a histone solution and subsequent TMV inoculation of its upper surface. This method excludes direct virus-histone contact.
4. Injecting histone solutions into the leaf (beneath the epidermis) and rubbing the upper surface of the leaf with a virus suspension. In this version, there was no direct virus-histone contact, either. As can be seen from the Table, the true antiviral effect of the histones is revealed only in the first version. Evidently, in addition to the contact with the virus, the histones must act on the cell for a definite period so as to preclude virus reproduction. A pool of free histones in the cell can be built up by the action of DNA-ase, which causes the dissociation of the nucleohistone. Besides, it is well known that histone synthesis occurs in the cytoplasm; the newly syntesized histones move into the nucleus. The pool of free histones may also be replenished by these histones. It is also known that histones, which possess definite physicochemical properties, are capable of complexing with polymer substances of the cell, particularly with nucleic acids.

II. *Blocking of the matrix function of DNA*. As is well-known, histones are capable of blocking the matrix function of the cell DNA in the synthesis of the cell RNA by forming complexes with molecules of the cell DNA:

histone - - - - cell DNA

as a result of which some of the functions of DNA necessary for virus replication are blocked. It has been shown that addition of histones disturbs the synthesis of structural virus proteins [6].

The following mechanism is not excluded, either:

III. *Indirect effect of histones on virus reproduction by changing the cell metabolism.* A number of papers demonstrate the effect of exogenous histones:

1. on energy-metabolism *processes* (oxidative phosphorylation, oxygen uptake, ATP-ase activity, etc.). Exogenous histones may inhibit or activate the energy-exchange processes, depending on the concentration [7—10].
2. *Effect on the activity of enzymes* (RNA-polymerase, alkaline phosphatase, etc.) by complexing with the protein portion of the enzyme [11, 12].
3. Histones are capable of *changing the membrane permeability and conformation*, and, thus they can act as regulators of the metabolic activity of the cell, which, in turn may substantially affect the virus reproduction [13].

The author's experiments and the data available in the literature indicate that plant, as well as animal, organisms have at their disposal a whole arsenal of protective reactions against virus reproduction resulting from joint evolution of the host plant and the pathogen. Among the similar protective mechanisms of plants and animals are evidently also the natural proteins — histones. The absence of specificity in plant and animal histones in the manifestation of their antiviral properties suggests the presence of some common protective reactions of plants and animals determined by natural substances of a protein nature.

Literature

[1] MILLER, J. E., WATSON, D. W., "Proceedings of the Society for Experimental Biology and Medicine", 131, No. 2, (1969), 339. "Biochemical Characterization of the Antimicrobial Histone Released by Deoxyribonuclease and Lactic Acid"

[2] CONNOLY, J. H., "Effect of Histones and Protamine on the Infectivity of Semlike Forest Virus and Its RNA", Nature, 212 (1966), 852

[3] ZHDANOV, V. M., MEL'NIKOVA, L. A., KOZLOVA, I. A. et al., "Repression of the Synthesis of the Pox Vaccine Virus by a Histone", DAN SSSR, 165, No. 1 (1965), 1182—1189

[4] DUBROVINA, T. YA., POLYAK, R. YA. et al., "Effect of Histones of a Normal and an Infected Cell on the Reproduction of some DNA- and RNA-containing Viruses", Voprosy virusologii, No. 6 (1969), 735—737

[5] BRESLER, V. M., VOROBIEV, V. I., "Inhibition of the Growth of a Transplanted Tumour by Nonfacrionated Histone Preparations", Tsitologiya, 5, No. 1 (1963), 69

[6] PRASKOW, D., TODOROV, S. et al., Zbl. Bacteriol., Parasitenk, Infektionskrankh. und Hyg. Abt. I — Orig. A-238, No. 1 (1977), 9—15

[7] NISHIMURA, C., NAKASUJI, K., NASHIGARI, N., Arch. Exp. Med. No. 4 (1974), 103

[8] DOVGII, I. E., KOROL', B. A., MINAYEV, P. F., "Effect of Histones on Some Energy-Exchange Processes in Liver Mitochondria of Totally Irradiated Animals", Radiobiologiya, 10, No. 6 (1970), 836

[9] DOVGII, I. E., MINAYEV, I. F., "To the Mechanism of the Effect of Histones on the Respiratory Chain of Mitochondria", Tsitologiya, 14, No. 9 (1972), 1156—1160

[10] TARSHCHIS, M. A., KUZIN, A. M., KOROL', B. A., "Effect of Histones on the Energy-Exchange Processes and on the Permeability of Isolated Goiter Nuclei", Tsitologiya, 12, No. 6 (1970), 794—797

[11] ZHDAN-PUSHKINA, S. M., AVENIROVA, E. L. et al., "Effect of Thymus Histones on the RN-ase, ATP-ase and Phosphatase Activity of *Staphylococcus* Cells", Zhurnal mikrobiologii, epidemiologii i immunologii, No. 6 (1977), 88

[12] SPELSBERG, T., HHILICA, L., "Studies on the RNA polymerase — histone Complexes", BBA, 195, No. 1 (1969), 55—59

[13] HILLER, M., WANG, D., STOLTZ, D., Mechanism of Histone Action on Proton and Cation Transport through the Mitochondrial Membrane", Biokhimiya, 42, No. 4 (1977), 589

P. Leroux and M. Gredt

Institut National de la Recherche Agronomique, Laboratoire de Phytopharmacie, CNRA, 78000 Versailles — France

Negatively Correlated Cross Resistance between Benzimidazole Fungicides and N-Phenyl Carbamate Herbicides

Introduction

The benzimidazole fungicides: benomyl, carbendazim, thiabendazole and the related thiophanate-methyl have a broad spectrum of biological activity. They inhibit the mycelial growth of fungi by blocking mitosis. These spindle toxicants bind to microtubule proteins (tubulins ?) and interfere with microtubule assembly. Davidse [1].
Various fungi can develop resistance towards the benzimidazole fungicides. In the case of *Penicillium expansum* and *Aspergillus nidulans*, Davidse [1] observes that tubulins isolated from resistant strains have a low affinity with carbendazim.
These modifications of properties of microtubules can affect the sensitivity to other spindle poisons. It is the reason why we investigated the effects of some antimitotic herbicides (Draber and Fedtke [2]) and griseofulvin on strains of *Aspergillus nidulans, Botrytis cinerea* and *Penicillium expansum* sensitive or resistant to benzimidazole fungicides. Some details of these experiments have already been published. Leroux et Gredt [3, 4, 5].

Material and Methods

Fungal strains

Aspergillus nidulans strains 003, Ben 13 and Ben 14 were generously supplied by J. M. van Tuyl (Laboratory of Phytopathology, Wageningen, The Netherlands) [7].
Botrytis cinerea strain S. Ver. 1 was isolated on experimental vines at Versailles. Strains B. Bor. and B. Col. came from vineyards near Bordeaux and Colmar where benzimidazole fungicides were used. Strain B. Eme. was isolated by Lacroix on genarium before the discosvery of benzimidazoles.
Penicillium expansum strains P. UB 4 and P. UB 5 descended from strain P. were selected on nutrient medium containing carbendazim (10 µg/ml). Strain P. NB was isolated on apples treated with thiophanate-methyl.

Chemicals (Fig. 1)

The fungicides used were: benomyl [1-(butyl-carbamoyl) benzimidazol-2-yl carbamate], carbendazim [benzimidazol-2-yl carbamate], thiabendazole [2-(thiazol-4-yl) benzimidazole] and thiophanate-methyl [dimethyl 4,4'-(0-phenylene) bis (thioallophanate)]. The herbicides used were: benfluralin [-butyl-N-ethyl-α,α,α-trifluoro-2,6-dinitro-p-toluidine], butralin [N-sec-butyl-4-tert-butyl-2,6-dinitroaniline], pendimethalin [N-(1-ethylpropyl)-

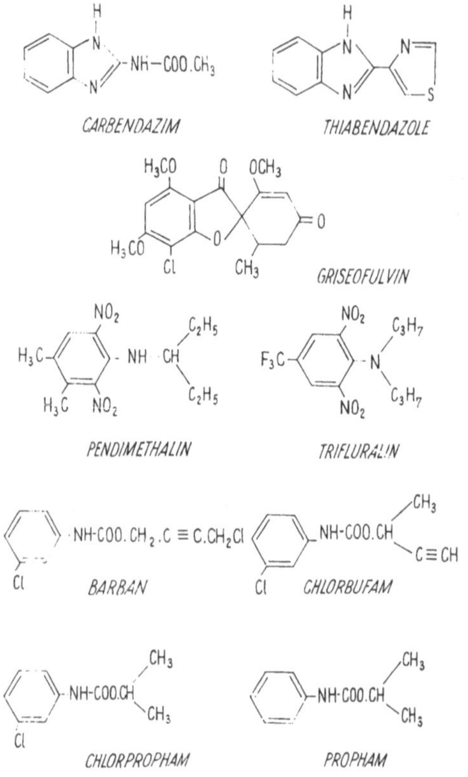

Fig. 1. Compounds that interact with microtubule formation and function

2,6-dinitro-3.4-xylidine], trifluraline [α,α,α-trifluoro-2,6-dinitro-N,N-dipropyl-p-toluidine], barban [4-chlorobut-2-ynyl 3-chloro phenylcarbamate], chlorbufam [1-methyl-prop-2-ynyl 3-chlorophenylcarbamate], chlorpropham [isopropyl 3-chlorophenylcarbamate] and propham [isopropyl phenylcarbamate].
We tested also the antibiotic griseofulvin.
All these chemicals were obtained as technical or pure forms.

Growth tests

Inhibitory effect of toxicants on mycelial growth was measured as described previously. LEROUX et al. [6]. All tests were carried out on solid medium and chemicals were added in methanol to molten agar; the final solvant concentration was 1%. Inoculation was realized by placing inverted agar disc with young mycelium on the agar surface. Incubation took place a 35 °C for *Aspergillus nidulans* and 25 °C for the two other fungi. Radial growth was measured every day during one week. For each condition the concentration (ED 50) which gave a growth rate inhibition of 50% was calculated.

Results

As mentioned in Table 1, cross resistance between benomyl, carbendazim, thiabendazole and thiophanate-methyl was always observed with *Botrytis cinerea*. Strains sensitive or resistant to these benzimidazole fungicides were inhibited by similar concentrations of griseofulvin, dinitroaniline herbicides (benfluralin, butralin, trifluralin

Table 1
Effects of toxicants on mycelial growth of Botrytis cinerea (ED_{50}-values in µg/ml)

Toxicants		*Botrytis cinerea* strains		
	S. Ver. 1	B. Eme.	B. Bor.	B. Col.
benomyl	0.03	>100	>100	—
carbendazim	0.02	>100	>100	>100
thiabendazole	0.1	8	15	—
thiophanate-methyl	0.1	>100	>100	>100
griseofulvin	0.3	0.5	0.2	0.4
benfluralin	300	—	100	200
butralin	>1000	—	>1000	>1000
pendimethalim	30	—	30	25
trifluralin	200	—	100	150
barban	20	20	0.2	0.2
chlorbufam	25	10	2	2
chlorpropham	40	20	2.5	3
propham	120	80	140	120

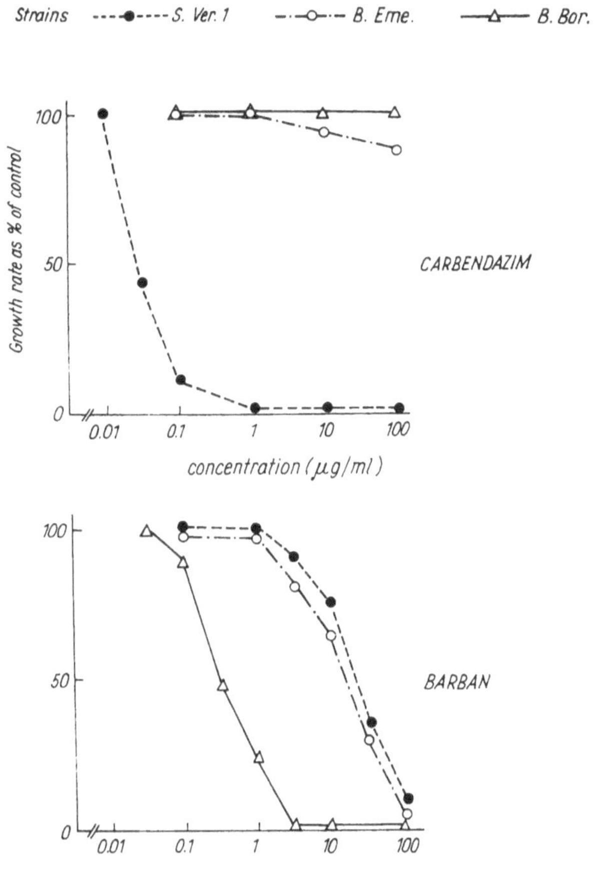

Fig. 2. Effects of carbendazim and barban on mycelial growth of *Botrytis cinerea*

or pendimethalin) and propham. With the other N-phenylcarbamate herbicides (barban, chlorbufam, chlorpropham) the ED_{50} values for strains B. Bor and B. Col. were approximatively 5 to 10 times lower than for B. Eme. and S. Ver. 1 (Table 1 and Fig. 2).

With *Penicillium expansum* strains P. UB 4, P. UB 5 and P. NB, resistance occured simultaneously with carbendazim and thiabendazole. The dinitroaniline herbicides and griseofulvin were not effective against this fungus. As observed with *Botrytis cinerea*, some strains tolerant to benzimidazole fungicides (P. UB 5 and P. BN) were hyper-sensitive to the N-phenylcarbamate herbicides (including propham) (Table 2).

The results reported in Table 3 showed that all the strains of *Aspergillus nidulans* had similar comportments towards griseofulvin, dinitroaniline and N-phenylcarbamate herbicides (Table 3).

Table 2
Effects of toxicants on mycelial growth of Penicillium expansum (ED_{50}-values in µg/ml)

Toxicants	*Penicillium expansum* strains			
	P.	P. UB 4	P. UB 5	P. NB
carbendazim	0.1	>100	>100	>100
thiabendazole	0.4	40	20	40
griseofulvin	>100	>100	>100	>100
benfluralin	>1000	>1000	>1000	>1000
butralin	>1000	>1000	>1000	>1000
pendimethalin	>1000	>1000	>1000	>1000
trifluralin	>1000	>1000	>1000	>1000
barban	50	60	15	4
chlorbufam	35	40	10	10
chlorpropham	40	40	3	5
propham	150	200	30	50

Table 3
Effects of toxicants on mycelial growth of Aspergillus nidulans (ED_{50}-values in µg/ml)

Toxicants	*Aspergillus nidulans* strains		
	003	Ben 13	Ben 14
carbendazim	0.3	>90	0.15
thiabendazole	2	>100	20
griseofulvin	5	7	7
pendimethalin	>1000	>1000	>1000
trifluralin	>1000	>1000	>1000
barban	8	10	12
chlorbufam	10	15	15
chlorpropham	30	30	20
propham	90	120	80

Discussion

The benzimidazole fungicides posses a biochemical mode of action different from the other antimitotic compounds tested because we never observed cross resistance between them.

The fact that the benzimidazole resistant strains have two types of comportment with N-phenyl carbamate herbicides indicates that, probably several mechanisms of resistance exist. It is also possible that the variations in sensitivity to N-phenylcarbamate herbicides are not correlated with the modification of properties of tubulin. Further biochemical investigation is needed to clarify this phenomenon.

However, the existance of such negatively correlated cross resistance between benzimidazole fungicides and some N-phenyl carbamate herbicides can be interesting in practice. The adjunction of a N-phenylcarbamate derivative (not phytotoxic!) to benzimidazole formulations can prevent the development of resistance of these fungicides in the field.

References

[1] DAVIDSE, L. C., The antimitotic properties of the benzimidazole fungicide carbendazim and a mechanism of resistance to this compound in *Aspergillus nidulans*, Thesis of Agricultural University, Wageningen, The Netherlands (1976), 1–84

[2] DRABER, W. and C. FEDTKE, Herbicide interaction with plant biochemical systems. in GEISSBÜHLER, H. (ed.)/Advances in Pesticide Science Pergamon press, Oxford and New York 1979, 475–486

[3] LEROUX, P. et M. GREDT, Effects du barbane, du chlorbufame, du chlorprophame et du prophame sur diverses souches de *Botrytis cinerea* et de *Penicillium expansum* sensibles ou resistantes au carbendazime et au thiabendazole, C. R. Acad. Sc. Paris, série D, **289** (1979), 691–693

[4] LEROUX, P. et M. GREDT, Phénomenes de résistance croisée négative chez *Botrytis cinerea* entre les fongicides benzimidazoles et des herbicides carbamates, Phytiatr. Phytopharm. **28** (1979), 79–86

[5] LEROUX, P et M. GREDT, Effets de divers herbicides sur la croissance mycélienne de souches de *Botrytis cinerea* et de *Penicillium expansum* sensibles ou resistantes à certains fongicides, Weed Res. (sous presse)

[6] LEROUX, P., R. FRITZ et M. GREDT, Etudes en laboratoire de souches de *Botrytis cinerea* résistantes à la dichlozoline, au dichloran, au quintozene, à la vinchlozoline et au 260 19 RP (ou glycophène), Phytopath. Z. **89** (1977), 347–358

[7] VAN TUYL, J. M., Genetics of fungicide resistance, Meded. Landb. Hogesch. Wageningen, **77 2** (1977), 1–236

H. LYR, L. BANASIAK und H. AURICH

Institut für Pflanzenschutzforschung der Akademie der Landwirtschaftswissenschaften, Kleinmachnow, DDR und Sektion Biowissenschaften, Wissenschaftsbereich Biochemie der Martin-Luther-Universität, Halle-Wittenberg, DDR

Wirkung und Wirkungsweise flüchtiger Alkenale als Fungizide

Alk-2-en-1-ale sind flüchtige, biozid wirksame Verbindungen, die bei Pflanzen weit verbreitet vorkommen und von einigen Arten auch in größeren Mengen über die Blätter gasförmig ausgeschieden werden. Nach ersten Hinweisen bei *Ailanthus glandulosa*, *Ginkgo biloba* und *Robinia pseudoacacia* [8, 9, 13, 15] zeigt sich, daß derartige Verbindungen, wie Hex-2-en-1-al, neben anderen Aldehyden und niedermolekularen Alkoholen Bestandteile des Aromas von Pflanzen und deren Früchten (Tomate, Gurke, Bohnen, Äpfel, Bananen u. a.) sind [1, 19].
Nach neueren Erkenntnissen [3, 5—7, 10] entstehen Alkenale durch Einwirkung des Enzyms Lipoxygenase auf ungesättigte Fettsäuren (Linolsäure, Linolensäure) in Gegenwart von Sauerstoff über eine intermediäre Hydroperoxid-Zwischenstufe nach folgendem Reaktionsschema für Hexenal:

Linolsäure
↓ Lipoxygenase + O_2
13-Hydroperoxy-Linolsäure
↓
(Z)-Hex-3-en-1-al $\overset{\text{ADH}}{\rightleftharpoons}$ (Z)-Hex-3-en-1-al
↓
(E)-Hex-2-en-1-al $\overset{\text{ADH}}{\rightleftharpoons}$ (E)-Hex-2-en-1-al

Lipoxygenase kommt membrangebunden in Chloroplasten-Lamellen und auch in Mitochondrien vor, wobei die Fettsäureoxidation verstärkt in alternden Zellen auftritt [7] und möglicherweise sogar ein Trigger-Prozeß zur Auslösung von Zellalterungsprozessen ist.
Über die Bedeutung dieser Verbindungen als potentielle Abwehrstoffe bei Pflanzen (Phytonzide nach TOKIN) ist bisher nur wenig bekannt. Es wird angenommen, daß diese flüchtigen Blattbestandteile im interzellulären Raum auftreten [4]. Eingehende Untersuchungen zur fungiziden Wirkung, auch im Verhältnis zur Phytotoxizität, fehlen bisher. Das Arbeiten mit flüchtigen Fungiziden bringt auch besondere methodische Probleme. Um den Grund der Wirksamkeit abschätzen zu können, synthetisierten wir einige Alkenale (C_3-C_7) und deren Acetale und unterzogen sie einer vergleichenden Testung gegen einige Pilzarten und zwei Wirt-Parasit-Kombinationen.

Die Pilze bzw. infizierte Getreidepflanzen wurden einer Gasphase dieser Verbindungen in geschlossenen Glaszylindern bei einre Einwirkungsdauer von 24 bis 72 Stunden und einer Temperatur von 25 °C ausgesetzt.
Testobjekte waren *Pythium* spec., *Rhizoctonia solani* und *Trametes versicolor*, die auf Agarplatten (3% Malzagar) kultiviert und dann in geschlossenen Glasgefäßen der Gasphase

der Verbindungen ausgesetzt wurden. Jede Konzentration wurde in siebenfacher Wiederholung geprüft. Als Maß der Wirkung diente die Reduktion des Radialwachstums. Die Werte für die ED_{50} und ED_{95} wurden durch grafische Interpolation erhalten. Junge Gersten- und Weizenpflanzen wurden in Quarzsand angezogen, mit Sporen von *Erysiphe graminis* bzw. *Puccinia triticina* infiziert und 24 Stunden in einer Klimakammer der Gasphase der Verbindungen ausgesetzt. Nach 6 bis 8 Tagen (Mehltau) bzw. 10 bis 12 Tagen (Rost) erfolgte die Auszählung der gebildeten Pusteln. Die Mittelwerte wurden in Beziehung zu den unbehandelten Kontrollen gesetzt und daraus die Hemmwirkung in Prozent errechnet. Bei Applikation über die Wurzel wurden 0,5 bis 3,0 ml der Testsubstanzen 24 Stunden vor bzw. nach der Infektion dem Quarzsand als Kultursubstrat zugesetzt. Zu Ermittlung der phytotoxischen Wirkung der Verbindungen wurden die Blätter von jeweils 10 Testpflanzen mit 1 ml der Lösungen fein besprüht. Die Phytoxizität wurde entweder an der Zahl abgestorbener Pflanzen oder nach dem Schädigungsgrad (Boniturskala 9 bis 1) bestimmt.

Tabelle 1
Fungizide Wirkung von Alk-2-en-1-alen und deren Acetale bei Applikation über die Gasphase gegenüber drei Testpilzen im Vergleich zu bekannten fungiziden Wirkstoffen (Agar-Platten-Test).
ED_{50}-Werte (ED_{95}-Werte) in µl/l

Fungizide	Pythium	Rhizoctonia	Trametes	⌀ ED_{50}
Formaldehyd	0,8 (>20)	4,5 (100)	3,6 (>100)	3,0
Propenal	0,8 (>20)	0,7 (1,9)	1,4 (18)	1,0
Butenal	4,8 (25)	1,3 (5,6)	2,2 (70)	2,8
Pentenal	1,4 (>3)	2,7 (7,0)	3,0 (28)	2,7
Hexenal	3,5 (>30)	6,6 (>30)	14,7 (320)	8,3
Heptenal	2,1 (>25)	1,4 (11)	6,1 (34)	3.2
Mittelwerte	2,5	2,5	5,5	3,5
Propenaldiethylacetal	1,3 (4)	0,5 (1,3)	0,8 (4)	0,9
Butenaldiethylacetal	3,3 (50)	6,3 (48)	2,5 (48)	4,0
Pentenaldimethylacetal	12,3 (54)	2,9 (16)	60,0 (>160)	25,0
Hexenaldiethylacetal	32,7 (>65)	19,0 (>60)	37,8 (>170)	29,8
Heptenaldiethylacetal	5,7 (60)	6,8 (>60)	27,9 (>150)	13,4
Mittelwert	11,0	7,1	25,8	14,6
Benomyl	>100	5	>100	
Carboxin	100	7	8	
Triarimol	>10	>10	0,3	
Tridemorph	60	40	6	

In Tabelle 1 sind für drei Pilzarten die ED_{50}- und ED_{95}-Werte von einigen Alkenalen und deren Acetale dargestellt. Eine Entfernung der Wirkstoffe aus der Gasphase führt nicht zur Wiederaufnahme des Wachstums, d.h. die Wirkung ist fungizid. Auffallend in Tabelle 1 sind die sehr niedrigen Werte im Vergleich zu konventionellen Fungiziden, die dem Agar-Nährmedium in üblicher Weise zugesetzt wurden. *Trametes versicolor*, ein baumbewohnender Holzzerstörer ist deutlich unempfindlicher als die beiden bodenbewohnenden Pilzarten. Im Mittel aller Werte nimmt die Wirksamkeit mit steigender Zahl der C-Atome ab, wobei sich hier zwei Faktoren überlagern: sinkender Dampfdruck mit zunehmender Kettenlänge — steigende Lipophilie. Bei Hexenal wird oft ein Minimum der Wirkung erreicht. Verbindungen mit ungerad-

zahliger Anzahl von C-Atomen haben im allgemeinen eine relativ stärkere Wirkung als solche mit geradzahligen C-Atomen. Dieser Befund könnte mit dem Abbau über einen β-Oxidationsmechanismus zusammenhängen. Die Acetale sind bei Gasphasenapplikation im Durchschnitt etwas schwächer wirksam, wofür der geringere Dampfdruck und der für die Wirkung erforderliche Zerfall in die freien Alkenale verantwortlich sind. Formaldehyd hat zum Vergleich eine teilweise sehr geringe Wirksamkeit. Bei Zugabe der Verbindungen zum Nährmedium sinkt deren Effektivität um das 32- bis 610-fache (Propenal) bzw. 6- bis 70-fache (Butenal). Reaktionen mit Nährbodenbestandteilen und damit verbundener Inaktivierung der Verbindungen sind mögliche Ursachen für das Absinken der Aktivität.
Sehr interessant ist die starke Wirksamkeit gegenüber obligaten Parasiten (Tab. 2). Die Substanzen sind 24 Stunden nach der Infektion zur Einwirkung gekommen, haben also einen deutlich kurativen Effekt. Weizenrost reagiert dabei empfindlicher als der

Tabelle 2
Fungizide Wirkung von Alk-2-en-1-alen und deren Acetale auf die Wirt-Parasit-Kombinationen Puccinia triticina (Weizen und Erysiphe graminis) Gerste bei Applikation über die Gasphase (24 Stunden Einwirkzeit).
Q = Quotient aus LD_{50} Wirtspflanze/ED_{50} Parasit ED_{50}-Werte in µl/l

Alkenale	*Puccina*	Q	Erysiphe	Q
Formaldehyd	0,21	4,8	0,25	2,2
Propenal	0,01	11,2	0,02	15,2
Butenal	0,09	4,5	0,08	4,8
Pentenal	0,02	110,0	0,77	3,4
Hexenal	0,09	52,6	0,22	6,3
Heptenal	0,06	19,6	0,04	35,7
Mittelwerte	0,04	40,3	0,22	13,2
Propenaldiethylacetal	0,04	5,5	0,04	5,5
Butenaldiethylacetal	0,50	9,1	0,26	16,6
Pentenaldiethylacetal	0,64	14,3	1,41	11,1
Hexenaldiethylacetal	0,16	33,3	0,94	5,9
Heptenaldiethylacetal	0,80	4,3	0,18	14,3
Mittelwerte	0,43	13,3	0,56	10,7

Gerstenmehltau. Da die Verbindungen in höheren Konzentrationen phytotoxisch sind, war es interessant, einen phytotherapeutischen Index in Form eines Wertes Q (Quotient aus LD_{50} Wirtspflanze/ED_{50} Parasit) zu ermitteln. Dabei zeigt sich, daß besonders Pentenal und Hexenal bzw. deren Acetale günstige Indices ergeben, was auf eine Bedeutung als negative Fungizide hinweist. Im Gegensatz zu Getreide erweisen sich andere Pflanzenarten (Bohne, Tomate, Gurke, Begonia) leider empfindlicher gegenüber Alkenalen, so daß ein direkter Einsatz als Pflanzenschutzmittel über Blattapplikation kaum in Betracht kommt. Die phytotherapeutische Spanne zur Bekämpfung von Blattkrankheiten ist zu gering.
Bemerkenswert und wegen der phytotoxischen Nebenwirkung etwas unerwartet, ist ein eindeutiger systemischer Effekt einiger Verbindungen gegen *Puccinia triticina* (Tab. 3). Der ED_{50}-Wert für Propenal und Butenal liegt im Bereich von 6 bis 14 µl/l.

Tabelle 3
Fungizide Wirkung von Alk-2-en-1-alen und deren Acetale bei Wurzelapplikation auf die Wirt-Parasit-Kombination Puccina triticina/Weizen Hemmwerte in Prozent

Alkenale	protektiv 10^{-3}	10^{-4}	kurativ 10^{-3}	10^{-4} (Mol/l)
Propenal	100	100	100	38,3
Butenal	88,4	61,8	100	100
Propenalglycerinacetal	100	100	100	100
Butenalglycerinacetal	100	100	100	66,7

Mehltau reagiert unempfindlicher und erfordert höhere Dosen. Bei 60 µl/l treten unter diesen Bedingungen noch keine phytotoxischen Schäden auf. Die in Tabelle 3 aufgeführten, nicht flüchtigen Acetale zeigen erst über 1000 µl/l eine Schädigung, wobei an der Verlagerung der Schadzonen in akropetaler Richtung der Xylemtransport erkennbar ist. In anderen Versuchen ergeben sich auch deutliche insektizide und akarizide Wirkungen, die bei Milben und einigen Dipteren stärker als bei Colepteren sind.
Bei der relativ breiten bioziden Wirksamkeit dieser Verbindungsklasse ist ein relativ unspezifischer Wirkungsmechanismus wahrscheinlich. Die Wirkung beruht vor allem auf der hohen Reaktivität der Alkenale. Als „vinologe" Verbindungen liegen im Unterschied zu den weniger wirksamen analogen gesättigten Aldehyden [12] zwei elektrophile Reaktionszentren vor, die mit nucleophilen Reaktionspartnern, wie Aminen, Aminosäuren, Alkohlen, Phenolen, Mercaptanen u. a., in Wechselwirkung treten. Zur Erklärung der inhibierenden Wirkung der Alkenale ist besonders die Reaktion mit Sulfhydryl-Gruppen enthaltenden Verbindungen, wie Glutathion, Cystein interessant, weil das eine Ursache der Inaktivierung von wichtigen Enzymen sein kann [2].
Die Wirkung verschiedener Aminosäuren auf Alkenale wurde experimentell untersucht, indem die Veränderung der Absorptionsmaxima gemessen wurde (Abb. 1). Aus dieser Reaktion läßt sich folgern, daß hierdurch ein wesentliches Unterscheidungsmerkmal zu gesättigten Aldehyden gegeben ist, die bei einer Konzentration von 80 µmol/l (8 bis 12 µl/l) nach NYMAN [14] keine fungizide Wirkung aufweisen, sondern das Wachstum von *Dipodascus aggregatus* mehr oder weniger stimulieren. Gleiches zeigen auch Versuche von SCHILDKNECHT und RAUCH [15] an Ciliaten, wo Hexanal im Gegensatz zu Hexenal biologisch inaktiv ist. Hexadienal wirkt hier schwächer, während andere Aldehyde nur wenig wirksam sind.

Da die unterschiedliche Empfindlichkeit der einzelnen Pilzarten (vgl. Tab. 1) auch in einem differenzierten oxidativen Abbau der Alkenale begründet sein kann, ist die Angreifbarkeit von Pentenal und Hexenal durch die Aldehyddehydrogenase von *Acinobacter calcoaceticus* untersucht worden. Das Enzym ist membrangebunden, NADP-abhängig und oxidiert langkettige Aldehyde. Es wird nach Anzucht der Bakterien auf n-Alkanen gebildet [17]. Das Enzym, das gesättigte Aldehyde (Nonanal) gut oxidiert, greift Pentenal und Hexenal jedoch nicht an. Diese Alkenale haben bis zu einer Konzentration von 0,45 mmol/l keinen kompetitiven Hemmeffekt gegenüber Nonanal. Eine partikuläre, akzeptorabhängige Alkoholdehydrogenase aus *Pseudomonas putida* (Anzucht auf n-Alkanen) oxidiert aliphatische Alkohole, wie Hexanol, Octanol, mit Dichlorphenolindophenol als Akzeptor zu den entsprechenden Aldehyden [18]. Messungen mit Hexa-2,4-dien-1-ol (Sorbinalkohol) als Substrat ergaben bei Zusatz von 1,25 mmol/l Hexenal eine Hemmung von 57%, während Pentenal bei 1,0 mmol/l keine Wirkung zeigt. Eine NADH-abhängige Alkoholdehydrogenase aus *Saccharomyces cerevisiae* blieb durch Alkenale unbeeinflußt.

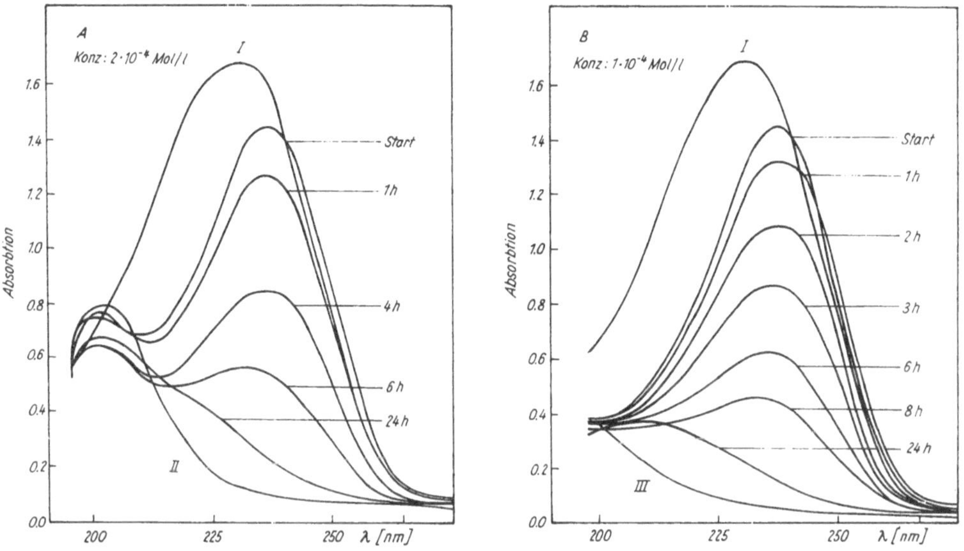

Abb. 1. Verlauf der Adduktbildung von Alk-2-en-1-alen (I) mit Glutathion (II) [A] bzw. α-Alanin (III) [B] in 50%igem n-Propanol

Diese ersten Ergebnisse zeigen, daß Alkenale resistent gegen Aldehyddehydrogenasen sind, die am Abbau von n-Alkanen mitwirken. Ob spezifische, induzierbare Alkenal-Dehydrogenasen durch Mikroorganismen gebildet werden, ist gegenwärtig noch nicht geklärt. Der Metabolismus in der Pflanze verläuft reduktiv von Hexenal über Hexenol zu Hexanol [16], der dann durch akzeptorabhängige ADH oxidiert werden kann. Orientierende Versuche an ETP aus Rinderherzmitochondrien ergeben keinen Hinweis auf eine Hemmung der Atmungskette durch Alkenale [11].

Schildknecht und Rauch [15] fanden in der Umgebungsluft von *Robinia pseudoacacia* eine Konzentration an Hexenal von 3 μg/m³. Das entspricht einer Konzentration von $3 \cdot 10^{-6}$ μl/l. Eine Rostinfektion kann bei 24-stündiger Einwirkung von 0,3 μl/l Hexenal bzw. 0,08 μl/l Pentenal eliminiert werden. Da man annehmen kann, daß die Hexenal-Konzentration im Interzellularsystem von *Robinia* höher als in der Umgebungsluft ist, könnte eine solche Konzentration ausreichen, um einen Pilzbefall zu verhindern. Für andere Pflanzenarten liegen ähnliche Messungen nicht vor. Möglicherweise ist die Bildung von Alkenalen geringer und die Eigenempfindlichkeit höher. Immerhin ist nicht auszuschließen, daß Alkenalen eine innertherapeutische Funktion zur Keimfreihaltung des Pflanzeninneren zukommt. Alkenale werden besonders intensiv nach Gewebeverletzungen gebildet, was wahrscheinlich mit der Freisetzung von Fettsäuren durch Phospholipasen und der Aktivierung von Lipoxygenasen zusammenhängt, die besonders in Chloroplasten und Mitochondrien lokalisiert sind. Dadurch ist ein zusätzlicher Abwehrmechanismus gegen pilzliche oder tierische Schädlinge denkbar, was nach Funktion und Bildungsweise den Phytoallexinen entspricht.

Literatur

[1] Drawert, F. u. a., Über die Biogenese von Aromastoffen bei Pflanzen und Früchten, Ann. Chem. **694** (1966), 200–208

[2] ESTERBAUER, H., H. ZOLLNER and N. SCHOLZ, Reaction of glutathione with conjugated carbonyls, Z. Naturforsch. **30** c (1975), 466–473
[3] GARDNER, H. W., Decomposition of linolenic acid hydroperoxides. Enzymatic reactions compared with nonenzymatic, J. Agric. Food. Chem. **23** (1975), 129–136
[4] GRAMBOV, H. J., The influence of volatile leaf constituents on the in vitro differentiation and growth of *Puccinia graminis* f. sp. *tritici*, Z. Pflanzenphysiol. **85** (1977), 361–372
[5] GROSCH, W. and G. LASKAWY, Differences in the amount and range of volatile carbonyl compounds formed by lipoxygenase isoenzymes from soybeans, J. Agric. Food. Chem. 23 (1975), 791–794
[6] HATANAKA, A. u. a., Formation of 12-oxo-trans-10-dodecenic acid in chloroplasts from *Thea sinensis* leaves, Phytochemistry **16** (1977), 1828–1829
[7] MACLEOD, J. A. and H. E. PIKK, Formation of (E)-hex-2-enal and (Z)-hex-3-en-1-ol by fresh leaves of *Brassica oleracea*, J. Agric. Food Chem. **27** (1979), 469–475
[8] MAJOR, R. T., P. MARCHINI and T. SPROSTON, Isolation from *Gingko biloba* L. of an inhibitor of fungus growth, J. Biol. Chem. **235** (1960), 3298–3299
[9] MAJOR, R. T., P. MARCHINI, and A. J. BOULTON, Observation on the production of α-hexenal by leaves of certain plants, J. Biol. Chem. **238** (1963), 1813–1816
[10] MATTHEW, J. A. and T. GALLIARD, Enzymatic formation of carbonyls from linolenic acid in leaves of *Phaseolus vulgaris*, Phytochemistry **17** (1978), 1043–1044
[11] MÜLLER, W., Persönliche Mitteilung, 1979
[12] NANDI, B., Effect of volatile aldehydes, ketones, esters and terpenoids on growth and development of fungi associated with wheat grains in the field and in storage, Z. Pfl. Krankh. **84** (1977), 11–128
[13] NYE, W. and H. A. SPOEHR, The isolation of hexenal from leaves, Arch. Biochem. Biophys. **2** (1943), 23–35
[14] NYMAN, B., The effect of various aliphatic aldehydes and related compounds on the growth of *Dipodascus aggregatus*, Physiol. Plant **22** (1969), 1322–1328
[15] SCHILDKNECHT, H. und G. RAUCH, Die chemische Natur der „Luftphytoncide" von Blattpflanzen, insbesondere von *Robinia pseudoacacia*, Z. Naturforsch. **16** b (1961), 423–429
[16] STONES, E. J., R. M. HALL and, S. J. KAZENIAC, Formation of aldehydes and alcoholes in Tomato fruits from U-^{14}C labeled linolenic and linoleic acid, J. Food. Sci. **10** (1975), 1138–1141
[17] TAUCHERT, H. u. a., Pyridinnucleotid-unabhängige Oxydation von längerkettigen aliphatischen Alkoholen durch ein Enzym aus *Acinobacter calcoaceticus*, Z. Allg. Mikrobiol. **15** (1975), 457–460
[18] TAUCHERT, H., M. GRUNOW und H. AURICH, Regulation und einige Eigenschaften einer partikulären, akzeptorabhängigen Alkoholdehydrogenase aus *Pseudomonas putida* beim Wachstum auf n-Alkanen, Z. Allg. Mikrobiol. **18** (1978), 675–680
[19] TRESSL, R. and DRAWERT, F., Biogenesis of banana volatiles, J. Agric. Food. Chem. **21** (1973), 560–565

R. MAXMIOWA, N. PALMOWA, T. SCHARKOWA, B. CHURATOWA, M. LEJKINA

Staatliche Lomonossow-Universität Moskau, UdSSR

Wirksamkeit von Trichothezin gegen Myzelien von Fungi imperfecti und Gewebekulturen

Trichothezin- ein fungizides Antibiotikum — wird aus *Trichothecium roseum* gewonnen. Seine Summen-Formel ist $C_{19}H_{24}O_5$. Chemisch ist es ein Esther des Trichothecolons und der Isokrotonsäure vom Molekulargewicht 332, hat den Schmelzpunkt 118° und ist in organischen Lösungsmitteln gut löslich und optisch aktiv ($[\alpha]D^{18} = +44$). Eine fungizide Aktivität seiner Nährlösung wurde von BRIAN und HEMMING bereits 1947 beschrieben. Isolierung und Identifizierung gehen auf FREEMAN und MORRISON [4] zurück.

Trichothezin wirkt gegen Pyrenomyceten, Actinomyceten und Basidiomyceten, sowohl gegen einige phytopathogene als auch tierpathogene Arten (BECKER et al. [3], FREEMAN und MORRISON [4], SILAEW [6] VÖRÖS [7]). Trichothezin hemmt pflanzen- und tierpathogene Viren, wirkt aber nicht gegen Bakterien.

Trichothezin gehört, wie auch ökologisch wichtige Mykotoxine und andere Antibiotika, zu den Sesquiterpenen. Diese Substanzen mit sehr ähnlicher Biogenese (über Mevalonsäure) haben auch einen ähnlichen Wirkungsmechanismus, der bei eukaryotischen Zellen in einem Angriff auf die ribosomale Proteinsynthese besteht (BAMBERG [2]). Für den Wirkungsmechanismus der Sesquiterpene ist die Struktur der C-15-Position sehr wichtig. Substanzen, die in dieser Position eine sauerstoffhaltige Gruppe besitzen, wie z.B. Verrucarin verursachen einen schnellen Abbau der Polysomen. Stoffe, die in dieser Position eine Methylgruppe haben, wie z.B. Trichodermin, hemmen zwar die Verlängerung der Peptidkette, zerstören aber die Polysomen nicht. Vermutlich hat auch Trichothezin, das in dieser Position ebenfalls eine Methylgruppe besitzt, einen solchen Wirkungsmechanismus (Abb. 1). Die Gruppe der Sesquiterpene besitzt gegen Pilze, Protozoen und verschiedene Säugetierzellen hohe biologische Aktivität.

Trichothezin wird in der Sowjetunion produziert und im Pflanzenschutz angewendet, insbesondere gegen Gurken-Mehltau in Treibhäusern [1].

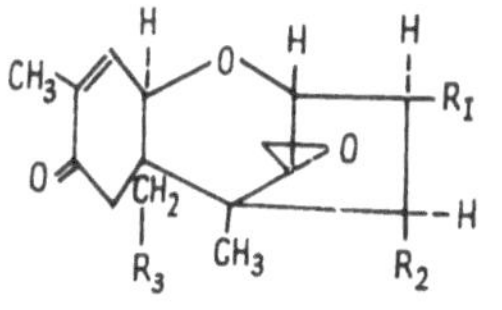

$R_1 = H$

$R_2 = CH_3CH{=}CHOCO$

$R_3 = H$

Fig. 1. Strukturformel von Trichothezin

Im Laboratorium der Staatsuniversität Moskau untersuchen wir die Biologie, Variabilität und Physiologie von *Trichothecium roseum* und die Biosynthese von Trichothezin sowie auch anderer, von diesem Organismus gebildeter, biologisch aktiver Verbindungen (MAXIMOWA et al. [5]).

Für den Produzentenstamm stellt Trichothezin auch keinen neutralen Stoff dar. Die Empfindlichkeit von *T. roseum* gegen Trichothezin liegt bei 200—400 mkg/ml. In 0,1—1,0 mkg/ml-Konzentrationen hemmt es die Keimung von Konidien von *T. roseum*, im Nährmedium das Myzelwachstum schon in den ersten Phasen der Entwicklung. Nach einer Adaptationsperiode nimmt die Wachstumsintensität wieder zu. 500 mkg/ml bewirken einen stärkeren Hemmeffekt (Abb. 2). Zugabe des Antibiotikums zu einem vorher 48 Stunden gewachsenen Myzel verursacht stärkere Verminderung der Biomasse. Eine Lyse des Myzels kann festgestellt werden.

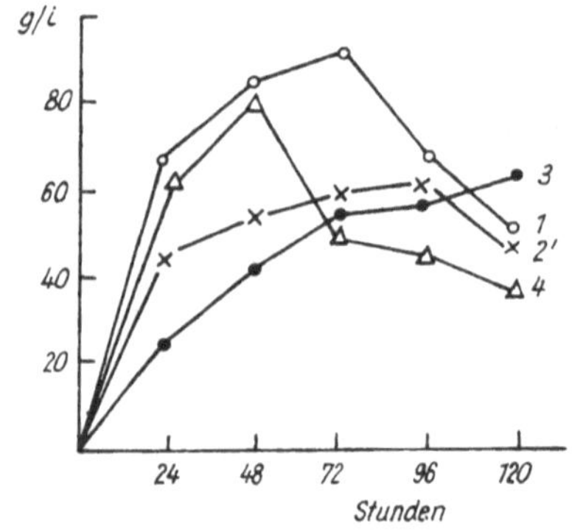

Fig. 2. Einfluß von Trichothezin auf das Wachstum von *T. roseum*
1 — Kontrolle, 2 — 200 mkg/ml Trichothezin, 3 — 500 mkg/ml, 4 — 200 mkg/ml Trichothezin nach 48 Stunden

Trichothezin hemmt die biosynthetische Aktivität von *T. roseum* stark. Davon ist auch die Biosynthese des Antibiotikums betroffen. Diese kann in den ersten 72 Stunden der Kultur beobachtet werden. Danach steigt die Biosyntheseaktivität etwas an; wird das Antibiotikum 48 Stunden alten Pilzkulturen zugesetzt, ist der Einfluß auf die Biosynthese des Antibiotikums weniger ausgeprägt. In jedem Fall wird nicht nur die Trichothezin-Biosynthese gehemmt, sondern auch ein Abbau des Antibiotikums in der Kultur bewirkt. Tricholysin",, ein Enzymkomplex ,der Enzyme mit proteolytischer und Esteraseaktivität enthält, bewirkt diesen Abbau (Abb. 3).

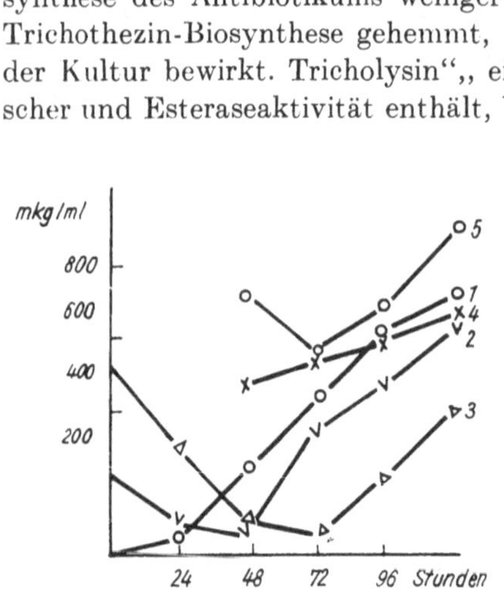

Fig. 3. Einfluß von Trichothezin auf die biosynthetische Aktivität von *T. roseum*
1 — Kontrolle, 2 — 200 mkg/ml Trichothezin vor der Saat, 3 — 500 mkg/ml vor der Saat, 4 — 200 mkg/ml nach 48 Stunden Wachstum, 5 — 500 mkg/ml nach 48 Stunden Wachstum

Es hat den Anschein, daß die Verminderung der biosynthetischen Fähigkeiten der Pilzzelle das Resultat einer Störung wichtiger Stoffwechselprozesse ist. Betroffen sind Energiehaushalt und Eiweißsynthese, was sich am Wachstum wiederspiegelt. In Gegenwart von Trichothezin war die Dehydrogenaseaktivität des Myzels vermindert, besonders auffällig in den ersten Phasen der Kultur. Die Dehydrogenaseaktivität wird nach Adaptation des Myzels an das Antibiotikum wiederhergestellt (Abb. 4).

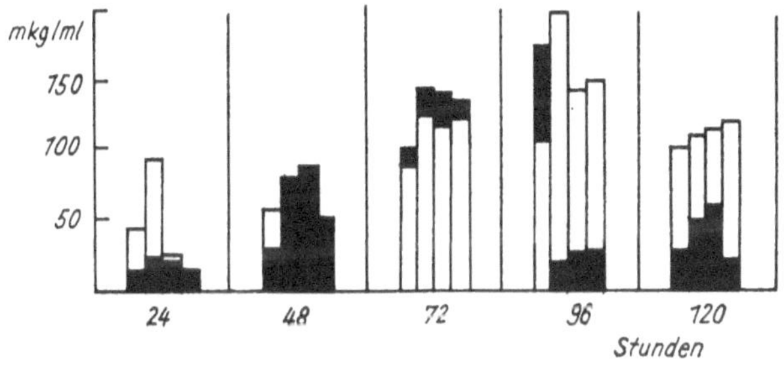

Fig. 4. Einfluß von Trichothezin auf die Dehydrogenaseaktivität
1 — Glukose, 2 — Pyruvat, 3 — Zitrat, 4 — Sukzinat. Helle Säulen — Kontrolle, dunkle Zugabe von 200 mkg/ml Trichothezin

Trichothezin vermindert auch bei anderen empfindlichen Pilzen, z. B. *Verticillium dahliae* und *Fusarium moniliforme* die Dehydrogenaseaktivität. Resistenz gegen Trichothezin ist bei diesen Pilzen von einer Abnahme der Wirkung des Antibiotikums auf das Dehydrogenasensystem begleitet.

Trichothezin hat auch auf die Morphologie des Submersmyzels von *T. roseum* Einfluß. Es beschleunigt die Myzelentwicklung, begünstigt die „Tiefenkonidienbildung", die Bildung von Initialzellen, von knospendem Myzel und von Chlamydosporen. Letzteres führt zur Herausbildung neuer, antibiotikumresistenter und zu intensiverer Trichothezin-Bildung befähigter Myzelgenerationen.

Der Einfluß des Trichothezins auf den die Verbindung produzierenden Pilz wurde nach kurzfristigem Kontakt des Antibiotikums mit dem Pilz in einer 200—500 mkg/ml-Konzentration bei 48 Stunden altem Myzel, bei pH 4,0 und 7,0 untersucht. Es findet eine Adsorption an der Myzeloberfläche statt (Abb. 5). Die Zellmembranpermeabilität

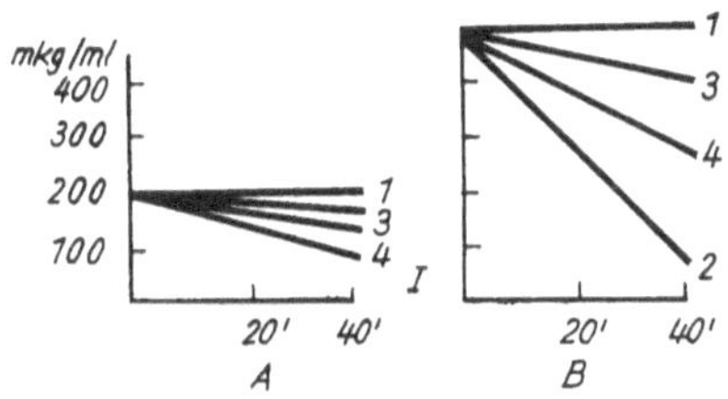

Fig. 5. Trichothezinadsorption auf der Myzeliumoberfläche
A — Trichothezinkonzentration 200 mkg/ml
B — 500 mkg/ml
1 — Kontrolle, pH = 7,0, 2 — im Kontakt mit Myzelium, pH = 7,0, 3 — Kontrolle, pH = 4,0, 4 — dasselbe im Kontakt mit Myzelium

wird erhöht. Als Folge kommt es zum Austritt wichtiger Zellkomponenten in das Außenmedium. Betroffen sind z. B. Phosphorverbindungen und bei 260 nm adsorbierende Substanzen (Nucleotide) (Abb. 6). Adsorption des Wirkstoffes und Permeabilitätsstörungen in den Zellen sind bei pH = 7,0 und einer Konzentration von 500 mkg/ml intensiver als bei pH 4,0, bzw. niederen Konzentrationen. In Anwesenheit von NH_4^+ und Cholesterin ist die toxische Wirkung von Trichothezin herabgesetzt. Das Wachstum setzt neu ein, der Verlust an Phosphorverbindungen ist verlangsamt und vermindert.

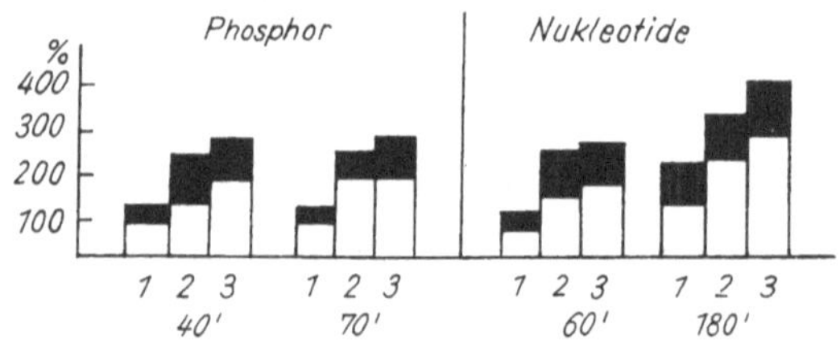

Fig. 6. Verlust der Zellen an Phosphor und an bei 260 nm UV-absorbierenden Stoffen
1 — Kontrolle, 2 — 200 mkg/ml Trichothezin, 3 — 500 mkg/ml, helle Säulen — pH = 4,0, dunkle — pH = 7,0

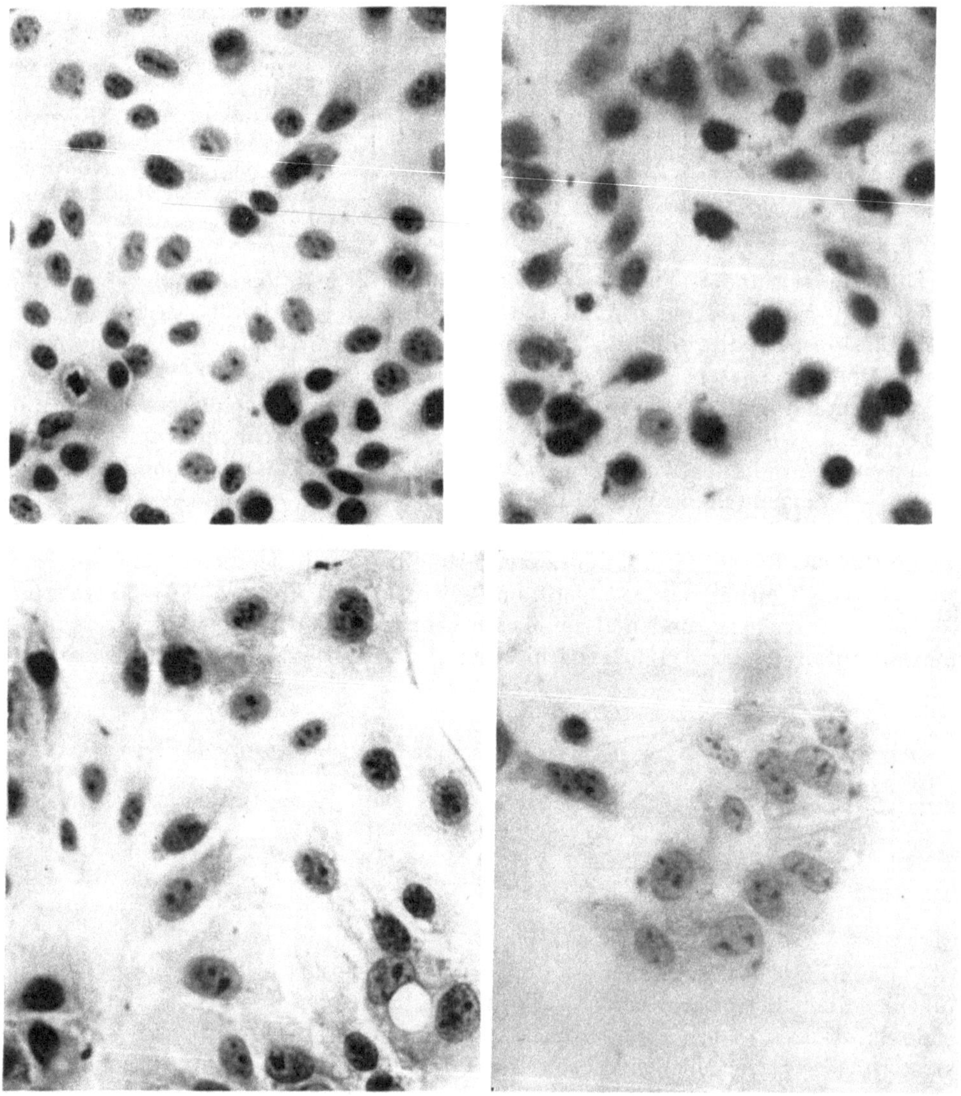

Fig. 7. Einfluß von Trichothezin auf Gewebekulturen
1 — Monoschicht von „SPEW" (Kontrolle), 2 — Monoschicht von SPEW in Gegenwart von 0,2 mkg/ml Trichothezin 3-2SPEW Kulturen nach 48 Stunden [Kontrolle], 4 — dasselbe in Gegenwart von Imkg/ml Trichothezin

Von Monoschichten von Schweinsembryonierenepithel (SPEW) wurde Trichothezin in hoher, zytostatisch wirksamer Konzentration, gebunden. Trichothezin hemmt das Wachstum der Zellen der SPEW bei Zellpassage in Konzentrationen von 0,2—0,3 mkg/ml und im Kontakt mit einer in 48 Stunden gewachsenen Kultur in einer solchen von 0,6—1,0 mkg/ml (Abb. 7). Nach 24 bzw. 48 Stunden Wachstum wird eine Hemmung der Zellteilung beobachtet; die mitotische Aktivität ist vermindert. Es kommt zu keiner Schichtbildung in der Gewebekultur. Veränderungen der Zellform und Zellhäufungen wurden beobachtet. Zugabe von Trichothezin zur reifen Gewebekultur führt zur Zerstörung der Zellschicht. Die Zellen werden rundlich, die Kerne schrumpfen. Dann entsteht eine Zellpyknose und als Endresultat folgt Lyse, charakterisiert durch Membranzerstörung und Ausfließen des Zellinhaltes. Die toxische Wirkung von Trichothezin steigt mit seiner Konzentration und Einwirkzeit.
Es wurde die hemmende Wirkung von Trichothezin für Pilze und seine hohe Aktivität bei Zellmonoschichten von Gewebekulturen gezeigt. Trichothezin besitzt ein breites Wirkungsspektrum. Es kann das Wachstum hemmen, morphogenetische Wirkungen hervorrufen, energetische und biosynthetische Prozesse hemmen und auch die mitotische Aktivität. Diese Stoffwechelstörungen sind Folgen der spezifischen Einwirkung des Antibiotikums auf die Zellpermeabilität, die zu einem Mangel an Stoffen in den Zellen führt. Der Wirkungsmechanismus von Trichothezin, mit seinem membranotropen Effekt, ist dem Wirkungsmechanismus der Polyen-Antibiotika ähnlich.

Literatur

[1] Annotierung von Pestiziden. Werke der Staatskomission für chemische Kampfmittel gegen Pflanzenschädlinge und Unkraut (russisch), Moskau, 1973

[2] Bamberg, James, R., Chemische und biochemische Forschungen der Trichothezenmykotoxinen. Mykotoxine und andere mit Fungi verbundene Nahrungsprobleme (englisch).

[3] Becker, S. E., Lisina, E. S., Maximowa, R. A., Die Eigenschaften des antibiotischen Spectrums und der Wirkungsmechanismus einiger Antibiotika aus Pilzen. Intern. Symposium Eberswalde, Academieverlag Berlin, 1967, 115—119 (deutsch)

[4] Freeman, G. G., Morrison, R. J., Trichothezin — ein fungizides Metabolitenprodukt aus *Trichothecium roseum*, Nature, **162** (1948), 4106—4110 (englisch)

[5] Maximowa, R. A., Silaew, A. B., Palmowa, N. P., Scharkowa, T. S., Churatowa, B. G., Poch, L. I., Gruschina, W. A., Minaewa, T. A., Biologische Eigenschaften eines imperfekten Fungus, *Trichothecium roseum* des Trichotezinantibiotikumsproduzenten. Sammlung: „Die Antibiotika und ihre Produzenten“, Verlag „Nauka“, Moskau, 1975, 232—240 (russisch)

[6] Silaew, A. B., Maximowa, R. A., Scheklanow, E. D., Leschenko, W. M., Rukawischnikowa, W. M., Nagibin, W. W., Trichothezin und seine klinische Anwendung in der Behandlung von einigen Dermatomykoseformen. Sammlung: Chemotherapie von Infektionen und die Heilmittelresistenz pathogener Mikroorganismen, Moskau, 1973, 123—124 (russisch)

[7] Vörös, I. A., Trichothecin antibioticum alkalmazasa hovenyi karekozok ellen Növenytermeles, **4** (1955), 233—240 (Hung.)

R. MÜLLER und U. BURTH

Institut für Pflanzenschutzforschung der Akademie der Landwirtschaftswissenschaften, Kleinmachnow, DDR

Benzimidazol-Tensid-Formulierungen mit Wirksamkeit bei Benomylresistenz

1. Einleitung

Pilzkrankheiten der Kulturpflanzen verursachen jährlich hohe Ernteverluste. Neben züchterischen Maßnahmen kommt dem Einsatz von Fungiziden eine entscheidende Bedeutung bei der Ausschöpfung des Ertragspotentials unserer Kulturpflanzen zu. Mit dem weltweiten Einsatz systemischer Fungizide seit etwa 12 Jahren trat das Problem der Resistenzentwicklung bei pflanzenpathogenen Pilzen gegenüber einigen Wirkstoffgruppen in den Vordergrund. Speziell für die Gruppe der Benzimidazol-Fungizide (Benomyl, Carbendazim, Thiabendazol, Fuberidazol und Thiophanatmethyl), wurden bei verschiedenen pilzlichen Schaderregern Resistenzerscheinungen mit Gruppen-Resistenz für diese Wirkstoffe beobachtet. Ziel der vorliegenden Untersuchungen war es, die fungizide Wirksamkeit von Benzimidazol-Tensid-Formulierungen (gegen Populationen von *Erysiphe graminis* DC. und *Sphaerotheca fuliginea* (SCHLECHT.) SALMON) zu prüfen, bei denen eine ausgeprägte Benomylresistenz nachgewiesen worden war.

2. Material und Methoden

In Tabelle 1 werden die Handelspräparate, Prüfmuster, Tenside und Lösungsmittel aufgeführt, die für die Versuche verwendet wurden.
Die Bestimmung der fungiziden Wirkung der Präparate gegen *Erysiphe graminis* DC. erfolgte an Pflanzen der Sommergerstensorte „Elgina" im Einblattstadium. Die Applikation der Fungizide wurde mittels Tropfnaß-Behandlung in den Konzentrationen 250, 50 und 10 ppm Wirkstoff durchgeführt. Nach Infektion und Inkubation der Versuchspflanzen wurde die Auswertung durch Blattbonitur in den Boniturstufen 9 bis 1 nach STEPHAN [6] vorgenommen.
Die Untersuchungen zur Penetration in Anlehnung an einen Penetrationstest nach DRANDAREWSKI und MAYER [2] (an Epidermisstücken von *Allium porrum*) durchgeführt. Als Testorganismus diente *Aspergillus niger*. Als Penetrationszeit wurde 24 Std. gewählt. Die Auswertung erfolgte durch Messen der Hemmhöfe.
Die Wirksamkeit der Präparate bei Benomylresistenz wurde an einem benomylresistenten Stamm von *Sphaerotheca fuliginea* (SCHLECHT.) SALMON an Gewächshausgurken (SORTE Quix) und *Erysiphe graminis* DC. an Sommergerste (Sorte „Eligna") untersucht. Zur Selektion der benomylresistenten Stämme wurden Gewächshausgurken in wöchentlichem Rhythmus mit 50; 100; 250 und 500 ppm Benlate und Sommergerstenpflanzen mit 50; 100; 500; und 1000 ppm Benlate tropfnaß behandelt. Im Anschluß an die Selektion benomylresistenter Stämme wurde ein Kreuzresistenztest durchgeführt. Die Auswertung erfolgte durch eine Blattbonitur.

Tabelle 1
Fungizide Wirkung von Benomyl- bzw. Carbendazim-Tensid-Formulierungen und Vergleichspräparaten gegen einen benomylsensiblen und einen benomylresistenten Stamm

Fungizide	Wirkstoffgehalt	Wirkstoff
Benomyl-Wirkstoff	–	–
Carbendazim-Wirkstoff	–	–
Benlate	50%	Benomyl
Derosal	60%	Carbendazim
Calixin	75%	Tridemorph
Sickosul	88%	Schwefel
Prüfmittel 1	5%	Benomyl
Prüfmittel 2	5%	Carbendazim
Prüfmittel 3	7,5%	Carbendazim
Prüfmittel 4	10%	Carbendazim
Prüfmittel 5	15%	Carbendazim
Prüfmittel 6	25%	Carbendazim

Tenside	chemische Zusammensetzung
Tensid I	Alkylphenyläthoxilat 7 ... 8 AeO/mol[1]
Tensid II	Alkylphenyläthoxilat 4 ... 5 AeO/mol
Tensid III	Alkylphenyläthoxilat 50 ... 52 AeO/mol
DMSO	Dimethylsulfoxid
DMF	Dimethylformamid

[1] AeO = Äthylenoxidzahl

3. Ergebnisse und Diskussion

Die fungizide Wirkung von Tankmischungen der Handelspräparate Benlate und Derosal mit Tensiden sowie Lösungsmitteln ist aus Tabelle 2 ersichtlich. Ein Vergleich der fungiziden Wirkung der verschiedenen Varianten zeigt, daß bei einer Anwendungskonzentration von 50 ppm Wirkstoff mit den Handelspräparaten Benlate ein Wirkungsgrad von 66% und Derosal von 3% erreichbar ist. Durch den Zusatz von Lösungsmitteln und Tensiden in die Spritzbrühen konnte in jedem Fall eine Steigerung der fungiziden Wirkung erreicht werden. Es traten jedoch deutliche Unterschiede in Abhängigkeit vom Wirkstoff und von den Zusätzen auf. In den Tankmischungen mit Lösungsmitteln kam es nur in der Kombination von DMSO mit Benlate zu nennenswerten Effekten, die zu erwarten waren, da DMSO ein gutes Durchdringungsvermögen für biologische Membranen besitzt und die Aufnahme anderer Substanzen in die Zelle fördert (Martin und Hauthal [4]). Einen weitaus stärkeren Einfluß üben Tenside auf die fungizide Wirkung beider Handelspräparate aus. Bei Verwendung von Benlate in Kombination mit Tensid I und II wurde der Wirkungsgrad von 66% auf 99% erhöht. Deutliche Unterschiede in der Effektivität des einzelnen Tensids waren nur in Tankmischungen mit Derosal feststellbar. Tensid I und II mit Äthylenoxidzahlen von 7 ... 8 bzw. 4 ... 5/mol hatten einen erheblichen höheren Effekt als Tensid III mit 50 ... 52 AeO/mol. Ein vergleichbarer Effekt durch Mischung von Thiophanatmethyl mit einem Tensid wurde von Cole und Cox [1] bei der Bekämpfung von *Sphaerotheca fuliginea* und *Botrytis cinerea* und von Mercer u. a. [5] bei der Bekämpfung

Tabelle 2
Fungizide Wirkung von Tankmischungen der Handelspräparate Benlate und Derosal mit Tensiden gegen Erysiphe graminis DC. *an Sommergerste nach kurativer Anwendung*

Variante	Wirkstoff 50 ppm	Lösungsmittel 500 ppm	Tensid 250 ppm	Wirkungsgrad %
Benlate	Benomyl		—	66
Benlate	Benomyl		+ Tensid I	99
Benlate	Benomyl		+ Tensid II	99
Benlate	Benomyl		+ Tensid III	98
Benlate	Benomyl	+ DMF	—	71
Benlate	Benomyl	+ DMSO	—	95
Derosal	Carbendazim		—	3
Derosal	Carbendazim		+ Tensid I	96
Derosal	Carbendazim		+ Tensid II	93
Derosal	Carbendazim		+ Tensid III	82
Derosal	Carbendazim	+ DMF	—	10
Derosal	Carbendazim	+ DMSO	—	14
—	—		Tensid I	74
—	—		Tensid II	56
—	—		Tensid III	31
—	—	DMF	—	0
—	—	DMSO	—	0

von *Podosphaera leucotricha* erreicht. KLOPPING u. a. [4] weisen auf eine beträchtliche Wirkungssteigerung durch Tankmischungen von Carbendazim mit den Tensid „Trem 014" gegen *Sphaerotheca fuliginea* hin.

Die Ursachen für die Wirkungsverbesserung bei Fungizid-Tensid-Kombinationen lassen sich mit Hilfe eines Penetrationstestes klären. Für diesen Test wurden flüssigformulierte Prüfmuster der Wirkstoffe Benomyl und Carbendazim mit einem Wirkstoff-Tensid-Verhältnis von 1:1,2 bis 1:7 verwendet (Tab. 3). Es zeigen sich deutliche Unterschiede in der fungiziden Wirkung und im Penetrationsvermögen zwischen den

Tabelle 3
Kurative Wirkung von Handelspräparaten und Prüfmustern gegen Erysiphe graminis DC. *an Gersteund Ergebnisse zum Penetrationsvermögen an Allium porrum* L.

Variante	Wirkstoff	Wirkstoff-Tensid-Verhältnis	Wirkungsgrad (%) 50 ppm Wirkstoff	Penetrationsvermögen (Hemmhofdurchmesser mm)
Benomyl	Benomyl	—	21	24
Carbendazim	Carbendazim	—	0	8
Benlate	Benomyl	—	56	25
Derosal	Carbendazim	—	2	8
Prüfmuster 1	Benomyl	1:7	98	54
Prüfmuster 2	Carbendazim	1:7	96	51
Prüfmuster 3	Carbendazim	1:2	82	36
Prüfmuster 4	Carbendazim	1:6	95	41
Prüfmuster 5	Carbendazim	1:4	92	38
Prüfmuster 6	Carbendazim	1:1,2	75	30

einzelnen Varianten. Für den Wirkstoff Carbendazim wurde erwartungsgemäß ein geringeres Penetrationsvermögen beobachtet als für Benomyl (Drandarewski und Mayer [2]. Ähnliche Ergebnisse wurden für die Handelspräparate (Benlate und Derosal) ermittelt. Erst durch Flüssigformulierung mit einem hohen Tensidanteil wird eine erhebliche Wirkungs- und Penetrationsverbesserung erreicht. Für einen optimalen Effekt scheint das Wirkstoff-Tensid-Verhältnis der Formulierung von Bedeutung zu sein. Die carbendazimhaltigen Prüfmuster haben bei einer Einteilung sowohl nach dem Penetrationsvermögen als auch nach dem Wirkstoff-Tensid-Verhältnis die gleiche Rangordnung:
Prüfmuster 2 (Verhältnis 1:7) > Prüfmuster 4 (Verhältnis 1:6) > Prüfmuster 5 (Verhältnis 1:4) > Prüfmuster 3 (Verhältnis 1:2) > Prüfmuster 6 (Verhältnis 1:1,2).
Nach den Ergebnissen des Penetrationstestes wird die ausgezeichnete fungizide Wirkung der Flüssigformulierungen mit einem hohen Tensidgehalt durch eine erhöhte Wirkstoffaufnahme bewirkt.
Benomylresistenz konnte in Gewächshausversuchen an *Sphaerotheca fuliginea* nach 18 und an *Erysiphe graminis* nach 25 Behandlungen nachgewiesen werden. Mit beiden benomylresistenten Erregern wurde ein Kreuzresistenztest durchgeführt, für die die Muster 1 und 2 mit dem höchsten Penetrationsvermögen ausgewählt wurden. Die Ergebnisse des Kreuzresistenztests sind in Abbildung 1 dargestellt. Der Abbildung 1 ist zu entnehmen, daß Gruppenresistenz bei beiden benomylresistenten Testpilzen nur für die Handelspräparate Benlate und Derosal auftrat. Die Vergleichsmittel Sicksul und Calixin hatten erwartungsgemäß eine unveränderte fungizide Wirkung. Die Prüfmuster 1 und 2 (Wirkstoff-Tensid-Verhältnis 1:7) auf der Wirkstoffbasis von Benomyl und Carbendazim wirkten überraschenderweise an beiden benomyl-

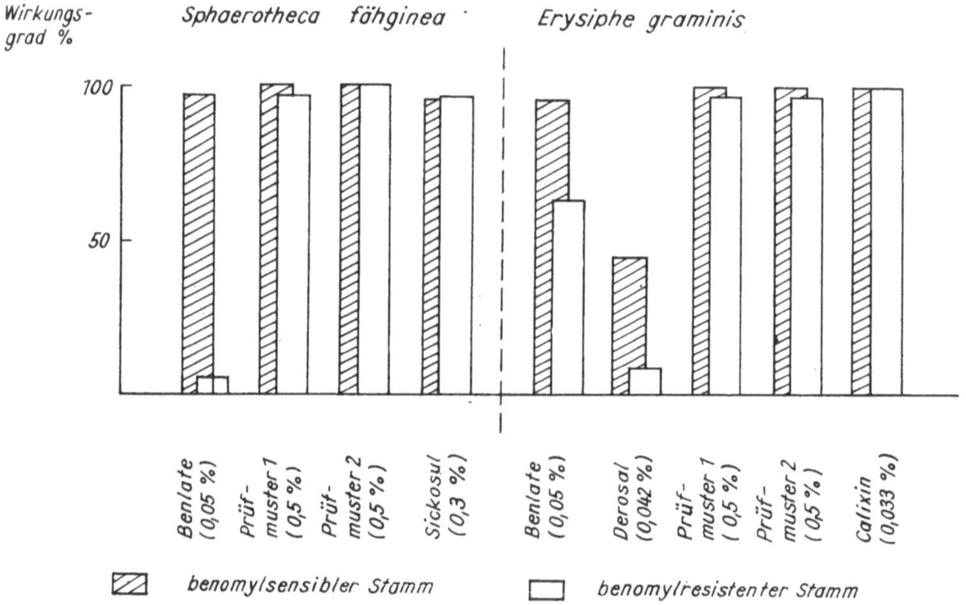

Abb. 1. Fungizide Wirkung von Benomyl- bzw. Carbendazim-Tensid-Formulierungen und Vergleichspräpartaen gegen einen benomylsensiblen und einen benomylresistenten Stamm

resistenten Testpilzen resistenzbrechend. Sowohl das benoyml- als auch das carbendazimhaltige Prüfmuster erreichten gegen den benomylresistenten Stamm von *Sphaerotheca fuliginea* einen Wirkungsgrad von 99% sowie gegen den benomylresistenen Stamm von *Erysiphe graminis* einen Wirkungsgrad von 96%. Dieser Effekt konnte auch mit dem Prüfmuster 5 (Wirkstoff-Tensid-Verhältnis 1:4) an einem benomylresistenten Stamm von *Venturia inaequalis* nachgewiesen werden.

4. Zusammenfassung

In Tankmischungen ergeben Tenside (Alkylphenyläthoxylate) mit einer Äthylenoxidzahl von 4 ... 8/mol in Kombination mit Benlate und Derosal eine erhebliche Verbesserung der fungiziden Wirkung. In Flüssigformulierungen ist das Wirkstoff-Tensid-Verhältnis für die fungizide Wirkung gegen *Erysiphe graminis* DC. von Bedeutung. Die bessere fungizide Wirkung der Benomyl- bzw. Carbendazim-Tensid-Formulierungen wird durch eine Erhöhung des Penetrationsvermögens verursacht.
Die Benomyl- bzw. Carbendazim-Tensid-Formulierungen (Verhältnis 1:4 und 1:7) zeigen gegenüber benomylresistenten Stämmen von *Sphaerotheca fuliginea* (SCHLECHT.) SALMON und *Erysiphe graminis* DC. eine uneingeschränkte Wirkung.

Ю. М. Плотникова, Г. В. Сережкина

Главный ботанический сад АН СССР, Москва, СССР

Особенности Ультрасруктуры Совместимой и Несовместимой Комбинации гриба *Puccinia graminis tritici* и Пшеницы

Для исследования изменений, происходящих на субклеточном уровне в устойчивых и восприимчивых растениях, ыла использована комбинация биотипа расы 21 *Puccinia graminis tritici*, возбудителя стеблевой ржавчины пшеницы, и двух линий сорта Маркиз, различающихся одним геном устойчивости. Маркиз Sr_6 восприимчив к этому биотипу. Маркиз Sr_{11} высокоустойчив. Исследование проводилось на 6-ой, 10-ый и 22-ой день после заражения.

Проникая в листья пшеницы, гифы *P. graminis tritici* распространялись в межклеточном пространстве, посылая гаустории внутрь клеток мезо-

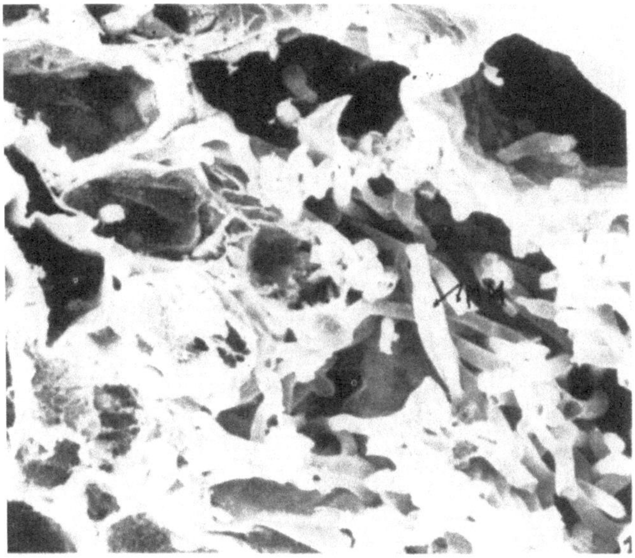

Рис. 1. Развитие гриба *P. graminis tritici* в ткани пшеницы (совместимая комбинация) на 10-ый день после заражения СЭМ × 1500

В — вакуоль
Г — гаустория
Л — липиды
М — митохондрия
ММ — межклеточный мицелий
МК — мембранный комплекс
Мт — микротело
О — оболочка
П — псевдопаренхима
Пл — плазмалемма
ПГ — пузырьки Гольджи
Хл — хлоропласт
ШГ — шейка гаустории
ЭР — эндоплазматический ретикулум
Я — ядро

филла (рис. 1). При совместимой комбинации гриб образовывал разветвленный межклеточный мицелий и многочисленные гаустроии (рис. 2), при несовместимой — формировал слабо развитый мицелий и единичные гаустории (рис. 3). При совместимой комбинации на ранних сроках инфекции наблюдались слабые структурные изменения пораженных клеток (рис. 4). Отмечалась явная стимуляция клеток растения с внедрившимися гаусториями, выражающаяся в увеличении объема протоплазмы и таксисе всех органелл: хлоропластов, ядра, митохондрий, микротел, эндоплазматического ретикулума, рибосом. При несовместимой комбинации уже на ранних сроках инфекции наблюдались сильные изменения всех клеточных органелл (рисс. 5). Происходило разушение внешней оболочки хлоропластов и уменьшение электронной плотности пластоглобул. Отмечались нарущения структуры митохондрий, выражающиеся в распаде крист и разрушении внешней мембраны. В обеих комбинациях был ярко выражен таксис ядра лкетки хозяина в сторону гаусторий паразита (рис. 6, 7). Изменения касались не только формы ядра, которая из округлой превращалась в чашевидную или лопастную, обеспечивающую максимальный контакт с гаусторией патогена, но и его содержимого

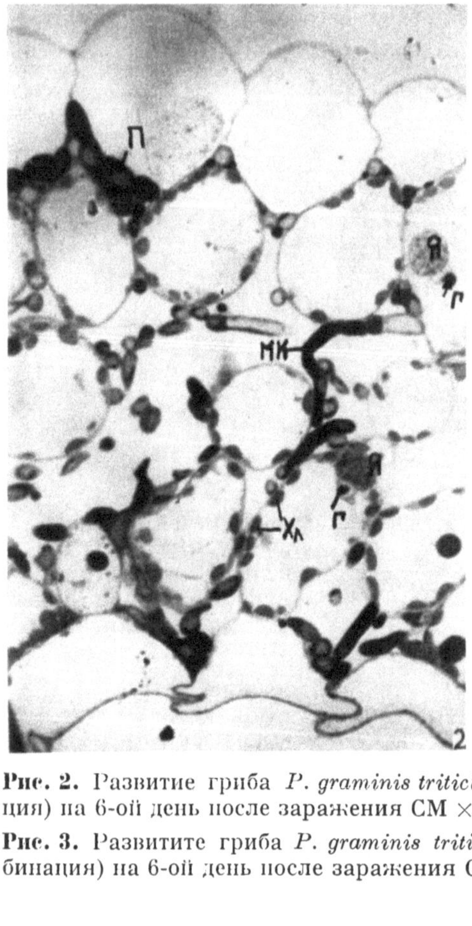

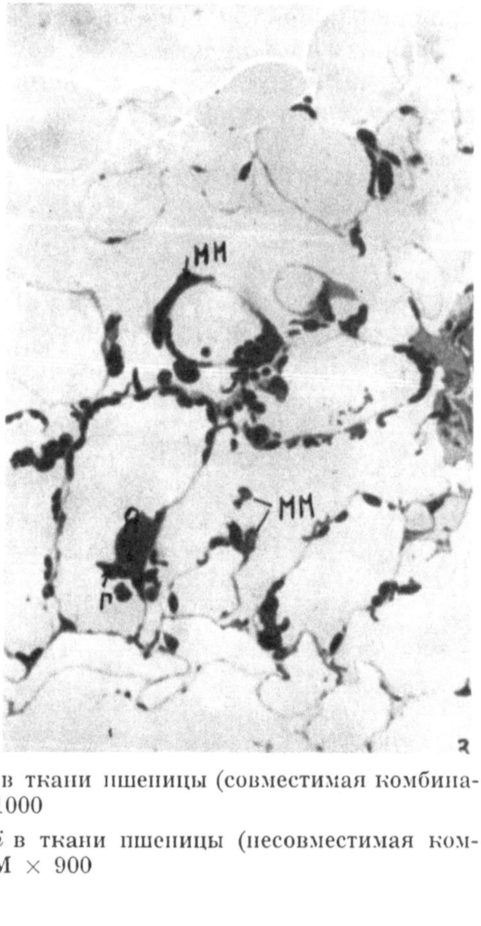

Рис. 2. Развитие гриба *P. graminis tritici* в ткани пшеницы (совместимая комбинация) на 6-ой день после заражения СМ × 1000

Рис. 3. Развитите гриба *P. graminis tritici* в ткани пшеницы (несовместимая комбинация) на 6-ой день после заражения СМ × 900

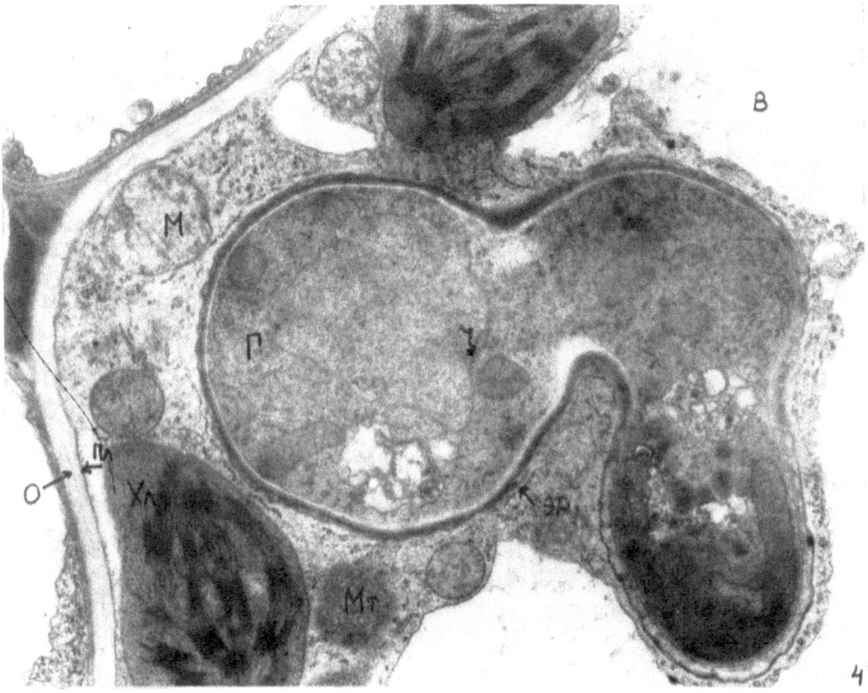

Рис. 4. Клетка пшеницы Маркиз Sr_6 с внедрившейся гаусторией гриба *P. graminis tritici* 6-ой день после заражения ТЭМ × 60000

происходило увеличение электронной плотности нуклеоплазмы и позднее резкое снижение электронной плотности зон хроматина (рис. 7). Наблюдалось постепенное увеличение размеров ядер под влиянием инфекции, причем при совместимой комбинации оно происходило быстрее (табл. 1). На более поздних сроках инфекции при совместимой комбинации происходили изменения и пластидного аппарата клетки хозяина, выражающиеся в сокращении ламелл гран и увеличении пластоглобул хлоропластов (рис. 7, 8), по сравнению со структурой хлоропластов в неповрежденных

Таблица 1
Параметры хлоропластов и ядер в здоровых и пораженных листьях различных по устойчиьости линий пшеницы сорта Маркиз на разных сроках инфекции

Линии сорта Маркиз	Период развития ржавчины в днях после инокуляции	Объем хлоропласта мкм³	Объем ядра мкм³
	6	9,03 ± 1,12	407,4 ± 48,1
Sr_6	10	2,08 ± 0,30	554,7 ± 67,2
восприимчивая	контроль без инокуляции	11,05 ± 1,01	185,3 ± 19,2
	6	1,74 ± 0,14	274,7 ± 23,5
Sr_{11}	10	2,56 ± 0,30	541 ± 49,6
устойчивая	контроль без инокуляции	12,98 ± 1,13	178,5 ± 16,3

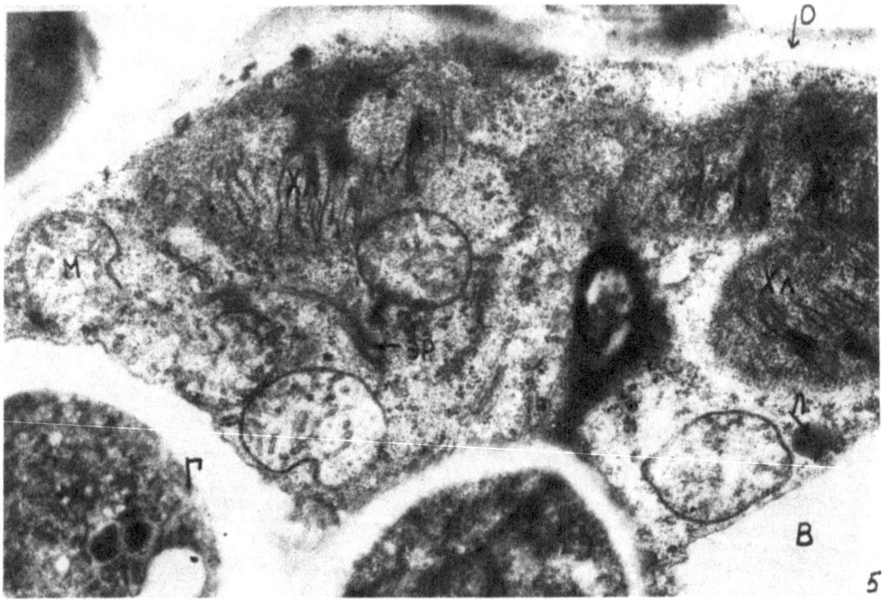

5

Рис. 5. Клетки пшеницы Маркиз Sr_{11} с внедрившейся гаусторией гриба *P. graminis tritici* 6-ой день после заражения ТЭМ × 70000

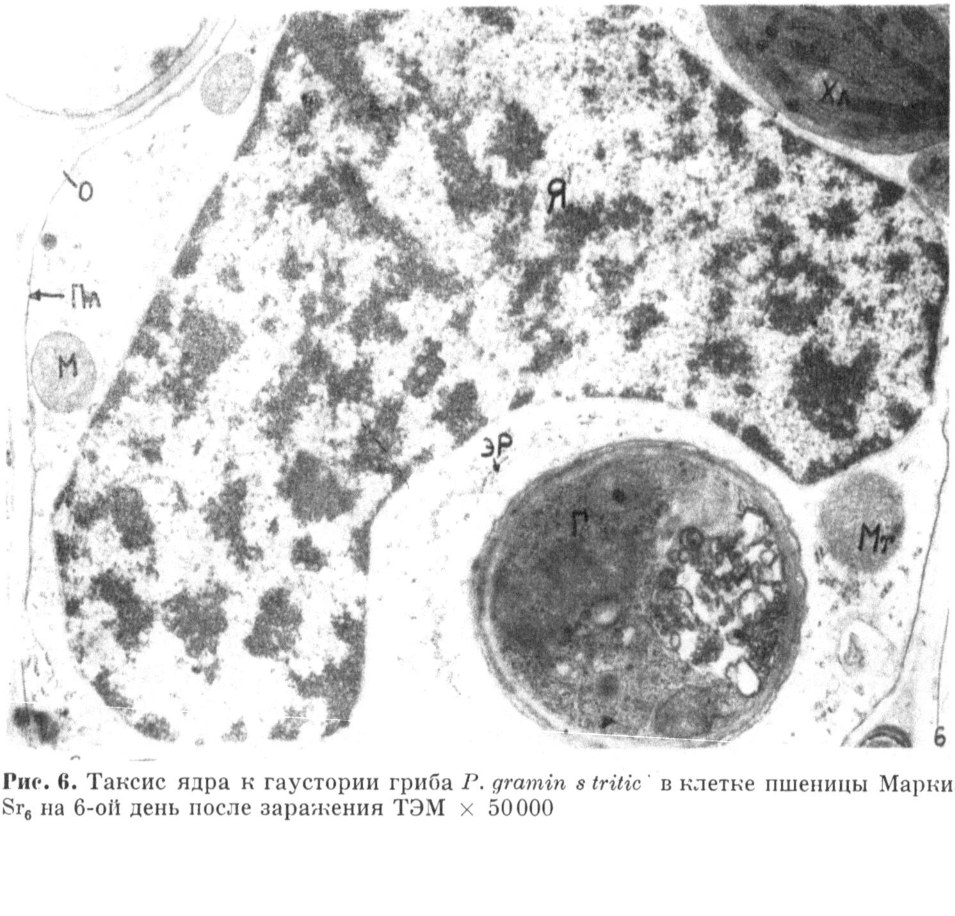

6

Рис. 6. Таксис ядра к гаустории гриба *P. gramin s tritic* в клетке пшеницы Маркиз Sr_6 на 6-ой день после заражения ТЭМ × 50000

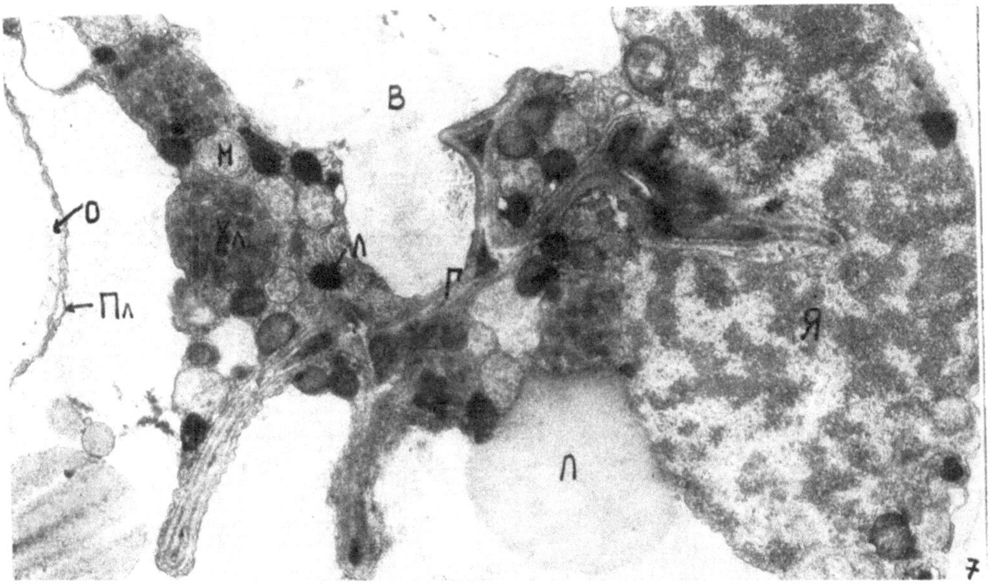

Рис. 7. Таксис и изменение ультраструктуры ядра в клетке пшеницы Маркиз Sr_6 на 22-ой день после заражения ТЭМ × 50000

Таблица 2

Толщина экстрагаусториального матрикса у гаусторий Puccinia graminis tritici, образующихся в клетках различных по устойчивости линий пшеницы сорта Маркиз на разных сроках инфекции (мкм)

Линии сорта Маркиз	Период развития ржавчины в днях после инокуляции	У физиологически активных гаусторий	У некротизированных гаусторий
Sr_6	6	0,116 ± 0,014	0,275 ± 0,028
восприимчивая	10	0,148 ± 0,015	0,256 ± 0,013
Sr_{11}	6	0,136 ± 0,014	0,210 ± 0,02
устойчивая	10	0,530 ± 0,34	0,540 ± 0,029

клетках пшеницы (рис. 9). Объем хлоропластов резко снижался при несовместимой комбинации и постепенно при совместимой (табл. 1). В ассоциации с гаусториями было обнаружено образование особого мембранного комплекса — производного эндоплазматического ретикулума, не встречающегося в здоровых клетках пшеницы (рис. 10). В тесном контакте с гаусториями гриба находились тяжи эндоплазматического ретикулума (рис. 11) и пузырьки аппарата Гольджи (рис. 12). Толщина экстрагаусториального матрикса, роль которого в обмене метаболитами между растением-хозяином и патогеном хорошо известна, значительно увеличивалась у устойчивых растений (табл. 2).

Таким образом, защитные реакции растений на субклеточном уровне выражались в более быстром изменении ультраструктуры устойчивых растений и изменении строения пограничной зоны между хозяином и патогеном.

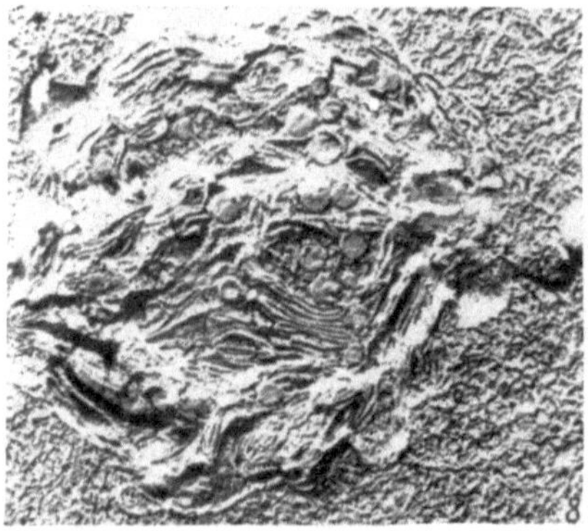

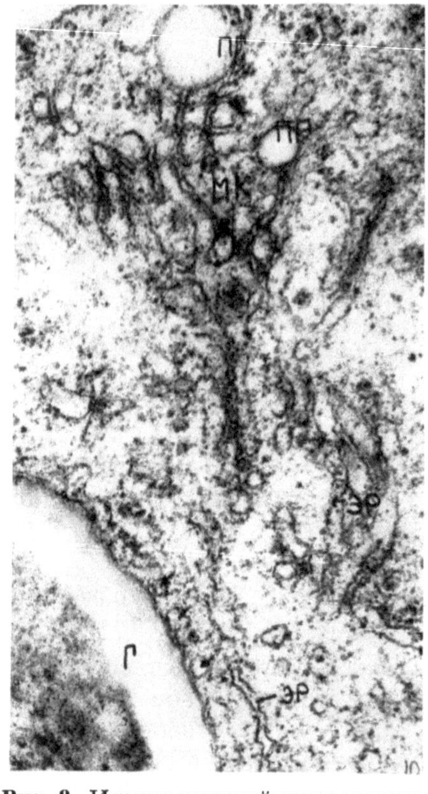

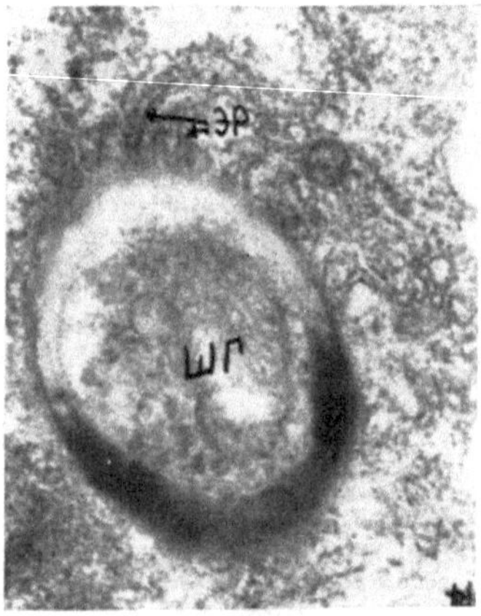

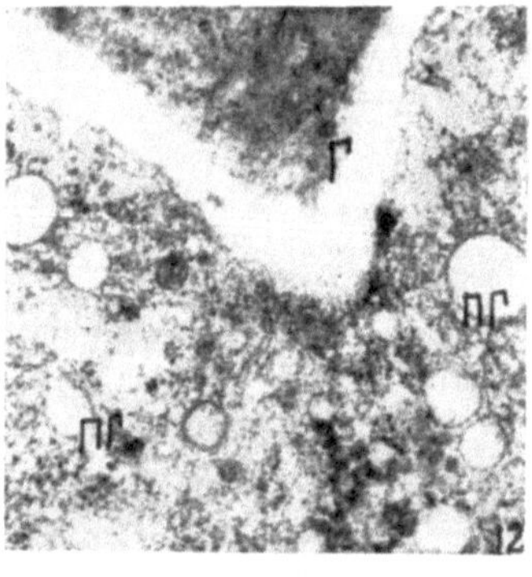

Рис. 8. Изолированный хлоропласт из пораженной клетки пшеницы Маркиз Sr_6. Замораживание — травление ТЭМ × 50000

Рис. 9. Изолированных хлоропласт из здоровой клетки пшеницы Маркиз Sr_6. Замораживание — травление ТЭМ × 40000

Рис. 10. Мембранный комплекс в контакте с гаусторией гриба *P. graninis tritici* в клетке пшеницы Маркиз Sr_{11} (10-ый день после заражения) ТЭМ × 80000

Рис. 11. Контакт тяжей эндоплазматического ретикулума с экстрагаусториальной мембраной шейки гаустории гриба *P. graninis tritici* в клетке пшеницы Маркиз Sr_6 (10-ый день после заражения) ТЭМ × 90000

Рис. 12. Контакт аппарата Гольджи с экстрагаусториальной мембраной *F. graninis tritici* в клетке пшеницы Маркиз Sr_6 (10-ый день после заражения) ТЭМ × 40000

H.-P. Schmauder und D. Gröger

Institut für Biochemie der Pflanzen der Akademie der Wissenschaften der DDR, DDR-4020 Halle/Saale, Weinberg 2

Zur Hemmung der Chitin-Synthetase von *Claciceps purpurea* durch ausgewählte Effektoren

Chitin ist ein wichtiger Bestandteil der *Claviceps*-Zellwand, der in ihr bis zu 48% (bestimmt als Glucosamin) enthalten sein kann [10, 11]. Es lag daher nahe, das Chitinsynthetisierende System von *Claviceps* näher zu untersuchen und zu charakterisieren. Die Einflüsse verschiedener fungizider Stoffe auf das Wachstum und Alkaloidbildung von *Claviceps* haben Schmauder und Gröger [12] beschrieben. Von den untersuchten Substanzen zeigten im submersen in-vivo-Test vor allem Hinosan, Nystatin, Parathionmethyl, Pentachlorphenol und Terrazol einen deutlichen Einfluß auf Wachstum und Alkaloidbildung. Polyoxin B, Polyoxin D wie auch Streptovirudin zeigten in diesem Testsystem nur geringe Wirkungen, die auf Permeationsprobleme zurückgeführt werden könnten [12].

Im in-vitro-System (zellfreier Überstand der 20000 g Fraktion in 0,05 M Imidazol/HCl-Puffer, pH 7,0–7,1, mit 30% Glycerin) zeigten die Effektoren Polyoxin B, Polyoxin D, Hinosan, Pentachlorphenol, Streptovirudin, UDP, UMP sowie einige ungesättigte Fettsäuren deutliche Hemmeffekte gegenüber der Chitin-Synthetase [13].

Es war überraschend, daß die Chitin-Synthetase von *Claviceps* durch Präparate, die Diflubenzoran enthielten, nicht signifikant gehemmt wird, wie es für andere Pilze u. a. Brillinger [3], von Eck [4] sowie Lyr und Seyd [8] beschrieben haben. Für Polyoxin D ist aufgrund der strukturellen Ähnlichkeit mit dem Substrat des Enzyms (Abb. 1) von verschiedenen Autoren eine kompetitive Hemmung beobachtet worden (z.B. [1, 2, 4]). Eine Reduzierung der Enzym-Aktivität durch Hinosan fanden Lyr und Seyd [8] bei *Mucor rouxii*.

Von den bereits erwähnten negativen Effektoren für das *Claviceps*-Enzym wurden die Polyoxine B und D, Hinosan, das in seiner Struktur noch unbekannte Streptovirudin [5, 14] und Pentachlorphenol näher auf einen möglichen Hemmtyp hin untersucht. Während es unmöglich war, die mit Pentachlorphenol erzielten Ergebnisse zu interpretieren, da die Substanz nur allgemein toxische Einflüsse zeigte, könnten für die Polyoxine B und D sowie das Streptovirudin-Gemisch eine kompetitive Hemmung wahrscheinlich gemacht werden. Hinosan zeigte dagegen einen Hemmtyp, in dem eine kompetitive Hemmung dominiert, aber von anderen Einflüssen des Hemmstoffs auf das Enzym überlagert wird [13].

Abbildung 2 zeigt für alle vier untersuchten Verbindungen die Hemmkurven und Dixon-Plots. Man kann ihr entnehmen, daß im Gegensatz zu einigen anderen Organismen (*Coprinus* [3], *Aspergillus* [1], *Mucor rouxii* [2]) der K_i-Wert für Polyoxin D relativ hoch ist. Er liegt, wie im Falle des bisher noch nicht als Chitin-Synthetase-

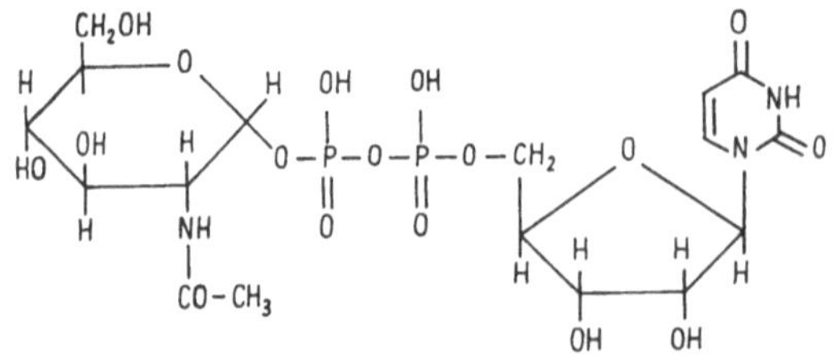

UDP-N-acetylglucosamin

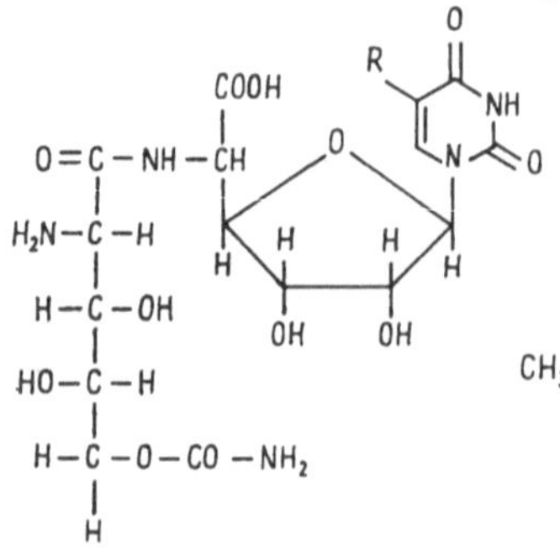

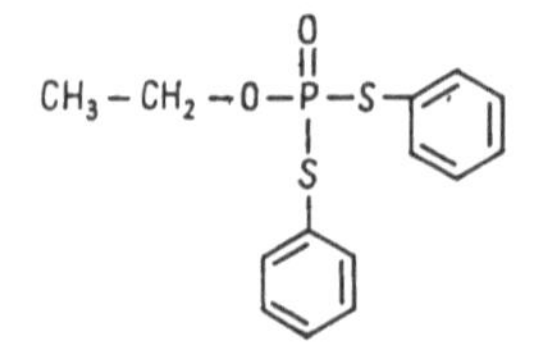

Polyoxin B: R=CH_2OH O-Ethyl-S,S-diphenylphosphorodithiolat
Polyoxin D: R=COOH "Hinosan"

Abb. 1. Formeln des Substrates der Chitin-Synthetse (UDP-N-acetylglucosamin) und einiger Hemmstoffe dieses Enzyms. (Vergl. hierzu [6])

Hemmstoffes beschriebenen Polyoxin B, im Bereich des K_m-Wertes für das Substrat ($\approx$ 0,13 bzw. $\approx$ 0,1 mM). Die anderen getesteten Hemmstoffe zeigen analoge Konzentrationsabhängigkeiten für die Größe von K_i: Streptovirudin $K_i \approx$ 0,18 mM; Hinosan $K_i \approx$ 0,13 mM, so daß auffällig ist, daß für die Chitin-Synthetase von *Claviceps purpurea* bislang kein hochwirksamer kompetitiver Hemmstoff, der im µ Molbereich wirkt, gefunden werden konnte. Die erhaltenen Dixon-Plots zeigen bei hohen Hemmstoffkonzentrationen eine Abweichung von der Linearität, die der für UDP beschriebenen entspricht [7, 9]. Dieses Phänomen deutet auf das Vorliegen einer komplexen Kinetik hin, d. h., in hohen Konzentrationen beeinflußt der Hemmstoff nicht nur die Bindungsstelle für das Substrat im Enzym, sondern übt weitere, noch nicht näher charakterisierte, Wechselwirkungen auf das Enzym bzw. Substrat aus.

Danksagung

Wir danken Herrn Prof. Dr. sc. H. LYR, Institut für Pflanzenschutzforschung Kleinmachnow, für die Überlassung einiger Fungizide sowie Herrn Dr. sc. K. ECKARDT, Zentralinstitut für Mikrobiologie und experimentelle Therapie der AdW, Jena, für die Proben von Streptovirudin.

Literatur

[1] ARCHER, D. B., Chitin biosynthesis in protoplasts and subcellular fractions of *Aspergillus fumigatus*, Biochem. J. **164** (1977), 653–658

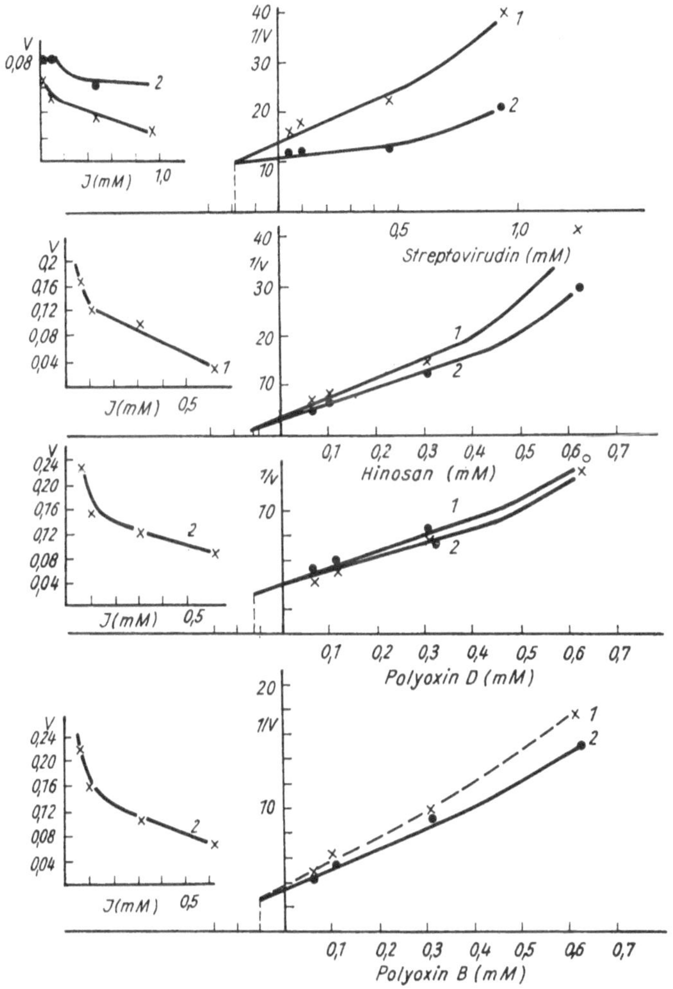

Abb. 2. Der Einfluß ausgewählter Hemmstoffe auf Präparationen der Chitin-Synthetase von *Claviceps purpurea*, Pepty 695, einem saprophytischen Ergotoxinstamm. Die Kurven entsprechen folgenden Konzentrationen an UDP-N-acetylglucosamin

Streptovirudin:	1 = 0,06 mM;	2 = 0,3 mM
Hinosan:	1 = 0,14 mM;	2 = 0,28 mM
Polyoxin D:	1 = 0,14 mM;	2 = 0,28 mM
Polyoxin B:	1 = 0,14 mM;	2 = 0,28 mM

[2] Bartnicki-Garcia, S. and E. Lippman, Inhibition of *Mucor rouxii* by Polyoxin D: effects on chitin synthetase and morphological development, J. Gen. Microbiol. **71** (1972), 301–309

[3] Brillinger, G. U., Metabolic products of microorganisms. 181. Chitin synthase from fungi, a test model for substances with insecticidal properties, Arch. Microbiol. **121** (1979), 71–74

[4] van Eck, W. H., Mode of action of two benzoylphenyl ureas as inhibitors of chitin synthesis in insects, Insect. Biochem. 9 (1979), 295–300

[5] ECKARDT, K., H. THRUM, G. BRADLER, E. TONEW and M. TONEW, Streptovirudines, new antibiotics with antibacterial and antiviral activity. II. Isolation, chemical characterization and biological activity of streptovirudines A_1, A_2, B_1, B_2, C_1, C_2, D_1 and D_2, J. Antibiotics **28** (1975), 274–279

[6] GLASBY, J. S., Encyclopaedia of Antibiotics, JOHN WILEY & Sons, Chichester, New York, Brisbane and Toronto, 1979, second Edition, pp 362–363

[7] LOPEZ-ROMERO, E. and J. RUIZ-HERRERA, Synthesis of chitin by particulate preparations from *Aspergillus flavus*, Antonie van Leeuwenhoek **42** (1976), 261–276

[8] LYR, H. und W. SEYD, Hemmung der Chitin-Synthetase von *Mucor rouxii* in vitro durch Fungicide und andere Wirkstoffe, Z. Allg. Mikrobiol. **18** (1978), 721–9

[9] RUIZ-HERREA, J., E. LOPEZ-ROMERO and S. BARTNICKI-GARCIA, Properties of chitin synthetase in isolated chitosomes from yeast cells of *Mucor rouxii*, J. biol. Chemistry **252** (1977), 3338–3343

[10] SCHMAUDER, H.-P. and D. GRÖGER, Chitinbestimmung in *Claviceps*, Biochem. Physiol. Pflanzen **173** (1978), 139–145

[11] SCHMAUDER, H.-P., D. GRÖGER, W. LÖSECKE, A. RUDOLPH and H. D. GRIMMECKE, Über die Zusammensetzung der Zellwand von *Claviceps*arten, Biochem. Physiol. Pflanzen **173** (1978), 295–310

[12] SCHMAUDER, H.-P. and D. GRÖGER, Einfluß ausgewählter fungizider Stoffe auf Wachstum, Alkaloidbildung und Zellwandsynthese von *Claviceps purpurea*, Biochem. Physiol. Pflanzen, in Vorbereitung

[13] SCHMAUDER, H.-P. and D. GRÖGER, The chitin-synthetase of *Claviceps* properties and inhibition, FEMS Microbiol. Letters, in Vorbereitung

[14] THRUM, H., K. ECKARDT, G. BRADLER, R. FÜGNER, E. TONEW and M. TONEW, Streptovirudins, new antibiotics with antibacterial and antiviral activity. I. Culture taxonomy, fermentation and production of streptovirudin complex, J. Antibiotics **28** (1975), 514–521

W. Schnaak

Zentralinstitut für Ernährung der Akademie der Wissenschaften der DDR, Potsdam-Rehbrücke, DDR

Neuere Erkenntnisse zum Abbau und Metabolismus von Benzimidazolcarbamidsäuremethylester in Pflanzen

1. Einleitung

Fungizide auf der Basis von Benzimidazol-Derivaten werden in zunehmendem Maße in der Landwirtschaft eingesetzt. Im Gegensatz zu den klassischen Kontakt-Fungiziden, die hauptsächlich als Oberflächenbeläge wirken, besitzen Benzimidazol-Derivate die Fähigkeit, in die Pflanzen einzudringen. Sie ermöglichen so einen innertherapeutischen Schutz der Pflanze einschließlich der nach Behandlung neu hinzuwachsenden Pflanzenteile und sind deshalb aus der Sicht des Pflanzenschutzes besonders vorteilhaft.

Mit der Aufnahme in das Innere der Pflanzen sind die Wirkstoffe aber verstärkt metabolischen Einflüssen ausgesetzt. Daraus resultiert, daß für die lebensmittelhygienisch-toxikologische Beurteilung von Behandlungsrückständen an den Erntegütern das Vorkommen von Metaboliten zunehmend zu beachten ist. Die Feststellung, daß der ursprünglich ausgebrachte Wirkstoff nicht mehr nachweisbar ist, gibt noch keine Gewähr dafür, daß eine Entgiftung stattgefunden hat. Zum Nachweis der Entgiftung eines Pestizides in der Umwelt ist es notwendig, den Weg der Substanz über die verschiedenen Abbaustufen zu verfolgen, bis er letztlich in den Kreislauf natürlich vorkommender Substanzen (CO_2, Wasser, Oxalsäure, Glycin usw.) endet. Dabei gilt es, rückstandsbildende Metabolite zu erkennen, deren Vorkommen aufgrund ihrer Toxizität oder anderer unerwünschter Nebenwirkungen limitiert und kontrolliert werden muß.

Unter den Benzimidazol-Fungiziden nimmt der Benzimidazolcarbamidsäuremethylester (MBC) eine zentrale Rolle ein. Es hat sich gezeigt, daß die fungizide Wirkung so bedeutender Präparate wie Benomyl und Thiophanat-methyl auf ihre Umwandlung zu MBC beruht [9, 10]. MBC wird deshalb selbst als Fungizid (Wirkstoffbezeichnung Carbendazim) eingesetzt. Die Struktur der Verbindungen ist aus Abbildung 1 ersichtlich.

Während Translokation und Rückstandsdynamik des MBC in Pflanzen mehrfach beschrieben wurden (vgl. [1, 2, 5]), gibt es in der Literatur nur wenige Angaben zum Abbau und Metabolismus dieses Fungizids. Die Bildung des Hydrolyseproduktes 2-Aminobenzimidazol (2-AB) wurde von Siegel und Zabbia [8] beschrieben. Rouchaud u. a. [6, 7] geben als weitere pflanzliche Metabolite Benzimidazol, o-Aminobenzonitril, o-Phenylendiamin und Anilin an.

Ziel der Arbeiten waren weitere Erkenntnisse zum Abbau und Metabolismus des MBC in Pflanzen als Grundlage zur Einschätzung der aus der Anwendung der Benzimidazol-Fungizide resultierenden Rückstandssituation.

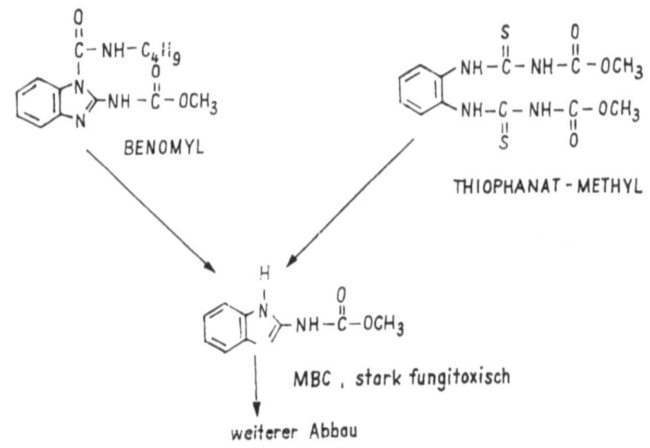

Abb. 1. MBC-bildende Fungizide

2. Versuchsdurchführung

Die Untersuchungen wurden durchgeführt an Bohnen- und Weizenpflanzen. Die Pflanzen wurden als Hydrokultur mit KNOP'scher Nährlösung gezogen, welche die zu untersuchenden Wirkstoffe enthielt. Getreidepflanzen wurden außerdem in 5 cm hoher Humuserde in Glasschalen gezogen. Die gewählten Modellsysteme sollten Aussagen erlauben über den Abbau und Metabolismus des MBC in einkeimblättrigen Pflanzen, in zweikeimblättrigen Pflanzen, in isolierten Pflanzen sowie im System Pflanze + Boden.
Zur Festlegung der Reaktionskette des Abbaus wurden nach einem Screening-Test mit ^{14}C-MBC in Weizenpflanzen die nicht markierten Metabolite 2-AB, Benzimidazolon ebenfalls über das Wurzelsystem an Pflanzen appliziert. In Abständen von 1—4 Wochen nach Zugabe der Wirkstoffe wurden Pflanzen und Boden extrahiert und auf vorhandene Folgeprodukte analysiert. Der Analysengang ist in Abbildung 2 dargestellt. Die chromatographische Analyse der einzelnen Fraktionen des Extraktes wurde nach der von ENGST und SCHNAAK [3] angegebenen Methode ausgeführt.

3. Ergebnisse

3.1. Verteilung und Abbau des MBC in Pflanzen

Die fungizidwirksame Komponente MBC wird über die Wurzeln aufgenommen und gelangt mit dem Saftstrom in die oberen Pflanzenteile. Hier ist in Abhängigkeit von der Einwirkungsdauer eine zunehmende Bindung an Pflanzeninhaltsstoffe feststellbar, während sich der Anteil des freien mit Lösungsmitteln extrahierbaren MBC verringert. Die Bindung erfolgt hauptsächlich an polymere unlösliche Pflanzenbestandteile. Wasserlösliche Konjugate, aus denen sich durch Hydrolyse MBC freisetzen läßt, wurden nur zu einem geringen Teil festgestellt. Der Anteil des in der wasserlöslichen Fraktion gebundenen MBC betrug in jedem Fall weniger als 5% der Gesamtmenge.
Neben der Bindung an Pflanzenbestandteile ist ein Abbau des Wirkstoffes feststellbar. Die Hydrolyse der Carbendazimsäuremethylester-Gruppe unter Bildung des 2-Aminobenzimidazols (Benzimidazolonimid) wurde als Hauptabbaureaktion ermittelt. In Weizenpflanzen tritt 2-AB als relativ stabiler Metabolit auf, der im Gegensatz zum MBC ausgeprägte phytotoxische Eigenschaften besitzt.

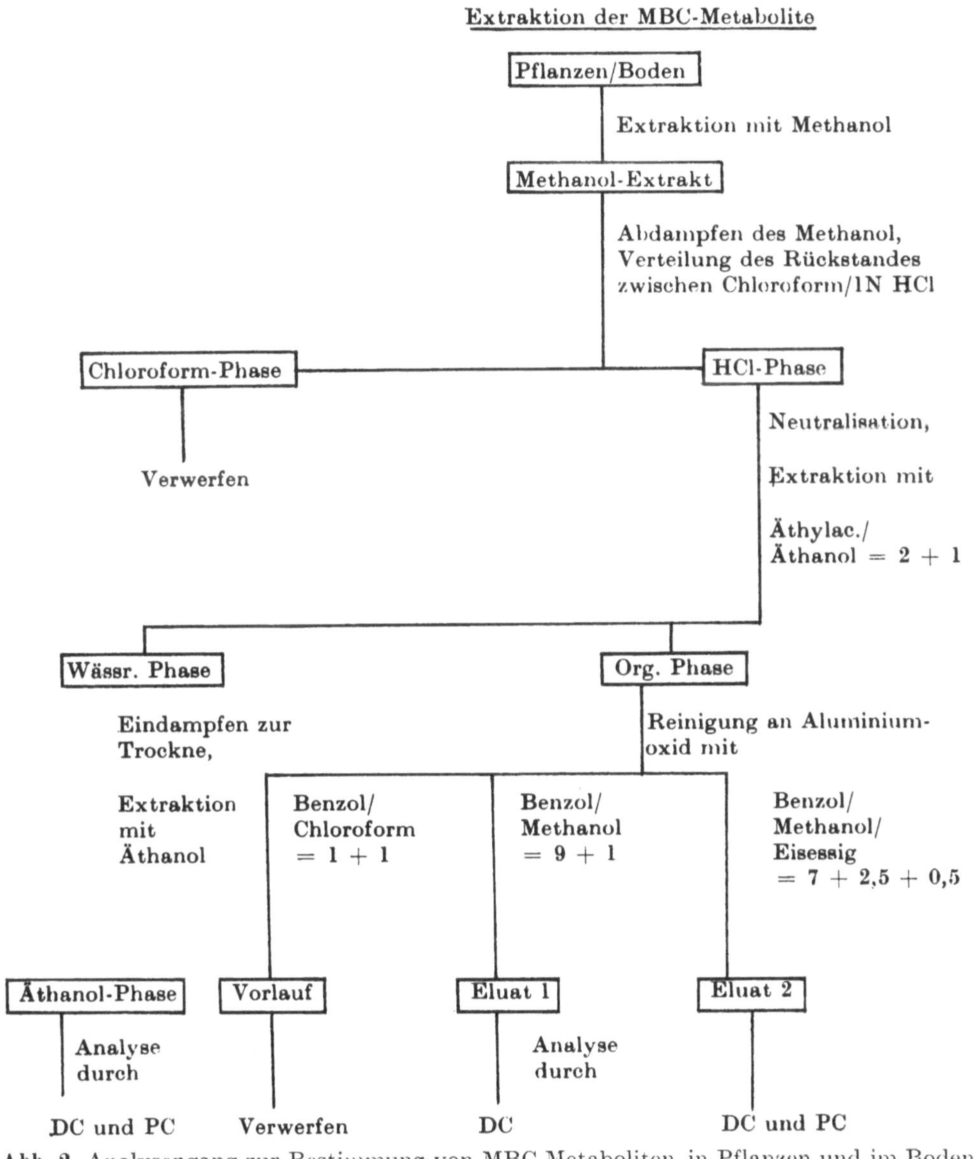

Abb. 2. Analysengang zur Bestimmung von MBC-Metaboliten in Pflanzen und im Boden

Der Metabolit 2-AB wird wie MBC selbst an Pflanzenbestandteile gebunden. Die Hauptmenge der gebundenen Benzimidazol-Derivate kann durch mehrtätige Extraktion mit stark polaren Lösungsmitteln gelöst werden, während ein geringer Rest nur durch Hydrolyse der Extraktionsrückstände freigesetzt werden kann. Aus diesem Verhalten resultiert, daß die Benzimidazole weitgehend adsorptiv an das Pflanzenmaterial gebunden werden. Es gibt keine sicheren Hinweise für einen chemischen Einbau in Pflanzenpolymere (z. B. als Fremdbasen in die Nukleinsäure-Fraktion).
Verteilung und Abbau von MBC in Weizenpflanzen in Abhängigkeit von der Einwirkungsdauer sind aus Abbildung 3 ersichtlich.

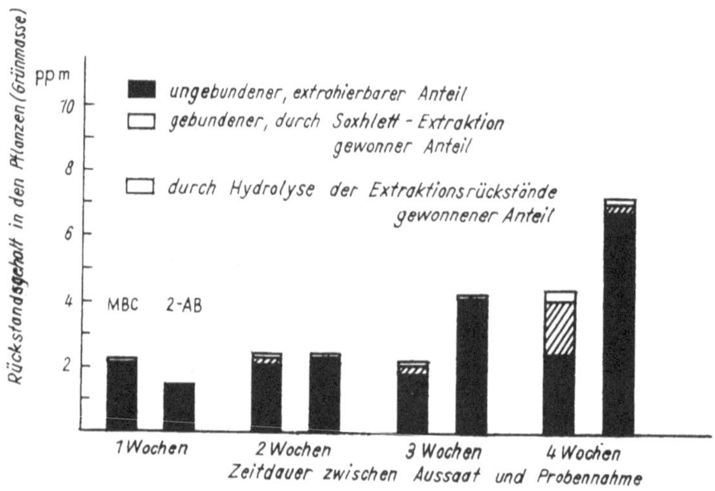

Abb. 3. Verteilung und Abbau von MBC in Weizenpflanzen

Die Auswertung der Chromatogramme durch biologische Detektion mit *Aspergillus niger* [4] ergab, daß neben MBC keine weiteren fungitoxischen Substanzen gebildet werden. Damit können auch MBC-Konjugate weitgehend ausgeschlossen werden. Bei Anwendung von im Imidazolring ^{14}C-markiertem MBC verblieb nach 4wöchiger Einwirkungsdauer etwa $^1/_3$ der in den Weizenpflanzen vorhandenen Aktivität in der Fraktion der wasserlöslichen Substanzen. Ein geringer Teil der Aktivität (unter 1%) wurde als $^{14}CO_2$ abgegeben. Damit ist ausgewiesen, daß die MBC-bildenden Benzimidazol-Fungizide im System Pflanze + Erde letztlich unter Aufspaltung des Imidazolringes abgebaut werden. Neben dem Abbau in der Pflanze sind dabei jedoch insbesondere mikrobielle Einflüsse des Bodens in Betracht zu ziehen.

Die Bildung des 2-AB aus MBC ist allerdings hauptsächlich dem pflanzlichen Metabolismus zuzuordnen. Der 2-AB-Gehalt im Boden blieb innerhalb der Versuchsdauer

Tabelle 1
Verteilung und Abbau von MBC in Bohnenpflanzen (MBC-Gehalt der Nährlösung = 0,5 mg = 0,001%)

	nach 2 Wochen	nach 4 Wochen
ungebundener, extrahierbarer Anteil [ppm]	MBC: 5,6 2AB: 0,5	MBC: 6,0 2-AB: 0,9
gebundener, durch Soxhlet-Extraktion gewonnener Anteil [ppm]	MBC: 1,0 2-AB: n. n.	MBC: 2,2 2-AB: n. n.
durch Hydrolyse der* Extraktionsrückstände gewonnener Anteil [ppm]	MBC + 2AB = 0,1	MBC + 2-AB = 0,3
MBC-Gehalt der Nährlösung [mg]	0,2	0,08

* nach alkalischer Hydrolyse als 2-AB bestimmt
n. n. = nicht nachweisbar (Nachweisgrenze: 0,1 ppm)

von 4 Wochen erheblich niedriger als in Pflanzen. Der höchste Gehalt wurde nach 3 Wochen mit einem Verhältnis 2-AB/MBC = 0,03 im Gegensatz zu 1,95 in den Pflanzen gefunden.
In Bohnenpflanzen war die Bildung von 2-AB weniger deutlich ausgeprägt, obwohl der MBC-Gehalt der Nährlösung innerhalb einer Versuchsdauer von 4 Wochen um 84% abnahm (Tab. 1). Durch Zugabe von 2-AB in die Nährlösung wurde festgestellt, daß diese Substanz in Bohnenpflanzen schneller abgebaut wird als in Weizen. Benzimidazol war jedoch nur in Spuren feststellbar.

3.2. Reaktionsfolge des Abbaus von MBC

Durch Abbau von ^{14}C-MBC wurden 4 Wochen nach Zugabe des Wirkstoffes im Boden 7, in Weizenpflanzen 3 extrahierbare Folgeprodukte festgestellt. Die in Weizenpflanzen nachgewiesenen Metabolite entsprachen qualitativ den im Boden gebildeten. Es zeigte sich jedoch, daß von den bisher beschriebenen pflanzlichen Metaboliten [6, 7] nur 2-AB sowie Spuren von Benzimidazol und o-Phenylendiamin gebildet wurden. o-Aminobenzonitril und Anilin waren nicht nachweisbar.
Zur Festlegung der Reaktionskette des Abbaus wurden neben MBC auch die Metabolite 2-AB, Benzimidazol sowie das durch Oxidation (in-vitro-Abbau) gebildete Benzimidazolon als vermutlicher Metabolit über das Wurzelsystem an Pflanzen appliziert. Die Analyse der gebildeten Folgeprodukte erfolgte ebenfalls nach dem in Abbildung 2 dargestellten Schema.

Die eingesetzten Metabolite zeigten folgendes Abbauverhalten:

- 2-AB, Benzimidazol und Benzimidazolon besitzen in Pflanzen eine dem MBC vergleichbare hohe Persistenz. Sie waren selbst 4 Wochen nach der Zugabe (3 mg/100 ml Nährlösung) noch nachweisbar.
- 2-AB und Benzimidazol bilden neben der Bindung an polymere Pflanzenbestandteile auch lösliche Konjugate. Dieser Effekt war bei Bohnenpflanzen kräftiger ausgeprägt als bei Weizenpflanzen.
- Die Benzimidazol-Derivate werden nur langsam unter Aufspaltung des Imidazol-Ringes und gleichzeitigem Austritt von Stickstoff (vermutlich als NH_3) abgebaut. Der dabei gebildete Metabolit x ist wesentlich polarer als die Ausgangsverbindungen und deshalb mit unpolaren organischen Lösungsmitteln nicht aus der wäßrigen Phase extrahierbar.
- Die Bildung von Benzimidazol aus 2-AB tritt nur als Nebenweg des Abbaus in Erscheinung.
- Im Boden sind als weitere Nebenwege des Abbaus die Kernhydroxylierung des MBC zu 5-OH-MBC sowie der oxidative Abbau des Benzimidazolring-Systems zu Imidazoldicarbonsäure feststellbar.

Die Ergebnisse der Untersuchungen sind in dem folgenden Abbauschema (Abb. 4) zusammengefaßt:
Aus den Untersuchungen zur Abbaudynamik geht hervor, daß alle Folgeprodukte des MBC, in denen das Benzimidazolringsystem noch erhalten ist, eine relativ hohe Persistenz im Boden und Pflanzen zeigen. Die metabolische Bildung dieser Produkte ist deshalb mit ihrer zwischenzeitlichen Anreicherung verbunden. Dies trifft insbesondere für 2-AB zu. Diese Substanz ist deshalb als potentieller Rückstandsbildner in Betracht

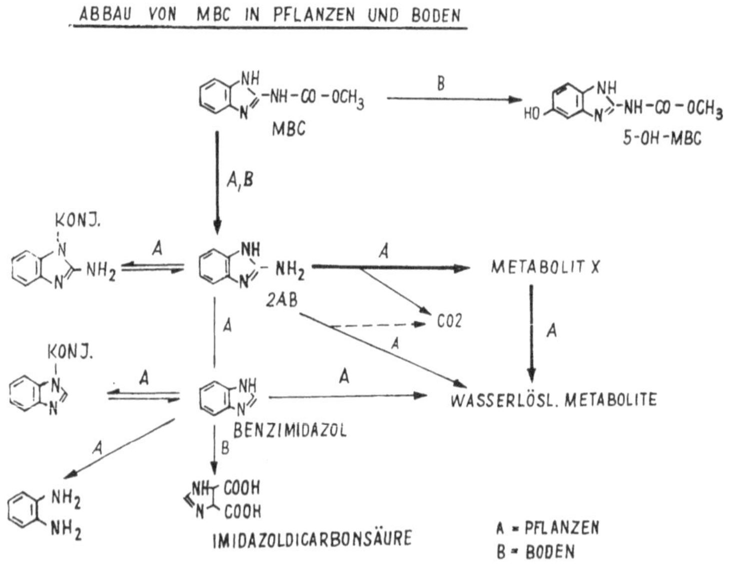

Abb. 4. Metabolismus von MBC in Pflanzen und im Boden

zu ziehen. Mit der Aufspaltung des Benimidazol-Ringes werden Produkte gebildet, die demgegenüber schneller weiter verstoffwechselt werden. Eine Anreicherung dieser Metabolite in der Umwelt ist deshalb nach dem gegenwärtigen Stand der Kenntnisse nicht zu erwarten.

Literatur

[1] BEN AZIN, A., u. N. AHARONSON, Pestic. Biochem. Physiol. **4** (1974), 120

[2] BODROVA, M. R., S. S. KUKALENKO u. E. A. DVOJČENKOVA, Chim. selsk. choz., Moskau **167** (1978), 52

[3] ENGST, R., u. W. SCHNAAK, Nahrung **23** (7), (1979), 701

[4] ENGST, R., u. W. SCHNAAK, Pharmaz. Z. Halle **106** (1967), 382

[5] NETZER, D., u. J. DISHON, Phytoparasitica **1** (1973), 33

[6] ROUCHAUD, J. P., J. R. DECALLONNE u. J. A. MEYER, Phytopathology **64** (12), (1974), 1513

[7] ROUCHAUD, J. P., J. R. DECALLONNE u. J. A. MEYER, Pestic. Sci. **8** (1977), 31

[8] SIEGEL, M. R., u. A. J. ZABBIA, Phytopathology **62** (1972), 630

[9] SIMS, J. J., H. MEE u. D. C. ERWIN, Phytopathology **59** (1969), 1775

[10] VONK, J. W., u. A. K. SIJPESTEIJN, A. Pestic. Sci. **2** (1971), 160

TH. STRUMPF, D. ZANKE und E. KLUGE

Institut für Pflanzenschutzforschung der Akademie der Landwirtschaftswissenschaften Kleinmachnow, DDR

Fungizide ß-Napthoxy-alkylhydroxamsäuren

In den letzten 15 Jahren wurde eine Reihe von Naturstoffen mit einer oder mehreren oxidierten Peptid (Amid)-bindungen, -CO-N(OH)-, in mikrobiellen Spezies aufgefunden, die sich als potente Wachstumsfaktoren, Antibiotika, antibiotische Antagonisten, Tumorhemmer oder Zellteilungsfaktoren erwiesen [3, 5].

Problem

Charakteristische Strukturelemente der in der Natur vorkommenden Hydroxamsäuren dienten als Synthesemodelle für die Darstellung neuer, potentieller Wirkstoffe. Es wird über die fungizide und bakterizide Wirkung von β-Naphthoxy-alkylhydroxamsäuren berichtet und der Einfluß der Hydrophobizität der Verbindungen auf ihre biologische Aktivität untersucht.

Resultate

Die chemische Darstellung der β-Naphthoxy-alkylhydroxamsäuren erfolgte durch Umsetzung der entsprechenden Carbonsäureester mit Hydroxylamin nach HAUSER und RENFROW [2]. Die Acylhydroxamsäuren wurden durch Reaktion von Alkalisalzen der ω(β-Naphthoxy)-alkylhydroxamsäuren mit Aroyl-/Alkyloylhalogeniden [1], die N-Arylhydroxamsäuren durch Umsetzung der β-Naphthoxy-alkanoyl-halogenide mit N-substituierten Hydroxylaminen erhalten [4].
Die Hydroxamsäuren wurden auf ihre Wirkung gegenüber Hefen, Pilzen und Bakterien untersucht. Der Einfluß von ω(β-Naphthoxy)-alkylhydroxamsäuren auf die Wachstumshemmung (WH) von *Torulopsis candida* BERLESE in Abhängigkeit von der hydrophoben Substituentenkonstante π ist in Abbildung 1 dargestellt. Im untersuchten Bereich steigt die Wirksamkeit mit wachsender Alkylkette an. Für einen linearen Regressionsansatz ergibt sich eine signifikante Abhängigkeit ($\alpha = 0{,}05$).

$$WH = 19\pi - 34{,}92 \qquad (n = 1, 2, 3 \ldots 10)$$

Bei $n \gg 10$ ist jedoch aufgrund der geringen Wasserlöslichkeit der Verbindungen ein starker Abfall der Wirksamkeit zu erwarten. Eine Verlängerung der Alkylkette in α-Stellung zur funktionellen Gruppe bei β-Naphthoxy-alkylhydroxamsäuren führte dagegen zu keiner Wirkungssteigerung.

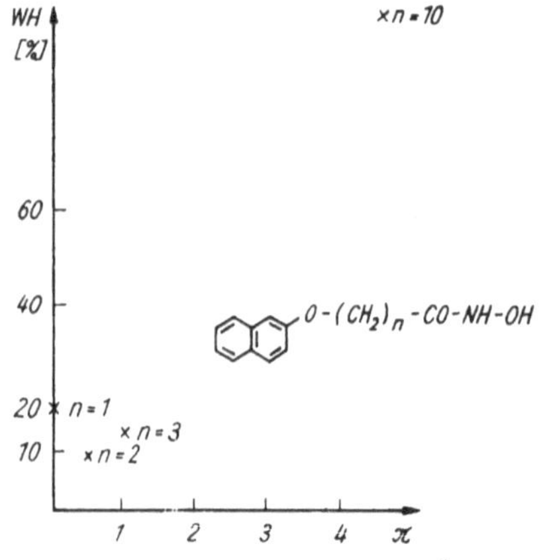

	n	B	r	α = 0,05	s_R^2
WH = 19π + 6,5	4	0,9304	0,965	ja	168,75

Abb. 1. Zusammenhang zwischen der Wirkung von ω(β-Naphthoxy)-alkylhydroxamsäuren gegenüber *Torulopsis candida* BERLESE (Wachstumshemmung bei 10^{-4} Mol/l) und der Substituentenkonstanten π

Untersuchungen zur bakteriziden Aktivität dieser Verbindungen an *Pektobakterium carotovorum* v. *atrosepticum*/(van HALL), DYE ergaben eine Umkehr dieses Effektes. Während bei ω(β-Naphthoxy)-alkylhydroxamsäuren mit steigender Lipophilie die Wirkung nicht anwächst, wurde in der Reihe der α-alkylsubstituierten β-Naphthoxyacethydroxamsäuren die Beziehung

$$WH = 13{,}57\pi - 24{,}69 \qquad (n = 0, 1, 2, 3, \ldots, 9)$$

gefunden (Abb. 2).

Die Unterschiede in der biologischen Aktivität von strukturisomeren Hydroxamsäuren bei der Hefe und dem Bakterium lassen sich nur auf sterische Einflüsse zurückführen; entweder ist die Raumerfüllung des Moleküls für das Passieren von Barrieren entscheidend oder Unterschiede in der Rezeptorstruktur erfordern zur Wechselwirkung eine bestimmte sterische Anordnung des Effektors.

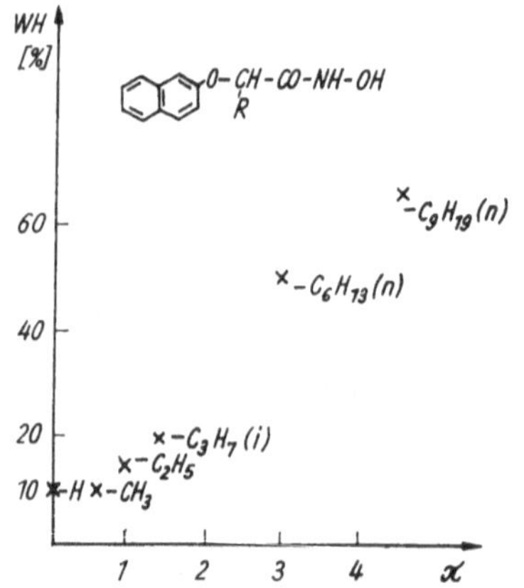

	n	B	r	α = 0,05	s_R^2
WH = 13,56π + 4,89	6	0,973	0,987	ja	18,32

Abb. 2. Zusammenhang zwischen der Wirkung von α(β-Naphthoxy)-acethydroxamsäuren gegenüber *Pectobakterium carotovorum v. atrosepticum* van HALL, DYE (Wachstumshemmung bei 10^{-4} Mol/l) und der Substituentenkonstanten π

Tabelle 1
Fungitoxizität ausgewählter ω(β-Naphthoxy)-alkylhydroxamsäuren

Wirkstoff	ED_{50} in Mol/l Wirkstoff		
	Phytophthora cactorum Zoosporenkeimtest	*Phytium debaryanum* Myzelwachstumstest	*Erysiphe graminis* Wirtspflanzentest (Gerste)
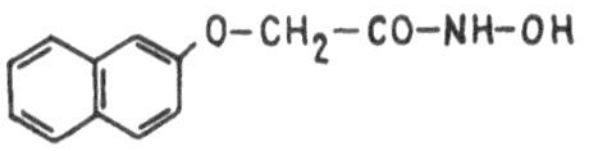	$9{,}2 \cdot 10^{-5}$	$9{,}2 \cdot 10^{-5}$	$9{,}2 \cdot 10^{-5}$
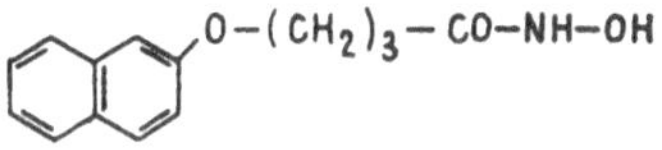	$16{,}3 \cdot 10^{-5}$	$8{,}2 \cdot 10^{-5}$	$4{,}1 \cdot 10^{-5}$
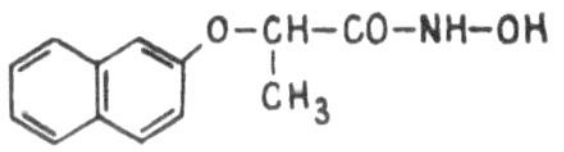	$10{,}8 \cdot 10^{-5}$	$13{,}0 \cdot 10^{-5}$	$>25{,}0 \cdot 10^{-5}$
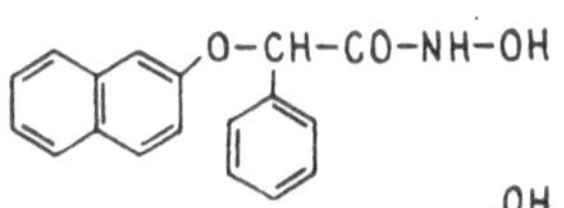	$6{,}8 \cdot 10^{-5}$	$10{,}2 \cdot 10^{-5}$	$17{,}0 \cdot 10^{-5}$
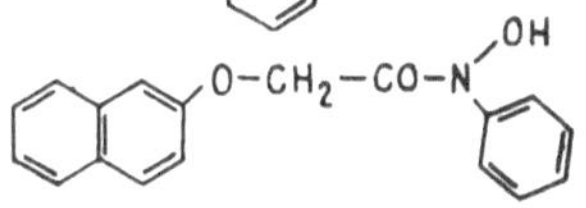	$10{,}2 \cdot 10^{-5}$	$3{,}4 \cdot 10^{-5}$	$0{,}7 \cdot 10^{-5}$
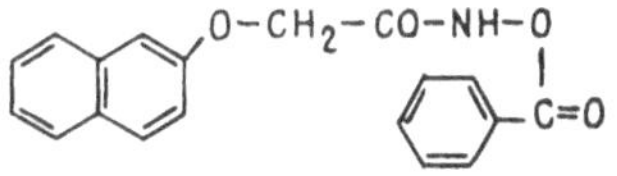	$6{,}2 \cdot 10^{-5}$	$1{,}9 \cdot 10^{-5}$	$0{,}6 \cdot 10^{-5}$
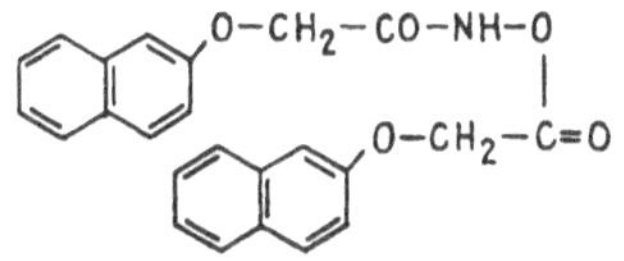	$5{,}0 \cdot 10^{-5}$	$1{,}2 \cdot 10^{-5}$	$2{,}0 \cdot 10^{-5}$

Bei pilzlichen Spezies wurde eine gute Wirkung gegenüber einigen Oomyceten und *Erysiphe graminis* Marchal (Tab. 1) gefunden. Die fungizide Wirkung nimmt mit steigender Lipophilie ab (Abb. 3); es ist somit für die Aktivität relativ bedeutungslos, ob sich der β-Naphthoxy-Rest in der α- oder ω-Stellung von Alkylhydroxamsäuren befindet. Es wurde eine lineare Beziehung zwischen den ED_{50}-Werten und der hydrophoben Substituentenkonstanten π gefunden ($\alpha = 0{,}05$).

$$ED_{50} = 3{,}12 \cdot 10^{-5}\,\pi + 15{,}45 \cdot 10^{-5} \qquad (n = 1, 2, 3, \ldots, 10)$$

Danach ist die β-Naphthoxy-acethydroxamsäure (n = 1) die wirksamste Verbindung in der Reihe der β-Naphthoxy-alkylhydroxamsäuren mit freier funktioneller Gruppe, und eine Wirkungssteigerung durch Verlängerung der Alkylkette ist nicht erreichbar.

In weiteren chemischen Synthesen wurde der Wasserstoff am N- und O-Atom der Hydroxamsäurefunktion substituiert und die fungizide Aktivität dieser Verbindungen untersucht. Eine Veränderung im pilzlichen Wirkungsspektrum trat nicht ein; es

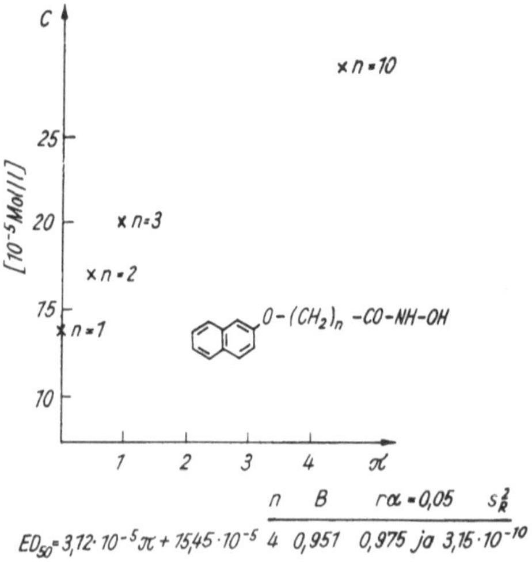

Abb. 3. Zusammenhang zwischen der Wirkung von ω(β-Naphthoxy)-alkylhydroxamsäuren gegenüber *Phytophthora cactorum* (LEN. et COEN) SCHROET. (Myzelwachstumstest — ED_{50}) und der Substituentenkonstanten π

wurde aber eine deutliche Steigerung der Aktivität beobachtet (Tab. 1). Zwischen der Lipophilie und der Wirkung konnte keine signifikante Korrelation gefunden werden, so daß bei den an der funktionellen Gruppe substitutierten Verbindungen neben dem hydrophoben Term elektronische oder sterische Einflüsse beteiligt sein müssen.

Schlußfolgerungen

Die Wirksamkeit von β-Naphthoxy-alkylhydroxamsäuren wird wesentlich durch den hydrophoben Term beeinflußt, was auf eine relativ unspezifische Wirkung schließen läßt.

Die Verbindungen vereinigten in sich eine günstige Kombination von fungiziden und bakteriziden Eigenschaften, ohne jedoch die Aktivität von Spitzenwirkstoffen zu erreichen. Ihre Phytotoxizität ist gering.

Literatur

[1] ARTEMENKO, A. J., Synthese von 0-Acylhydroxamsäuren (russ.), Zh. org. Khim. 7 (1971), 714—721

[2] HAUSER, C. R. and W. B. RENFROW jr., Benzohydroxamic acid, in BLATT, A. H. (Hsg) Organic Synthesis, John Wiley & Sons, New York 1943, Vol. II, 67

[3] MAEHR, H., Antibiotics and other naturally occuring Hydroxamic acids and Hydroxamates, Pure Appl. Chem. 28 (1971), 603—636

[4] PRIYADARSHINI, U. and S. G. TANDON, Preparation and properties of some N-Aryl Hydroxamis acids ,J. Chem. Eng. 12 (1967), 143—144

Л. Н. Андреев

Главный ботанический сад Академии наук СССР, Москва, СССР

Роль Витаминов в Устойчивости Растений к Ржавчине

Значение физиологически-активных веществ в устойчивости растений к инфекционным заболеваниям установлено рядом исследователей. Еще в 30-х годах работами К .Т. Сухорукова с сотрудниками [1, 2] было показано значение веществ биоса в устойчивости моркови к *Sclerotinia libertiana* и арбуза к *Fusarium wolgense*. Несколько позже К. Т. Сухоруковым [3] было сформулировано положение о том, что взаимоотношения между паразитом и растением-хозяином в отделбных случаях зависят от содержания в растении фракторов роста, в частности, бактериальных витаминов необходимых паразиту. Ингибирование развития раразита может быть вызвано несоответствием между его питательными потребностями и возможностью удовлетворения их клетками растений. Впоследствии Гербером [4] была предложена питательнотормозящая гипотеза, согласно которой одним из условий восприимчивости растений является соответствие между питательными потребностями паразита и адекватной концентрацией соответствующих соединений, необходимых патогену, в растении. На определенную роль витаминов, в частности, мезоинозита, биотина, пантотеновой кислоты, тиамина и др., в развитии возбудителей инфекционных болезней, особенно вызываемых сапрофитными организмами, указывается в ряде работ [5—7]. Учитывая большую приспособленность облигатных паразитов к обмену веществ растения-хозяина, можно предположить, что при облигатном паразитизме роль витаминов в устойчивости растений к инфекционным заболеваниям значительно возрастает. В связи с этим изучение роли веществ дополинтельного питания при облигатном паразитизме имеет большое значение для выявления факторов устойчивости растений к инфекции, тем более, что экспериментальные исследования в этом отношении очень немногочисленны [8—16]. Наши исследования были посвящены изучению роли витаминов группы В — биотина, пантотеновой кислоты и тиамина в устойчивости растений пшеницы к желтой *(Puccinia striiformis)* и стеблевой *(P. graminis* f. sp. *tritici)* ржавчине. Как показало изучение содержания витаминов в здоровых и больных растениях, под влиянием инфекции в листьях пшеницы содержание витаминов возрастает (табл. 1). Особенно возрастает в больных растениях содержание биотина. Уже через сутки уровень биотина под влиянием инфекции возрастает на 30—40%, а на 12—15 сутки — его содержание увеличивается в 2—2,5 раза по сравнению со здоровыми, не-

Таблица 1
Содержание биотина и пантотеновой кислоты в листьях различных по устойчивости сортов, инокулированных P. graminis f. sp. tritici раса 21 (в мкг/гр сухого веса)

Сорт	Степень устойчовости	Состояние растений	Время после инокуляции, в сутках			
			0	6	12	18
		Биотин				
Капли	устойчивый	здоровое	0,48	0,39	0,40	0,40
		больное	—	0,73	1,02	1,04
Грекум 114	средне-устойчивый	здоровое	0,46	0,43	0,40	0,34
		больное	—	0,42	1,02	0,83
Литтл Клаб	восприимчивый	здоровое	0,75	0,67	0,70	0,55
		больное	—	1,17	1,34	1,31
		Пантотеновая кислота				
Кпли	устойчивый	здоровое	41,0	49,0	50,0	48,0
		больное	—	68,0	68,0	54,0
Грекум 114	средне-устойчивый	здоровое	13,5	13,5	12,5	15,0
		больное	—	15,0	9,0	22,5
Литтл Клаб	восприимчивый	здоровое	40,0	40,0	35,0	40,0
		больное	—	47,0	66,0	68,0

инокулированными растениями. Содержание пантотеновой кислоты и тиамина под влиянием инфекции также возрастает, хотя и в меньшей степени. Наибольшее увеличение содержания витаминов наблюдается в период перехода гриба в репродуктивную фазу развития. Значительное повышение содержания витаминов обнаруживается в листьях пшеницы у восприимчивых к ржавичине сортов по сравнению с устойчивыми. Обращает на себя внимание высокое содержание витаминов в уредоспорах стеблевой, желтой и бурой ржавчины. По сравнению с тканями растений содержание биотина в них в 7—10 раз выше. Такое высокое содержание витаминов в уредоспорах, видимо, является важным для первых этапов развития патогена в фазе инфекционной капли. Прорастающие споры, выделяя во внешнюю среду витамины, создают благоприятные условия для своего развития. Изучение влияния витаминов на прорастание и развитие уредоспор показало, что введение в среду экзогенных витаминов увеличивало количество прорастающих уредоспор и, что особенно важно, количество спор, образующих при прорастании полные инфекционные структуры. Учитывая определенную роль витаминов в устойчивости растений к ржавчине, а также имеющиеся литературные данные о том, что многие грибы и, в частности, возбудители ржавчины, не способны к синтезу ряда витаминов, в том числе биотина, мы испытали ряд неактивных структурных аналогов биотина и его антиметаболитв для борьбы с ржавчинными болезнями. Введение в среду экзогенных неактивных структурных аналогов биотина, по нашему предположению, должно было создать дефицит витаминов в среде и привести к неполноценному питанию патогена. В связи с этим мы испытали действие ряда неактивных аналогов биотина (дестиобиотин, изодестиобиотин, нордестиобиотин, гомодестиобиотин и др.). Испытание препаратов проводили путем учета ингибирующего действия препаратов при проращивании уредоспор стеблевой ржавчины в растворах испытываемых соединений разной концентрации. Было установлено, что препарат изодестиобиотин ингибирует прорастание уредоспор в концентрации 100 мг/л. Сходным действием обладал изомер нордестиобиотина и гомодестиобиотин. Полученные данные демонстрируют перспективность дальнейших исследований с целью использования неактивных структурных аналогов биотина для борьбы с ржавчиной.

Резюме

Установлена определенная роль витаминов — биотина, пантотеновой кислоты и тиамина в устойчивости растений пшеницы к ржавчине. Устойчивые растения, как правило, характеризуются пониженным содержанием витаминов по сравнению с восприимчивыми растениями. Под влиянием инфекции содержание витаминов в тканях растений возрастает, причем это увеличение достигает больших размеров у восприимчивых растений. Испытание различных препаратов неактивных структурных аналогов биотина в целях ингибирования развития инфекции показало, что эти препараты являются перспективными химическими соединениями, создающими неполноценную питательную среду для патогена.

Литература

[1] Сухоруков, К. Т., Э. Х. Гербер, Г. П. Барабанова, Н. А. Бородулина К биохимии растительного иммунитета Ученые записки Саратовского университета, X (1933) № 1

[2] Вернер, А. Р., Е. Г. Клинг, К иммунитету плодов. Причины устойчивости плодов арбуза к „черной гнили". Труды комиссии по ирригации АН СССР, 3 (1934)

[3] Сухоруков, К. Т., Физиология иммунитета растений Изд. АН СССР, Москва 1952

[4] Garber, E. D., A nutrition — inhibition hypothesis of pathogenicity, Amer. Naturalist **90** (1956), 183—194

[5] Mandels, G. R., Biotin and interrupted growth of *Myrothecium verrucaria*, Amer. J. Bot. No. **42** (1955), № 10

[6] Lukeric, F. L., J. E. De Vay, Effect of myoinositol in host tissues on parasitism of *Bromus domestica* var. President by *Rhodosticta querciana*, Phytopathology **54** (1964), 697

[7] Сухоруков, К. Т., В. В. Филиппов, Н. А. Шадманова, Физиологически активные вещества при вертициллезе хлопчатника Доклады АН ССР 188 (1966), 470—472

[8] Defago, G., Effect de l'aneurine et de ses composants et de l'heteroauxine sur la croissance de trois parasites du ble, Phytopathol. Zeitschr. **13** (1940), 292—299

[9] Pryor, D. E., The influence of vitamin B′ on the development of cantaloupe powdery mildew, Phytopathology **32** (1942), № 10

[10] Gottlieb, D., H. Hart, Growth substances and the rust fungi, Phytopathology **33** (1943), No. 8

[11] HASSEBRAUK, K., Untersuchungen über die Einwirkung von Sulfonamiden und Sulfonen auf Getreideroste. 3. Untersuchungen über den Wirkungsmechanismus von Sulfonamiden und Sulfonen, Phytopath. Z. **19** (1952), 56–72
[12] HOTSON, H. H., The inhibition of wheat stem rust by sulfadiazine and the reversal of this effect by paraaminobenzoic acid and folic acid, Phytopathology, **42** (1952), No. 1
[13] YARWOOD, C. E., A. P. HALL, M. M. NELSON, Nutritive value of rust-infected leaves, Science **117** (1952), 326–327
[14] HOBBS, C. D., M. C. FUTRELL, Influence of vitamins and other compounds on the degree of infection of wheat by *Puccinia graminis tritici*, Phytopathology **53** (1963), 230–231
[15] HUSAIN, M., C. D. HOBBS, M .C. FUTRELL, Effect of infection by *Puccinia graminis tritici* on the vitamin B_6, niacin and pantothenic acid contents of wheat plants, Phytopathology **54** (1964), 502–505
[16] SHAW, M., N. COLOTELO, The physiology of host-parasite relations. XII. The effect of stem rust on the nitrogen and amino acids in wheat leaves, Can. J. Bot. **39** (1961), 1351–72

Беккер, З. Э., И. О. Селях, Е. С. Лобакова, О. В. Камзолкина, И. С. Кулаев и С. Э. Мансурова

Кафедры Микробиологии, Низших растений и Биохимии Биологического факультета московского Государственного Университета, Москва, СССР

Биоритмы в Онтогенезе Возбудителя Вилта и Действие Системых Антибиотиков

Предпринятые ранее рядом авторов этой работы исследования биоритма при развитии девяти видов грибов в погруженной и агаровой культурах [5, 6] показали, что он имеется у всех изученных представителей разных таксономических групп *(Peronosporales, Saccharomycetales, Pyrenomytales, Plectascales, Polyporales, Agaricales* и *Hyphomycetales)* даже, если он не выявляется по зональности роста на агаре. Эти ритмы различаются только по длительности их периодов (от 5-ти-6-ти сусточного у представвителя пероноспоровых *Phytophthora infestans* до 4-х-5-ти часового у дрожжей), причем отдельным таксономическим группам свойственна определенная продолжительность периода. Например, гифомицетам от суток до 2-х суток, базидиомицетам от 3-х до 4-х суток и т. д.. Возбудитель фузариозного вилта *Fusarium oxysporum* f. sp. *vasinfectum* имеет ритм развития от 2-х суток до суток в зависимости от сезона года (весной и осенью учащенный). Наиболее подробно биоритм грибов был нами изучен на этом объекте.

Выявление биоритма в погруженной культуре удается лучше всего с помощью количественных цитохимических методов, из которых основным является подобранный нами для этой цели метод окраски азур — эозином по Майгрюнвальд [6, 11], дополняемый витальной окраской нейтраль рот и люминесцентной микроскопией с окраской акридин-оранж, а также реакциями на некоторые ферментные системы (дегидрогеназы, полифенолоксидазы, сульфгидрильные соединения и состояние ред-окс системы цитоплазмы).

В результате этих исследований было установлено, что биологический ритм у грубов заключается в смене периода деления заблокированных основым белком ядер (адсорбирующим анион красителя, эозин) и периода их освобождения от блока (адсорбции ядрами катиона красителя, азура), выделения в цитоплазму рибосом (красно-оранжевая окраска гиф акридин-оранж), синтеза белка и роста гиф новой генерации. В следующие пол периода происходит дифференциация клетки с формированием митохондрий (зеленая окраска цитоплазмы гиф акридин-оранж), волютина (красные гранулы в вакуолях при той же окраске) и образованием адаптивных приспособлений (например, фермента глюкоамилазы у *Aspergillus awamori* [6, 7] или вилтотоксина у *Fusarium oxysporum*).

Результаты биохимических исследований, проведенных на погруженной культуре *Fusarium oxysporum* согласуются с приведенной трактовкой хода биологического ритма на основании цитохимических параметров, полученных для этого вида. Осцилляции скорости изменений количества нуклеиновых кислот в мицелии, выделяемых методом Шмидта и Тангаузера и определенных спектрофотометрически в мкг фосфора на мг веса сухого мицелия, согласуются с пиками и понижениями кривых, построенных на основе цитохимического анализа поведения ядер (рис. 1). Так ход

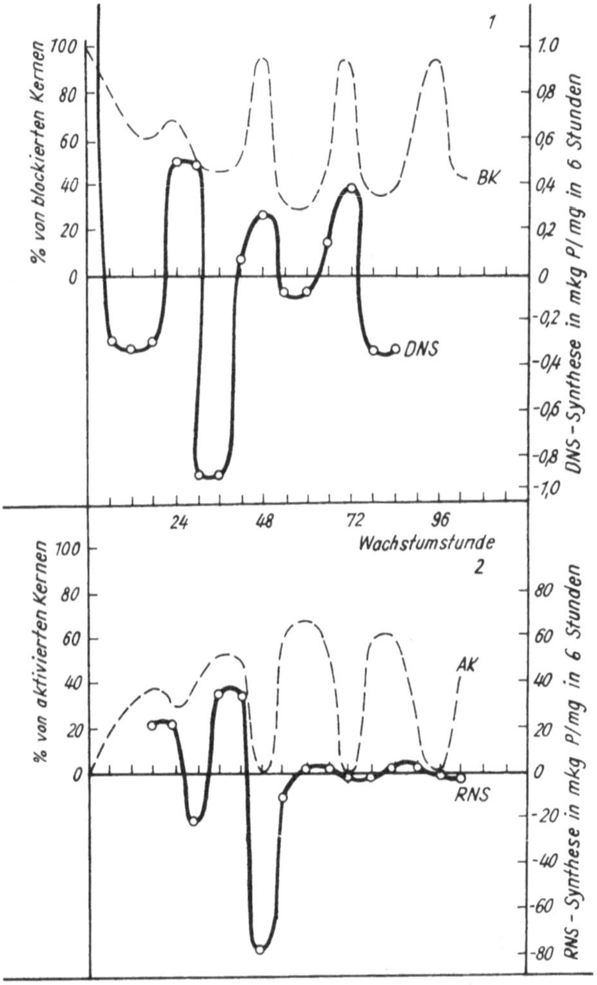

Рис. 1. Согласование осцилляции синтеза ДНК [1] и и РНК [2] в гифах *Fusarium oxyrporum* соответственно с периодами блока [БЯ] и активации [АЯ] клеточных ядер

2 Phase	DNS-Synthese (Kernteilung) RNS-Synthese (Ribosomenreiche Hyphen)
2 Antiphase	Eiweiß-Synthese (Wachstum und Erniedrigung der Kern/Plasma Verhältnisse)
3 Phase	Differenziation der Zellen (Mitochondrien- und Volutinformation, adaptive Enzyme, Antibiotika- und Toxinformation und andere adaptive Anpassungen)

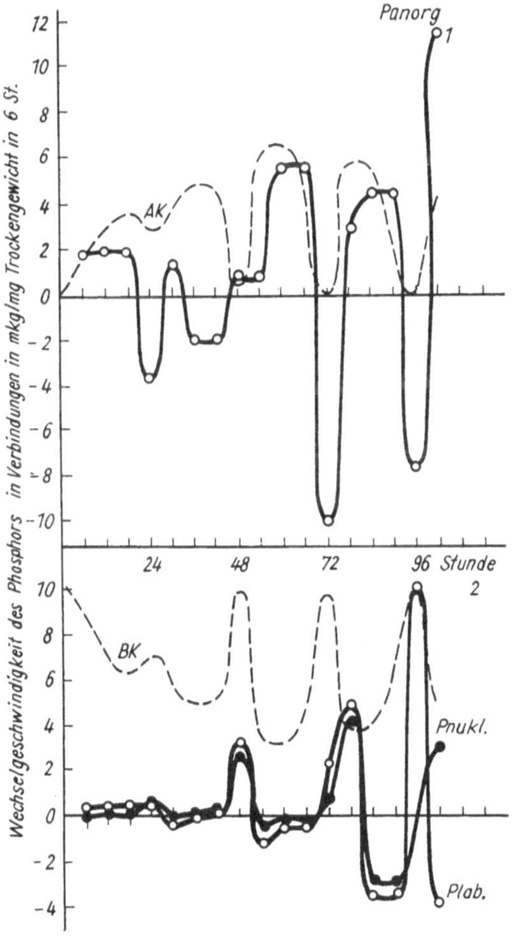

Рис. 2. Сопряжение осцилляций накопления неорганического фосфора [$P_н$] с активацией ядер [АЯ] и ростом [1] и трансформации его в лабильные органические фосфаты [$P_{лаб}$] и нуклеотиды [$P_{нукл}$] с блокированием [БЯ] и делением [2] ядер в мицелии *Fusarium oxysporum*

осцилляции кривой количества заблокированных основным белком ядер хорошо совпадает с осцилляцией кривой изменения количества ДНК (рис. 1 [1]), а кривая количества активированных ядер, находящаяся в противофазе к первой, с кривой изменения количества РНК, особенно в первой половине наблюдаемого онтогенеза (до 72 часов роста) (рис. 1 [2]).

Последнее состояние ядер, т. е. снятие белкового блока и их активация, оказалось совпадающим с кривой изменения в мицелии неорганического ортофосфата (рис. 2 [1]), что может свидетельствовать одновременно и о ходе его поступления в клетку из окружающей среды и о скорости его высвобождения из макроэргов в процессе идущих в эти репиоды биосинтезов РНК и белка.

Наоборот, изменение количества лабильных фосфоорганических соединений и нуклеотидов оказывается в противофазе к изменению неорганического ортофосфата (рис. 2 [2]), совпадая начиная со второго цикла осцилляции с блоком ядра.

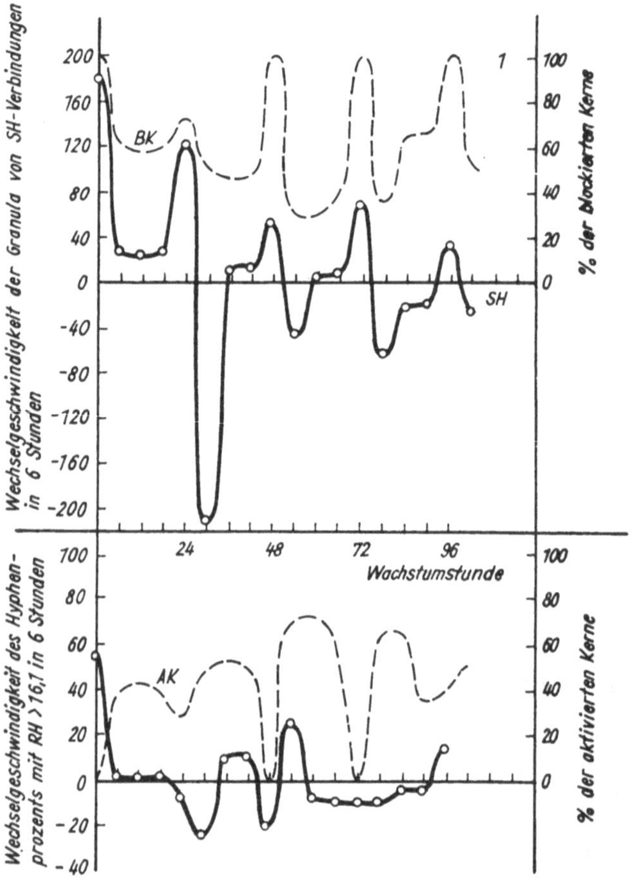

Рис. 3. Синхронизация процента ядер в состоянии блока [БЯ] и активации [АЯ] с осцилляцией количества SH-групп [1] и гиф с уровнем RH-цитоплазмы > 16,1 [2] в мицелии из погруженной культуры *Fusarium oxysporum*

С наблюдаемыми явлениями хорошо согласуются также определенные цитохимическими методами (метод Мазон и витальная окраска янусгрюн [1]) изменения некоторых общих параметров обмена (рис. 3). Так накопление количества сульфгидрильных групп, обычно сопровождающее деление ядер, хорошо согласуется с кривой изменения количества заблокированных ядер (рис. 3 [1]), а активация ядра и рост согласуются с кривой изменения количества гиф с высоким уровнем RH, оркашивающихся янусгрюн (рис. 3 [2]), что является сигналом о повышении интенсивности дыхания в процессе роста.

Исходя из предшествовавших [5, 6], а также из приведенных здесь дополнительных биохимических данных видимо следует принять следующие точки зрения на механизм хода биоритма в развитии грибов:

1. Ритмы развития грибов инициируются в области клеточного ядра и, видимо, наследственно запрограммированы.
2. Передача этой инициации на другие клеточные структуры протекает по известной молекулярно-биологической последовательности: синтез ДНК → синтез РНК → синтез белка и рост → дифференциация клеточных структур и синтез адаптивных продуктов обмена.

3. Биологический ритм в онтогенезе грибов имеет непосредственную связь с адаптивным онтогенезом [3], выраженным ранее в форме двустадийного [1, 2] и полициклического [12] развития продуцентов антибиотиков в погруженной культуре, являющимся основой универсальных механизмов приспособления грибов к их среде обитания.

Все приведенные здесь три определения и в особенности положение о этапности приспособления к среде в онтогенезе [3], принятое сейчас многими фитопатологами, в приложении к приспособлению к паразитизму, которое включается в ход биоритма начиная со второго его цикла [6], привлекает особое внимание с точки зрения возможности затормаживания или исключения формирования паразитической стадии в онтогенезе грибов возбудителей болезней растений. С таким случаем нам пришлось встретиться несколько лет назад при сравнительном исследовании механизмов защиты хлопчатника от фузариозного вилта агентами его естественного иммунитета, как формируемые им факторы регуляции проницаемости (диоксифенилаланин ДОФА) и β-индолилуксусная кислота (ИУК)) и образуемый ризосферными почвенными грибами антибиотик, гризеофульвин, имеющими эффект системного типа, в основом ограниченный влиянием на паразитическую стадию в растении.
Анализ методом окраски по Майгрюнвальд состояния ядер *Fusarium oxysporum* в стеблях растения хозяина в контроле и после обработки гризеофульвином показал увеличение числа ядер в мицелии (рис. 4) и в особенности крупных ядер, заблокированных основным белком, тогда как при обработке ДОФА общее количество ядер резко падало без преимущества для заблокированных гистоном [4, 10]. Обработка хлопчатника ИУК, также как и ДОФА в дозе уплотняющей клеточные мембраны хозяина и создающей голодание патогена, давала сходное с ДОФА состо-

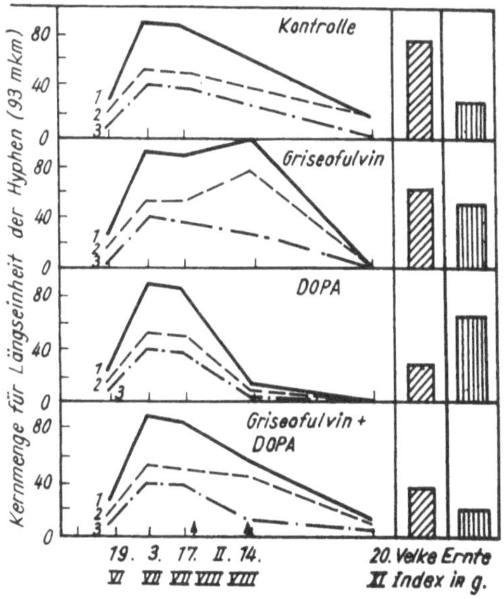

Рис. 4. Влияние обработки хлопчатника гризеофульвином, диоксифенилаланином [ДОФА] и их сочетанием в максимальных дозировках [200 мг] л и 40 мг/л соответственно] на изменение числа ядер в мицелии [1], степени их блокировки основным белком [2], наличия мелких ядер [3] при сопоставлении с уровнем вилта и урожаем на куст. Стрелки — даты обработки

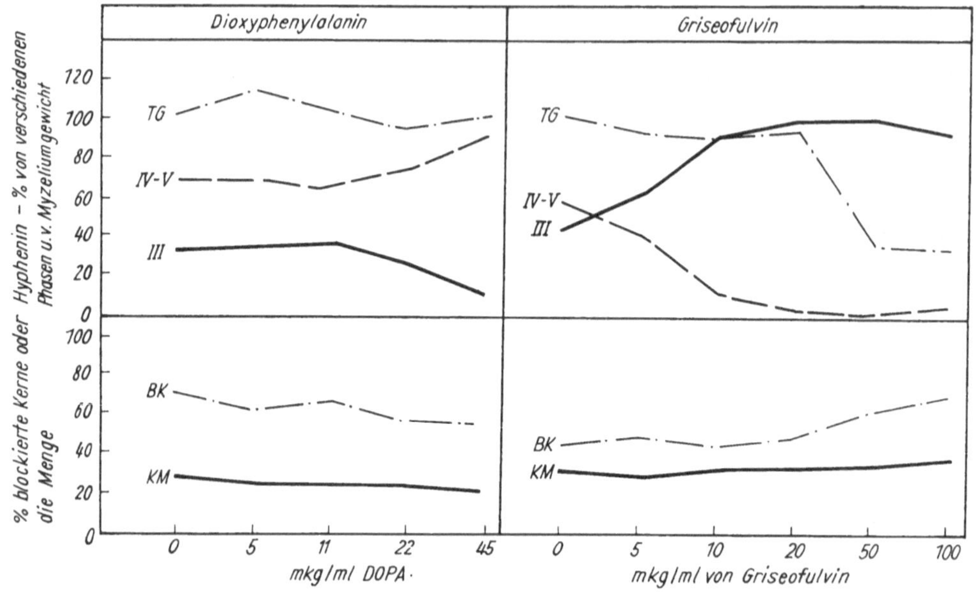

Рис. 5. Состояние развития мицелия *Fusarium oxysporum* на 4 сутки роста в погруженной культуре при разных дозировках ДОФА и гризеофульвина. СВ — вес сухого мицелия; III и IV—V — мицелий III и IV—V возрастных фаз; БЯ — блокированные ядра; КЯ — количество их на 100 мкм длины гиф

яние ядер [10]. Наоборот, в условиях свободного доступа питательных веществ среды в лабораторной погруженной культуре добавка угнетающих этот гриб в условиях патогенеза доз ДОФА практически совершенно не сказывалась на количестве и степени блокирования его ядер (рис. 5) [4] и на ходе роста и развития мицелия, которое даже несколько ускорялось.

В противоположность этому при добавке гризеофульвина в лабораторную культуру патогена, даже в дозах еще не влияющих на рост мицелия (20 мкг/мл), уже наблюдался сильный эффект в отношении влияния антибиотика на его развитие. Способная к паразитизму генерация мицелия (т. е. IV—V возрастные фазы) полностью угнеталась (рис. 5) и он оставался в ранней, недифференцированной III фазе. Это сопровождалось возрастанием количества заблокированных цитоплазматическими гистонами ядер, хотя общее количество их практически не менялось (рис. 5).

Поскольку нам было уже известно на основании цито- и биохимических анализов из более ранних работ [8, 9], что действие гризеофульвина связано с торможением продукции РНК, осцилляция которой по ириве денным здесь данным совпадает с периодами активации ядра в ритме онтогенеза, мы продолжили наши исследования в плане анализа хода биоритма данного патогена. Для сравнения с гризеофульвином на том же объекте, *Fusarium oxysporum*, нами был применен несколько задерживающий рост грибов полимиксин, действие которого считается связанным с биосинтезом клеточных мембран и оболочек. Исследование проводилось методом количественной цитохимии с применением окраски по Майгрюнвальд с дополнением ее люминесцентной микроскопией с окраской акридин-оранж.

Полученный результат показал, что, как и ожидалось, гризеофульвин показывает очень сильное воздействие на биоритм развития *F. oxysporum* (рис. 6), которое состоит в торможении перехода от состаяния блока ядра к его активации, приводя к почти полному блокированию его в период

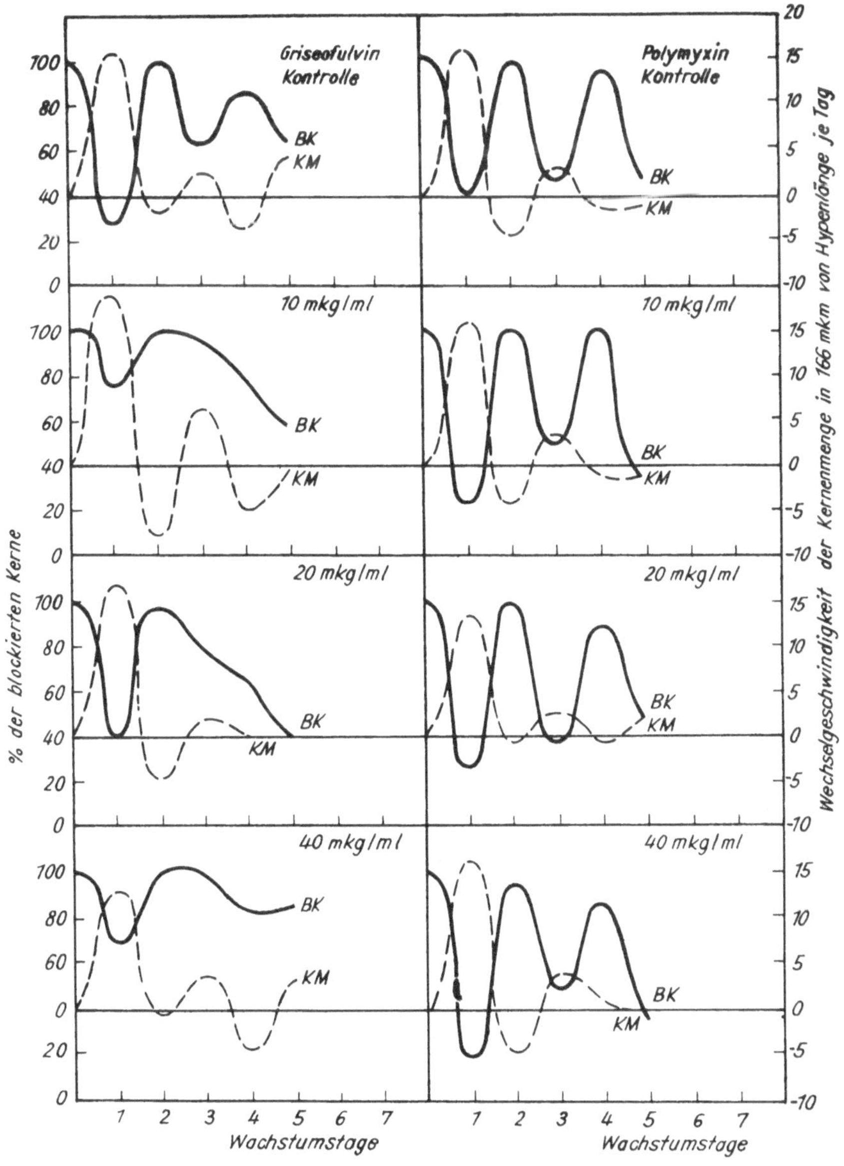

Рис. 6. Осцилляция процента заблокированных ядер [БЯ] и их количества [КЯ] при различных дозировках гризеофульвина и полимиксина

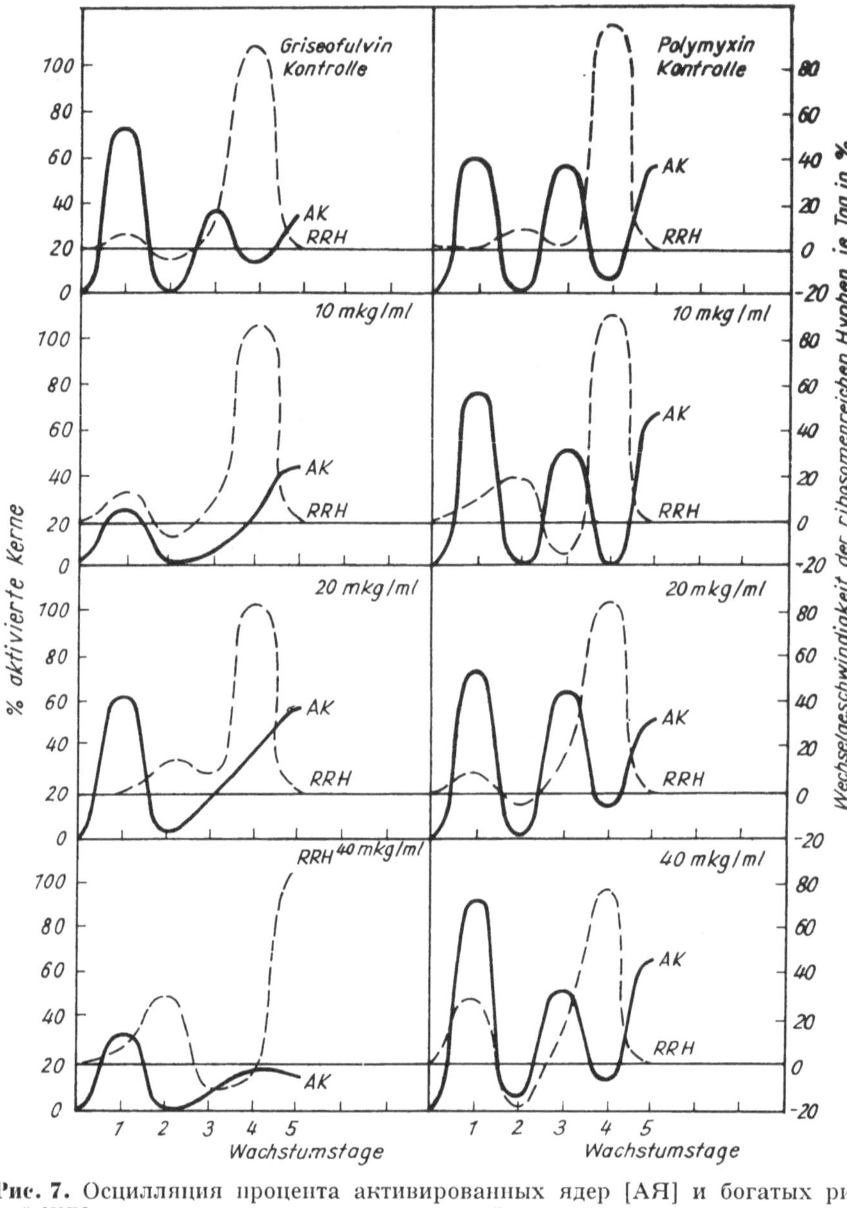

Рис. 7. Осцилляция процента активированных ядер [АЯ] и богатых рибосомами гиф [ГР] при различных дозировках гризеофульвина и полимиксина в погруженной культуре *Fusarium oxysporum*

второго цикла при дозе антибиотика в 40 мкг/мл, при которой обычно наблюдается начало снижения роста у этого организма. [4] Осцилляция увеличения количества ядер при этом не изменяется.

Продление периода блокировки ядер естественно ведет к процессу уменьшения и замедления их активации (рис. 7), что приводит в конечном счете к замедлению и угнетению развития новой генерации гиф, обладающих адаптацией к паразитизму, хотя процент богатых рибосомами в их числе осциллирует на уровнях практически не отличающихся от контроля.

Из этого можно предположить, что основной механизм системного эффекта гризеофульвина, свяван не столько с самим торможением процесса биосинтеза РНК, не превышающим при дозе в 100 мкг/мл 30% от контроля [8, 9] сколько с поддержанием ядра в состоянии блокирования ретикулумом из основного белка, видимо препятствующим формированию или высвобождению информационной фракции этой нуклеиновой кислоты в форме и-РНК или полисом. В таком случае оказывается понятным отсуствие появления при действии гризеофульвина мицелия с дифференциацией, обеспечивающей его адаптацию к паразитическому существованию.

К вышесказанному можно добавить, что весьма вероятно подобный же механизм действия через торможение биоритма онтогенеза свойствен не одному гризеофульвину, а и многим друлим веществам, обладающим системным эффектом, например, геодину из *Aspergillus terreus*.

Литература

[1] Беккер, З. Э., ,,Возрастные явления у *Penicillium chrysogenum* Thom.". Диссертация д-ра биологических наук. Московский Гос. Университет, Москва 1956

[2] Беккер, З. Э., ,,Физиология грибов и их практическое использование." Издат. Московского Гос. Университета, Москва 1963

[3] Беккер, З. Э., О конвергентных чертах сходства грибов с некоторыми группами животных. Журнал Общей Биологии, **36**, № 5 (1975), 670—687

[4] Беккер, З. Э., Р. М. Азимова и И. Д. Пушкарева, Влияние мембранноактивных веществ и антибиотиков с системным эффектом на *Fusarium oxysporum* Schlecht f. sp. *vasinfectum* (Robin) Berkh. Микология и фитопатология, **9**, № 5 (1975), 369—376

[5] Беккер, З. Э., Э. И. Бурцева, И. О. Селях и Е. С. Рогова. Биоритмы грибов в погруженной культуре. Сборник ,,Экологические особенности низших растений Советской Прибалтики,,. Тезисы докладов. Изд. Академии Наук Литовской СССР, Вильнюс 1977, 13—14

[6] Беккер, З. Э., Бурцева, Э. И., Е. С. Рогова и И. О. Селях. Биоритмы грибов в погруженной культуре. Журнал Общей Биологии, **39**, № 4 (1978), 521—533

[7] Беккер, З. Э., Э. И. Бурцев а Е. А. Двадцатова. Углеводы среды и биоритм онтогенеза *Aspergillus awamori* T. — I в связи с биосинтезом глюкоамилазы. Микробиология, **49**, № 3 (1980)

[8] Беккер, З. Э., С. В. Дмитриева, Т. П. Супрун, И. А. Трофимова и С. М. Иванчикова, О связи аномального морфогенеза конидиеносцев с обменом нуклеопротеидов, выявляемом методами флуорохромирования и биохимии. Микология и фитопатология, **6**, № 2 (1972), 105—110

[9] Becker, S. E. und S. M. Iwantschikowa, Über die Zusammenhänge morphogenetischer und biochemischer Effekte von Nukleinsäure angreifenden Antibiotika. Sammlung ,,Wirkungsmechanismen von Fungiziden und Antibiotika", Akademie-Verlag, Berlin 1967, 257—260

[10] Беккер, З. Э., В. Ф. Полетаева и И. Д. Пушкарева, Эффективность регуляторов проницаемости и РНК-ингибиторов против болезней хлопчатника

и некоторые данные о механизме их действия. Сборник 3: Эколого физиологические методы в борьбе с фузариозным вилтом хлопчатника. Изд. ,,ЫЛЫМ", Ашхабад 1977, 141—166
[11] Беккер, З. Э., и И. Д. Пушкарева, Онтогенез мицелия и трансформация ядерного аппарата при развитии возбудителя фузариозного вилта а погруженной культуре. Сборник 2: Эколого физиологические методы в борьбе с фузариозным вилтом хлопчатника. Изд. ,,ЫЛЫМ", Ашхабад 1973, 59—72
[12] Дмитриева, С. В., и Г. Д. Пестерева, Сравнительное изучение процесса развития штаммов *Penicillium chrysogenum* № 194 и ,,Новый гибрид". Антибиотики, **9** (1962), 783—786

H. Bergmann, R. Müller und W. Neuhaus

Institut für Pflanzenschutzforschung der Akademie der Landwirtschaftswissenschaften, Kleinmachnow, DDR

Bestimmung der phytotoxischen Wirkung von Calixin auf Kulturpflanzen

Einleitung

Die schnelle und rationelle quantitative Einschätzung der phytotoxischen Wirkung biologisch aktiver Wirkstoffe ist für die Entwicklung derartiger Substanzen und ihren Einsatz in der Praxis von großer Bedeutung. In der Regel wird die Phytotoxizität an Keimlingen bzw. Ganzpflanzen unter bestimmten Kulturbedingungen oder im Freiland bestimmt. Dabei erfolgt die Einschätzung äußerlich sichtbarer Schadwirkungen an Blättern und Sprossen bzw. an der Gesamtpflanze [3, 12, 17, 18, 19, 8, 14] oder es wird die Beeinflussung des Längenwachstums, des Frisch- und Trockengewichtes von Sprossen und Wurzeln ermittelt [6, 5, 11]. Derartige Untersuchungen sind z. T. zeitaufwendig, wobei diesbezügliche Ergebnisse und Aussagen unter Feldbedingungen zusätzlich durch nicht immer exakt erfaßbare Einflußfaktoren modifiziert werden. Bei der Entwicklung einer labormäßig zu handhabenden Methode ist die Auswahl der nach Einwirkung potentiell toxischer Substanzen zu bestimmenden Parameter problematisch. Veränderungen an Zellstrukturen, z.B. der Chloroplasten [2, 4], bzw. des Gehaltes metabolisch wichtiger Verbindungen, z.B. reduzierende Zucker [2], Stärke [10], Membranlipide [7] als entsprechender Ausdruck der pflanzlichen Reaktion, sind bezüglich des methodischen Aufwandes und der erreichbaren quantifizierbaren Aussage nicht unbedingt geeignet. Vorliegende Arbeit beschreibt ein Verfahren, mit dem auf der Basis des 2,3,5-Triphenyltetrazoliumchlorid (TTC)-Tests die phytotoxische Wirkung von Calixin quantitativ bestimmt werden kann.

Material und Methoden

Die Untersuchungen erfolgten an Pflanzen von *Hordeum vulgare* L., *Triticum aestivum* L. und *Cucumis sativus* L.. Die Pflanzen wurden unter klimatisierten Bedingungen bei 20 °C angezogen.

Die TTC-Methode geht davon aus, daß zwischen der Aktivität von Oxidoreduktasen und der Vitalität pflanzlicher Gewebe ein Zusammenhang besteht [15, 1]. Die farblose TTC-Lösung wird dabei zum rot gefärbten Formazan reduziert. Verminderte Formazan-Bildung bedingt durch verminderte Aktivität der Oxidoreduktasen als Folge der Calixin-Entwicklung, zeigt entsprechende phytotoxische Wirkung an. Die quantitative Bestimmung der Formazanmenge erfolgt kolorimetrisch.

Zu Blattsegmenten des zweitjüngsten Blattes 12 Tage alter Pflanzen (je Probe 100 mg Frischmasse) wurden ins Reagensglas je 6 ml der zu prüfenden Wirkstofflösung gegeben. Anschließend erfolgte einstündige Vakuumfiltration. Nach insgesamt 2 stündiger Einwirkzeit bei 20 °C wurde die Lösung abgegossen, die Blattsegmente mit Wasser gespült und je Probe 6 ml TTC-Lösung (0,25%) zugegeben. Nach erneuter einstündiger Vakuuminfiltration blieben die Proben 16 Stunden bei 30 °C im Dunkeln stehen. Danach wurde die TTC-Lösung abgegossen, das pflanzliche Material mit Wasser gespült und anschließend 16 bis 20 Stunden mit je 2 ml 96%igem Äthanol bei Zimmertemperatur extrahiert. Die Bestimmung der Extinktion der äthanolischen Formazan-Extrakte erfolgte bei 530 nm am Spekol (VEB Carl Zeiss Jena).

Jeder angegebene Versuchswert entspricht dem Mittel aus 3 Einzelwerten. Die Versuche wurden mehrmals wiederholt. Mit Hilfe des t-Tests wurde die Signifikanz der Differenzen eingeschätzt. Die Ergebnisse wurden mit an Ganzpflanzen unter Gewächshaus- oder Freilandbedingungen nach Applikation des Calixins über die Blätter erhaltenen, verglichen.

Ergebnisse und Diskussion

Die phytotoxische Wirkung von Calixin war abhängig von der einwirkenden Konzentration. Ein Vergleich der untersuchten Pflanzenarten ergab, in Übereinstimmung mit Untersuchungen an Ganzpflanzen unter Gewächshausbedingungen, steigende phytotoxische Wirkung in folgender Reihenfolge: Gerste < Weizen < Gurke (Abb. 1). Die verwendeten Calixin-Konzentrationen (10, 100, 1000 ppm) führten zu signifikant unterschiedlicher Formazanbildung. Die durch 100 bzw. 1000 ppm Calixin bedingten unterschiedlichen Formazanmengen der Blattsegmente der einzelnen Arten waren ebenfalls signifikant. Die erhaltenen Ergebnisse stimmen mit Angaben überein, die aufgrund des praktischen Einsatzes von Calixin bei einzelnen Kulturpflanzen-Arten erhalten wurden [3, 12, 19].

Vergleichende Untersuchungen an Winterweizen-Sorten zeigten zunehmende Empfindlichkeit der Sorten in folgender Reihenfolge: Almus = Alcedo < Mironowskaja 808 < Mironoswskaja Jubilejnaja (Abb. 2). Die Sorte M. Jubilejnaja reagierte auf alle 3 verwendeten Konzentrationen (100, 300, 500 ppm) signifikant am empfind-

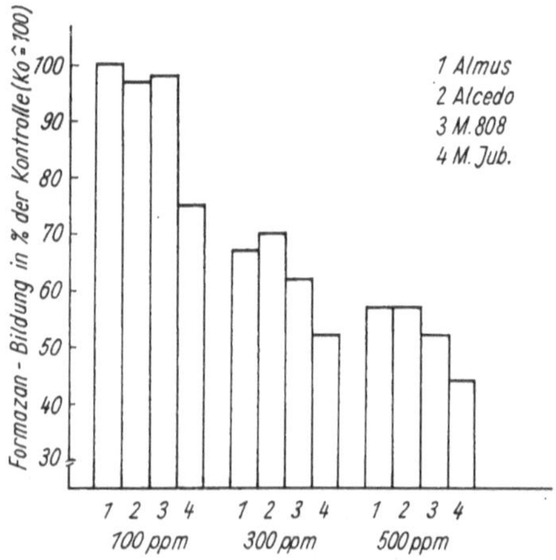

Abb. 1. Einfluß von Calixin (10, 100, 1000 ppm) auf die Bildung von Formazan in Blattsegmenten von Gerste-, Weizen- und Gurkenpflanzen

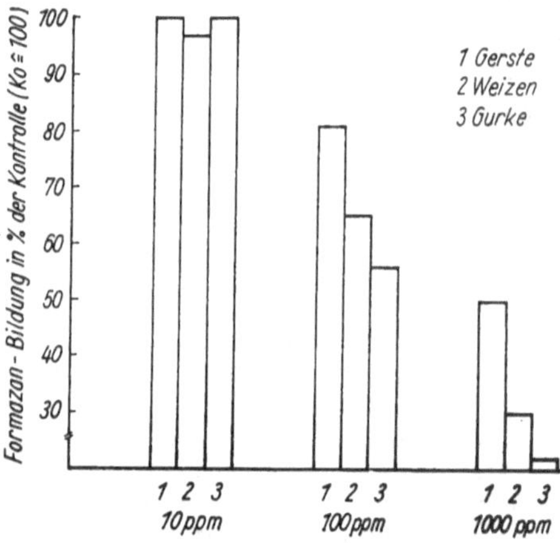

Abb. 2. Einfluß von Calixin (100, 300, 500 ppm) auf die Bildung von Formazan in Blattsegmenten von Winterweizen-Pflanzen der Sorten Almus, Alcedo, M. 808, M. Jubilejnaja

Tabelle 1
Phytotoxizität von Calixin an Winterweizen im Einblattstadium (Gewächshausversuch)

Sorte	Konzentration (ppm Wirkstoff)	Blattzerstörung (%)
Mironowskaja Jubilejnaja	40	5
	60	18
	80	31
Iljitschowka	40	2
	60	3
	80	13
Mironowskaja 808	40	4
	60	8
	80	8
Awrora	40	2
	60	7
	80	18
Alcedo	40	1
	60	4
	80	21
Fakir	40	5
	60	4
	80	7
Kawkas	40	2
	60	9
	80	14

lichsten. Nach Einwirkung von 300 bzw. 500 ppm Calixin ergaben sich auch zwischen den Sorten Alcedo, M. 808 und M. Jubilejnaja signifikante Abstufungen.

Gewächshausversuche an Ganzpflanzen eines breiten Sortenspektrums von Winterweizen wiesen die Sorte M. Jubilejnaja ebenfalls als empfindlichste der untersuchten Sorten aus (Tab. 1). Das Vorkommen sortenspezifisch unterschiedlicher Reaktion auf Calixin wurde bei Winterweizen bereits beim praktischen Einsatz von Calixin beobachtet [13, 17, 18], wobei die Ermittlung signifikanter Unterschiede unter Freilandbedingungen relativ aufwendig ist. Im Zusammenhang mit dem Einsatz verschiedener Herbizide wurde auch an anderen Kulturpflanzen-Arten sortenspezifisches Verhalten festgestellt [14, 11]. Unsere mit dem TTC-Test erhaltenen Ergebnisse lassen den Test zur Ermittlung geringer sortenspezifischer Unterschiede hinsichtlich der Reaktion auf Calixin als geeignet erscheinen.

An der Sommerweizen-Sorte „Hatri" durchgeführte Untersuchungen zeigten, daß mit zunehmendem Alter der Pflanzen die Empfindlichkeit gegenüber Calixin anstieg und zum Zeitpunkt der Ausbildung der Ähren, am Ende des Versuchszeitraumes, besonders hoch war (Abb. 3). Dieser Befund steht in Übereinstimmung mit unter Freilandbedingungen gemachten Beobachtungen, wonach Getreidepflanzen zum Zeitpunkt des Ährenschiebens besonders empfindlich gegenüber Präparaten zur Mehltau-Bekämpfung reagieren [9]. Die im Vergleich zu nicht ausgewachsenem Gewebe und jüngeren Blättern höhere toxische Wirkung anderer Pflanzenschutzmittel bei reifem

Stammgewebe von Baumwolle [16], weist ebenfalls auf Phytotoxizitätsunterschiede in Abhängigkeit vom Alter pflanzlicher Gewebe oder Organe hin.
Insgesamt wird eingeschätzt, daß der TTC-Test die schnelle, rationelle quantitative Bestimmung der phytotoxischen Wirkung von Calixin ermöglicht und relativ geringe Wirkungsunterschiede bzw. Unterschiede der pflanzlichen Reaktionsnorm erkennbar macht.

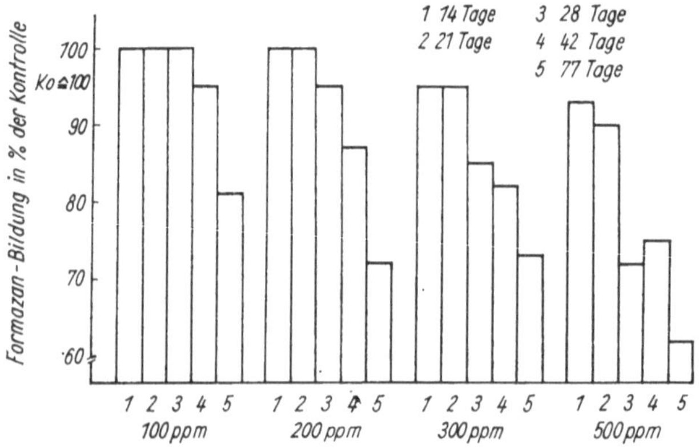

Abb. 3. Einfluß von Calixin (100, 200, 300, 500 ppm) auf die Bildung von Formazan in Blattsegmenten unterschiedlich alter Pflanzen (14, 21, 28, 42, 77 Tage) der Sommerweizen Sorte „Hatri"

Literatur

[1] Bergmann, H., Untersuchungen zur Aktivität reduzierender Enzyme in Blättern von Gerstenpflanzen verschiedenen Alters mit Hilfe des Triphenyltetrazoliumchlorid-Tests, Biol. Zbl. **93** (1974), 65–75

[2] Gur, A., Y. Gil and B. Bravdo, The efficacy of several herbicides in vineyard and their toxicity to grapevines, Weed Res. (Oxford) **19** (1979), 109–116

[3] Hampel, M., Versuchsergebnisse der Getreidemehltaubekämpfung mit ®Calixin 1969, Badische Anilin- und Soda-Fabrik AG, Landwirtschaftliche Abteilung, Ergebnisbericht 1970

[4] Kleudgen, H. K., Veränderungen der Pigment- und Prenylchinongehalte in Chloroplasten von Gerstenkeimlingen nach Applikation des Wuchsstoffherbizids MCPA (4-Chlor-2-methylphenoxyessigsäure), Z. Naturforschung **34** c (1979), 106–109

[5] Močalkina, K. I., G. S. Spiridinova, J. J. Spiridonov, Unterschiedliche Herbizidempfindlichkeit der Keimlinge verschiedener Arten, Chim. selsk. choz. (Moskau) **17** (1979), 42–44 (russ.)

[6] Murray, D. S., J. E. Street and J. K. Soteres, Growth inhibition of cotton *(Gossypium hirsutum)* and soybean *(Glycine max)* roots and shoots by three dinitroaniline herbicides, Weed Sci. (GENEVA, N. Y.) **27** (1979), 336–342

[7] Muschinek, G., G. I. Garab, L. A. Mustardy, The mechanism of linuron phytotoxicity in maize, Weed Res. (Oxford) **19** (1979), 101–107

[8] Mydlinova, E. und J. Zemanek, Sortenempfindlichkeit des Winterweizens gegen Herbizide bei Gewächshausversuchen, Ochrana Roslin (Praha) **11** (1975), 59–68 (tschechisch)

[9] Neuhaus, W., Unveröffentlichte Ergebnisse 1979

[10] Pool, R. A. F., A rapid bioassay for pesticide phytotoxicity, J. Agric. and Food Chem. (Wash.) **25** (1977), 1216–1218

[11] Richard, E. P. and J. B. Baker, Response of selected rice *(Oryza sativa)* lines to molinate, Weed, Sci. (Geneva, N. Y.) **27** (1979), 219–223

[12] Siddiqui, K. A., Resistance in some wild and cultivated grasses to the phytotoxicity of a systemic fungicide, Experientia (Basel) **28** (1972), 1239–1240

[13] Siddiqui, K. A. and V. Haahr, Different reactions of wheat mutants to a systemic fungicide, Naturwissenschaften **58** (1971), 415–416
[14] Souza Machado, V., I. L. Nonnecke and S. C. Phatak, Bioassay to screen tomato seedlings for tolerance to metribuzin, Can. J. Plant. Sci. (Ottawa) **58** (1978), 823–828
[15] Steponkus, P. L. and F. O. Lanphear, Refinement of the triphenyl tetrazolium chloride method of determining cold injury, Plant Physiology **42** (1967), 1423–1426
[16] Wills, G. D., Factors affecting toxicity and translocation of glyphosate in cotton *(Gossypium hirsutum)*, Weed. Sci. (Geneva, N. Y.) **26** (1978), 509–513
[17] Zwatz, B., Getreidemehltau: Erfahrungen 1973, Pflanzenarzt (Wien) **27** (1974), 5–6
[18] Zwatz, B., Anwendung von fungiziden Spritzmitteln im Getreideanbau In: Bericht über die Arbeitstagung 1973 der „Arbeitsgemeinschaft der Saatzuchtleiter", Gumpenstein 1973, 244–256
[19] Zwatz, B., Mehltaubekämpfung im Getreide: 1974–1975, Pflanzenarzt (Wien) **28** (1975), 8–11

G. Bischoff

Institut für Pflanzenschutzforschung der Akademie der Landwirtschaftswissenschaften, Kleinmachnow, DDR

Tridemorph-Resistenz bei *Botrytis cinerea*

Tridemorph (N-tridecyl-2,6-dimethylmorpholin) ist ein Wirkstoff aus der Gruppe der N-substituierten Tetrahydro-1,4-Oxazine auf der Basis der Morpholinstruktur, der vorrangig zur Bekämpfung von Gerstenmehltau *(Erysiphe graminis)* eingesetzt wird [3, 6].

Tridemorph wird als „multisite-inhibitor" betrachtet. Er besitzt wahrscheinlich drei Angriffspunkte im Hauptweg der Atmungskette [1, 8] sowie eine Membranaktivität [2]. Eine Hemmwirkung von Tridemorph auf die Proteinsynthese wurde bisher nur von Fisher [4] postuliert. Nach Polter und Casperson [10] beeinflußt Tridemorph die Lipid-Biosynthese bei *Botrytis cinerea*. Experimente von Kerkenaar [5] mit reinem Tridemorph ohne Formulierung weisen auf eine Wechselwirkung mit biosynthetischen Prozessen hin. Bisher wurde weder im Labor noch in der Praxis ein Fall von Tridemorph-Resistenz bekannt, obschon z. B. an *Colletotrichum lindemuthianum* intensive Versuche hierzu von Meyer [7] unternommen wurden. In einigen Untersuchungen sollte die Möglichkeit geprüft werden, ob unter Laborbedingungen Tridemorph-resistente Mutanten selektionierbar sind.

Wegen der obligat-parasitischen Lebensweise von *Erysiphe graminis* wurden die Untersuchungen an dem gegenüber Tridemorph hochempfindlichen Pilz *Botrytis cinerea* durchgeführt unter Berücksichtigung der unterschiedlichen Tridemorphempfindlichkeit von Sporen und Myzel [10].

Es wurde die übliche Selektionsmethode der Sporenaussaat auf Malzagar in Petrischalen verwendet. Zur Steigerung der Mutationsrate bestrahlten wir die Sporen mit UV-Licht. Die Wachstumsgeschwindigkeit, die fungizide Wirksamkeit von Tridemorph und die Kreuzresistenz der verschiedenen *Botrytis cinerea*-Stämme bestimmten wir im Agarplattentest bei +25 °C durch vergleichende Messung des radialen Zuwachses. Punktförmige Impfflecken erzielten wir durch Auftupfen von Sporensuspension mittels einer Kapillare und vermieden so ein Verstreuen von Sporen beim Auflegen von Impfstücken.

Es konnten zwei Tridemorph-resistente Stämme von *Botrytis cinerea* isoliert werden, die untereinander verschieden sind im Resistenzgrad, in der Wachstumsgeschwindigkeit und im Kreuzresistenzverhalten. Der R_1-Stamm hat einen Resistenzfaktor von ~ 20, seine Wachstumsgechwindigkeit in Emerskultur unter den oben genannten Kulturbedingungen ist um 60% reduziert, bezogen auf den Koloniedurchmesser nach 5 Tagen aus 11 Versuchen. Der R_2-Stamm besitzt einen Resistenzfaktor von 4, seine Wachstumsgeschwindigkeit ist gegenüber der des Wildstammes nur um 15% eingeschränkt, bezogen auf den Koloniedurchmesser nach 5 Tagen aus 9 Versuchen. Die Wachstumskurven des W, R_1 und R_2 Stammes ohne und mit Tridemorph gibt Abbildung 3 wieder.

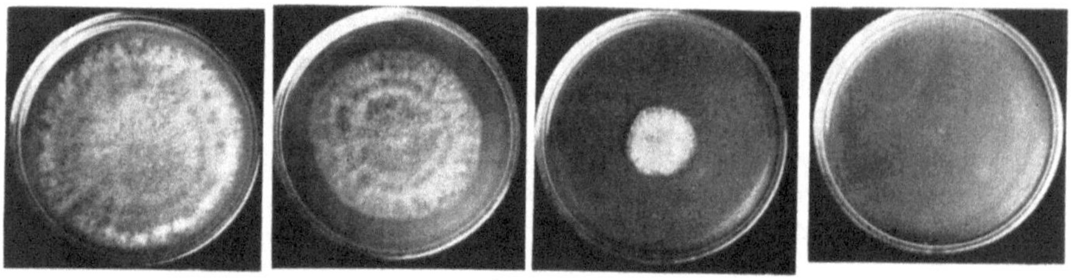

Abb. 1. Wuchsbild des Ausgangsstammes W von *Botrytis cinerea* nach 6 Tagen
1 Kontrolle
2 + 0,3 ppm Tridemorph
3 + 1,0 ppm Tridemorph
4 + 10,0 ppm Tridemorph

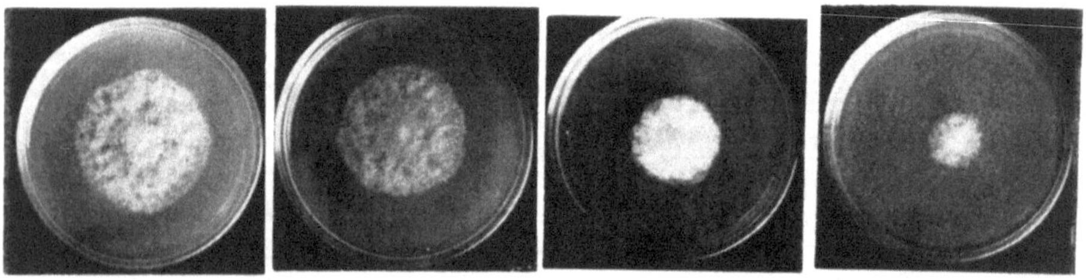

Abb. 2. Wuchsbild des Tridemorph-resistenten Stammes R_2 von *Botrytis cinerea* nach 6 Tagen
1 Kontrolle
2 + 0,3 ppm Tridemorph
3 + 1,0 ppm Tridemorph
4 + 10,0 ppm Tridemorph

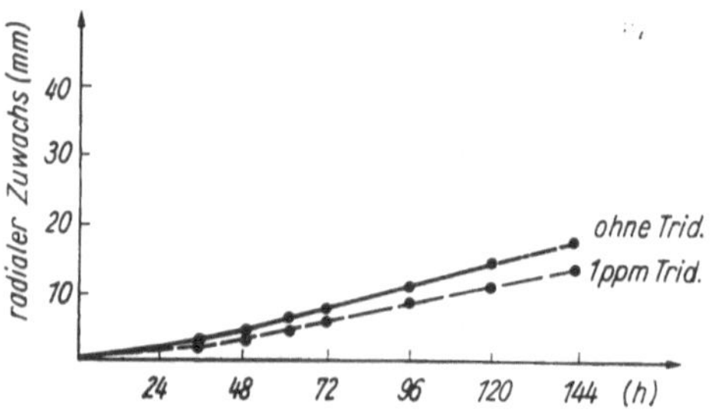

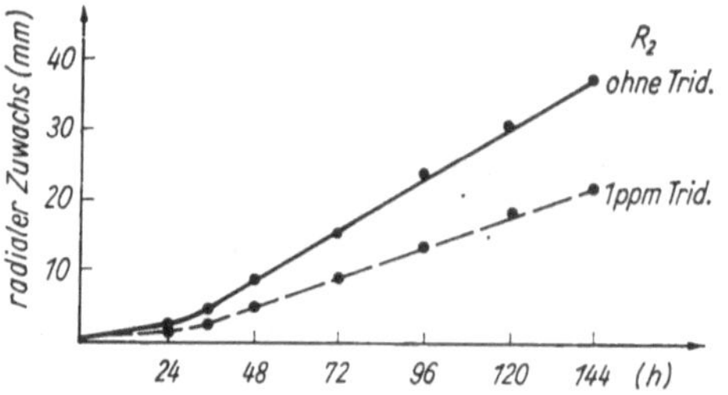

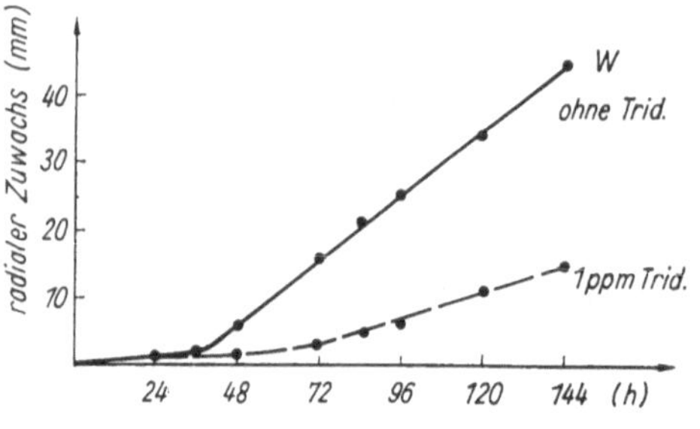

Abb. 3. Wachstumsverlauf von *Botrytis cinerea* W-, R_1- und R_2-Stamm ohne Tridemorph und in Gegenwart von 1,0 ppm und 10,0 ppm Tridemorph

Wie bei POLTER [9] dargestellt, erfolgt eine rasche Adsorption von Tridemorph in den Zellwänden und in den Schleimschichten gleichermaßen in einem Tridemorph-sensiblen und in einem Tridemorph-resistenten Pilz. Für die Unterschiede zwischen den sensiblen und resistenten Stämmen sei vor allem die Geschwindigkeit des Tridemorph-Metabolismus verantwortlich.

Die Resistenz der Stämme von *Botrytis cinerea* ist über einen Zeitraum von einem Jahr bei Kultivierung auf Malzagar mit und ohne Tridemorphzusatz stabil aufgetreten, so daß die Tridemorphresistenz als Folge genetischer Veränderungen in den Pilzzellen zu betrachten ist. Einzelsporenkulturen erwiesen sich als einheitlich.

Kreuzresistenz besteht gegenüber Dodine (Faktor 6) und den Atmungshemmstoffen Carboxin, Rotenon und Amytal, jedoch nicht gegenüber Antimycin A und Drazoxolon (Tab. 1). Dies könnte ein Hinweis darauf sein, daß die Tridemorph-Resistenz bedingt ist durch Veränderungen in der Atmungskette an Site I. Der Antimycin A-Locus

Tabelle 1
Ergebnisse zur Kreuzresistenz von Botrytis cinerea W-, R_1- *und* R_2-*Stamm (Angaben in ppm)*

Substanz	ED_{50} W	ED_{50} R_1	ED_{50} R_2	Resistenzfaktor R_1	Resistenzfaktor R_2
Tridemorph	0,45	8	1,7	18	4
Dodine	20	120	40	6	2
PCNB	1	1	1	–	–
Triarimol	3	3	3	–	–
Triforine	50	50	50	–	–
Carboxin	0,5	1,5	1,0	3	2
Drazoxolon	5	5	5	–	–
Antimycin A	3	3	3	–	–
Amytal	~100	>300	~100	>3	–
Rotenon ED_{20}	<10	100	10	>10	–

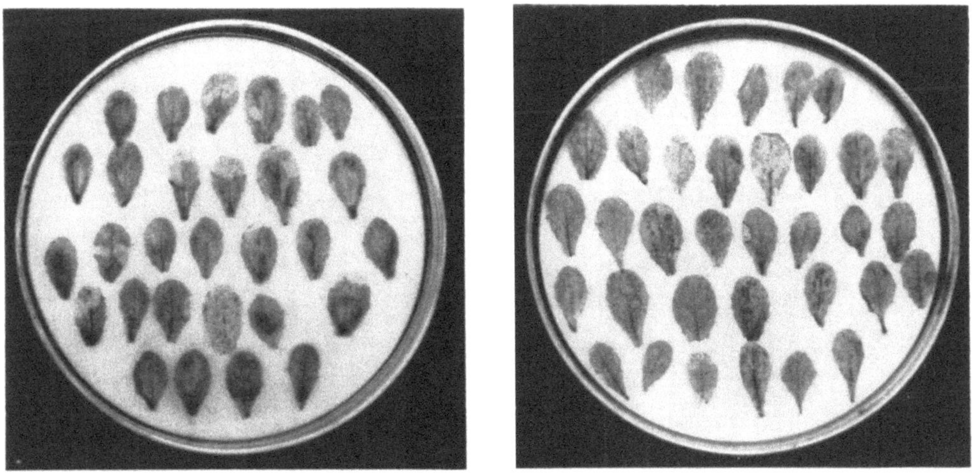

Abb. 4. Pathogenität des W und R_1-Stammes von *Botrytis cinerea* auf Salatblättern (dunkle Flecken: nekrotisches Gewebe)

scheint nicht mutiert zu sein, während Darzoxolon als Entkoppler einen andersartigen Wirkungsmechanismus haben dürfte.

Die Pathogenität der beiden Tridemorp-resistenten Stämme von *Botrytis cinerea* bleibt im Salatblattest erhalten (Abb. 4), so daß mögliches Auftreten von resistenten Populationen im Freiland [3] nicht mehr als völlig abwegig erscheint. Ob allerdings die allgemeine Fitness infolge der verringerten Wachstumsrate zu einer stabilen Durchsetzung in einer Wildpopulation ausreicht, läßt sich nicht vorhersagen.

Literatur

[1] Bergmann, H., Die Wirkung von Tridemorph auf *Torulopsis candida*, Symposiumbericht Systemfungizide, Reinhardsbrunn (1977), 215–222

[2] Buchenauer, H., Various response of cell membranes of plant and fungal cell to different systemic fungicides, Proc. Int. Plant Prot. Congr. Moscow **3** (1975), 94–109

[3] Fehrmann, H., Systemische Fungizide – ein Überblick. II. Fungizidresistenz phytopathogener Pilze, Phytopath. Z. **86** (1976), 144–185

[4] Fischer, D. J., A note on the mode of action of the systemic fungicide tridemorph, Pestic. Sci. **5** (1974), 219

[5] Kerkenaar, A. and A. Kaars Sijpesteijn, On a difference in the antifungal activity of tridemorph and its formulated product calixin, Pestic. Biochem. Physiol. **12** (1979), 124–129

[6] Marsh, R. W. (ed.), Systemic fungicides Longman 1977, London

[7] Meyer, E., Resistenzbildung gegen systemische Fungizide (Benzimidazolderivate) bei *Colletotrichum lindemuthianum* (Sacc. et Magn.), Bri et Cav. Mitt. Biol. Bundesanstalt **166** (1976), 135

[8] Müller, W. und T. Schewe, Das Systemfungizid Tridemorph als Hemmstoff der Atmungskette von Elektronentransportpartikeln aus Rinderherzmitochondrien, Acta Biol. Med. Ger. **35** (1976), 693

[9] Polter, C., Aufnahme und Metabolismus von trans-2,6-Dimethyl-N-dodecylmorpholin bei unterschiedlich empfindlichen Pilzen, Symposiumsbericht Systemfungizide, Reinhardsbrunn 1980

[10] Polter, C. und G. Casperson, Einfluß von Tridemorph auf Ultrastruktur und Lipidmetabolismus von *Botrytis cinerea*, Symposiumsbericht Systemfungizide, Reinhardsbrunn 1977, 225–240

L. Ferenczy, E. Hegedüs*, Zs. Jászay*, M. Szegedi, F. Kevei, T. Pfliegel*, A. Franko*

Department of Microbiology, József Attila University, Szeged, Hungary
*CHINOIN Pharmaceutical and Chemical Works Ltd., Budapest, Hungary**

1-(Xanthene-9-carbonyl)-4-phenyl-semicarbazide (MKF 199), a New Anti-*Phytophthora* Compound

A superselective anti-*Phytophthora* compound (code name: MKF 199) was discovered among new substances synthetized in the CHINOIN Pharmaceutical and Chemical Works Ltd., Budapest, and analyzed microbiologically at the Department of Microbiology, Attila Jozsef University of Szeged.

It is well known that a number of active pesticides are thiocarbamates, benzimidazole carbamates, mehtyl carbamates, etc. The primary aim of our work was to develop original, patent-independent molecules. First of all xanthoxim-carbamates, xanthone-hydrazone and xanthin-urea, later xanthene-9-carbonicacid derivates were synthesized. The inclusion of a xanthene moiety into the molecule was considered advantageous in the literature (Adams et al. [1], Buchenauer et al. [2, 3, 4] Farbenfabriken Bayer [5], Tamura [6]).

Synthesis of Molecules

There are two different ways of preparing xanthene-9-carbonyl semicarbazide derivates:

a) By the reaction of the proper acid hydrazide (eg. xanthene-9-carbonic acid hydrazide) and different carbonic acid derivates (e.g. isocyanate, N-phenylcarbamoyl chloride, N-phenylurethanes, etc.) (Fig. 1)

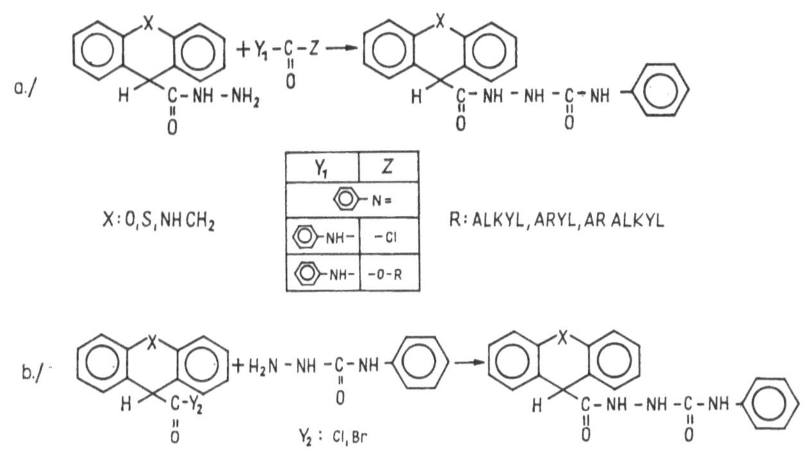

Fig. 1

b) By the reaction of proper acid halides (e. g. xanthene-9-carbonylchloride) and 4-phenylsemicarbazide.

Antimicrobial Activity

In vitro antimicrobial activities were determined by using serial dilution method. Media contained 1% glucose and 0.5% yeast extract or green peas extract (2% dry weight). These media were solidified with 2% agar. The test compounds were dissolved in dimethylsulfoxide and were added to the warm nutrient media. The concentration of dimethylsulfoxide did not exceed 1% by volume of agar. Agar plates containing different concentrations of test compounds were inoculated with the suspension or the mycelia of the test organisms. The minimum inhibitory concentration (µg/ml) was determined. The following list contains the test microorganisms with the code numbers of Szeged Microbiological Collection (Tab. 1, 2).

Table 1

Actinomucor repens 0487
Alternaria tenuis 0048
Ascochyta pisi 0039
Aspergillus niger 0050
Botrytis cinerea 0036
Candida albicans 0056
Candida clausseni 0816
Candida crusei 0779
Candida natalensis 0872
Candida parapsilosis 0814
Candida pseudotropicalis 0786
Candida pulcherima 0798
Candida guillermondi 0799
Candida solani 0783
Candida tropicalis 0787
Candida utilis 0788
Cercospora beticola 0038
Colletotrichum lini 0164
Debaromyces globosum 0722
Debaromyces tyrocola 0721
Dekkaromyces fragilis 0734
Dekkaromyces lodderi 0735
Endomycopsis febuliger 0866
Endomycopsis wickernamii 0723
Epidermophyton floccosum 0051
Fusarium moniliforme 0040
Geotrichum candidum 0055
Hansenula sp. 0915
Hansenula anomala 0731
Hansenula beijerinckii 0727
Hansenula subpelliculata 0728
Kluyveromyces polysporus 0733

Table 2

Nematospora coryli 0750
Nematospora gossypii 0751
Nigrococcus nigricans 0826
Phytophthora drechsleri 0394
Phytophthora infestans 0001
Pichia saitoi 0726
Plasmopara viticola
Procandida albicans 0774
Procandida tropicalis 0771
Pythium ultimum 0395
Saccharomyces bayanu 0717
Saccharomyces cerevisiae 0425
Saccharomyces chevalieri 0704
Saccharomyces chodati 0716
Saccharomyces diastaticus 0643
Saccharomyces heterogenicus 0709
Saccharomyces logos 0642
Saccharomyces mellis-acidi 0705
Saccharomyces oviformis massandra 0713
Saccharomyces oviformis xeres 0711
Saccharomyces paradoxus 0715
Saccharomyces pretoriensis 0714
Saccharomyces transwalensis 0708
Saccharomyces ludwigii 0720
Stemphylium radicinum 0037
Torulopsis cantarelli 0817
Torulopsis stellata 0818
Torulopsis vanzylii 0820
Torulopsis westerdykii 0821
Trichophyton mentagrophytes 0004
Trichophyton rubrum 0157

Antimicrobial activity of 58 compounds was tested (Tab. 3—7). The compound 1-(xanthene-9-carbonyl)-4-phenylsemicarbazide (MKF 199) proved to be the most active. The compound prevented growth of *Phytophthora infestans* de BARY at 0,3 to 0,7 µg/ml but proved to be inactive not only against taxonomically distant micro-

Table 3a

$$R^1-\underset{O}{\underset{\|}{C}}-NH-\underset{R^2}{\underset{|}{N}}-\underset{X}{\underset{\|}{C}}-NH-R^3$$

NO	R^1	R^2	R^3	X	EFFECT µg/ml
1	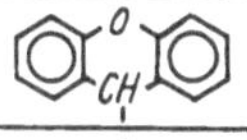	H	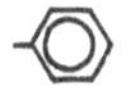	0	0,75
2	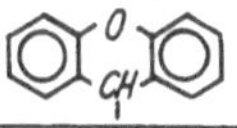	H	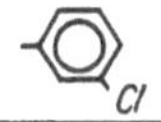	0	no
3	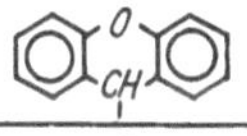	H		0	no
4	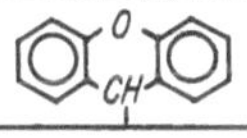	H	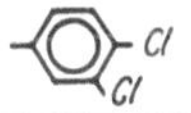	0	no
5	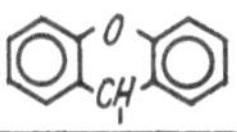	H	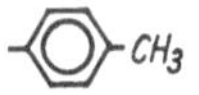	0	no
6	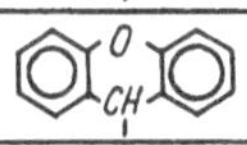	H	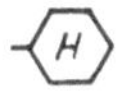	0	no
7	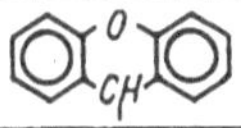	H	$-CH_3$	0	no
8	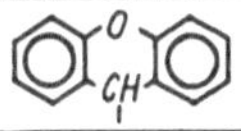	H	$-(CH_2)_3CH_3$	0	no
9	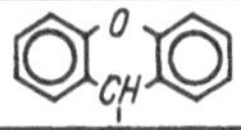	H	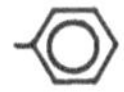	S	no
10	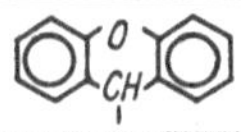	H	$-CH(CH_3)_2$	S	no
11	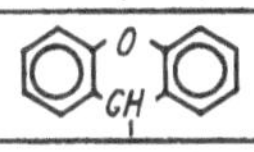	H	$-C(CH_3)_3$	S	no
12	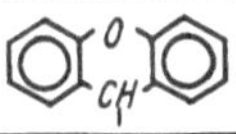	$-(CH_2)_2-O-\underset{O}{\underset{\|}{C}}-NH-PH$	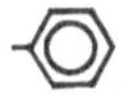	0	no
13	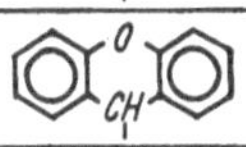	$-(CH_2)_2-O-CH_3$	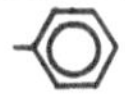	0	no
14	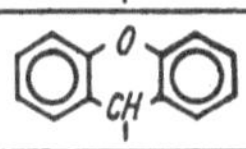	$-CH_3$		0	no
15	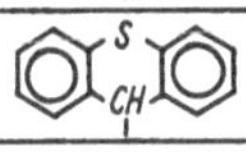	H	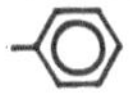	0	6,25p 12,5t
16	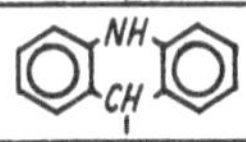	H	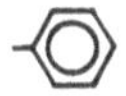	0	25

organisms (Gram-positive and Gram-negative bacteria, yeasts and a rather wide spectrum of filamentous fungi) but even against such taxonomically closely related microbes as the *Pythium* species.

Table 3b

NO	R^1	R^2	R^3	X	EFFECT µg/ml
17	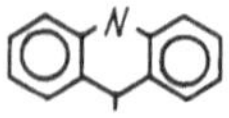	H	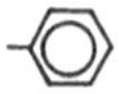	0	50p
18	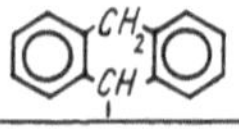	H	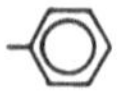	0	25
19	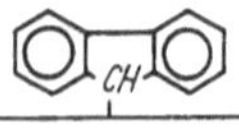	H	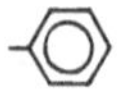	0	0,7p 1,5t
20	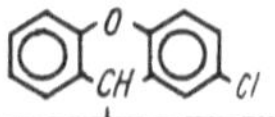	H	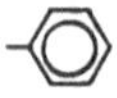	0	no
21	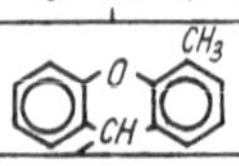	H	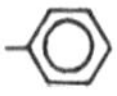	0	no
22	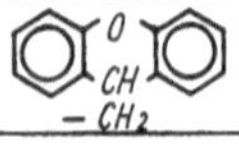	H	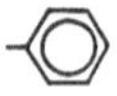	0	no
23	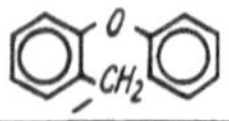	H	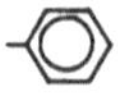	0	no
24	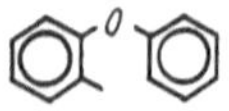	H	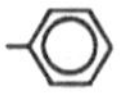	0	no
25	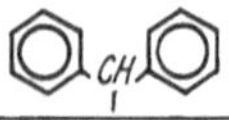	H	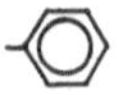	0	no
26	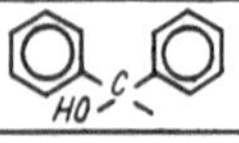	H	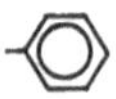	0	no
27	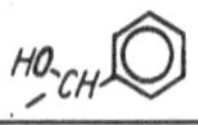	H	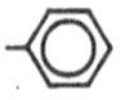	0	no
28	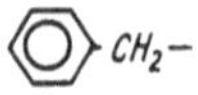	H	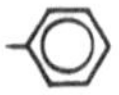	0	no
29	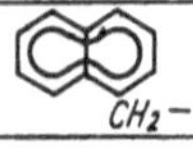	H	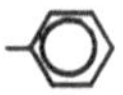	0	no
30	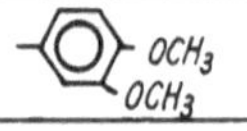	H	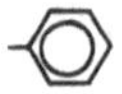	0	no
31	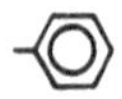	H	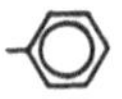	0	no
32		H	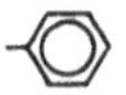	0	no
33	$-CH_3$	H	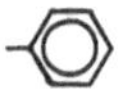	0	no
34	$-OC_2H_5$	H	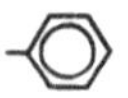	0	no

We find this result (superselectivity, high activity) most interesting, because *Phytophthora infestans* is a widely spread and highly pathogenic disease of tomatoes and potatoes which cannot be controlled with certainty.

Table 4

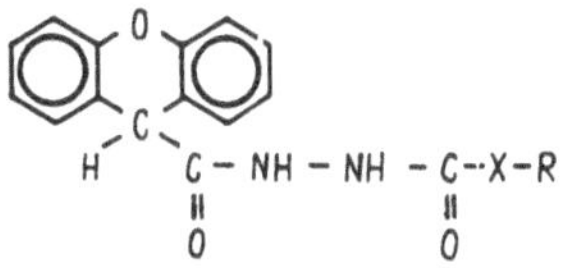

NO	R	X	EFFECT µg/ml
1	$-C_2H_5$	0	no
2		0	no
3		S	no
4	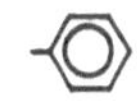	$-CH_2-$	no
5	$-N(C_2H_5)_2$	$-CH_2-$	no

Table 5

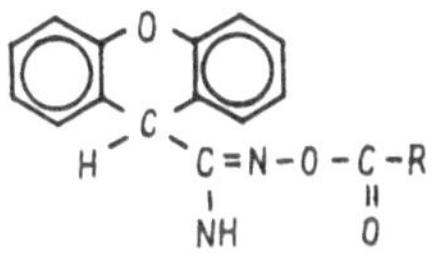

NO	R	EFFECT µg/ml
1		no
2	$-CH_3$	no
3	$-CH_2-Cl$	no
4	$-O-C_2H_5$	no
5	$-NH-$	no
6	$-NH-CH_3$	no

In vivo investigation of MKF 199 is very difficult because of its physical properties, first of all of its very low solubility in water. It is practically insoluble in many of the current organic solvents (methanol, ethanol, chloroform, benzene, acetone etc.) but readily soluble in dimethylsulfoxide (6% at 25 °C) and dimethylformamide (10% at 25 °C).

Another difficulty is its strong idioeletric property which makes formulation extremly difficult. Probably these serious problems are responsible for the unsatisfactory results in preventive protection of the tomato against *Phytophthora infestans* in our field experiments. A number of derivatives was synthesised to improve these unfavourable characteristics. Unfortunately the derivatives with better physical features were moderately active.

Relation between Activity and Structure

According to our experiments a 3-ring structure with two aromatic rings on both sides (xanthene, thioxanthene, fluorene, dihydroacridine, dihydroanthracene) is necessary for activity. Activity is moderate when the whole 3-ring structure is aromatic. Activity completely disappears with any substitution of the 3-ring structure, e.g. if the 3-ring structure is changed to a similar ring-structure or any kind of single ring or aliphatic groups (Tab. 3).

Only molecules in which the 4-N of the semicarbazide moiety is attached to a phenyl group without any substitution are active. Similarly the 2-N substituted derivates are inactive, too (Tab. 3). Carbazates, thiocarbazates and diacylhydrazides synthesized by isoster substitution of the 4-N group are inactive although they have a very similar steric structure to MKF 199.

Table 6

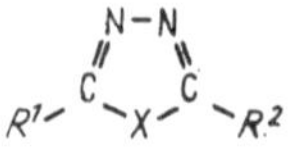

NO	R^1	R^2	X	EFFECT µg/ml
1	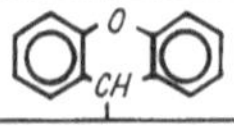	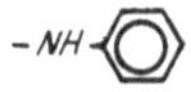	O	12,5
2	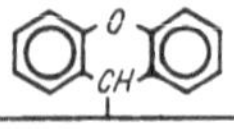	$-NH-(CH_2)_3-CH_3$	O	no
3	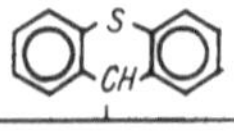	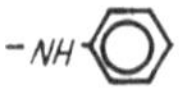	O	no
4	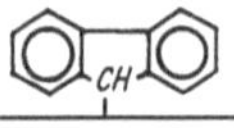	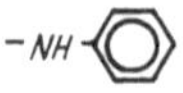	O	25
5	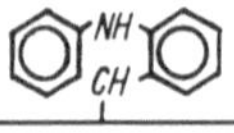	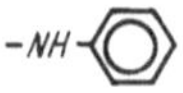	O	no
6	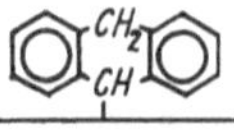	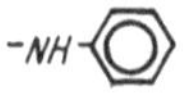	O	no
7	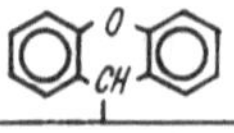	$-OH$	O	no
8	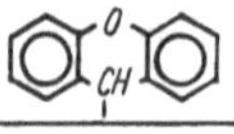	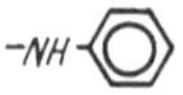	S	no
9	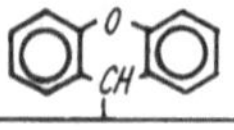	$-SH$	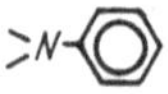	no

Table 7

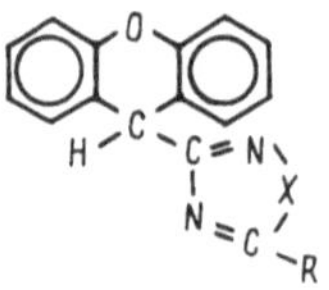

NO	R	X	EFFEKT µg/ml
1	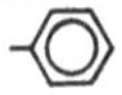	O	no
2	$-CH_2-Cl$	O	3
3	$-OH$	O	pH7 no pH5 5,5
4	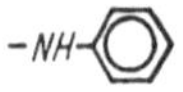	S	no

References

[1] Adams, S., S., Armitage, B. J., Bristow, N. W., Heatcote, B. V., Xanthenyl semicarbizides, G. B. Patent 39, 284 (1966)
[2] Buchenauer, H., Grossmann, F., Therapeutic effect of fluorene, phenanthrene and xanthene derivates of fungal diseases of tomato, Nature (London) **227** (1970), 1267 to 1268
[3] Buchenauer, H., Grossmann, F., Morphoregulatory effect of xanthene derivates, Planta **93/1/** (1970), 86–88
[4] Buchenauer, H., Effect of phenanthrene and xanthene derivates on *Fusarium* and *Verticillium*-wilt of tomato, Phytopathol. Z. **72/4/** (1971), 291–304
[5] Farbenfabriken Bayer A.G., New N-substituated imidazole preparations and application as antimycotic agents, D.P.B. P.2053080.7, (1970)
[6] Tamura, S., Okuda, S., Sanai, S., Kimura, Y., Oksuka, Y., Nakata, T., Fluorenoncarboxylic acids, plant growth regulators, Jpn. Kokai Tokyo Koho 78.91.129 (1978)

E. KARTASHOVA and S. GORJACHEV

Moscow State University, Departments of Plant Physiology and Biophysics, Moscow, U.S.S.R.

Effect of Zinc upon Viability and Virulence of the Fungus *Fusarium Oxysporum* f. sp. *Vasinfectum*

The use of microelements is one of the possible ways to exert an influence on pathogenic organisms. Microelements, in particular zinc, are capable to determine direction of metabolism of biological systems. In this study the research of responses of the pathogenic organism to high doses of zinc is presented.

The fungus *Fusarium oxysporum* f. sp. *vasinfectum*, especially its pathogenic forms, are characterized by high resistance to drying. The fungus survives high temperatures up to 80 °C and in a dryed state does not lose its viability for several years (KLIMOVA, KOSHKELOVA [4]). On reviewing the literature it is known that maintenance of viability of eucaryotic organisms is impossible when they completely lose their water. In experiments with some objects (yeast, moss, lichen, nematode) correlation between the cell resistance to drying and in a dryed state maintenance of some isolated compartments where free water is "trapped" water can again interact with environment only when the lipid-protein structures are damaged or when the permeability is changed by moistening from outside (AKSJONOV et al. [1, 2]).

We were interested in the ability of the fungus *Fusarium oxysporum* f. sp. *vasinfectum* of various virulence to retain free water in a dryed state and in the effect of zinc upon this ability.

In the dryed mycelium (freeze and oven drying at 55 °C was used) the presence of water was estimated with the help of the method of nuclear magnetic resonance spin echo (NMR). The use of the spin echo method to study water conditions in biological objects is based on a marked difference between signals of water protons and signals of immobile or slightly mobile molecules such as proteins and other high polymerous compounds. In the study the virulent strain No 155904 (No 15) and the avirulent strain No 619123 (No 61) were used. The fungus was grown on CZAPEK's medium. Zinc was added in the following quantities: 0,2 mg/l — the normal dose and 30 mg/l — the excessive dose.

The analysis of NMR spin echo signals of the dryed samples of the avirulent and virulent strains of the fungus (normal zinc) have shown differences in decrease of the signal curves. On the Figure 1 the curves of spin echo decay are presented. In case of the virulent strain the decrease is slower, the amplitude of the signal is higher, the time of spin-spin relaxation is longer comparing to the same parameters of the avirulent strain. The experiments with deuterium were performed to clarify the nature of the slowest component of the echo decay.

The frequency of NMR for deuterium is in several times lower, than that of the common proton and the resonance and the resonance from deuterium does not add anything to the proton one. The dryed samples of the fungus were treated with 10-fold volume of heavy water D_2O during 50 hours at 20 °C. During incubation the heavy

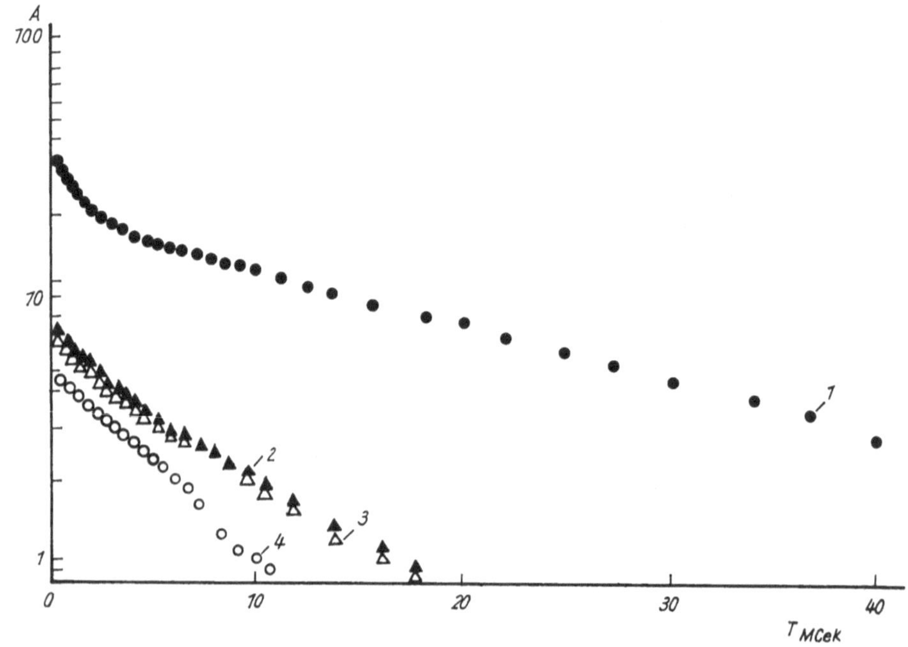

Fig. 1. Curves of spin echo decay, obtained on dryed samples of mycelium tissue
1 virulent strain
2 avirulent strain
3 avirulent strain + D_2O
4 virulent strain + D_2O

water does not accomplish isotopic exchange with any giving a signal cell component except water. After the deuterium-hydrogen exchange the samples were dryed again and then measured (Fig. 1).

During the measurements of the avirulent strain samples we did not find difference before deuterium treatment and after it. In the virulent strain the slow component of the proton resonance signal disappeared (Fig. 1). Thus, in the virulent strain the slow component belongs to protons of the heavy water, retaining in the cells after drying. In the avirulent strain this component obviously belongs to lipid protons.

The resistance to drying and the viability under high temperature conditions is probably due to the retention of the mobile water by the pathogenic forms.

The addition of excessive doses of zinc to the medium of the virulent strain led to the slight decrease of water content in the mycelium after drying (Fig. 2). These results coincide with the data on determination of viability of the fungus virulent strain after drying at 55 °C for 77 hours. Regeneration of the fungus growth was accomplished by the following way: the fungus mycelium was suspended in a liquid medium and following cultured on the medium with 2% agar. The growth of the virulent strain fungus, grown on various doses of zinc before oven drying, differed unadequately statistically. However, the content of the fungus toxin, that is fusaric acid, in the virulent strain, grown on the medium with the excessive dose of zinc, was at an average by 40% lower than the same strain, grown on the medium with the normal dose of zinc. In the experiments when the virulent strain of the fungus was grown in the presence of the exces-

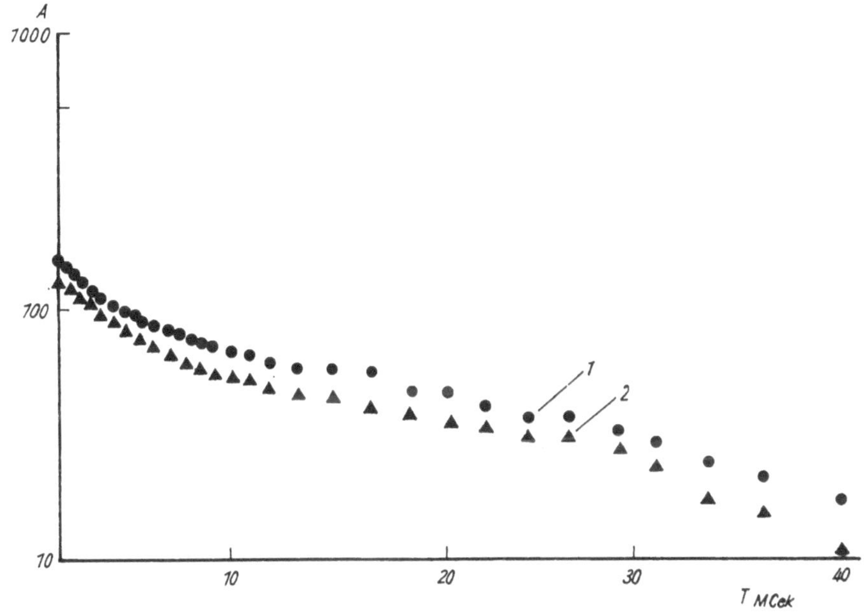

Fig. 2. Curves of spin echo decay, obtained on dryed samples of the virulent strain mycelium tissue
1 virulent strain, normal zinc
2 virulent strain, excess zinc

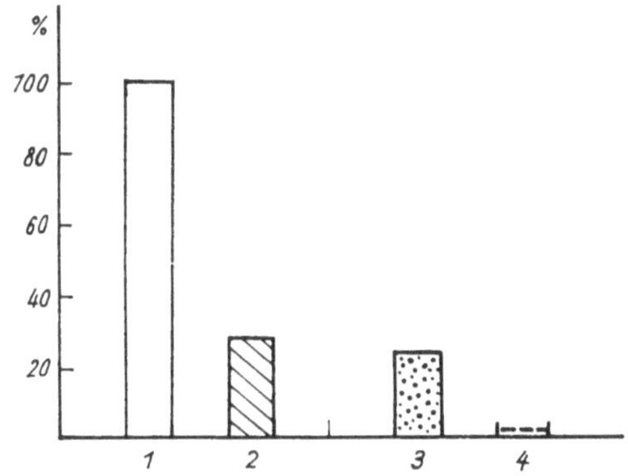

Fig. 3. Relative content of fusaric acid in mycelium tissue (in per cent to the virulent strain, grown on the medium with the normal dose of zinc)
1 virulent strain, normal zinc
2 virulent strain, excess zinc
3 avirulent strain, normal zinc
4 avirulent strain, excess zinc

sive dose of zinc, the content of fusaric acid in its mycelium tissue approached to that of the avirulent strain (Fig. 3). The content of fusaric acid in the avirulent strain, grown on the medium with the excessive dose of zinc, was not estimated. In all experiments fusaric acid was determined chromatographically with following spectrophotometry at 268 nm.

The rapid decrease of fusaric acid content as a result of competition for zinc between components-activators of dehydrogenase, as they need nicotinamidadeninenucleotides, and the components responsible for the synthesis of tryptophane precusor of fusaric acid (Bekker [2], Fig. 4). This hypothesis is supported by our experiments on estimation of the total dehydrogenase activity and isoenzymic spectra of hydrogenase. In

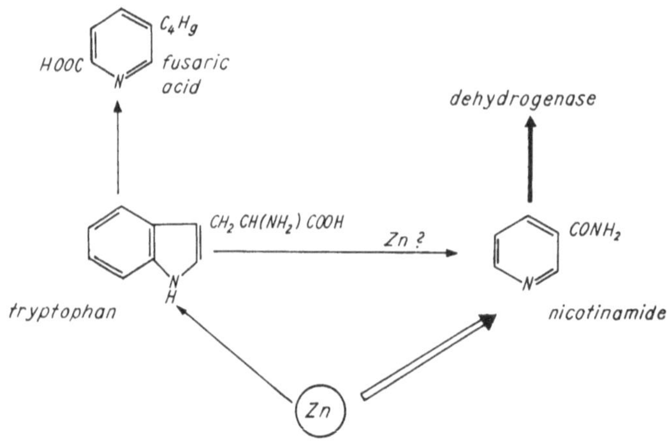

Fig. 4. Scheme of zinc participation in activation of dehydrogenase and in fusaric acid synthesis (according to BEKKER Z. E., 1973)

the virulent strain, grown on the dose of zinc favourable to fusaric acid synthesis — 0,2 mg/l, the total dehydrogenase activity was more than 3 times lower than in the virulent strain grown on the excessive dose of zinc. For example, in case of malate dehydrogenase (MDH) we showed that the fungus mycelium growth on the medium with the excessive dose of zinc for 4 hours led to marked changes in isoenzymic spectrum. Effect of the excessive dose of zinc upon isoenzymic spectrum of MDH increased with the prolongation of its action. During culturing of the virulent strain on the medium with the excessive dose of zinc the decrease of heterogeneity of the spectrum is observed which is followed by the increase in activity of individual protein fractions

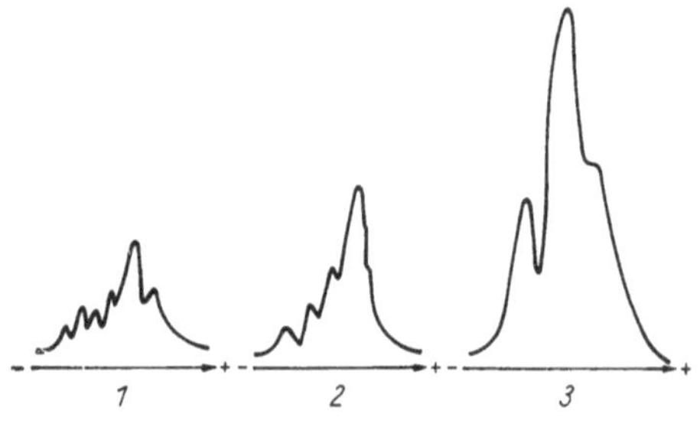

Fig. 5. Enzymograms of malate dehydrogenase of the virulent strain mycelium tissue
1 normal zinc, 7 days
2 normal zinc, 7 days + excess zinc, 4 hours
3 excess zinc, 7 days

(Fig. 5). Thus, the excessive dose of zinc, probably, changes the metabolic pathway, associated with fusaric synthesis, into activation of dehydrogenase system (Fig. 4).
So, the main effect of the addition of the excessive doses of zinc to the medium of the fungus growth is not the loss of viability by pathogenic forms but the loss of their virulence. This conclusion is in agreement with our earlier studies which showed that biosynthetic activity of the fungus grown, on the medium with excessive doses of zinc, is easily suppressed by the host-plant (KARTASHOVA [3]). Special vegetative experiments showed that the excessive dose of zinc (30 mg/kg of soil) had favourable effect on growth and development of cotton. We suggest that excessive doses of zinc should be used to decrease virulence of the fungus *Fusarium oxysporum* f. sp. *vasinfectum* without producing unfavourable effect upon cotton.

References

[1] AKSJONOV, S. I., GORJACHEV, S. N., FATEEVA, M. V., NIKITINA, T. N., The study of yeast various strains of *Saccharomycodes* with different resistance to freeze drying by the spin echo, NMR method. Izvestia AN SSSR, „Biologia", 5 (1973)

[2] AKSJONOV, S. I., GORJACHEV, S. N., SHERMAN, F. B., BABJEVA, I. P., RESHETOV, I. S., On isolated water in dryed microorganisms, Mikrobiologia, 47, 3 (1978), 549–553

[3] BEKKER, Z. E., Theoretical basis of pathogenesis, ecological background and study of a complex of measures against cotton fusaric wilt. In: "Eco-physiological methods of measures against cotton fusaric wilt", Ashchabad, 1973, 233–254

[4] KARTASHOVA, E. R., Some Pathogen – Host Plant Interaction Mechanisms. Systemfungizide, V. Internationales Symposium veranstaltet durch die Sectionen Mikrobiologie und Phytopathologie der Biologischen der DDR vom 9.–13. Mai 1977 in Schloß Reinhardsbrunn, Akademie-Verlag Berlin, 1979

[5] KLIMOVA, A. P., KOSHKELOVA, E. N., The effect of a period of culture storage of *Fusarium oxysporum* f. *vasinfectum* upon their pathogenic properties. In: "Eco-physiological methods of measures against the cotton fusaric wilt", Ashchabad, 1973, 33–37

E. KLUGE und H. LYR

Institut für Pflanzenschutzforschung der Akademie der Landwirtschaftswissenschaften, Kleinmachnow, DDR

Vergleichende Untersuchungen über die Wirksamkeit von Systemfungiziden gegen *Erysiphe graminis* DC

In den letzten Jahren wurden hochwirksame Systemfungizide gegenüber Pilzen aus der Ordnung der Erysiphales entwickelt. In Gewächshaus- und Freilandversuchen wurde die praktische Anwendbarkeit der Präparate bestätigt. Es fehlen jedoch exakt vergleichbare Angaben über die Wirksamkeit und Art der Wirkung der einzelnen Fungizide gegenüber den Mehltaupilzen. Wir haben deshalb in Labortests unter definierten Bedingungen 12 bekannte Systemfungizide hinsichtlich ihrer protektiven, kurativen und systemischen Wirksamkeit gegenüber *Erysiphe graminis* an Gersten-Keimpflanzen untereinander verglichen.

Methoden

Fungizide

In der Regel wurden für die Untersuchungen reine Wirkstoffe verwendet. Sie wurden in Dimethylformamid (DMF) aufgenommen und mit destilliertem Wasser zu Konzentrationsreihen verdünnt. Untersuchte Wirkstoffe:

Chloraniformethan
(1-(3,4-Dichloroanilino)-1-formylamino-2,2,2-trichloroethane
Triforine
(1,4-bis-(2,2,2-Trichlor-1-formamidoäthyl)-piperazin)
Triadimefon
(1-(4-Chlorphenoxy)-3,3-dimethyl-1-(1,2,4-triazol-1-yl)-2-butanon)
Fenarimol
(α-(2-Chlorphenyl)-α-(4-chlorphenyl)-5-pyrimidinmethanol)
Ethirimol
(2-Äthylamino-4-hydroxy-6-methyl-5-butyl-pyrimidin)
Tridemorph
(N-Tridecyl-2,6-dimethyl-morpholin)
Trimorphamid
(N(1-Formamido-2,2,2-trichlorethyl)-morpholin)
Benomyl
(1-(N-Butylcarbamoyl)-2-(methoxy-carboxamido)-benzimidazol)
Thiabendazol
(2-(4-Thiazolyl)-benzimidazol)
Thiophanat M
(1,2-bis-(3-Methoxycarbonyl-2-thioureido)-benzol)

Pyrazophos
2-(0,0-Diäthylthionophosphoryl)-5-methyl-6-carbäthoxypyrazolo(1,5-2)-yprimidin
Ditalimfos
(0,0-Diäthylphthalimido-phosphonothiat)

Protektivtest

Für diese vergleichenden Versuche entwickelten wir einen speziellen Test, der gut reproduzierbare Werte lieferte. Zur Verwendung gelangten 5 cm lange Stücke von Primärblättern im Gewächshaus angezogener junger Pflanzen der Sommergerstensorte „Astacus". Fungizidbehandlung und Infektion erfolgten gleichzeitig, indem die jeweilige Wirkstoffkonzentration, die Mehltausporensuspension und 10 Blattstücke je Variante in Kolben zusammengefügt wurden. Um ein vorzeitiges Absetzen der Wirkstoffe und Mehltausporen zu verhindern, wurden die Kolben 2 Stunden langsam geschüttelt. Die gut benetzten Blätter wurden danach in feuchten Kammern übertragen und bei 18 °C und Beleuchtung inkubiert. Nach 5 bis 7 Tagen erfolgte die Auswertung durch Auszählung der angegangenen Infektionen (Pusteln, grüne Inseln). Mit den Ergebnissen aus den Konzentrationsreihen wurden sowohl hier als auch in den anderen Tests die ED_{50}-Werte grafisch ermittelt. Durch diese Methodik kommt ein sehr guter Kontakt zwischen Mehltausporen und Fungizid zustande. Die ED_{50}-Werte liegen daher relativ niedrig und lassen sich als absolute Zahlen nicht mit solchen vergleichen, die unter Gewächshaus- oder Freilandbedingungen erzielt wurden. Erfahrungsgemäß entsprechen sie am ehesten den ED_{50}-Werten, die mit anderen Pilzen im Agarplattentests erzielt werden.

Kurativtest

Der Test unterscheidet sich von der vorhergenannten Methode lediglich darin, daß die Fungizidbehandlungen 48 Stunden nach der Infektion durchgeführt wurden, wobei die Blattstücke in den feuchten Kammern mit den Wirkstoffsuspensionen überschichtet wurden. Überstehende Flüssigkeit wurde nach 2 Stunden abgegossen.

Systemtest

In Anlehnung an die Sandröhrchenmethode nach MÜLLER [3] wurden Pflanzen der Sommergerstensorte „Astacus" in Röhrchen mit Quarzsand angezogen. Nach der Bildung des 2. Blattes erfolgte Fungizidzugabe bis zur Sättigung des Sandes. Nach einer Aufnahmezeit von 48 Stunden bei 25 °C wurden die Pflanzen durch Besprühen mit Mehltausporensuspension infiziert. Die Auswertung erfolgte 6 Tage später durch Auszählung der Pusteln je Blattlängeneinheit. Die hierbei ermittelten ED_{50}-Werte betreffen nicht die Wirkstoffkonzentration in der Pflanze, sondern die Konzentration der Suspension, aus der die Pflanzen die Substanzen entnehmen. Der systemische Effekt, der sich in den ED_{50}-Werten ausdrückt, ist daher die Resultante aus absoluter Fungizidwirksamkeit, Grad der Aufnahme und Transport in der Pflanze sowie Metabolismus der Substanzen.

Ergebnisse

Bei allen Wirkstoffen ließ sich eine gute protektive Wirksamkeit gegen *Erysiphe graminis* nachweisen (Tab. 1). Lediglich Ditalimfos hatte einen etwas ungünstigeren ED_{50}-Wert. An der Spitze der Rangfolge stehen die Präparate Triadimenol, Ethirimol und Fenarimol. Die besonders gute Wirkung von Triadimefon in allen Tests kann durch die Dampfphase bedingt sein, die bei den angewendeten Methoden nicht ausgeschlossen werden konnte.
Der kurative Effekt durch Blattbehandlung 48 Stunden nach der Infektion ist bei allen Fungiziden wesentlich geringer als der protektive. Unter den Versuchsbedingungen

Tabelle 1
Vergleich der protektiven, kurativen und systemischen Wirksamkeit von 12 Fungiziden gegen Gerstenmehltau (Erysiphe graminis)

Fungizid	ED_{50}-Werte in µg/ml Protektiv	Kurativ	Systemisch
Chloraniformethan	0,05	unwirksam	100*
Triforin	0,08	1	7
Triadimefon	0,0006	0,4	0,005
Fenarimol	0,003	5	0,4
Ethirimol	0,005	unwirksam	0,02
Tridemorph	0,02		3
Trimorphamid	0,5	unwirksam	0,4
Benomyl	0,03	5	0,5
Thiabendazol	0,5	2	10
Thiophanat	0,02		5
Pyrazophos	0,2	unwirksam	100*
Ditalimfos	5	10	100

* Konzentration phytotoxisch

erwiesen sich Chloraniformethan, Ethirimol, Trimorphamid und Pyrazophos als kurativ unwirksam.

Im Vergleich zur kurativen war die systemische Wirksamkeit bei den meisten Präparaten bereits in niedrigeren Konzentrationsbereichen nachweisbar, was auf die bessere Aufnehmbarkeit der Substanzen durch die Wurzel als durch die Blattoberfläche hindeutet. Bei allen geprüften Fungiziden ließ sich eine systemische Aktivität feststellen, die jedoch bei Chloraniformethan, Pyrazophos und Ditalimfos nur sehr gering war. Da die ED_{50}-Werte bei Chloaniformethan und Pyrazophos außerdem im phytotoxischen Bereich lagen, erscheinen besonders diese Fungizide als wenig geeignet für eine systemische Anwendung. Die günstigsten ED_{50}-Werte im Systemtest wiesen Triadimefon, Ethirimol, Trimorphamid und Benomyl auf.

Diskussion

Bei der Einwirkung von Fungiziden auf den Infektionsvorgang bei *Erysiphe graminis*, der durch die Stadien Sporenkeimung — Appressorienbildung — Infektionshyphenbildung — Haustorienbildung gekennzeichnet ist, wird ein kurativer Effekt frühestens im Stadium der Haustorienbildung erforderlich. Dieses Stadium liegt bei optimalen Temperaturbedingungen etwa 13 bis 18 Stunden nach dem Sporenanflug. Zu diesem Zeitpunkt beginnt sich die Infektion zu stabilisieren; das Fungizid muß über ein gewisses Eindringvermögen in die Epidermiszellen verfügen, um die sich innerhalb der Wirtszellen befindlichen Pilzstrukturen zu erreichen. In Übereinstimmung hiermit zeigen die vorliegenden Ergebnisse, daß alle Fungizide, die als spezifische Haustorienhemmer beschrieben wurden (Schlüter u. Weltzien [4], Buchenauer [1]), auch kurativ wirksam sind (Triforin, Triadimefon, Fenarimol, Tridemorph). Darüber hinaus erwiesen sich aber auch Benzimidazole und Thiophanat M als kurative Fungizide. Es muß daher vermutet werden, daß diese Wirkstoffe nicht nur die Appressorienbildung (Schlüter u. Weltzien [4]), sondern auch spätere Entwicklungsstadien hemmen.

Zwar wirken alle hier genannten Fungizide auch sporulationshemmend, wie aus der oben zitierten Literatur und eigenen Beobachtungen hervorgeht, doch wurde in unseren Versuchen dieser Effekt von der protektiven und kurativen Wirkung nicht getrennt, da auch die nicht sporulierenden grünen Inseln als angegangene Infektionen gewertet wurden.

Der Grad der protektiven, kurativen und systemischen Wirkungen und ihr Verhältnis zueinander geben Hinweise auf die günstigste Form der praktischen Anwendung der einzelnen Fungizide. Ein Vergleich der Ergebnisse aus dem Protektivtest und dem Systemtest läßt deutliche Differenzdn zwischen den ED_{50}-Werten erkennen. So kommt z. B. bei Trimorphamid die potentielle Fungitoxizität, die sich in den Protektivwerten ausdrückt, bei systemischer Anwendung voll zur Auswirkung. Daraus muß auf besonders gute Aufnehmbarkeit der Substanz durch die Wurzeln und Weiterleitung im Gefäßsystem geschlossen werden. In dieser Beziehung steht Trimorphamid an der Spitze der Rangfolge, gefolgt von Ethirimol und Triadimefon. Nicht immer ist ein guter systemischer Effekt mit einer entsprechenden kurativen Wirksamkeit verbunden. Das trifft vor allem für Ethirimol und Trimorphamid zu, die bei Blattbehandlung keinen kurativen Effekt aufwiesen. Für diese beiden Präparate ist deshalb eine Zuführung über das Wurzelsystem günstiger, zumal sie auch recht gute systemische Eigenschaften haben. Das entspricht den Erfahrungen aus der Praxis, denn Ethirimol wird überwiegend als Beizmittel angewendet, und Trimorphamid zeigt seine Wirkung mehr als Gießmittel, denn als Spritzmittel (Hudecova u. a. [2]). Andererseits erscheint bei Wirkstoffen, die besser kurativ als systemisch wirken, eine Blattbehandlung erfolgversprechender (Triforin, Tridemorph, Thiabendazol, Ditalimfos). Präparate mit fehlender Kurativwirkung und ungenügender Systemwirkung sind praktisch nur als protektive Mehltaufungizide einsetzbar (Chloraniformethan, Pyrazophos). Eine weitere Gruppe von Fungiziden mit guter Wirkung bei allen durchgeführten Tests ist demgegenüber vielseitiger anwendbar (Triadimefon, Fenarimol, Benomyl). Die Möglichkeit, mit Formulierungshilfsstoffen Wirkungen zu verstärken, soll nur angedeutet werden; wir haben diese Problematik nicht in unsere Untersuchungen einbezogen.

Form und Umfang des Einsatzes dieser Fungizide wird neben ökonomischen Erwägungen in der Praxis auch von anderen Eigenschaften wie Phytotoxizität, Wirkungsspektrum und Neigung zur Resistenzbildung bestimmt. So ist bekannt, daß der Anwendungsbereich von Tridemorph durch seine Phytotoxizität begrenzt wird. Die systemische Anwendung von Chloraniformethan und Pyrazophos verbietet sich aus den gleichen Gründen. Die untersuchten Wirkstoffe unterscheiden sich wesentlich im Umfang ihres Wirkungsspektrums. Nach eigenen Untersuchungen können nur Chloraniformethan, Ethirimol, Trimorphamid und Pyrazophos als spezifisch wirksam gegen Echte Mehltaupilze angesehen werden. Von den übrigen Fungiziden ist auch eine Wirksamkeit gegenüber anderen Pilzen bekannt; sie sind daher umfassender einsetzbar. Das Auftreten von Resistenz ist nach bisherigen Erkenntnissen vor allem gegenüber Benomyl, Thiabendazol, Thiophanat M und Ethirimol zu erwarten.

Zusammenfassung

In Labortests wurden 12 bekannte Fungizide hinsichtlich ihrer protektiven, kurativen und systemischen Wirksamkeit geprüft. Alle Präparate zeigten einen guten protektiven Effekt, lediglich Ditalimfos fiel ab. Auch eine systemische Wirksamkeit war überall nachweisbar, sie war jedoch bei Chloraniformethan, Pyrazophos und Ditalimfos nur sehr gering und lag teilweise auch im phytotoxischen Bereich. Nicht immer ist gute systemische Wirksamkeit mit einem entsprechenden kurativen Effekt verbunden, wie bei Ethirimol und Trimorphamid. Kurativ bei Blattbehandlung 48 Stunden nach Infektion wirkten die Haustorienhemmer Triforin, Triadimefon, Fenarimol und Tridemorph sowie die Benzimidazole und Thiophanat M. Der Grad der protektiven, kurativen und systemischen Wirkungen und ihr Verhältnis zueinander geben Hinweise auf die günstigste Form der praktischen Anwendung der einzelnen Fungizide.

Literatur

[1] Buchenauer, H., Mode of action and selectivity of fungicides which interfere with ergosterol biosynthesis, Proceed. Brit. Crop. Protection Conference. Pests and Diseases 1977, 699–711

[2] Hudecová, D., J. Kovač, S. Gahér, Biologicka ucinnost trimorfamidu proti *Erysiphe cichoracearum* na uhorkách, Agrochemia, Bratislava 18 (1978), 361–363

[3] Müller, H. J., Eine Methode zur Prüfung von Systemfungiziden, Zentralbl. f. Bakteriologie, Parasitenkunde, Infektionskrankheiten und Hygiene, II. Abt., 120 (1966), 720–725

[4] Schlüter, K., H. C. Weltzien, Ein Beitrag zur Wirkungsweise systemischer Fungizide auf *Erysiphe graminis*, Meded. Fac. Landbouwetensch. Rijksuniv Gent 36 (1971), 1159–1164

M. LENGIES, G. SCHELLENBERG, H. LYR und D. ZANKE

Institut für Pflanzenschutzforschung der Akademie der Landwirtschaftswissenschaften, Kleinmachnow, DDR

Wirksamkeit des Wachstumsregulators Pydanon gegen Mehltauarten

Pydanon (3,6-Dioxo-4-hydroxy-hexahydro-pyridazinyl-4-essigsäure) ist als Wachstumsregulator aus der Literatur bekannt, KNOEVENAGEL u. a. [2], ROLLI [3]. Diese Verbindung und ihre Derivate wurden als Wachstumshemmer in verschiedenen Anwendungsgebieten beschrieben. Durch SINHA und WOOD [4] wurde der Einfluß von Wachstumsregulatoren auf die Verticilliumwelke von Tomatenpflanzen untersucht.

Tabelle 1
Wuchshemmung von Winterweizen und Wintergerste durch Pydanon-Derivate (Wirkstoffkonzentration: 10 mg/ml)

Substanz	Wuchshemmung in % nach 10 Tagen			
	Wintergerste		Winterweizen	
	Blatt- applikation	Wurzel- applikation	Blatt- applikation	Wurzel- applikation
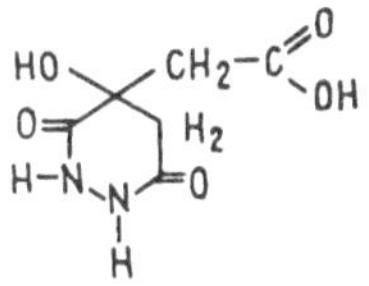	88	66	65	80
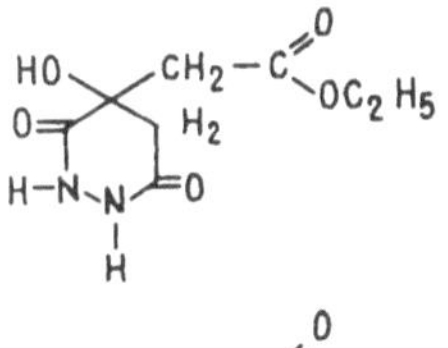	83	74	33	58
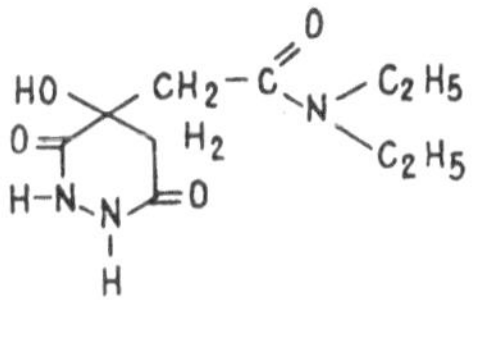	60	58	28	60
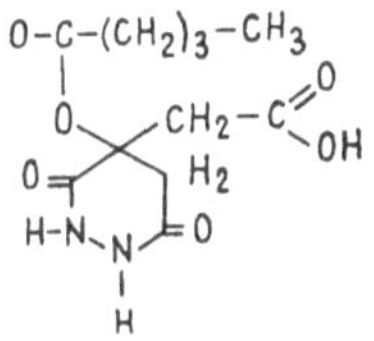	63	60	23	80

BUCHENAUER und ERWIN [1] demonstrierten, daß durch Blatt- oder Bodenbehandlung von Tomaten- und Baumwollpflanzen mit Pydanon vor der Inoculation mit einem mikrosklerotischen Stamm von *Verticillium dahliae* die Schwere der Blatt- und Gefäßverfärbungen gemindert werden kann. Hieraus resultierte unser Interesse, die Wirksamkeit dieser Verbindungen gegen Echte Mehltauarten im Zusammenhang mit ihrer wachstumsregulierenden Wirkung näher zu untersuchen.

Zur Ermittlung der wuchshemmenden Eigenschaften von Pydanon und seinen Derivaten wurde Wintergerste der Sorte „Vogelsanger Gold" zur Blattbehandlung in Töpfen auf Erde und zur Wurzelbehandlung in Kulturröhrchen auf Quarzsubstrat angezogen. Nachdem die Pflanzen eine durchschnittliche Höhe von 8 bis 10 cm erreicht hatten, wurden die Substanzen in Form einer 1%igen wässrigen Lösung über die Blätter als Tropfnaßspritzung und über die Wurzeln durch Gießen appliziert. 10 Tage nach der Behandlung wurde der Längenzuwachs der Pflanzen bestimmt und daraus die prozentuale Wuchshemmung im Vergleich zu den behandelten Kontrollen ermittelt. Analog wurde mit Winterweizen der Sorte „Alcedo" verfahren.

Es zeigte sich, daß Pydanon und seine Derivate gute wachstumsretardierende Effekte bei den untersuchten Pflanzenarten bewirkten, die sowohl bei Blatt- als auch bei Wurzelapplikation zu beobachten sind. Dabei fiel außerdem eine gewisse Sortenabhängigkeit in der Wirkung auf (Tab. 1).

Tabelle 2
Befallsunterdrückung von Erysiphe graminis MARCHAL *durch Pydanon-Derivate (Wirkstoffkonzentration: 100 µg/ml)*

Substanz	Mehltaubefallsunterdrückung	
	in % nach 7 d protektiv	kurativ
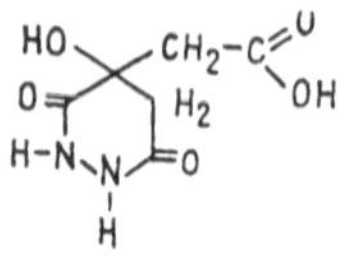	44	50
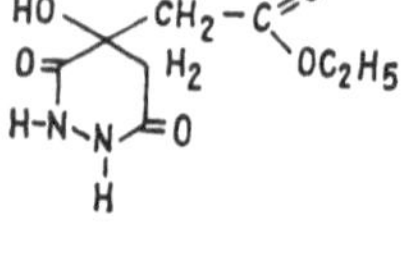	50	57
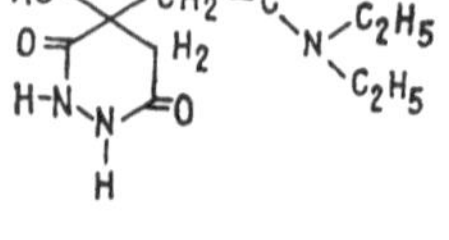	26	78
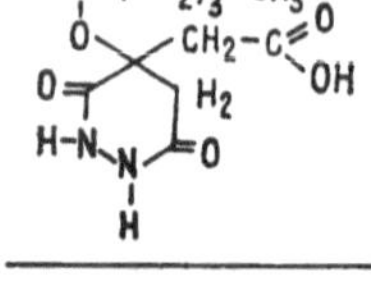	37	61

Tabelle 3
*Befallsunterdrückung von Podosphaera leucotricha (*Ell. *et* Ev.*)* Salm *durch Pydanon Derivate (Wirkstoffkonzentration: 500 µg/ml)*

Substanz	Behandlungsart	Mehltaubefallsunterdrückung in % nach 8 d	15 d
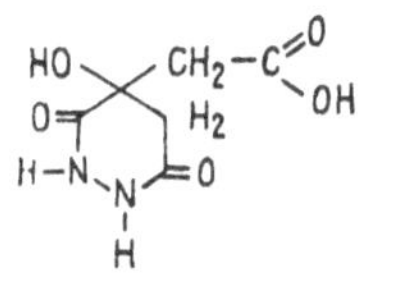	protektiv	100	83
	kurativ	93	83
	systemisch-protektiv	86	69
	systemisch-kurativ	86	79
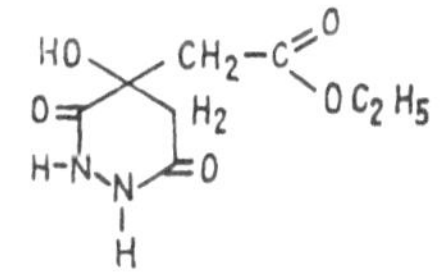	protektiv	100	79
	kurativ	78	69
	systemisch-protektiv	86	65
	systemisch-kurativ	78	51
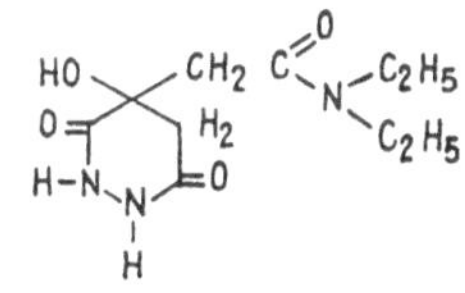	protektiv	71	62
	kurativ	100	69
	systemisch-protektiv	49	20
	systemisch-kurativ	93	41
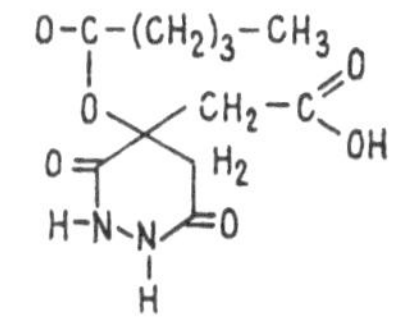	protektiv	100	97
	kurativ	100	79
	systemisch-protektiv	71	79
	systemisch-kurativ	71	62

Die fungizide Wirkung der Verbindungen wurde an Sommergerste geprüft. Eine empfindliche Sorte wurde auf Quarzsubstrat angezogen, die Primärblätter nach 10 Tagen auf eine geeignete Länge beschnitten. Die künstliche Infektion erfolgte mit *Erysiphe graminis* Marchal am 12. Tag. Die Applikation des Pydanons und seiner Derivate erfolgte sowohl protektiv (24 Stunden vor der Infektion) als auch kurativ (72 Stunden nach der Infektion) über die Blätter durch Tropfnaßspritzung mit wässrigen Wirkstofflösungen.
Ähnlich wurden getopfte Apfeltypenunterlagen vom Typ A II behandelt, die in einem Entwicklungsstadium von 6 bis 10 Blättern pro Trieb künstlich mit *Podosphaera leucotricha* (Ell et Ev.) Salm infiziert wurden. Die Konzentration der wässerigen Wirkstofflösung betrug hierbei 500 µg/ml. Zur Ermittlung der systemischen Wirkung wurden die Töpfe mit wässerigen Wirkstofflösungen gegossen. Die prozentuale Unterdrückung des Mehltaubefalls wurde durch den Vergleich von behandelten und unbehandelten Pflanzen bestimmt.

Es zeigte sich, daß Pydanon und seine Derivate eine deutliche Verringerung des Befalls von *Erysiphe graminis* und *Podosphaera leucotricha* bewirken (Tab. 2 und 3). Besonders bei Gerstenmehltau läßt sich erkennen, daß die kurative Wirkung deutlich stärker ausgeprägt ist als die protektive; jedoch läßt sich diese Tendenz nicht verallgemeinern, da der Effekt bei Apfelmehltau nur vereinzelt auftritt. Die durch Wurzelapplikation erzielte Befallsunterdrückung von *Podosphaera leucotricha* unterscheidet sich nur unwesentlich von den durch Blattapplikation erreichten Ergebnissen, was auf eine gute systemische Wirkung schließen läßt.

Die durchgeführten Untersuchungen ergaben, daß der Wachstumsregulator Pydanon und seine Derivate im Gegensatz zu anderen Wachstumsregulatoren bereits in relativ geringen Konzentrationen einen antifungalen Effekt nicht nur bei Tracheomykosen, sondern auch bei obligaten Parasiten bewirken. Die protektive, kurative und systemische Wirkung gegenüber Mehltauarten kann in Verbindung mit ihrer wachstumsregulierenden Wirkung als gut eingeschätzt werden. Der Wirkungsmechanismus der Verbindungen ist noch unbekannt.

Literatur

[1] BUCHENAUER, H. and D. C. ERWIN, Effect of the Plant Growth Retardant Pydanon on *Verticillium* Wilt of Cotton and Tomato, Phytopathology **66** (1976), 1140–1143

[2] KNOEVENAGEL, K., D. ROLLI und R. HIMMELREICH, Wirkung von Pyridazinyl-Derivaten auf das Pflanzenwachstum, Die Naturwissenschaften **57** (1970), 395–396

[3] ROLLI, K., Pydanon, eine Möglichkeit zur chemischen Kultursteuerung, Der Erwerbsgärtner 26 (1970), 1254

[4] SINHA, A. K. and R. K. S. WOOD, The Effect of Growth Substances on *Verticillium* Wilt of Tomato Plants, Ann. Appl. Biol. 59 (1967), 117–128

H. LYR und M. SUNKEL

Institut für Pflanzenschutzforschung der Akademie der Landwirtschaftswissenschaften, Kleinmachnow, DDR

Zur Wirkungsweise von Metanaxin

Metanaxin-DL-N-(2,6-Dimethylphenyl)-N-(2'-methoxyacetyl)-alanin-methylester ist ein neuartiger Wirkstoff von sehr starker Aktivität und mit einer extremen Selektivität gegen Peronosporales, wie sie bisher bei nur wenigen Fungiziden festgestellt werden konnte. Der formulierte Wirkstoff Ridomil wird in verschiedenen Ländern zur Bekämpfung von *Phytophthora-*, *Pythium-*, *Sclerospora-*, *Peronospora-*, *Plasmopara-* und *Bremia-*Arten mit gutem Erfolg eingesetzt.
Die Selektivität von Metanaxin konnte durch eigene Untersuchungen bestätigt werden (Tab. 1 und 2).

Tabelle 1
Wirkung auf das Myzelwachstum

Peronosporales	ED_{50} (mol/l)
Pythium debaryanum	10^{-8}
Phytophthora cactorum	8×10^{-8}
Phytophthora cactorum (submers)	3×10^{-7}
Phytophthora infestans (submers)	$<10^{-7}$

Tabelle 2
Wirkung auf das Myzelwachstum

	Pilzgruppe	ED_{50} (mol/l)
Achlya racemosa *Achlya ambisexualis* *Achlya ambisexualis* *Saprolegnia turfosa* *Saprolegnia literalis*	Saprolegniales	$>10^{-3}$
Mucor mucedo	Zygomycetales	$>10^{-4}$
Botrytis cinerea	Helotiales	
Rhizoctonia solani	Tulasnellales	

Der Wirkstoff wurde aus dem Präparat Ridomil 50 WP (Ciba-Geigy) durch Extraktion mit Hexanol, anschließender Vakuumdestillation und Umkristallisierung gewonnen.
Die Wirkung auf das Myzelwachstum der in den Tabelle 1 und 2 untersuchten Pilze wurde in einem Myzelwachstumstest auf Agarplatten bestimmt.

Zu einem Standardnährboden (2,5% Malz, pH 5—6) werden vor dem Plattenguß die gewünschten Wirkstoffkonzentrationen zugegeben. Die Petrischalen werden mit Myzelscheiben des Testpilzes beimpft. Nach 8 Tagen Wachstumsdauer bei 20 bis 25 °C wird der radiale Myzelzuwachs gemessen und zur Kontrolle ohne Wirkstoff in Beziehung gesetzt. Die ED_{50} wurde grafisch ermittelt.
Zur Untersuchung der Wirkungsweise von Metanaxin bei *Phytopthora cactorum* wurde mit einer Submerskultur gearbeitet. Dazu wird eine Kultur von *Phytopthora cactorum*, die auf SISLER-Medium und Glasperlen angezogen wird (SISLER-Original + 2 g Hefeextrakt/l, pH 5—6) mit SISLER-Medium abgeschwemmt. Die dabei abgelösten Myzelstückchen werden in festgesetzten Volumina auf Schüttelkolben verteilt. Metanaxin, das in Dimethylformamid gelöst ist, wird nach entsprechender Verdünnung hinzugegeben (Inkubationszeit 24 h). Die Temperatur betrug 25 °C. Proben wurden nach 24, 48 und 72 h entnommen. Von den Proben wird das Trockengewicht bestimmt. Die Auswertung erfolgte durch Angabe der prozentualen Wachstumshemmung im Vergleich zur unbehandelten Kontrolle.

Die Ursachen der starken Wirkung und Selektivität sind z. Z. noch unbekannt. Im Unterschied zu zahlreichen anderen Fungiziden tritt die Wirkung zeitlich verzögert ein (Abb. 1). STAUB u. a. [1] stellten einen ähnlichen Effekt beim Einsatz von Ridomil gegen *Plasmopara viticola* an Weinreben fest. Sie beobachteten, daß die Wirkung erst nach der Ausbildung der ersten Haustorien eintritt.

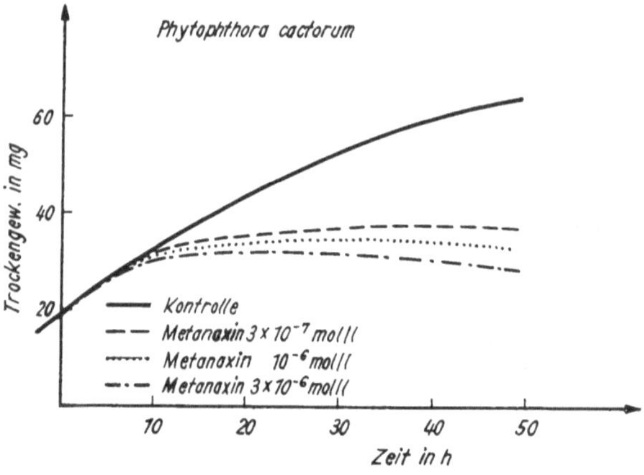

Abb. 1. Einfluß auf den Wachstumsverlauf
Kontrolle Metanaxin 3×10^{-7} mol/l, Metanaxin 10^{-6} mol/l, Metanaxin 3×10^{-6} mol/l

Um Hinweise auf mögliche Angriffsorte zu finden, wurde versucht, durch Zusätze die Hemmwirkung aufzuheben. Dazu wurden der Submerskultur sowohl Metanaxin als auch entsprechende andere Substanzen zugesetzt.

Es wurden Lecithin (SERVA-International, Chemie- und Handels GmbH & Co, Heidelberg), Cholesterin (FERRAK Chemikalien, Berlin), RNS aus Hefe 85%ig (VEB Arzneimittelwerk Dresden), die Vitamine D 2, B 2 und B 12 (VEB Jenapharm) und Kaseinhydrolysat eingesetzt. Letzteres wurde aus Kasein (SERVA Entwicklungslabor, Heidelberg) gewonnen (man erhitzt Kasein 8 h mit 2 n Salzsäure am Rückfluß, entfernt anschließend das Wasser im Rotationsverdampfer und gibt sooft kleine Portionen destilliertes Wasser hinzu und entfernt es wieder, bis der Salzsäuregeruch verschwunden ist).

Wie Tabelle 3 und Abbildung 2 und 3 zeigen, konnte bei den eingesetzten Zusatzstoffen nur ein sehr mäßiger Effekt erzielt werden.
Der Hemmeffekt von Metanaxin (100% Wachstumshemmung bei 10^{-6} mol/l) ist bei Übertragung des Myzels auf wirkstofffreien Agar reversibel (0% Wachstumshemmung, so daß unter diesen Bedingungen ein fungistatischer Effekt vorliegt. Es wurde der

Tabelle 3
Einfluß auf die Wachstumshemmung

Zusatzstoffe	Wirkung
Lecithin	–
Cholesterin	–
Vitamin B 2	–
Ribonukleinsäure	0
Kaseinhydrolysat	+
Vitamin D 2	+
Vitamin B 12	+

\+ Wachstumshemmung wird schwach verstärkt
– Wachstumshemmung wird schwach vermindert
0 Wachstumshemmung wird nicht beeinflußt

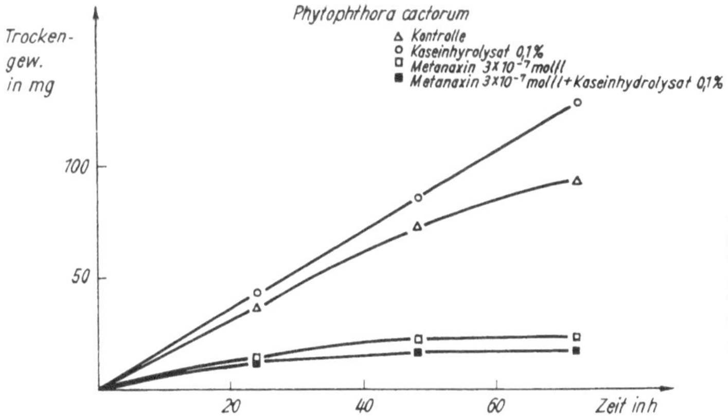

Abb. 2. Einfluß von Kaseinhydrolysat auf die Wachstumshemmung Kontrolle, o Kaseinhydrolysat 0,1%, Metanaxin 3 × 10^{-7} mol/l, Metanaxin 3 × 10^{-7} + Kaseinhydrolysat 0,1%

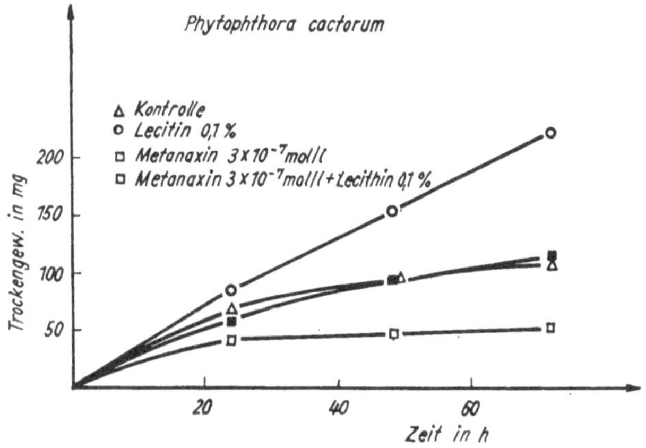

Abb. 3. Einfluß von Lecithin auf die Wachstumshemmung Kontrolle, o Lecithin 0,1%, Metanaxin 3 × 10^{-7} mol/l, Metanaxin 3 × 10^{-7} mol/l + Lecithin 0,1%

Agarplattentest angewendet, wobei das 8 Tage lang unter Wirkstoffeinfluß gehaltene Myzel auf Wirkstoff-freien Agar übertragen wird und nach weiteren 8 Tagen zu einer Kontrolle ohne Wirkstoff in Beziehung gesetzt wird. Der Testpilz war *Phytophthora cactorum*.

Die Ursachen für die sehr spezifische und mit einer lag-Phase einsetzende Wirkung sind noch ungeklärt. Als mögliche Ursache werden eine metabolische Umwandlung (Toxifikation), eine Interaktion mit Kernteilungsprozessen oder die Hemmung einer sehr spezifischen Produktsynthese im Pilz diskutiert.

Literatur

[1] Staub, T., H. Dahmen and F. J. Schwinn, Effect of Ridomil on the development of target pathogens on their host plants, Abstracts of papers, 3rd Int. Cong. Plant Pathol., München (1978), 366

Z. MACIEJOWSKA-POKACKA

Institute of Plant Protection, Poznań, Poland

The Influence of Systemic Fungicides on the Development of *Rhizoctonia* on Wheat

Introduction

The causal organism of sharp eyespot of cereals in the literature indicated as *Rhizoctonia solani* KÜHN [2, 3, 4, 5, 7, 8, 9, 10 and 11], and recently also as *R. cerealis* VAN DER HOEVEN spec. nov. [1, 12].

Earlier observations on the incidence of sharp eyespot of cereals in field cultures sprayed with benzimidazole fungicides suggested, that the incidence of the disease was related to the sprays performed [2, 10]. At the same time an observation was made that the increase of sharp eyespot was also related to the control or to the absence of *Pseudocercosporella herpotrichoides* (FRON) DEIGHTON [10,12]. On the basis of several hundred field experiments REINECKE and FEHRMANN [12] came to the conclusion that the occurrence of *Rhizoctonia* infection of cereals was negatively correlated with *Pseudocercosporella* infection, this being the case on untreated fields as well as on fields treated with MBC-fungicides. Parallel it was shown by these authors that the control of *Rhizoctonia* on winter wheat and rye with fungicides resulted in the increase of *Pseudocercosporella* index.

In experiments conducted in vitro the adaptive ability of isolates of *R. solani* to systemic and non-systemic preparations was reported [3]. Laboratory experiments were not always in agreement with greenhouse and field results. It was recently suggested that a reversal of toxicity of some systemic and non-systemic fungicides in laboratory conditions may be due to the presence in the medium of metal ions, phenolics and some other components [5].

Results obtained by RASHID and SCHLÖSSER [11] showed that benomyl resistant strains of *R. solani* were present in a number of fields, which had never been treated with benzimidazole fungicides. However mutations to resistance to these fungicides may occur. According to VAN TUYL [13] induced mutations to resistance to benomyl in fungi were of a practically recessive condition and they occurred relatively infrequently.

In the view of cited literature it may be concluded that the variation of results concerning *Rhizoctonia* on cereals is due to several interrelated factors, of which the antagonistic relationship between *R. solani* and *P. herpotrichoides* may play a major role.

In this study are presented data concerning the incidence of sharp eyespot on winter wheat in field experiments on chemical control of diseases, performed in the years 1978—1979 in Plant Protection Institute.

Material and Methods

Total number of experimental objects: 45
The number of sprays performed: 1 or 2

Dates of treatments:

one treatment
growth stage according to FEEKES scale:
5 to 7 or 10.1 to 10.5

two treatments
growth stages according to FEEKES scale:
5 to 7 and 10.1 to 10.5

Fungicides used:

Groups of preparations	Active ingredients	Number of objects
Preparations of Epidor type	benzimidazole + mankozeb	18
Benzimidazoles	benomyl	
	carbendazim	13
	tiophanate-methyl	
Other preparations containing	carbendazim + captafol	6
benzimidazoles	carbendazim + sulphur	
Triazoles	triadimefon	6
Sulphur	sulphur	1
Others unspecified		1

The incidence of sharp eyespot was determined in majority of cases on 100 culms, exceptionally on 50 culms, as percent of culms with disease symptoms before harvest. No grades of infection were distinguished, because disease symptoms were rather of a superficial character. The results of fields experiments were compared using U test, on the level of significance 0.01.

In laboratory experiments was used an isolate of *R. solani* from winter wheat, highly pathogenic to winter wheat seedlings. The following fungicides were used in laboratoy tests:

Product name	Active ingredient	% a.i.
Benlate	benomyl	50
Topsin M	tiophanate-methyl	70
Funaben 50	carbendazim	50
Bayleton 25 WP	triadimefon	25
Vitavax	carboxin	75
Dithane M-45	mankozeb	80
Zaprawa Nasienna T	thiram	50

Laboratory tests were carried out using standard techniques consisting of mixing the fungicides with sterile PDA cooled to the temperature of 40°. After solidifying the plates were inoculated with small discs overgrown with a test fungus from one week-old culture on PDA. Beside of standard laboratory tests two other experiments were performed. The first one consisted of inoculating with a suspension of mycelium non-solidified, cooled to 40° PD agar, to which fungicides were added just before inoculation. The second test was of a rather qualitative nature and it consisted of placing small amounts of fungicides on the surface of PDA, or on N-deficient and C-deficient mineral media. These plates were inoculated centrally with the test organism. A similar auxanographic test is widely used in screening yeast cultures for the ability of utilization of certain carbon and nitrogen sources [14].

Inoculated plates were incubated at room temperature. The radial growth of colonies of the fungus was measured at various time intervals. The number of colonies on plates inoculated with a suspension of mycelium was counted after one week.

Results

Field experiments

Because of differences in the number of fungicides tested in various experiments, for statistical interpretation U test was used (Fig. 1). Mean values for the occurrence of sharp eyespot in various experimental combinations are presented in Table 1. For once sprayed objects at various growth stages the incidence of the disease was nearly the same as for unsprayed control plots. A great difference was observed between experimental objects sprayed once or twice with fungicides. When compared with

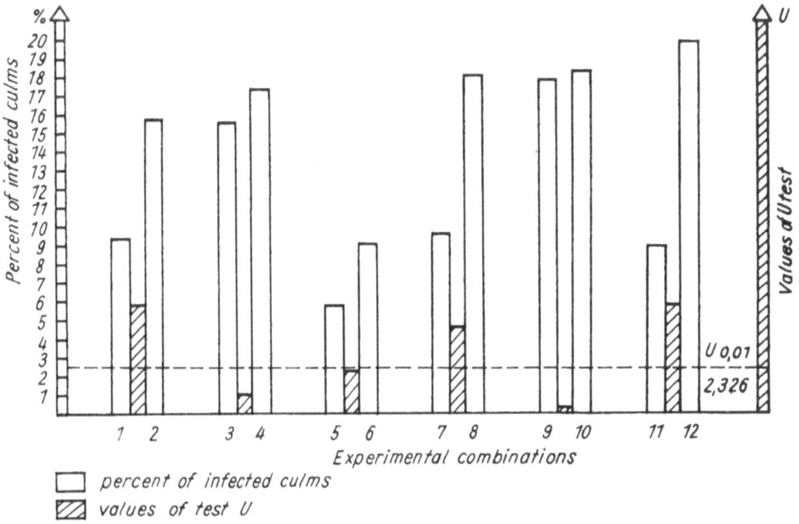

Fig. 1. Effects of treatment of winter wheat with fungicides on the occurrence of sharp eyespot caused by *Rhizoctonia solani* Kühn [a legend of experimental combinations 1 – 12 is given in Tab. 1]

control, the incidence of sharp eyespot being over 60% greater on plots where only one spray was performed. The difference between these two groups of objects was highly significant. On experimental plots sprayed once with fungicides at the stage of growth 10.1 to 10.5 in most cases a good control of sharp eyespot took place. On plots sprayed once at the stage 5 to 7 an increase of the disease occurred. Observed differences between these two groups of objects were highly significant at the level of probability 0.01.

Fungicides of Epidor containing mankozeb and benzimidazoles were better for control of sharp eyespot than all others, not containing mankozeb. However these differences were insignificant at the level of probability 0.01. In the case of two treatments they were significant at the level of 0.02. Benzimidazole fungicides not containing mankozeb

Table 1

Effects of treatment of winter wheat with fungicides on the occurrence of sharp eyespot (Rhizoctonia solani KÜHN*) before harvest, 1978–1979*

The number and dates of treatments as related to growth stages (Feekes scale)		Sprayed plots % culms with disease symptoms	Incidence of disease as % of control	Unsprayed plots % culms with disease symptoms
1 Two treatments	5 to 7 and 10.1 to 10.5	9.4	40.1	23.5
2 One treatment	5 to 7 or 10.1 to 10.5	15.7	101.4	15.5
3 One treatment with mankozeb containing fungicides	5 to 7 or 10.1 to 10.5	15,6	83,6	18,7
4 One treatment with fungicides not containing mankozeb		17,2	126,4	13,6
5 Two treatments with mankozeb containing fungicides	5 to 7 and 10.1 to 10.5	5,9	23,4	25,1
6 Two treatments with fungicides not containing mankozeb		9,1	41.9	21.8
7 One treatment	10.1 to 10.5	9.6	59.3	16.2
8 One treatment	5 to 7	18.1	117.8	15.4
9 One treatment with mankozeb containing fungicides	5 to 7	17.9	101.1	17.7
10 One treatment with fungicides not containing mankozeb		18.3	131.7	13.9
11 Two treatments with fungicieds containing benzimidazoles but not mankozeb	5 to 7 and 10.1 to 10.5	9.0	40.5	22.2
12 One treatment with fungicides containing benzimidazoles but not mankozeb	5 to 7 or 10.1 to 10.5	19.9	129.2	15.4

were highly effective in controlling sharp eyespot in the case of two treatments. When applied as a single spray, they favoured a higher incidence of sharp eyespot, in contrast to mankozeb containing fungicides.

Laboratory tests

When the test fungus was inoculated on the surface of PDA containing increasing doses of fungicides (Table 2), the concentration of 10 mg/kg of carboxin, carbendazim, benomyl and tiophanate-methyl strongly inhibited the hyphal growth. The fungicidal action of triadimefon was relatively weaker. The concentration of 100 mg/kg was sublethal, with the exception of triadimefon. In the case of the later fungicide some growth was observed even on concentrations of 1000 and 2000 mg/kg. As it can be seen from the Table 2, 10 mg/kg of triadimefon inhibited the radial growth of the

Table 2
Inhibition of mycelial growth of Rhizoctonia solani KÜHN *on PD agar with funigcides after 7 days*

Active ingredient	Percent of inhibition in comparison to control			
	Superficial inoculation		Suspension inoculation	
	10 μg/ml	100 μg/ml	10 μg/ml	100 μg/ml
Carbendazim (MBC)	86,4	98,6	95,3	100
Benomyl	87,8	99,6	91,3	100
Tiophanate-methyl	79,7	95,9	88,7	100
Carboxin	93,2	99,9	97,4	100
Triadimefon	48,6	89,2	89,7	100

fungus by 48,6%. On plates with 10 mg/kg of carboxin inhibition was of a highest range (93.2%). Benzimidazole fungicides used at the same rate were also inhibitory (79.7 to 87.8% of inhibition). The character of growth of the fungus was different on the medium with carboxin, than on media containing other fungicides. In spite little radial growth, the fungus grew in vertical position, reaching the cover lid.
When PDA plates containing 10 mg/kg of fungicides were inoculated using the suspension method, on plates with carboxin the radial growth of the fungus was also of a lowest range (Table 2). At the concentration of 100 mg/kg of fungicides no growth occurred on plates after 7 and 14 days. Carbendazim showed a better fungicidal action than other benzimidazole preparations. Triadimefon, less effective on plates inoculated superficially with the test fungus, exhibited a satisfactory fungicidal action when the inoculum was incorporated into the medium in a suspension form.
There are not sufficient data on mechanisms of detoxification observed in laboratory tests with *R. solani* and various fungicides. In this study qualitative auxanographic tests were performed on N-deficient and C-deficient mineral media in order to investigate the ability of *R. solani* to decompose and to utilize fungicides as nitrogen and carbon source. Negative results of these tests are presented on Figures 2 and 3. In addition to these results, a reaction of the fungus to fungicides on auxanographic plates could be observed. The plates shown on Figures 2 and 3 were inoculated with two subcultures of the test organism, exhibiting a different reaction to benomyl. This fungicide confined the hyphal growth of one subculture only. Well defined inhibition

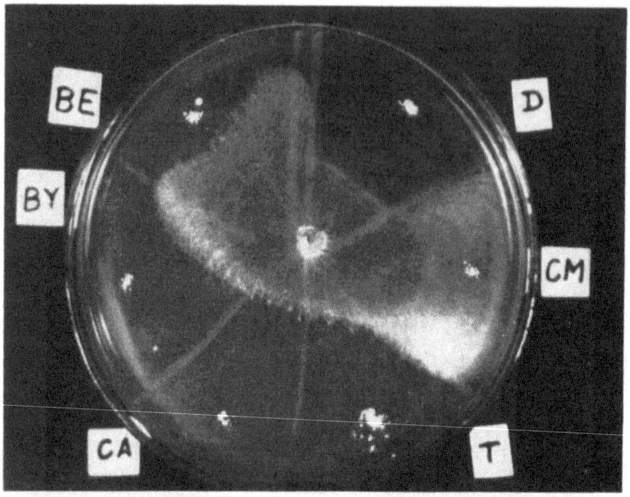

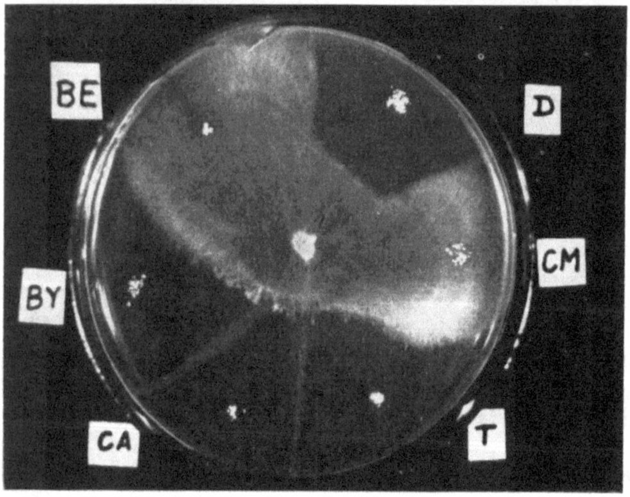

Fig. 2. and 3. The growth of *Rhizoctonia solani* KÜHN on N-deficient mineral medium in the presence of fungicides: BE – Benlate, D – Dithane M-45, CM – Funaben 50, T – Zaprawa nasienna T, CA – Vitavax, BY – Bayleton 25 WP

zones were formed by both subcultures as a result of fungicidal activity of mankozeb carboxin and thiram. Less inhibition occurred in the case of triadimefon. No wide inhibition zones were observed in the case of carbendazim and tiophanate-methyl. Similar results were obtained on C-deficient mineral medium, but the growth of the fungus was much weaker and difficult to study.

Auxanographic tests on PDA where only one fungicide was applied on a single plate, gave somewhat different image. The inhibiton of fungal growth was very strong in the case of carboxin, strong in the case of makozeb, thiram and benomyl. Less inhibition occurred in the case of carbendazim, tiophanate-methyl and triadimefon (Fig. 4–7).

Discussion

Results obtained in this study are in agreement with conclusions presented in the paper by REINECKE and FEHRMANN [12] who studied the occurrence of sharp eyespot on

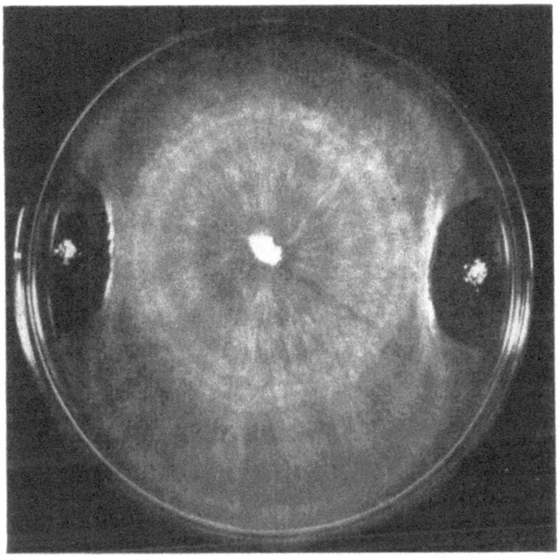

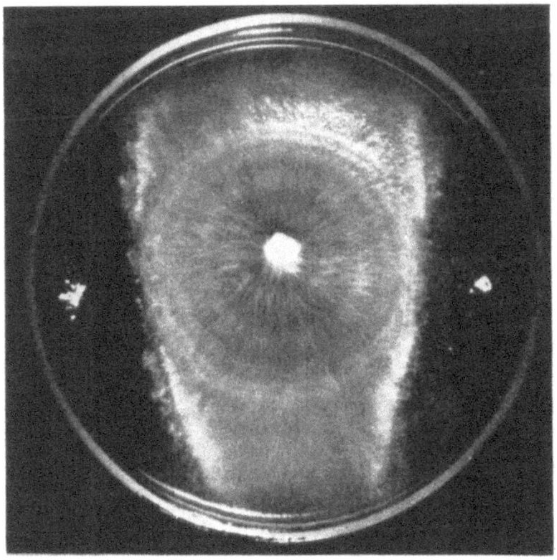

Fig. 4. and 5. The growth of *Rhizoctonia solani* KÜHN on PDA in the presence of fungicides: 4 — Vitavax, 5 — Dithane M-45

cereals in relation to chemical control and to the incidence of eyespot caused by *P. herpotrichoides*. Similar results were obtained by some other authors [7, 10]. The increase of sharp eyespot on fields sprayed with benzimidazole fungicides at early stages of growth may be related both to the control of *P. herpotrichoides* and to increased sensitivity of wheat to *Rhizoctonia* at earlier stages of growth [8]. Consequently, a better efficiency of a single spray performed at the stage of growth 10.1 to 10.5 may be related to a lower sensitivity of older plants to both diseases. A high degree of control was obtained by the application of two sprays on experimental plots. These results suggests that the development of sharp eyespot is a slow process, which however may reach at the end of vegetation a significant level, if appropriate measures of control are not applied. In experimental conditions, a second spray was applied about two weeks after a first spray, this prooved to be effective against *Rhizovtonia*. Average

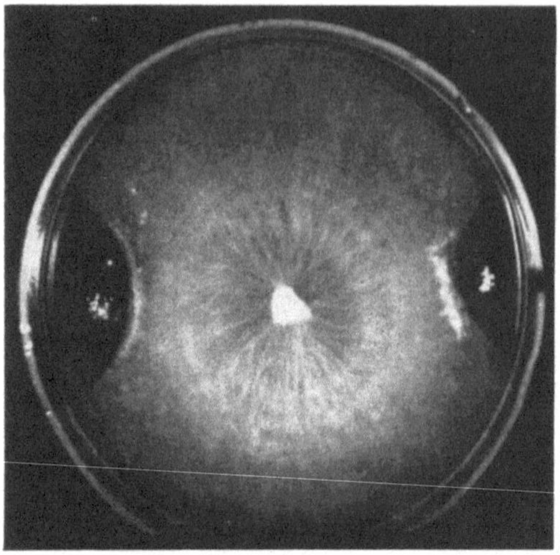

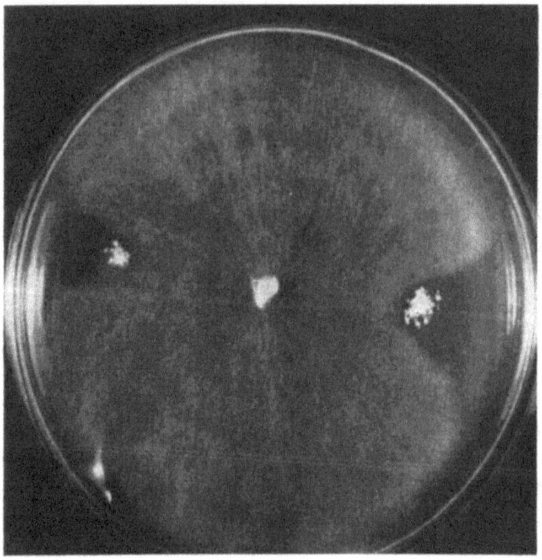

Fig. 6. and 7. The growth of *Rhizoctonia solani* KÜHN on PDA in the presence of fungicides: 6 — Benlate, 7 — Topsin M

results indicated that the application of two sprays mankozeb containing preparations of Epidor type gave best results. When compared with objects sprayed twice with fungicides not containing mankozeb, the significance at the level of 0.02 was observed mankozeb containing fungicides applied at the stage of growth 5 to 7 only slightly influenced the occurrence of sharp eyespot, when compared with control (Table 1). In contrast, a single spray performed at the same stage with benzimidazole fungicides not containing mankozeb markedly stimulated the development of *R. solani*. This may be explained on the base of competition between *R. solani* and *P. herpotrichoides*, the fungus particularly susceptible to benzimidazoles. On the other hand, mankozeb might be more effective against *R. solani*.

Laboratory experiments were conducted in order to elucidate some phenomena observed in the field. Fungitoxic action of benzimidazole fungicides against *R. solani* could be observed best on PDA plates mixed with fungicides (Table 2). A relatively weaker inhibition of the fungus by tiophanate-methyl could be related to the time requird for the transformation to carbendazim, a toxic principle of benzimidazoles. On auxanographic plates with mineral media and with PDA, a direct contact of growing mycelium with fungicides did not occur. In these conditions no distinct inhibition zones were observed with tiophanate-methyl (Fig. 2, 3 and 7) and with carbendazim (Fig. 2 and 3). However two types of reaction to benomyl were observed on auxanographic plates with N-deficient mineral medium (Fig. 2 and 3). Well-defined inhibition zones were formed only in the case of one subculture. On auxanographic plates with PDA inhibition zones were always present (Fig. 6), in contrast to thiophanate-methyl and carbendazim. On the base of experiment performed it was not possible to explain the reaction of the test fungus to benomyl. However it was evident that the mechanism of inhibition by benomyl might depend upon some other factors, not only upon carbendazim. As it was reported in the literature [6], in the case of benomyl an additional fungitoxic product, butyl-isocyanate may be formed, and possibly it might cause a different reaction of the fungus.

Carboxin was highly effective against the test fungus in vitro. As it was stated in earlier investigations, carboxin, commonly used for the control of loose smuts, exhibits a toxic action against other Basidiomycotina, and also against *R. solani* [6]. A very high degree of inhibition of fungal growth, especially on plates inoculated with a mycelial suspension, a vertical growth of mycelium confined to the area of inoculum discs placed on PDA plates with carboxin, and largest inhibition zones formed on artificial media in the presence of this fungicide (Fig. 2, 3 and 4) seem to confirm the fact, that the mechanism of fungicidal action of carboxin differs from the action of other preparations studied.

In tests on PDA mixed with triadimefon a good inhibition was observed only when the inoculum was incorporated into the medium. Differences obtained with two methods of inoculation were less distinct in the case of other preparations (Tab. 2). This indicated that the mode of action of triadimefon differed from the mode of action of other fungicides. Triadimefon was not effective against *Rhizoctonia* on plots sprayed once at the stage of growth 5 to 7. This was probably related to little effectiveness of this preparation against *P. herpotrichoides*. Mankozeb and thiram were inhibitory to *R. solani* in vitro (Fig. 2, 3 and 5). In the case of mankozeb these results were in agreement with field experiments.

Differences in mechanisms of action of fungicides discussed in this study were stated and elucidated in earlier findings, according to which carboxin is an inhibitor of mitochondrial respiration, while benzimidazoles are inhibitors of mitosis, and triadimefon inhibits ergosterol synthesis [6].

Laboratory experiments on control of *R. solani* with fungicides were not always consistent and not always in agreement with experiments conducted in vivo [4, 5]. Since the present time there are not sufficient data for explanation of these phenomena. According to KATARIA and GROVER [5] detoxification of certain fungicides may occur in the presence of some inorganic and organic components in artificial media. This however does not give a full explanation of differences between field and laboratory results. In the view of discussed facts, conclusions concerning field experiments should be based primarily on repeated field results.

Literature

[1] Boerema, G. H. and A. A. Verhoeven, Check-list for scientific names of common parasitic fungi. Series 2b. Fungi of field crops: cereals and grasses, Neth. J. Pl. Path. **83** (1977), 165–204

[2] Bollen, G. J. and E. P. van der Hoeven, Effect of benomyl on pathogens causing foot rot in rye, 2nd Intern. Congr. Plant Pathol. Abstr. 0196, Minneapolis 1973

[3] Kataria, H. R. and R. K. Grover, Adaptation of *Rhizoctonia solani* to systemic and non-systemic fungitoxicants, Z. PflKrankh. PflSchutz **81** (1974), 472–478

[4] Kataria, H. R. and R. K. Grover, Some factors affecting the control of *Rhizoctonia solani* by systemic and non-systemic fungicides, Ann. appl. Biol. **82** (1976), 267–278

[5] Kataria, H. R. and R. K. Grover, Reversal of toxicity of some systemic and non-systemic fungitoxicants by chemicals, Z. PflKrankh. PflSchutz **85** (1978), 76–83

[6] Leroux, P. and R. Fritz, Les modes d'action et de la séléctivité des fongicides, Phytiat.-Phytopharm. **27** (1978), 163–188

[7] Mielke, H., Beobachtungen über den natürlichen Befall mit *Rhizoctonia solani* Kühn an Winterweizenstämmen ohne und mit künstlicher Infektion durch *Cercosporella herpotrichoides* Fron, Nachrichtenblatt des Deutschen Pflanzenschutzdiensts **30** (1978), 67–68

[8] Pitt, D., Studies on sharp eyespot disease of cereals. I. Disease symptoms and pathogenicity of isolates of *Rhizoctonia solani* Kühn and the influence of soil factors and temperature on disease development, Ann. appl. Biol. **54** (1964), 77–89

[9] Pokacka, Z. and D. Wojtaszek, Z badan nad patógenicznością *Rhizoctonia solani* Kühn na pszenicy i życie, Materiały XVII Sesji Naukowej I. O. R., Poznań (1977), 182–191

[10] Prew, R. D. and A. H. McIntosh, Effects of benomyl and other fungicides on take-all, eyespot and sharp eyespot diseases of winter wheat, Pl. Path. **24** (1975), 67–71

[11] Rashid, T. and E. Schlösser, Cereal foot rot pathogens. II. Occurrence and distribution of benomyl-tolerant strains, Med. Fac. Landbouw. Rijksuniv. Gent **42** (1977), 1057–1065

[12] Reinecke, P. and H. Fehrmann, *Rhizoctonia cerealis* van der Hoeven an Getreide in der Bundesrepublik Deutschland, Z. PflKrankh. PflSchutz **86** (1979), 190–204

[13] Tuyl, J. M. van, Genetics of fungal resistance to systemic fungicides, Med. Landbouwhogeschool Wageningen **72** (1977), 1–136

[14] Walt, P. J., van der, Criteria and methods used in classification in Lodder, J. (ed.), The yeasts, North-Holland, Amsterdam 1970, 34–113

Hanna M. Müller und Horst Lyr

Institut für Pflanzenschutzforschung Kleinmachnow der Akademie der Landwirtschaftswissenschaften der DDR

Morphologische und zytologische Veränderungen bei *Phytophthora infestans* (Mont.) de By. und *Phytophthora cactorum* (Leb. et Cohn) Schroet. unter dem Einfluß von Metalaxyl

1. Einleitung

Mit der Entwicklung des DL-N-(2,6-Dimethylphenyl)-N-(2'-methoxyacetyl)alaninmethylesters (CGA-48988, Metanaxin, Metalaxyl; Präparat: Ridomil) als hochwirksames Systemfungizid spezifisch gegen Oomyzeten (Schwinn, Staub and Urech [16]; Urech, Schwinn and Staub [19]; Mitchell, Kannwischer and Sanden [15]; Biehn, Bseifried, Snow and Young [5]; Staub, Dahmen and Schwinn [17]) eröffnen sich neue Aspekte für die Bekämpfung dieser Krankheitserreger bei vielen Kulturpflanzen. Die ausgezeichnete Wirksamkeit gegenüber den Arten der Gattung *Bremia*, *Peronospora*, *Phytophthora*, *Plasmopara*, *Pythium* und *Sclerospora* wird noch begünstigt durch den akropetalen und nachfolgenden apoplastischen Transport, die Translokalisation von der Blattbasis zur Blattspitze (Staub, Dahmen and Schwinn [18]; Cohen, Reuveni and Eyal [7]) und den kurativen Effekt gegenüber den falschen Mehltaupilzen (Kannwischer and Mitchell [11]; anonym, [1, 2]; Johnson, Davies and O'Brien [9]).
Nach Buchenauer [6] ist der Wirkungsmechanismus der bisher vorliegenden Oomyzetenfungizide (Prothiocarb, DPX3217, Aluminiumäthylphosphit, Pyroxychlor) einschließlich der Acylalanin-Derivate Metalaxyl und Furalaxyl noch weitgehend unbekannt. Ausgehend von der geringen oder fehlenden Effektivität anderer Systemfungizide zur Bekämpfung der falschen Mehltaupilze und der hochspezifischen Wirkung der oben genannten Mittel gegen die Oomyzeten sind bisher nicht bekannte Wirkprinzipien für den chemotherapeutischen Effekt zu erwarten. Mit der Kennzeichnung der morphologischen Variationen von *Phytophthora* und der zytologischen Lokalisation von Veränderungen in der Hyphe unter dem Einfluß von Metalaxyl können erste Hinweise für den Eingriff der chemischen Substanz in den Stoffwechsel des pilzlichen Organismus gegeben werden.

2. Methodik

Untersuchungen zum morphologischen und zytologischen Effekt des Wirkstoffes Metalaxyl von Lyr und Sunkel [13] wurden an Bouillon- und Agarkulturen von *Phytophthora infestans* und *Phytophthora cactorum* durchgeführt.
Bei der Kombination Mikroorganismus-Fungizid wurden Konzentrationen von 0,5-1-2-5-10-25-30-50 und 100 ppm bei einer Einwirkungsdauer der in DMF gelösten Substanzen von 2–40 h an Bouillonkulturen auskeimender Zoosporen untersucht. In die histo- und

zytologische Auswertung wurden außerdem Agarkulturen von *Phytophthora cactorum*, die 8—10 Tage unter dem Einfluß von Metalaxyl (0,2—8 ppm) standen, einbezogen [13]. Die histologische Beurteilung erfolgte an Lebendmaterial oder nach herkömmlichen Fixierungs- und Färbemethoden, die Fluorochromierung erfolgte mit Wobital 3BTZ. Für die elektronenmikroskopische Untersuchung wurden die Hyphen und Blätter mit Glutaraldehyd (3%) und OsO_4 (2%) in Cacodylatpuffer pH 7,2 fixiert und in Durcupan oder Epon eingebettet. Die zytologischen Befunde wurden an mit Uranylacetat und Bleinitrat kontrastierten Ultradünnschnitten am SEM 3—2 und am JEM 100B ausgewertet.

3. Ergebnisse

Veränderungen im Habitus und in der Ultrastruktur der *Phytophthora*-Arten werden erst nach längerer Konfrontation der Pilze mit dem Fungizid sichtbar.

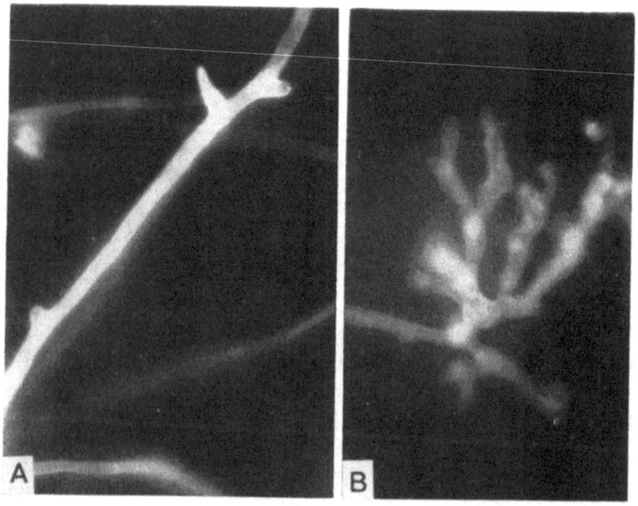

Abb. 1. *Phytophthora cactorum* nach Fluorochromierung mit Wobital 3 BTZ
A — Kontrolle, B — metalaxylbehandelt (1 Woche — 10^{-6} Mol)

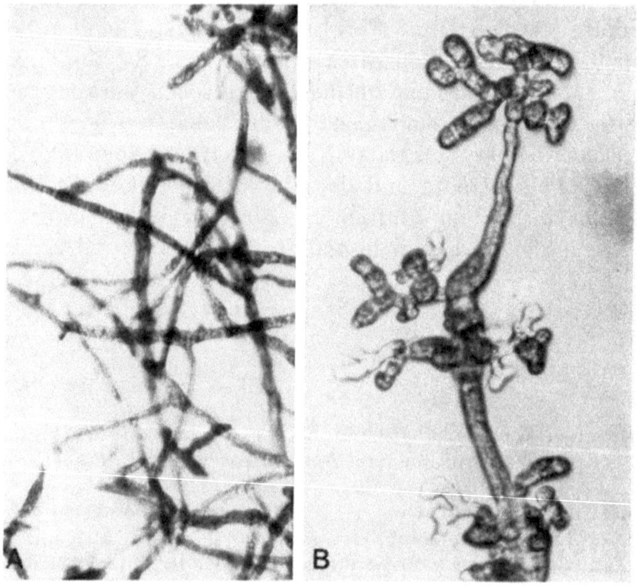

Abb. 2. *Phytophthora cactorum*, Übersicht des Hyphenwachstums
A — Kontrolle, B — metalaxylbehandelt (1 Woche — 10^{-6} Mol), Frischmaterial

Frühestens 3–4 Tage nach der Übertragung von Pilzmyzel auf metalaxylhaltigen Agar treten sowohl end- als auch seitenständig der fädig wachsenden Hyphen blasenförmige Sprossungen auf (Abb. 1–3).
Insbesondere die fluoreszenzmikroskopischen Abbildungen und Karminessigsäurefärbungen dieser Pilzhyphen verweisen auf ein verändertes Kern-Plasma-Verhältnis zugunsten des Kernanteils in der Zelle unter dem Einfluß von Metalaxyl (Abb. 1 und 3).
Im Elektronenmikroskop sind Veränderungen der Ultrastruktur jedoch schon frühestens 16–22 Stunden nach Applikation des Wirkstoffes in Abhängigkeit von seiner Konzentration feststellbar und verstärken sich bei längerer Einwirkdauer des Fungizides. Verstärkte Lomasomenbildung und Ausschleusung von Vesikeln verweisen auf eine erhöhte Stoffwechselaktivität im Bereich des Plasmalemmas. Die Membranen des endoplasmatischen Retikulums beginnen auseinanderzuweichen. Insbesondere sind Zellwandverdickungen lokal oder im Bereich der gesamten Zellwand unter Einlagerung von zytoplasmatischem Material (Abb. 5 und 6) beobachtet. Häufig ist der intraendoplasmatische Raum vesikulär aufgeweitet. Kern und Mitochondrien weisen

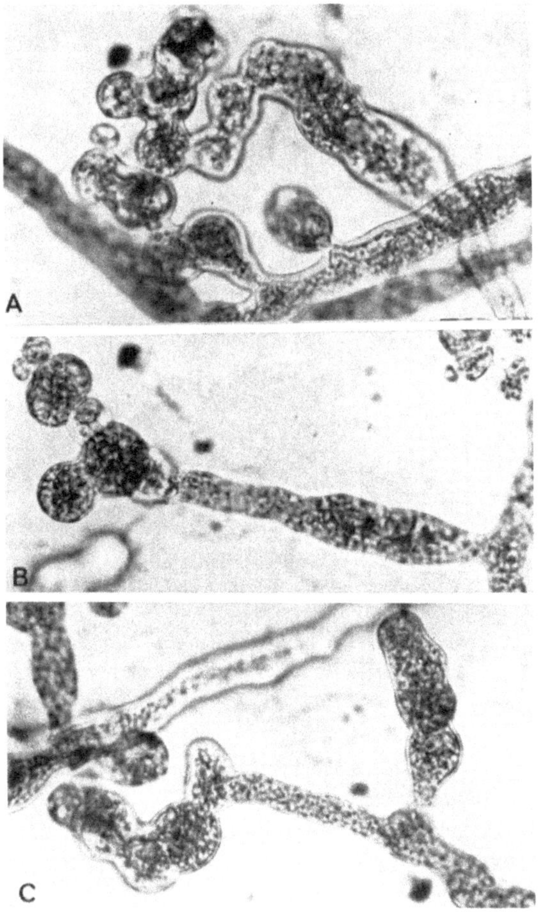

Abb. 3. Unter dem Einfluß von Metalaxyl (10^{-6} Mol – 1 Woche) entstandenen knollig verdickte Hyphenspitze von *Phytophthora cactorum*. Karminessigsäurefärbung

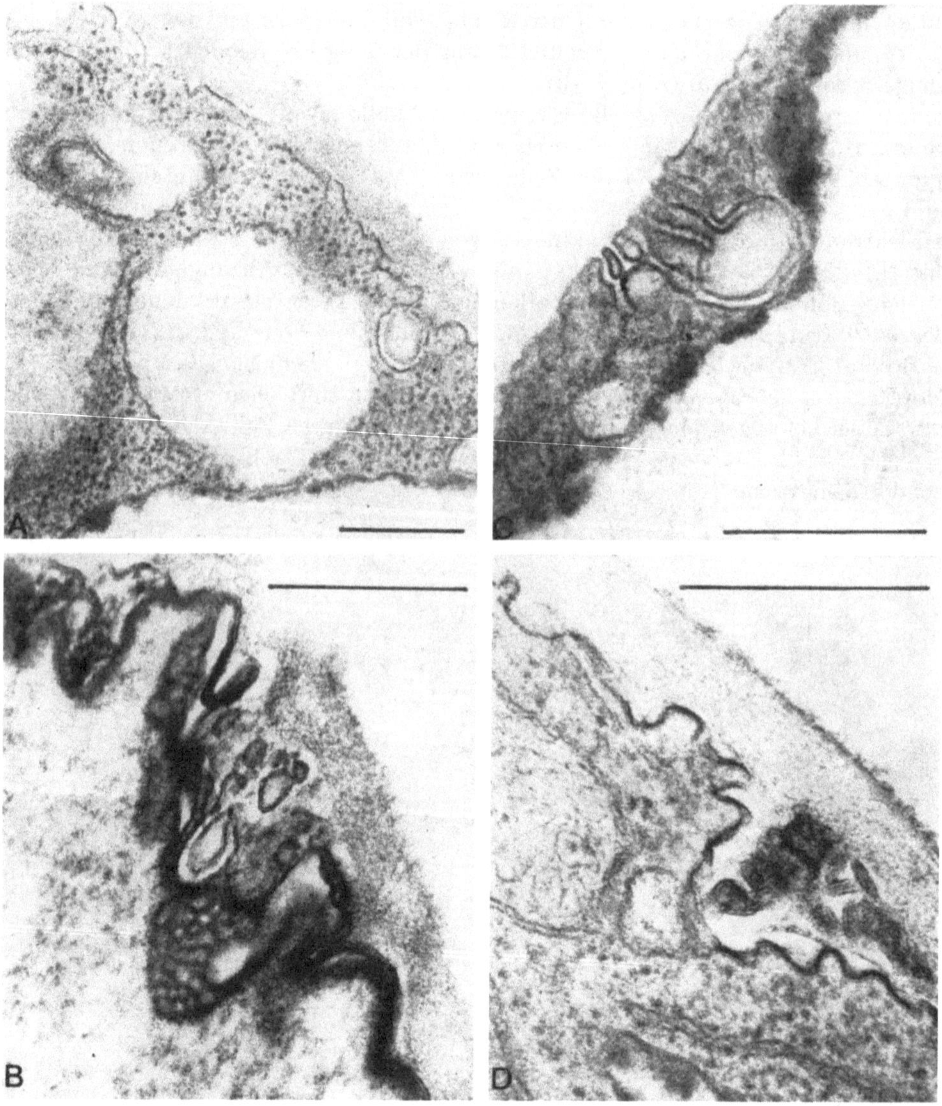

Abb. 4. Durch Metalaxyl hervorgerufene Aktivierung des Stoffwechsels im Bereich des Plasmalemmas, Ausschleusung von Vesikeln und Membranresten in den Raum zwischen Plasmalemma und Zellwand bei *Phytophthora infestans*
A – 0,5 ppm Metalaxyl 24 h
B – 0,5 ppm Metalaxyl 21 h
C – 1 ppm Metalaxyl 24 h
D – 5 ppm Metalaxyl 18 h

nur pathologische Veränderungen auf, wenn durch den Wirkstoffeinsatz von mehr als 50 ppm die normale Struktur der Zelle völlig zerstört wird.
Erste entsprechende Ergenisse zu den vorgenannten Veränderungen an den Pilzkulturen liegen auch aus den Untersuchungen zur Wirkung von Metalaxyl auf dem auf der Pflanze gewachsenen Mikroorganismus vor.

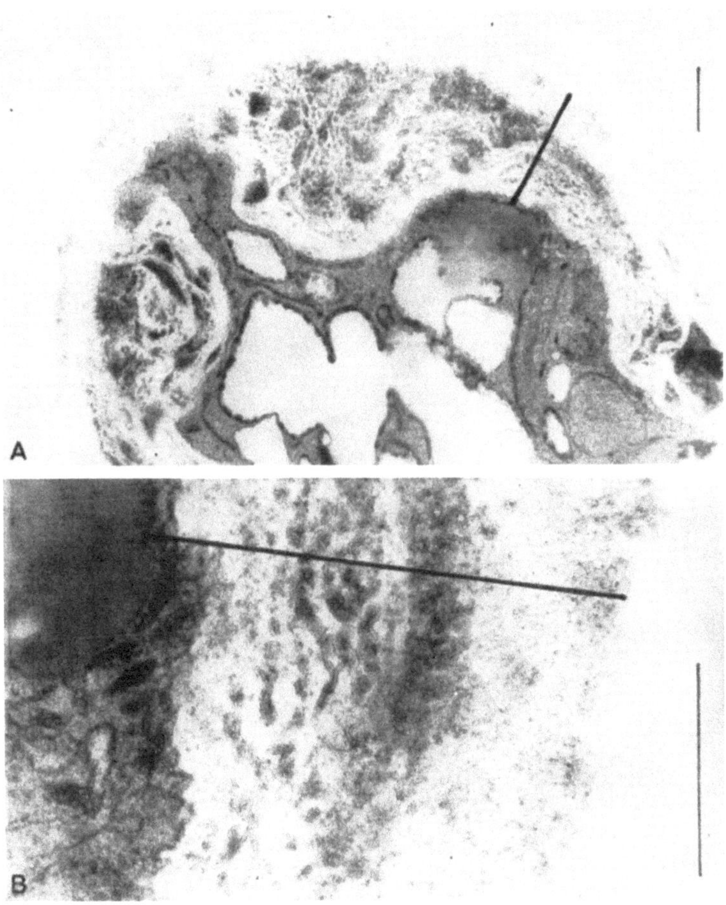

Abb. 5. In das Zellinnere erfolgende lokale Wandverdickung entsprechend dem „false septum" bei *Phytophthora infestans* unter dem Einfluß von 5 ppm Metalaxyl, 21 h.

A — Übersicht der lokalen Wandverdickung
B — Ausschnitt, den hohen Anteil an zytoplasmatischen Bestandteilen zwischen Plasmalemma und Zellwand zeigend

4. Diskussion

Aus den vorliegenden Ergebnissen erweist sich eindeutig eine beträchtliche zeitliche Verzögerung in der Ausprägung von pathologischen Merkmalen im Verhältnis zum Zeitpunkt der Wirkstoffapplikation. Die Ultrastrukturveränderungen sind abhängig von der Zeitdauer der Behandlung und der eingesetzten Substanzmenge. Da in den vorliegenden Versuchen die Wirkstoffkonzentration bei auftretenden pathologischen Effekten z. T. beträchtlich über den ED_{50}-Werten [13] liegen, können die im Bereich des endoplasmatischen Retikulums, des Plasmalemmas und der Zellwand lokalisierten Veränderungen nur sekundäre Effekte sein, die erst relativ langfristig (von 20 Stunden bis mehreren Tagen) nach einem umstimulierten Stoffwechsel der Pilzzelle sichtbar

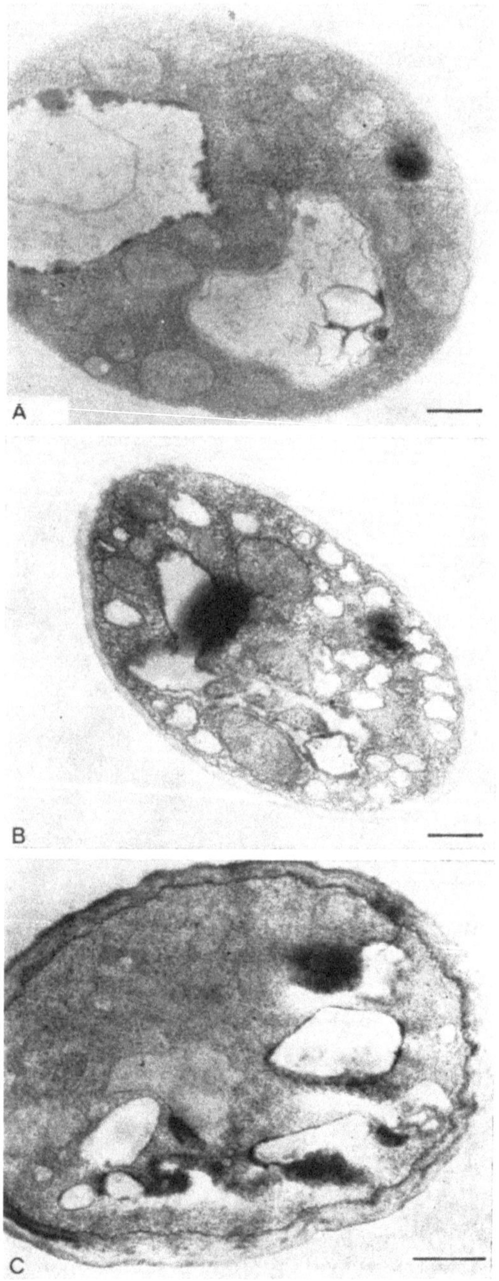

Abb. 6. Verdickung der gesamten Zellwand bei *Phytophthora infestans* unter dem Einfluß von Metalaxyl
A – Kontrolle
B – Beginnende Verdickung der Zellwand und Zunahme der Vesikelbildung peripher in der Hyphe, Metalaxyl 0,5 ppm, 24 h.
C – Relativ gleichmäßig verdickte, schwach gefaltete Zellwand, Metalaxyl 2 ppm, 16 h

werden. Mitochondrien werden erst durch eine hohe Dosis der Acylalanin-Derivate geschädigt, damit ist die für die Beeinflussung der Zellwand durch verschiedene Fungizide bei *Mucor* erarbeitete Hypothese zum veränderten Enzymstoffwechsel [12] auf diese Pilz-Fungizid-Kombination nicht übertragbar. Nach Hemmes und Hohl [10] und Bartnicki-Garcia [3, 4] ist für die Ausbildung der normalen Hyphenmorphologie und des Apikalwachstums der Pilze ein Regulationssystem des Zellwandmetabolismus verantwortlich zu machen, das sich aus parallel ablaufenden lytischen und synthetischen Prozessen zusammensetzt. Förderung der Vesikelbildung und häufiges Auftreten von Lomasomen sprechen für einen aktivierten Enzymtransport in Richtung auf die Zellwand. Nach MEYER, PARISH and HOWL [14] führt der Entzug von β-1,3-Glukan zu einem unkontrollierten Hyphenwachstum mit Anschwellen der Hyphen und apikaler Verzweigung. Jedoch ist eine Beeinflussung der zellulosehaltigen fibrillären Schicht der Zellwand ebenfalls nicht auszuschließen, da unter der Einwirkung von Cumarin, das sowohl die Synthese von Chitin und Zellulose fördert, ähnliche morphologische und zytologische Veränderungen bei *Trichophyton mentagrophytes* induziert werden (DALLOLIO and VANNIN [8]).

Die veränderte Kern-Plasma-Relation in den Hyphen von Metalaxyl behandelten Kulturen ist nicht unbedingt auf eine vermehrte Kernteilung zurückzuführen, sondern könnte ebenso auf einer Hemmung in der Bildung des zytoplasmatischen Anteils der Zelle beruhen, die Ultrastrukturuntersuchungen lassen gegenwärtig dazu keine Aussagen zu.

Weitere Untersuchungen zur Erfassung des primären Angriffspunktes der Acylalanine in der Oomyzetenzelle sind erforderlich.

Literatur

[1] ANONYM, Ridomil und seine Kombinationen, CIBA-GEIGY Agro-Division-Produktionsinformation, (1979a)

[2] ANONYM, Fungicide and Nematicide Tests, Results of 1978, American Phytopathology Society, 34 (1979b)

[3] BARTNICKI-GARCIA, S., Control of dimorphism in *Mucor* by hexoses: inhibition of hyphal morphogenesis, Journal of Bacteriology, 96, 5 (1968), 1586–1594

[4] BARTNICKI-GARCIA, S., Fundamental aspects of hyphal morphogenesis, Symp. Soc. Gener. Microbiology XXIII, Microbial Differentation, (1973), 245–267

[5] BIEHN, W. L., BSEIFRIED, B., SNOW, J. and YOUNG, T., A new type of systemic fungizide which its active against the phycomycetes, Proc. Am. Phytopathology Soc., 4 (Abstr.), (1977), 94

[6] BUCHENAUER, H., Entwicklungstendenzen in der Chemotherapie der Kulturpflanzen, Mitteilungsblatt der Biologischen Bundesanstalt, 191 (1979), 139–162

[7] COHEN, Y., REUVENI, M. and EYAL, H., The systemic antifungal activity of Ridomil against *Phytophthora infestans* on tomato plants, Phytopathology, 69, 6(1979), 645 to 649

[8] DALL'OLIO, G. and VANINI, G. L., Coumarin-induced disturbances of morphological development and cell wall formation in *Trichophyton mentagrophytes*, European Journal of Cell Biology, 18 (1979), 390–397

[9] JOHNSON, DAVIS, O'BRIEN, Soil application of CGA-48988 a systemic fungizide controlling *Peronospora tabacina* on tobacco, Pl. Dis. Rep., 63, 3 (1979), 212–215

[10] HEMMES, D. E. and HOHL, H. R., Ultrastructural aspects of encystation and cyst germination in *Phytophthora parasitica*, J. Cell. Sci., 9 (1971), 175–191

[11] KANNWISCHER, M. E. and MITCHEL, D. J., The influence of a fungicide on the epidemiology of Black shank of Tobacco, Phytopathology, 68 (1978), 1760–1765

[12] LYR, H. und CASPERSON, G., Pathologische Zellwandsynthese bei *Mucor mucedo* (L.) FRES. unter Einwirkung einiger Fungizide und anderer Verbindungen, Symposium Reinhardsbrunn, 1980

[13] LYR, H., SUNKEL, M., Zur Wirkungsweise von Metanaxin, Symposium Reinhardsbrunn, 1980

[14] MEYER, R., RARISCH, R. W. and HOHL, H. R., Hyphal tip growth in Phytophthora, Arch. Microbiol., 110 (1976), 215–224
[15] MITCHELL, D. J., KANNWISCHER, M. E. and SANDEN, G. E., Control of black shank of tobacco with an acylalanine analogue, Proc. Am. Phytopathology Soc., 4 (1977), 179–180
[16] SCHWINN, F. J., STAUB, T. and URECH, P. A., A new type of fungicide against diseases caused by Oomycetes, Meded. Fak. Landbouwhogesch. Gent, 42 (1977)
[17] STAUB, T., DAHMEN, H. and SCHWINN, F. J., Effect of Ridomil on the development of target pathogens on their host plants, 3rd. Intern. Congr. Plant Pathol. München 1978, Abstr. (1978), 366
[18] STAUB, T., DAHMEN, H. and SCHWINN, F. J., Biological characterisation of uptake and translocation of fungicidal acylalanins in grape and tomato plants, Z. Pflanzenkrankheiten und Pflanzenschutz, 85 (1978), 162–168
[19] URECH, P. A., SCHWINN, F. J. and STAUB, T., CGA-48988 a novel fungicide for the control of late blight downy mildew and related soilborne disease, Proc. Br. Crop. Prot. Conf., 42 (1977), 623–631

Anschriften der Verfasser:
Dr. sc. HANNA MÜLLER
Prof. Dr. H. LYR
Institut für Pflanzenschutzforschung
der Akademie der Landwirtschaftswissenschaften der DDR
1532 Kleinmachnow
Stahnsdorfer Damm 81

L. Orlikowski

Research Institute of Pomology and Floriculture, Skierniewice, Poland

Population Dynamics of *Phytophthora cryptogea* in Soil Drenched with Systemic Fungicides and Influences of Foot Rot of Gerbera

Occurrence of foot rot of gerbera (*Gerbera jamesonii* Bolus) incited by *Phytophthora cryptogea* was first described in the USA by Tompkins and Tucker [13]. Since that time disease has been recorded in many countries as the most serious and common disease of gerbera. Also in Poland the fungus is the main threat in many gerbera — cultivated greenhouses [7]. The incidence of the disease may be minimized by either substrate sterilisation with steam or the use of containers and strict sanitary cultural practices. Care should be taken not to re-infect beds with new transplanting gerberas. Also incorporation of effective fungicides into substratum at the time of gerbera planting or drenching the soil during plant cultivation may minimize disease losses.

Chemical control of *Phytophthora* foot rot of gerbera has been studied by several workers. Say [10] reported that dichlone or copper oxychloride applied into infested soil may stop the fungus spreading. Garibaldi [4] concluded that disease control can be achived by drenching weekly Dexon or Difolatan at the dosage of 1 g/m^2. Among systemic fungicides Schickedanz [11] used Previcur N, Bonnet et al. [1] Previcur and Aliette, whereas Rattink ,[9] Aliette, Fongarid and Trexal. The efficacy of those chemicals were conneted with the population density of *P. cryptogea* in soil, frequency of application and their concentration. The experiments reported here were undertaken to determine [1] effect of some systemic fungicides on population dynamic of *P. cryptogea* in peat and [2] development of gerbera foot rot.

Materials and Methods

Peat with pH of 5.7 was used for gerbera growing. The substrate was artificially infested with *P. cryptogea*, isolate 80 obtained from the rotted foot of gerbera. The fungus was grown for 3 weeks on rolled oats [5] at 25 °C in the dark. Cultures were blended with a minimum of water and the slurry mixed thoroughly with peat. After 2 weeks storage of infested substrate in plastic bags in the greenhouse, pots were filled with peat and 2-month-old gerberas were planted. Immediately after the transplanting of gerbera the following systemic fungicides were applied into substrate:

1. Aliette [aluminium tris (0-ethyl phosphonate)] in concentrations 0.1% and 0.2%
2. Fongarid 25WP [methyl-N-(2,6-dimethylphenyl)-N-furoyl-(2)-alaninate] in concentrations 0.001%, 0.005%, 0.01%, 0.05%
3. Previcur N [N-(3-dimethylaminopropyl) thiocarbamic acid S-ethylester hydrochloride] in concentration 0.15%

4. Ridomil 25WP [D, L-N-(2,6-Dimethylphenyl)-N-(2-methoxyacetyl)-alaninmethylester] in the same concentrations as Fongarid

Fifty ml of fungicide suspensions were incorporated into each pot. The plants were growing on greenhouse benches on gravel layers at temperatures varied from 15° to 27 °C. The disease incidence was recorded at 7-day intervals on the basis of 40 plants per treatment. Experiments were repeated 2 – 4 times.

Population densities of *P. cryptogea* in the infested peat was estimated before the gerbera planting (= initial population 389 propagules/g of air dry peat) during 45 days of gerbera cultivation at 15 day intervals. The propagule numbers per g of air dry peat were determined by the method of FLOWERS and HENDRIX [3].

Results

The fungicides incorporated into peat artifically infested with *P. cryptogea* differed significantly the fungus populations. At the time of gerbera planting 389 propagules per gram of air dry peat were detected, whereas, in the substrate non-treated with fungicides, the fungus density increased over 568 propagules after 15 and 30 days of plant cultivation. After the next 15 days the population density decreased to the level of 438 propagules [Fig. 1]. Application of Fongarid or Ridomil in concentration of 0.01% and 0.05% decreased sharply the fungus propagule numbers. From peat samples collected from the rhizospheres of gerbera drenched with both chemicals no/or few propagules were detected [Fig. 1, 2]. However, incorporation of those chemicals into peat even in concentration 0.05% did not eliminate the fungus completely. After 3 months of gerbera cultivation about 50% of plants wilted or died. The lower rates of Fongarid and Ridomil also affected population of *P. cryptogea*. The efficacy of Fongarid applied in the concentration of 0.001% was similar to Aliette 0.2% and better

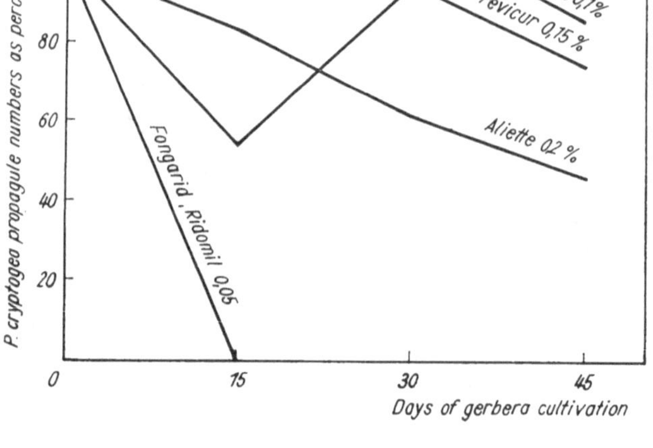

Fig. 1. Population dynamics of *Phytophthora cryptogea* in the rhizospheres of gerbera in relation to fungicides applied

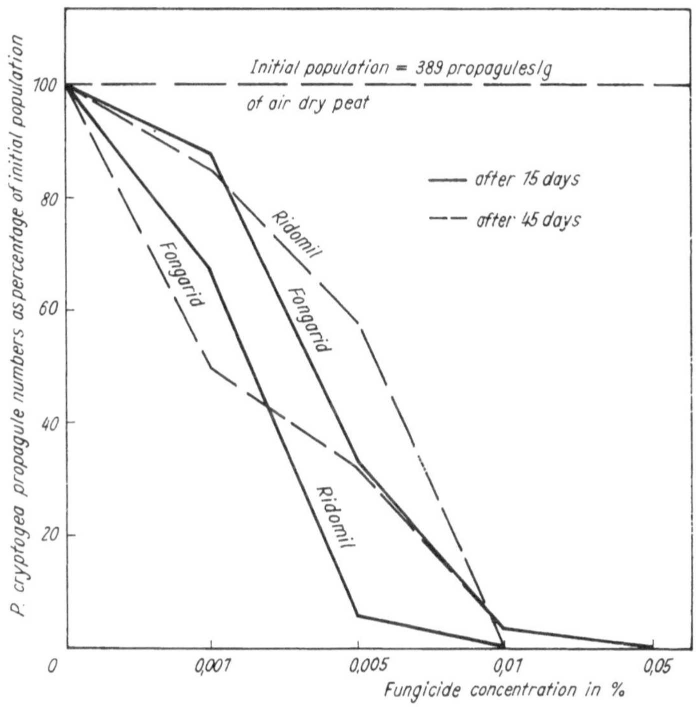

Fig. 2. Relationship between population dynamics of *Phytophthora cryptogea* and concentration of fungicides

than Previcur N 0.15% [Fig. 1]. Incorporation of Aliette in the rate of 0.1% only slightly decreased the level of the pathogen in peat [Fig. 1].

The apparent disease incidence on gerbera was determined by fungicides applied. Excellent protection from foot rot was provided by the treating of plants with Fongarid and Ridomil 0.05% at the time of planting [Tab. 1]. No gerberas were killed within 7 weeks of cultivation and only a few plants showed wilt symptoms after 63 days. Decrease of fungicides concentration to 0.01% was also very effective in disease control

Table 1
Percent of gerbera with Phytophthora foot rot symptoms

Fungicide	Concentration in %	Days after planting			
		14	28	49	63
Aliette	0,1	30	88	97	100
Aliette	0,2	18	76	97	100
Fongarid 25WP	0,001	27	47	54	92
Fongarid 25 WP	0,005	6	15	32	70
Fongarid 25WP	0,01	1	2	6	10
Fongarid 25WP	0,05	0	0	0	1
Precivur N	0,15	12	44	76	92
Ridomil 25WP	0,001	42	83	87	99
Ridomil 25WP	0,005	1	22	42	70
Ridomil 25WP	0,01	0	8	11	27
Ridomil 25 WP	0,05	0	0	0	3
Control	—	83	100	100	100

during the first 4—7 weeks. The lower rates of the both fungicides incorporated into infested peat, were as efficacious or better then Aliette or Previcur [Tab. 1].

Discussion

The established population of *P. cryptogea* in the peat used for gerbera growing, was almost the same as inoculum densities detected in rhizospheres of plants at the time of their wilting [8]. In greenhouse cultivation of gerbera, such high levels of the fungus may be detected only in places where plants show foot rot symptoms. Therefore, the efficacy of the systemic fungicides in gerbera protection, has been determined in conditions very favourable for plant infection.

The obtained results indicate on very high efficacy of Fongarid and Ridomil in foot rot control even at low concentrations. Beside their systemic and curative activity [12, 14] it was found here, that both chemicals influenced also the fungus population. KANNWISCHER and MITCHELL [6] reported almost complete inhibition of in vitro development of *P. cryptogea* at 5 µg/ml of Ridomil in V-8 juice agar. Afterwards, the authors concluded the lack of control of tobacco black shank with Ridomil beginging about 50 days after planting. In this study, the concentration of the both fungicides influenced disease incidence of gerbera. Complete control of foot rot on gerbera growing in peat drenched with chemicals in concentration 0.05% were observed for 8 weeks.

The inhibition of foot rot development on gerbera with Fongarid was earlier observed by RATTINK [9]. Incorporation of the fungicide into soil in rate 125 to 375 g/m³ gave excellent control of *P. nicotianae* var. *nicotianae* on saintpaulia and siningia [2]. DIRKSE and LANGHOUT [2] obtained also good control of the fungus using LS 74783 (Aliette) in concentration 0.25%. In this study, Aliette inhibited partly diseased development on gerbera only during the first two weeks. Also data obtained with the efficiency of Previcur N indicate, that it is possible to partially control *Phytophthora* foot rot by drenching substrate with this fungicide. However, my results do not confirm those of SCHICKEDANZ [11] indicated, that one treatment of gerbera with Previcur may keeps young plants healthy for about 3 months. Undoubtly, population density of the fungus, type of substrate and temperature, play an important role in disease development.

References

[1] BONNET, PH., P. RICCI, S., MERCIER and C. ANTONINI, Le deperissement du *Gerbera* du au *Phytophthora cryptogea* PETHYBR. et LAFF, Journee du Gerbera — INRA Fujus (1977), 43—52

[2] DIRKSE, F. B. and L. LANGHOUT, Bestrijding van *Phytophthora* in saintpaulia. Bestrijding van *Phytophthora* in siningia (gloxinia), Bloem onderzoek in Nederland over 1977, 157—160

[3] FLOWERS, R. A. and W. HENDRIX, Gallic acid in procedure for isolation of *Phytophthora parasitica* var. *nicotianae* and *Pythium* spp. from soil, Phytopathology, **59** (1969), 725—731

[4] GARIBALDI, A., Osservazioni sul merciume pedale della *Gerbera* causato da *Phytophthora cryptogea* PETHYBR. et LAFF, Riv. dell'Ortofl. Italiana **2** (1974), 1—13

[5] HENDRIX, J. W., Inhibition of growth of tobacco plants by *Phytophthora cryptogea*, Tob. Sci., **20** (1976), 152–153

[6] KANNWISCHER, M. E. and D. J. MITCHELL, The influence of a fungicide on the epidemiology of black shank of tobacco, Phytopathology, **68** (1978), 1760–1765

[7] ORLIKOWSKI, L., Przyczyny zamierania gerbery (*Gerbera jamesonii* BOLUS) w niektórych gospodarstwach ogrodniczych w Polsce, Prace Inst. Sad. s. B, **2** (1976/77), 197–201

[8] ORLIKOWSKI, L., The occurrence of *Phytophthora cryptogea* PETHYBR. et LAFF. in gerbera (*Gerbera jamesonii* BOLUS) growing sites, Bull. L'Acad. Pol. Sci., **26** (1978), 495–498

[9] RATTINK, H., Verwelkingsziekte bij gerbera, Bloem onderzoek in Nederland over 1978, 158

[10] SAY, J. S., Gerberas for pleasure and profit, N. Z. J. Agr. **106** (1963), 343–351

[11] SCHICKEDANZ VON F., Zur Bekämpfung von *Phytophthora cryptogea* an *Gerbera*, Nachr. Deut. Pflanzenschutzd., **28** (1976), 164–167

[12] STAUB, T., H. DAHMEN and F. J. SCHWINN, Biological characterization of uptake and translocation of fungicides acylalanines in grape and tomato plants, Z. Pflanzenkr. und Pflanzenschutz, **85** (1978), 162–168

[13] TOMPKINS, C. M. and C. M. TUCKER, Foot rot of Chine – aster, annual stock and transvaal daisy caused by *Phytophthora cryptogea*, J. Agric. Res., **55** (1937), 563–574

[14] URECH, P. A., F. SCHWINN and T. STAUB, CGA-48988, a novel fungicide for the control of late blight downy mildews and related soil-borne diseases, Proc. 1977 British Crop Prot. Conference, 623–631

G. Oros and S. Süle

Research Institute for Plant Protection
H-1525 Budapest, P. O. Box 102, Hungary

Antibacterial Activity of Calixin on Phytopathogenic Bacteria

The fungicid calixin is an emulsion of tridemorph (N-Tridecyl-2,6-diemthylmorpholin) and used widely as a foliar spray against powdery mildews and other fungal diseases [6, 7]. In field experiment it had a distinct fungicidal effect on *Helminthosporium* sp., *Septoria* sp., *Rhizoctonia* sp., *Penicillium* sp. [5]. In laboratory experiments it was active against representatives of nearly all taxonomic groups of fungi [3, 4, 1] and a few grampositive bacteria [4].
Previous work [8] from this laboratory had shown that calixin was very active against some phytopathogenic bacteria *(Corynebacterium michiganense, C. fascians)*, too. In this report we provide further evidence that calixin is effective against certain phytopathogenic bacteria and try to show its possible antibacterial mode of action.

Materials and Methods

Commercially available calixin was used over the experiments. Medium for bacterial cultures contained 1% bouillon extract (Human OEV), 0.1% yeast extract (Oxoid), 0.5% glucose, 0.1% K_2HPO_4, 0.05% KH_2PO_4, 0.05% $MgSO_4$ and Hoagland-Arnon's microelements. For solidification 2% agar-agar was added to this broth.
The minimal inhibitory concentrations (MIC) of calixin were determined by the common twofold dilution technique in Petri dishes using a multipoint inoculator. Calixin was added to the medium after autoclaving as an alcoholic solution. Bacterial growth was observed visually after 5 days incubation at 28 °C.
The used bacterial strains are listed in Table 1. Leakage of electrolytes were measured in 0.5% glucose solution or in distilled water using a Radelkis conductivity meter. Bacterial suspensions were prepared by harvesting bacterial cells by centrifugation from 24 hr broth shake culture and washing the cells in sterile 0.5% glucose solution. The concentration of cells was adjusted photometrically to contain 5×10^9 cells/ml. Calixin was added in 20 or 150 µg/ml active ingredient. The first measurements were made after 5 seconds and than recorded continually for 20 min.; after 1 and 20 hr the measurements were repeated. The conductance values (µS) obtained in each experiment are expressed as percent of the difference of maximum deviations (Δ%).

Results and Discussion

The minimal concentrations of calixin required to inhibit various plant pathogenic bacteria are summarized in Table 1. Bacteria differed considerably in sensitivity but, all of the *Corynebacteria* were sensitive. *C. michiganense* and *C. nebraskense* show very

Table 1
Antibacterial spectrum of calixin

Bacteria	Strain No.	Host plant	MIC ppm
AGROBACTERIUM			
A. tumefaciens	103	grapevine	>1000
A. tumefaciens	0	grapevine	>1000
A. tumefaciens	C-58	cherry	>1000
A. tumefaciens	B-6	tomato	>1000
A. radiobacter	54	soil	>1000
CORYNEBACTERIUM			
C. betae	101	sugar beet	100
C. fascians	27	pelargonium	50
C. flaccumfaciens	8	bean	25
C. michiganense	9	tomato	10
C. nebraskense	10	maize	10
C. oortii	11	tulip	50
PSEUDOMONAS			
P. lachrymans	31	cucumber	>500
P. mors-prunorum	32	pear	>1000
P. phaseolicola	67	bean	>500
P. pisi	68	pea	>500
P. syringae	33	apricot	>1000
P. syringae	48	Johnson gras	>1000
P. tabaci	41	tobacco	>1000
P. tomato	35	tomato	>1000
ERWINIA			
Erwinia carotovora			
pv. atroseptica	K459	potato	>1000
pv. carotovora	60	potato	>1000
XANTHOMONAS			
X. alfalfae	KX-1	*lucerne*	>500
X. campestris	2D 510	cabbage	200
X. pelargonii	58	pelargonium	>500
X. vesicatoria	53	pepper	>500
Bacillus subtilis			10
Staphylococcus aureus			10

low MIC values and this may suggest a promissing practical application against diseases caused by these bacteria. However our preliminary data (S. Süle and G. Oros unpublished) show that phytotoxicity of calixin may limit its practical applications. *Agrobacterium* and *Erwinia* were the most resistant; all of them were tolerant to 1000 µg/ml. In the case of *Pseudomonas* sensitivity depended from the strains but all had at least 500 µg/ml MIC value. *Xanthomonas* are moderately resistant, their MIC values were 200—500 µg/ml. According to these results we chosed a resistant *(Agrobacterium)* a semi-resistant *(Xanthomonas)* and a sensitive *(Corynebacterium)* strain for the electrolyte leakage measurements. The data concerning for the elctrolyte leakage of the first 20 min are presented in Figure 1. The concentrations of calixin in this experiment was 150 µg/ml. It is apparent that electrolyte leakage occurs very quickly in all of the cases. At the resistant and the semiresistant strains this values returned to the original level after 15—25 min, but at the sensitive strain it remained constantly high. This may show that cell membrane of the sensitive strain damaged irreversibly.

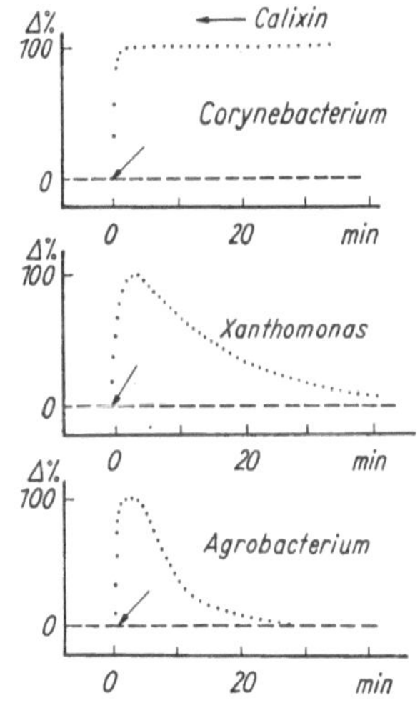

Fig. 1. Effect of Calixin [150 µg/ml] on electrolyte leakage of different bacteria

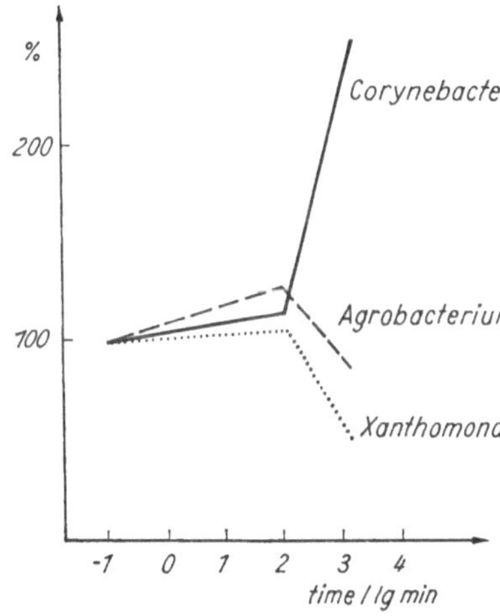

Fig. 2. Electrolyteleakage of bacteria treated with 20 µg/ml Calixin

In the second experiment (Fig. 2) 20 µg/ml calixin was used and measurements were continued till 20 hr. In this case, the electrolyte leakage of sensitive strain has increased steadily all over the experiment. However, at resistant and semi-resistant strains the initial increasing did not appear and there were a slight decreasing in electrolyte leakage of the suspension. This may show that this concentration apparently had no effect on these resistant bacteria.

The mechanism of calixin action on bacteria is unknown, but its antifungal mode of action is similar to the effects of inhibitors of sterol biosynthesis. This may not be an explanation for its antibacterial activity as bacteria do not contain or need sterols [4].

Our results suggest that calixin may primarily damage the cell membranes of sensitive bacteria irreversibly. Similar results were obtained with a few fungi [2], too.

Acknowledgements

The authors would like to express their thanks for helpful discussion to Dr. G. Josepovits.

References

[1] H. Bergmann, H. Lyr, E. Kluge and G. Ritter, Untersuchungen zur Wirkungsweise von Tridemorph, V. Int. Symp. Syst. Fungicide in Reinhardsbrunn (1975), p. 183, Academie-Verlag

[2] H. BUCHENAUER, Various response of cell membranes of plant and fungal cells to different systemic fungicides, VIIIth Intern. Plant Prot. Congress, Moscow (1975)
[3] M. GASZTONYI and T. CSERHÁTI, Die Komponenten von Tridemorph und ihre fungitoxische Aktivität, Arch. Phytopathol. u. Pflanzenschutz **14** (1978), 255–262
[4] A. KERKENAAR, D. BARUG and A. KAARS SIJPESTEIN, On the antifungal mode of action of tridemorph, Pest. Biochem. Physiol. **12**, 3(1979), 195–204
[5] J. KRADEL and E. H. POMMER, Beiträge zum Wirkungsspektrum von Tridemorph, Meded. Rijksfac. Landbouwwetensch. Gent. **36**, 1 (1971), 343–347
[6] S. OTTO, N. DRESCHER and S. HAHN, Rückstandsuntersuchungen mit Tridemorph, Meded. Rijksfac. Landbouwwetensch. Gent. **36**, 1 (1971), 419–423
[7] W. H. PHILLIPS and J. P. FROST, The use of carbendazim and tridemorph in the UK for the control of *Rhynchosporium secalis* in barley, Proc. 8th British Insecticide and Fungicide Conference (1975), 437
[8] S. SÜLE and G. OROS, A less known effect of Calixin, Scientific Days for Plant Protection of Hungarian Academy of Sciences, 3–4. May 1979, Budapest

C. POLTER

Institut für Pflanzenschutzforschung der Akademie der Landwirtschaftswissenschaften, Kleinmachnow, DDR

Aufnahme und Metabolismus von trans-2,6-Dimethyl-N-dodecylmorpholin bei unterschiedlich empfindlichen Pilzen

Trans-2,6-Dimethyl-N-dodecylmorpholin hat, wie andere N-alkylmorpholine, eine spezifische fungizide Wirksamkiet. Außer gegen echten Mehltau wirkt es z.B. im Labortest gut gegen *Botrytis cinerea* ($ED_{50} = 3 \times 10^{-7}$ mol) aber praktisch nicht gegen *Fusarium oxysporum* ($ED_{50} = 9 \times 10^{-5}$ mol). Auch gegen *Torulopsis candida* ist es wenig wirksam ($ED_{50} = 1 \times 10^{-5}$ mol). Geprüft werden sollte, ob diese Wirkunterschiede bei verschiedenen Pilzen durch Differenzen in der Stoffaufnahme bzw. Metabolisierung hervorgerufen werden, oder anders verursacht sind.
Eine Metabolisierung des Morpholinfungizids Tridemorph ist beschrieben worden, z.B. von OTTO u. DRESCHER [3], HAWKINS u. a. [1], WARING u. WOLFE [4]. Deshalb wurde geprüft, ob sich Hinweise für einen Metabolismus auch in oben genannten Pilzen finden lassen. Hinsichtlich einer Oxidation am Stickstoff des Morpholinringes trifft dies zu.

Methoden

Experimente wurden mit trans-2,6-Dimethyl-N-dodecylmorpholin sowie der entsprechenden 1-^{14}C-Dodecyl-Verbindung ausgeführt, die aus Isomerengemischen chromatographisch abgetrennt, gereinigt und auf Reinheit geprüft wurde. Dabei wurden DC-Trennungen an Kieselgel G (Merck) mit Gemischen von mehrfach destilliertem Hexan und Aceton in verschiedenen Verhältnissen (vorwiegend 3:1 und 1:3) eingesetzt. Diese Systeme wurden auch bei Metabolismus-Experimenten verwendet.
Aus chromatographisch gereinigtem fungizidem Wirkstoff wurde als Vergleichssubstanz und zur Biotestung das N-oxid nach JERCHEL u. JUNG [2] hergestellt und chromatographisch vom nicht umgesetzten Wirkstoff befreit. Im Gegensatz zum cis-2,6-Dimethyl-N-dodecylmorpholin und den entsprechenden 2,5-Dimethyl-N-dodecylmorpholinen, die in 16 Stunden vollständig zum N-oxid umgesetzt werden, mußte aus dem N-oxid der trans-2,6-Verbindung ein Rest von bis zu 20% nicht umgesetzter Verbindung abgetrennt werden.
Wirkstoff und Metabolite wurden aus Pilz und Nährlösung mit mehrfach destillierten Lösungsmitteln, vorwiegend mit Petroläther, Diäthyläther und Amylalkohol eluiert, aus Myzel auch mit Aceton. Die Radioaktivität wurde szintillationsspektrometrisch bestimmt, Myzel im Probenverbrennungegerät „Oximat" verbrannt. Überschlagswerte für Radioaktivitätsverteilungen wurden durch quantitative Auswertung von Radio-Scannogrammen gewonnen.
Verglichen wurden *Fusarium oxysporum*, *Botrytis cinerea* und *Torulopsis candida*. Die Hyphenpilze wurden unter sterilen Bedingungen in Nährlösung nach SISLER in Schüttelkulturen aus Sporen gezogen, die Hefe in gleicher Nährlösung kultiviert. Untersuchungen zur Metabolisierung des Wirkstoffes wurden unter gleichen Bedingungen wie Wachstums-

teste ausgeführt. Unterschiedlich waren die Größen der Kulturansätze (50, 200, 500 ml) und die Kulturzeiten. Radioaktive Substanzen wurden unter sterilen Bedingungen zum Ansatz gegeben.

Experimente und Resultate

Aufnahme von trans-2,6-Dimethyl-N-dodecylmorpholin in Pilze

Nach 0,5 bis 8 Stunden Aufenthalt in Lösung mit ^{14}C-trans-2,6-Dimethyl-N-dodecylmorpholin hatte Myzel von *Botrytis cinerea* einen höheren Gehalt an ^{14}C-Verbindungen als solches von *Fusarium oxysporum* (Abb. 1). Dabei wurde Radioaktivität im Myzel und in Nährlösung an Sterilkulturen (50 ml) bestimmt. Da die ^{14}C-Verteilung bilanzmäßig erfaßt wurde, sind durch Meßfehler bedingte Fehlbeurteilungen unwahrscheinlich. Trotz der Unterschiede im ^{14}C-Gehalt in beiden Pilzmyzelien war der Schluß auf entsprechende Unterschiede der Stoffaufnahme nicht zulässig, da bei beiden Pilzen schon nach kurzer Versuchszeit (30 min) pilztypische ^{14}C-Gehalte im Myzel erreicht wurden, die sich in 8 Stunden wenig änderten (Abb. 1).

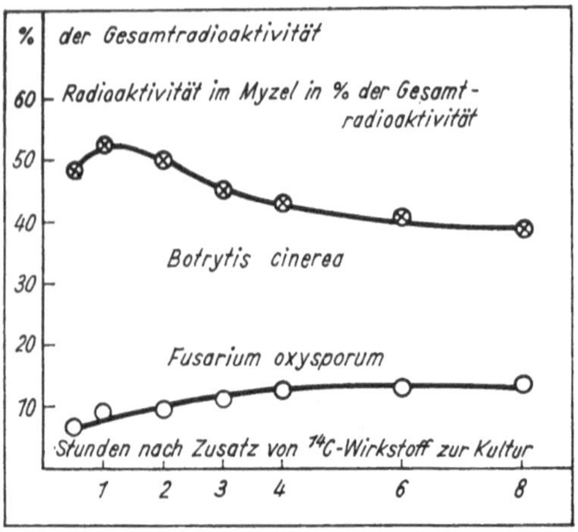

Abb. 1. Zeitabhängige Änderung des Radioaktivitätsgehaltes im Pilzmyzel bei steriler Schüttelkultur unter Zusatz von ^{14}C-trans-2,6-Dimethyl-N-dodecylmorpholin. Bestimmung des ^{14}C-Gehaltes der Myzelien szintillationsspektrometrisch nach automatisierter Verbrennung im Probenverbrennungsgerät „Oximat". Zur Bilanzierung der ^{14}C-Verteilung wurde auch die Radioaktivität der Nährlösung bestimmt (je Meßpunkt 3 Kulturen)

Danach wurden Ansätze mit verändertem Myzel-^{14}C-Verhältnis und unter etwas veränderten Versuchsbedingungen (Temperatur, Größe des Ansatzes) angesetzt. Um ausreichende Vergleichbarkeit bei kurzen Versuchszeiten zu erhalten, wurde aus je einem Ansatz jeweils ein Nährlösungsaliquot entnommen, das im Verhältnis zur Flüssigkeitsmenge des Ansatzes klein war. Dabei zeigte sich (Abb. 2), daß der ^{14}C-Gehalt der Medien beider Pilze fast gleich schnell abnimmt, das Fungizid also fast gleich gut von beiden Pilzen aufgenommen wird. Hierbei wird ein Teil wahrscheinlich zunächst von Zellwänden und Schleimschichten adsorbiert. Für das Einpegeln der ^{14}C-Konzentration im Pilz ist die Geschwindigkeit einer unmittelbar nach der Aufnahme bzw. Bindung des Fungizids an das Myzel, einsetzenden Wiederabgabe von radioaktiver Substanz an das Medium verantwortlich. Die an das Medium abgegebenen Substanzen sind nach chromatographischer Analyse polarer als der zum Versuch eingesetzte ^{14}C-Wirkstoff. Wahrscheinlich ist die Metabolisierung der die Stoffabgabe verursachende Prozeß. Da Metabolisierung und Stoffabgabe bei *Fusarium* intensiver ver-

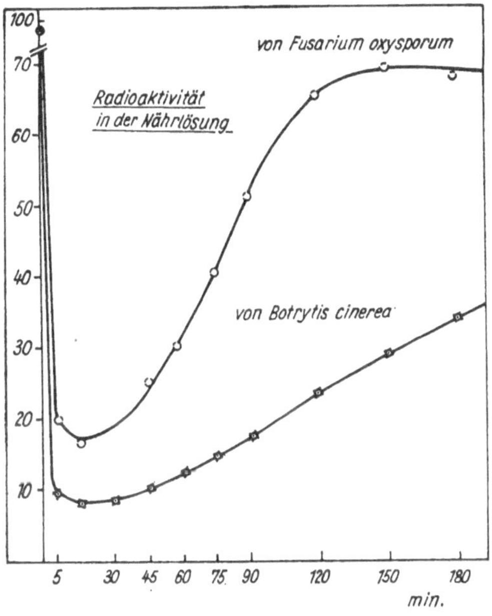

Abb. 2. Zeitabhängige Änderung des Radioaktivitätsgehaltes der Nährlösung von intensiv geschüttelten sterilen Pilz-Submerskulturen mit Zusatz von ^{14}C-trans-2,6-Dimethyl-N-dodecylmorpholin unter Berücksichtigung kurzer Versuchszeiten. Entnahme kleiner Nährlösungsproben unter halbsterilen Bedingungen und bei sehr kurzfristiger Unterbrechung der Agitation aus je einem großen Kulturansatz. Radioaktivitätsbestimmung szintillationsspektrometrisch

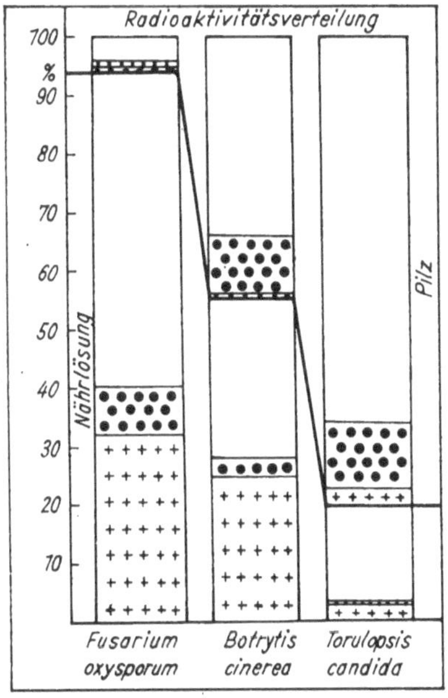

Abb. 3. Verteilung der Radioaktivität aus ^{14}C-trans-2,6-Dimethyl-N-dodecylmorpholin in Schüttelkulturen von *Fusarium oxysporum*, *Botrytis cinerea* und *Torulopsis candida* auf Pilz und Nährlösung einerseits und innerhalb dieser Gruppen auf mit Petroläther (weiß), nachfolgend mit Diäthyläther (Kreise) und danach mit n-Amylalkohol (Kreuze) extrahierbare radioaktive Substanzen. Radioaktivitätsbestimmung szintillationsspektrometrisch. Oberhalb der Treppenlinie Radioaktivität im Myzel, unterhalb in der Nährlösung

laufen als bei *Botrytis* kann dies die Ursache ihrer unterschiedlichen Empfindlichkeit gegenüber dem Wirkstoff sein.

Nachfolgend wurde geprüft, ob auch ein anderer schwach Wirkstoff-empfindlicher Pilz das Fungizid stark metabolisiert (Abb. 3). Unter erneut etwas veränderten Bedingungen wurde die Hefe *Torulopsis candida* in den Versuch einbezogen und die radioaktiven Substanzen aus Pilz und Nährlösung nacheinander mit Lösungsmitteln abgestufter Polarität extrahiert. Während die Radioaktivitätsverteilung auf Nährlösung und Pilz einerseits und auf verschieden polare Fraktionen bei *Fusarium* und *Botrytis* andererseits die bisherigen Ergebnisse bestätigen, wandelt *Torulopsis* den Wirkstoff nicht in nennenswertem Umfang in polare Metabolite um und hat deshalb nach vergleichbaren Versuchszeiten bei weitem die höchsten Konzentrationen an Wirkstoff in der Zelle (Abb. 3). Damit muß die geringe Empfindlichkeit der Hefe anders verursacht sein. Unzugänglichkeit oder Fehlen des Wirkstoffrezeptors könnte in Betracht kommen.

Metabolismus von trans-2,6-Dimethyl-N-dodecylmorpholin in Pilzen

Die dünnschichtchromatographische Analyse von radioaktiven Substanzen, die aus Myzel mit Aceton, Petroläther und Äther und aus Nährlösung mit Petroläther, Äther und Amylalkohol extrahiert wurden, ergab in Versuchen mit ^{14}C-Wirkstoff, wenn Extrakte ohne vorherige Reinigung chromatographiert wurden, meist eine Vielzahl von radioaktiven Flecken unterschiedlicher Lage. Hierbei handelt es sich hauptsächlich um Flecken, in denen Radioaktivität durch aus der Zelle stammende mitextrahierte Substanzen gebunden wurde. Bei Reinigungsschritten änderten sich die Rf-Werte dieser Flecke. Schließlich lag die Radioaktivität hauptsächlich in drei Flecken vor (Abb. 4), von denen derjenige mit dem höchsten Rf-Wert (Nr. 3) mit der Position des

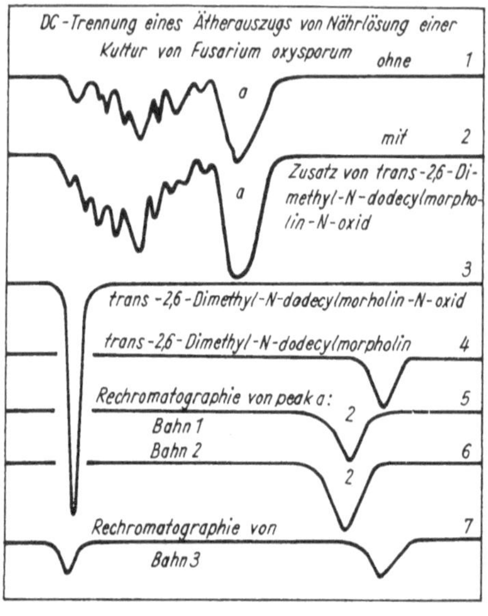

Abb. 4. Wiedergabe der Radioaktivitätsverteilung auf Scannogrammen von DC-Trennungen (Kieselgel G, Hexan: Aceton = 3:1) von ausgewählten typischen Extrakten aus Pilz und Nährlösung von Schüttelkulturen von *Fusarium oxysporum* mit ^{14}C-trans-2,6-Dimethyl-N-dodecylmorpholin. Die auftretenden radioaktiven Flecke wurden in Reihenfolge ihres Auftretens vom Start zur Front mit 1 bis 3 numeriert. Fleck 1 stimmt in der Lage dicht über dem Start mit der Position von ^{14}C-trans-2,6-Dimethyl-N-dodecylmorpholin-N-oxid und Fleck 3 mit der von ^{14}C-trans-2,6-Dimethyl-N-dodecylmorpholin überein. Insgesamt wurde das Pilzmyzel 3 mal mit 100 ml Aceton und die Nährlösung (500 ml) 5 mal mit je 150 ml Petroläther und 76 mal mit je 200 ml Diäthyläther ausgeschüttelt

Wirkstoffes übereinstimmte, derjenige mit dem niedrigsten Rf-Wert (Nr. 1) hatte die gleiche Position wie chemisch aus dem Wirkstoff durch Oxidation hergestelltes N-Oxid des Wirkstoffes. Bei *Fusarium* fand sich immer die Hauptmenge der Radioaktivität nach Fütterung von ^{14}C-trans-2,6-Dimethyl-N-dodecylmorpholin in diesem Fleck I (Abb. 5).

Der dritte Fleck (Nr. 2) hat, bezogen auf die Position des Wirkstoffes einen Rx-Wert von ca. 0,8. An dieser Stelle war, gemessen am ^{14}C-Gehalt, immer eine höhere Stoffkonzentration als an den anderen Positionen vorhanden. Aus Kulturen ohne radioaktiven Wirkstoff konnte durch entsprechende Chromatographie eine Substanz isoliert

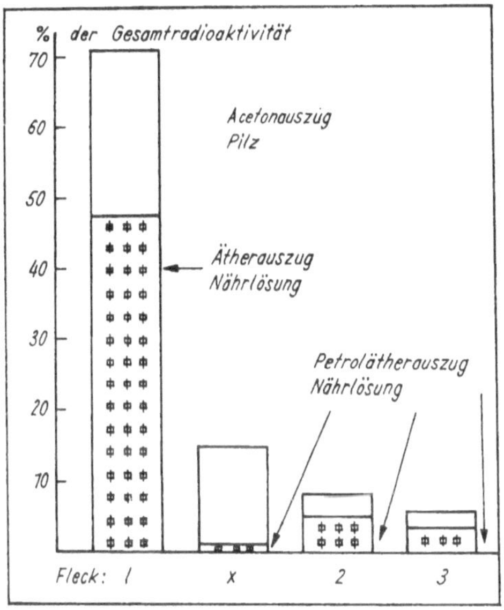

Abb. 5. Quantitative Verteilung der Radioaktivität aus ^{14}C-trans-2,6-Dimethyl-dodecylmorpholin nach Metabolisierung in einer Schüttelkultur von *Fusarium oxysporum* auf verschiedene Chromatogrammpositionen (Fleck 1, 2, 3 und verteilte Radioaktivität unsicherer Zuordnung = x) nach Elution mit drei verschiedenen Lösungsmitteln aus Pilz (Aceton-Extraktion) und Nährlösung (Extraktion mit Petroläther und Diäthyläther). DC-Trennungen an Kieselgel G mit Hexan:Aceton = 3:1

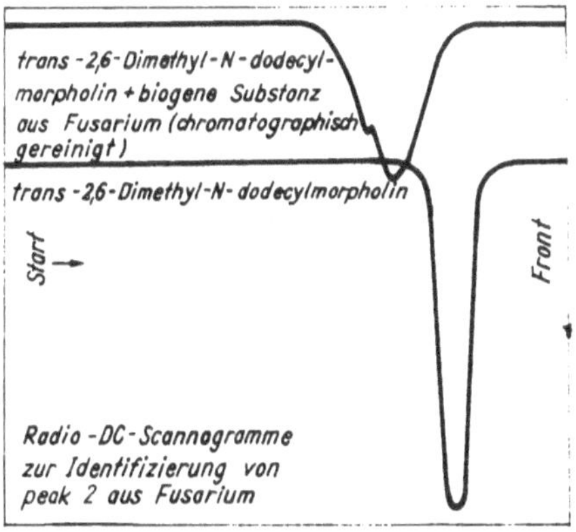

Abb. 6. Einfluß einer chromatographisch gereinigten biogenen Substanz aus einer *Fusarium*-Kultur auf das chromatographische Verhalten von trans-2,6-Dimethyl-N-dodecylmorpholin bei DC-Trennungen an Kieselgel G mit Hexan:Aceton = 3:1. Die Substanzpositionen bei Rechromatographie nach Elution sind unverändert

werden, die ^{14}C-Wirkstoff bindet und dann die Position des Fleckes 2 bei der Chromatographie einnimmt. Die gesamte Radioaktivität aus ^{14}C-trans-2,6-Dimethyl-N-dodecylmorpholin fand sich in diesem Fleck Nr. 2 ,wenn die unbekannte Substanz mit Wirkstoff chromatographiert wurde (Abb. 6). Auch in Chromatogrammen von Rohextrakten hob sich beim Scannen ein Fleck heraus, der bei Rechromatographie mit der Position 2 des gebundenen Wirkstoffes identisch war (Abb. 7, Bahn 1 u. 5). Diese Substanz bindet aber auch das Wirkstoff-N-Oxid (Abb. 7, Bahn 2 u. 6), wobei nach Rechromatographie ebenfalls nur Fleck 2 erhalten wird. Deshalb ist eine Entscheidung über die chemische Natur der gebundenen Radioaktivität danach nicht möglich, insbesondere, da sich gezeigt hat, daß unter den Bedingungen der DC-Trennungen aus dem ^{14}C-trans-2,6-Dimethyl-N-dodecylmorpholin-N-oxid bei Rechromatographie das trans-2,6-Dimethyl-N-dodecylmorpholin in gewissem Umfang zurückgebildet werden kann (Abb. 7, Bahn 3 u. 7).

Damit ergibt sich vorläufig als hauptsächliche Metabolisierung die Oxidation des fungiziden Wirkstoffes am Stickstoff. Auf die fungizide Wirksamkeit ist dieser Prozeß

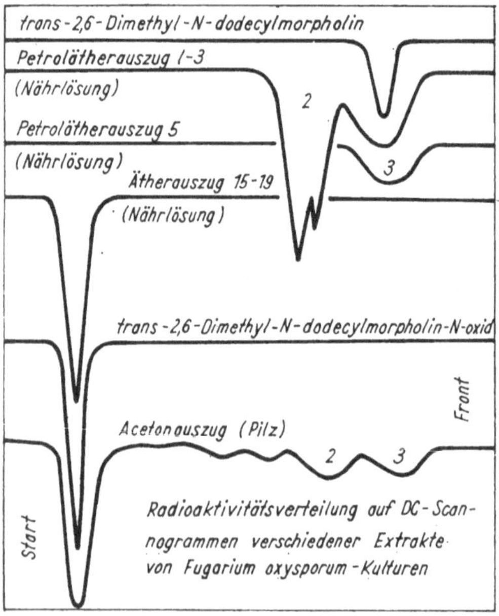

Abb. 7. Chromatographie von ^{14}C-trans-2,6-Dimethyl-N-dodecylmorpholin-N-oxid zusammen mit einem nicht gereinigten ätherischen Extrakt der Nährlösung einer *Fusarium*-Schüttelkultur unter Zusatz von ^{14}C-trans-2,6-Dimethyl-N-dodecylmorpholin, sowie Rechromatographie bestimmter eluierter Substanzflecke. DC-Trennungen an Kieselgel G mit Hexan:Aceton = 1:3 und 3:1

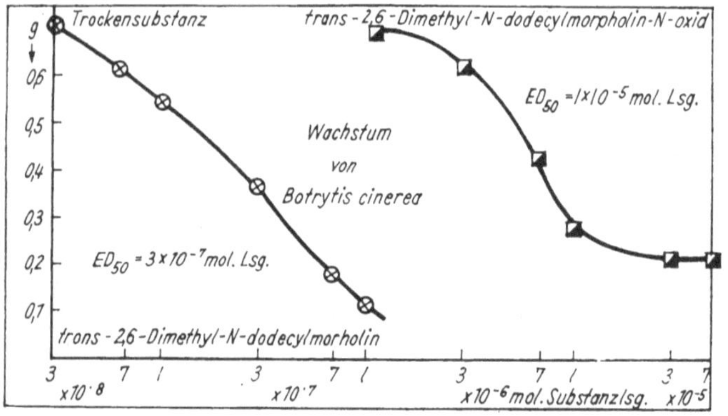

Abb. 8. Vergleich der als Trockenmassezuwachs-Hemmung gemessenen fungiziden Wirksamkeit chromatographisch gereinigter Substanzen von trans-2,6-Dimethyl-N-dodecylmorpholin und seinem N-oxid bei 65 Stunden Schüttelkultur von *Botrytis cinerea* in Nährlösung nach SISLER Die Wachstumshemmung ist bei beiden Substanzen in etwas unterschiedlichem Umfang mit Verminderung der typischen Pigmentbildung und Auftreten einer Myzelaggregation verbunden

von großer Bedeutung (Abb. 8). Gegenüber *Botrytis cinerea*-Kulturen ist die fungizide Wirksamkeit der oxidierten Verbindung erheblich geringer als diejenige des Wirkstoffes. N-Oxid-Bildung ist also ein wirksamer Entgiftungsmechanismus für den fungiziden Wirkstoff.

Literatur

[1] HAWKINS, D. D., W. H. DOWN, L. F. CHASSEAUD and J. D. LEWIS, The metabolism of tridemorph in rats, Pestic. Science **5** (1974), 535–542

[2] JERCHEL, O. und G. JUNG, N-Oxide und N,N'-Dioxide tertiärer Amine, Ber. Chem. Ges. **85** (1952), 1130–1138

[3] OTTO, S. and N. DRESCHER, The behaviour of tridemorph in soil, Proc. 7th Brit. Insect. Fung. Conf. 1973, 57

[4] WARING, R. H. and M. S. WOLFE, Distribution of tridemorph and its metabolites in barley plants grown to maturity, Pestic. Science **6** (1975), 169–172

Tröger, R., Blechschmidt, D. und G. Müller

Friedrich-Schiller-Universität Jena, DDR, Sektion Biologie

Verwendung von TMTD bei der Konservierung wertvoller Papiererzeugnisse

Zwischen 1964 und 1970 haben die Autoren eine Vielzahl von Untersuchungen zur Konservierung pilzgeschädigter Papiererzeugnisse durchgeführt. Neben dem Jenaer Papierspaltverfahren unter Anwendung desinfizierender und alkalisierender Heißwasserbäder sowie dem Einsatz von alkalischen Kleistern aus Cellin (Carboxymethylzellulose) durch den Restaurator (Müller) wurden zusätzlich die Möglichkeiten des Einsatzes von mikrobenvernichtenden Stoffen im Cellinkleister untersucht. An die Fungizide werden dabei hohe Anforderungen gestellt. Sie dürfen nicht humantoxisch, müssen ungefärbt, nicht flüchtig, nicht leim- bzw. kleisterverändernd und auch nicht schädlich gegen Druckfarben, Tinten und die Papierfasersubstanzen sein. Alle meist gegen Mikroben gut wirksamen Schwermetalle wie Hg, Ag, Cu, Cd, Zn und Pb beeinträchtigen die Leimung und die Kleisterkonsistenz oder wurden durch die Kleistersubstanzen unwirksam. Von den geprüften organischen Fungiziden hatten sich in den Kurzzeitversuchen (ca. 1–4 Jahre) am besten Tetramethylthiuramidsulfid (TMTD, 1–6 mg/ml Cellinkleister) und Captan (5 mg/ml Kleister) bewährt. 1966 und 1969 hatten die Autoren u. a. über die Anwendung und Beeinträchtigung des TMTD (Bestimmung mit modif. CuJ-Methode) auf und in Papier nach Normallager und auch bei künstlicher Alterung berichtet (Wiss. Z. d. Fr.-Schiller-Univ. 15, 465, 1966 und Zbl. Bibl. Wes. 83, 342, 1969). 1979 wurden wieder Analysen über die Beständigkeit des TMTD vorgenommen und die fungizidversetzten Papiere nach 10 bzw. 14 Jahren bezüglich Haltbarkeit und Vergilbung sowie Mikrobenfestigkeit bei Beimpfung und Feuchtlagerung getestet. Dabei zeigte sich, daß TMTD in Cellin mit Metallzusätzen *auf* Hadern- und Filterpapier nach 14 bzw. 10 Jahren noch mit meist mehr als 75% der Anfangsmenge vorhanden ist und die Papiere bei Beimpfung und Feuchtlager (6 Monate) nicht von Mikroben bewachsen und zerstört werden konnten (Kontrolle völlig verrottet). Folgender Ansatz hat sich am besten bewährt: TMTD bzw. Captan (5 mg/ml) gebrauchsfertiger Cellinkleister) wird in warmem Aceton gelöst und durch Wasserzugabe auch mit Zusätzen frisch feinst verteilt ausgefällt. In die milchigweiße Suspension wird die entsprechende Menge Cellin eingerührt. Der Kleisteransatz benötigt etwa 24 Std. Quellung für gut streichfähige Konsistenz. Wichtig für die Wirksamkeit des TMTD ist die feinste Verteilung an der Carboxymethylzellulose. Nicht gelöste Fungizid-Partikel verfärben sich braun und waren auf Filterpapierbögen festzustellen. Dicke Leimschichten auf dem Papier erscheinen nach 14 Jahren Lagerung leicht weißgelb verfärbt. Während der Papierfabrikation in den „Stoff" eingerührtes Handels-TMTD war im handgeschöpften Papier völlig abgebaut, und die Bögen waren vergilbt und schimmelten bei Beimpfung. Da TMTD sich über Dithiokarbamate zersetzt, kamen bei einigen Cellin-TMTD-Chargen geringe Zn- und Mn-Salz-Zusätze zur Anwendung, um in der Zersetzungsreaktion die wirksamen Fungizide Zineb (Ziram) bzw. Maneb in der Leimschicht entstehen zu lassen. Über ihre fungizide Wirkung kann noch nichts ausgesagt werden, da bei Anwesenheit von TMTD im Leim die fungizide Wirkung bereits gegeben war. Für die Erhaltung des TMTD scheinen diese Metallsalzzusätze aber sehr wesentlich zu sein.

(Die ausführliche Abhandlung erscheint im Zbl. Bibliothekswesen)

M. A. DE WAARD and A. FUCHS

Laboratory of Phytopathology, Agricultural University, Wageningen, The Netherlands

Resistance to Ergosterol Biosynthesis Inhibiting Fungicides

Introduction

Since 1967 fungicides have been developed which specifically interfere with ergosterol biosynthesis. For various reasons these fungicides occupy a rather special place among the systemic fungicides. Chemically, they comprise a heterogeneous group, however, with two common characteristics. They all have at least one nitrogen-containing heterocyclic ring; in addition, with two exceptions (buthiobate and prochloraz) they contain at least one asymmetric carbon atom.

On the basis of the chemical nature of the nitrogen-containing heterocyclic ring present the following sub-groups can be distinguished:

- piperazines: triforine
- morpholines: dodemorph, fenpropimorph, tridemorph
- pyridines: buthiobate
- pyrimidines: fenarimol, nuarimol, triarimol
- imidazoles: clotrimazole*, econazole*, imazalil, miconazole*, phenapronil, prochloraz
- triazoles: biloxazol, CGA 64.250, CGA 64.251, diclobutrazol, fluotrimazole, triadimefon, triadimenol

Due to the presence of (an) aysmmetric carbon atom(s) these chemicals occur in different steroisomeric forms which may often differ significantly in fungitoxicity. For instance, the (−) enantiomer of triforine is much more active than its (+) analogue (OST [15]). Fenpropimorph (POMMER and HIMMELE [16]) and triadimenol (BUCHENAUER [5]) present two other examples of which the activity has been found to depend on the stereoisomer involved.

Apart from the stereoisomerism due to the presence of the asymmetric carbon atoms in the side chains, in the triforine molecule another form of stereoisomerism originates from the nature of the piperazine ring. The chair form of the piperazine ring permits the 1-formamido-2,2,2-trichloroethyl side chains to occur in two different orientations at each of the ring nitrogen atoms, *viz.* equatorial or axial. According to JOSEPOVITS and GASZTONYI [14], with one side chain equatorial and the other axial, hydrogen bonding can be assumed to occur between the formamide group in one chain and the carbonyl group in the other. Alleged higher water solubility of this stereoisomer than

* Used in medicine against human fungal pathogens.

that of the one with both side chains oriented equatorially is believed to be responsible for its greater biological activity.
In general, the ergosterol biosynthesis inhibitors are broad spectrum fungicides, acting against representatives of *Ascomycetes*, *Basidiomycetes*, and *Fungi imperfecti* (FUCHS and DRANDAREVSKI [7]; BUCHENAUER [4]); *Oomycetes*, on the other hand, are much less sensitive (SIEGEL *et al.*, [20]; BUCHENAUER and RÖHNER [6]; DE WAARD and RAGSDALE [25]). In interpreting the literature data, it should be realized, however, that the reported sensitivities of fungi towards these fungicides — which even among taxonomically closely related fungal species often differ considerably — might depend on the predominant stereoisomer present in the formulations used (*cf.* OST [15]).
Ergosterol biosynthesis inhibitors affect spore germination generally less than germ tube elongation. Initial spore germination may be normal while germ tubes may become heavily distorted resulting in abnormal growth. Hyphae are frequently swollen and/or excessively branched. Nevertheless, dry weight increase, respiration activity mitosis and nucleic acid and protein synthesis are not or only slightly inhibited. However, in cultures treated with these fungicides often accumulation of free fatty acids occurs together with a decrease in C-4 desmethyl sterols (primarily ergosterol) and an increase in methyl and dimethyl sterols. Therefore, inhibition of C-14 demethylation in the ergosterol biosynthetic pathway is believed to be the primary site of action, although additional sites of attack are possible (*cf.* RAGSDALE [17]).

Resistance; Fitness and Virulence of Resistant Mutants

Practical application of fungicides which inhibit ergosterol biosynthesis has not so far led to any confirmed case of resistance*. In contrast, *in vitro*, resistant mutants can often be readily obtained by selection of untreated or mutagen-treated conidia on agar media with lethal concentrations of anyone of these fungicides.
Resistance in fungi to ergosterol biosynthesis inhibiting fungicides is often accompanied with a decrease in fitness or pathogenicity. This has been demonstrated first

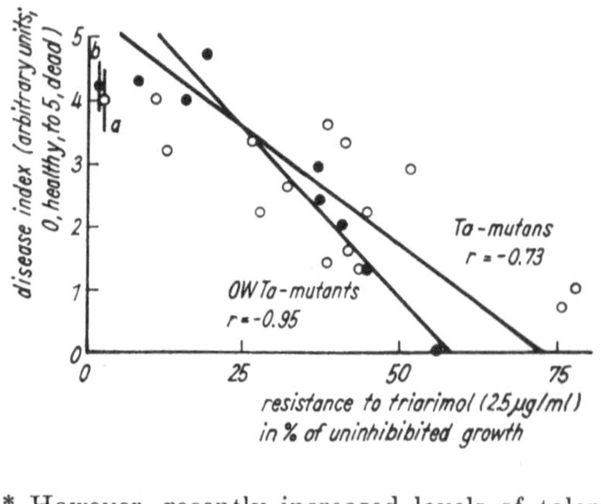

Fig. 1. Relation between virulence of some *Cladosporium cucumerinum* mutants in a cucumber seedling test and degree of resistance. a: green wild-type strain; b: off-white 'parent' strain (from: Fuchs *et al.*, 1977)

* However, recently increased levels of tolerance of *Erysiphe graminis* f. sp. *hordei* to tridemorph in both glasshouse and field experiments have been reported (WALMSLEY-WOODWARD et al., [27]).

with triarimol- and triforine-resistant mutants of *Cladosporium cucumerinum*; the degree of resistance was found to be inversely proportional to pathogenicity (FUCHS and VIETS-VERWEIJ [10]; FUCHS et al., [9]) (Fig. 1). Later, this finding has been confirmed in experiments with the same and other fungal species (BUCHENAUER [3]; VAN TUYL [21]). Imazalil resistance in *Penicillium expansum*, on the other hand, was not always coupled with decreased fitness (VAN TUYL [21]). Mutants of *Aspergillus nidulans* with identified genes for resistance displayed varying degrees of reduced fitness with respect to spore germination, germ tube elongation, mycelial growth or sporulation (DE WAARD and SISLER [26]; DE WAARD and GIESKES [22]). Germ tube elongation rate appeared to be negatively correlated with degree of resistance. All mutant strains resembled each other in impaired ability to produce spores. Mutant strains of *C. cucumerinum* also produced less spores, which at the same time were often inviable (SHERALD and SISLER [19]; FUCHS and DRANDAREVSKI [8]).
The above observations suggest that resistance in pathogenic fungi to ergosterol biosynthesis inhibiting fungicides might have significant epidemiological implications since the chance of such strains to survive in the absence of selection pressure by the fungicide is severely reduced. FUCHS and DRANDAREVSKI [8]), therefore, concluded that development of resistance to this type of fungicide under practical conditions is rather unlikely. The danger of development of resistant populations may once more be reduced inasmuch as resistant strains can be controlled with normal fungicide application rates (BROWN and HALL [2]).

Cross-Resistance

A summary of data on cross-resistant patterns to ergosterol biosynthesis inhibiting fungicides has been compiled by FUCHS et al. [9]. Usually, cross-resistance of fungi to these fungicides is reciprocal; however, exceptions to this rule have been noticed. For instance, imazalil resistance in *C. cucumerinum* (FUCHS et al. [9]), in *Phialophora cinerescens* (VAN TUYL [21]) and in *A. nidulans* strain J130 (DE WAARD and GIESKES [22]) was not always attended with resistance to other ergosterol biosynthesis inhibitors tested. Similarly, imazalil or tridemorph resistance in *Ustilago maydis* did not cause resistance to triadimefon (BARUG and KERKENAAR [1]). These data suggest that some of these fungicides — in particular imazalil — may have additional sites of attack besides C-14 demethylation. Moreover, different mutations, even within the same gene, might lead to different mechanisms of resistance or to slight differences in the same mechanism, for instance, to a differential uptake of these fungicides (DE WAARD and VAN NISTELROOY [24]). These hypotheses might also explain the varying degrees of resistance to a particular fungicide (e.g. FUCHS and VIETS-VERWEIJ [10]; VAN TUYL [21]; DE WAARD and GIESKES [22]; BARUG and KERKENAAR [1]).
The genetics of fungicide resistance has been thoroughly investigated by VAN TUYL [21] who described eight genes for multigenic resistance to imazalil and fenarimol in *A. nidulans*, the imaA or imaB genes conferring the highest degree of resistance to imazalil. Allelic imaA or imaB mutations, however, could also result in significantly lower levels of imazalil resistance. Upon combination of different ima genes for resistance in recombinant strains a higher degree of resistance was obtained, indicating additive effects (VAN TUYL [21]).

Resistance to sterol biosynthesis inhibiting fungicides in *A. nidulans* may also give rise to pleiotropic effects (cross-resistance, hypersensitivity) of fungitoxicants with a dissimilar mechanism of action such as acriflavin, chloramphenicol, cycloheximide, or neomycin (VAN TUYL [21]) and 8-azaguanine, p-fluorophenylalanine or D-serine (DE WAARD and VAN NISTELROOY [23]). Likewise, these data may be explained on the basis of a differential uptake of the toxicants.

Mechanism of Resistance

Resistance to fungicides with a specific mechanism of action may be based on a decreased affinity of the target site for the fungicide. Such a mechanism of resistance has been described for benzimidazoles and oxathiins (cf. GEORGOPOULOS [12]). Resistance to ergosterol biosynthesis inhibiting fungicides might be based on a decreased affinity of the enzyme(s) involved in C-14 demethylation of 24-methylene-dihydrolanosterol for this type of fungicides. Until now, it has not been feasible to test this hypothesis experimentally. Resistance may also be based on substitution of ergosterol by other more 'primitive' sterols. Such a mechanism has been described for polyene antibiotics (HAMILTON-MILLER [13]). However, resistant mutants of *C. cucumerinum* and *A. nidulans* showed almost normal concentrations of ergosterol (SHERALD and SISLER [19]; RAGSDALE and DE WAARD [18]). Factors such as decreased uptake or increased detoxification, which influence the intracellular concentration of a fungicide, may also result in decreased sensitivity. Both factors may play a role in natural resistance of fungi to triforine (GASZTONYI and JOSEPOVITS [11]). On the contrary, induced resistance to fenarimol in *A. nidulans* could not be ascribed to increased metabolism (DE WAARD and RAGSDALE [25])but rather to a differential mycelial uptake (DE WAARD and VAN NISTELROOY [23]).

These initial observations have led, in our laboratory, to elaborate investigations on the mechanism of uptake of fenarimol by *A. nidulans* which, therefore, are described here in some detail. The experiments were carried out with a wild-type strain and with 20 imazalil and fenarimol-resistant mutant strains, three of which were genetically characterized and shown to carry the imaB gene for resistance.

Mycelial uptake by the wild-type strain was characterized by a rapid initial accumulation during the first 10 min of incubation and a subsequent gradual release with time (Fig. 2). In further work (DE WAARD and VAN NISTELROOY [24] uptake by the wild-type strain appeared to be the resultant of influx and efflux. Influx of fenarimol could not be inhibited by low temperature, anaerobiosis, starvation of mycelium or incubation with several respirarory inhibtors and is, therefore, a passive process. Under identical test conditions efflux activity was severely inhibited and should. therefore, be regarded as an energy-dependent mechanism. After prolonged incubation (90 min) an equilibrium between influx and efflux was established, resulting in an energy-dependent steady state the efflux 'component' of which could be inhibited by addition of oligomycin or N,N'-dicyclohexylcarbodiimide causing uptake to be instantaneously enhanced. These obervations suggested that efflux activity is inducible; this hypothesiswas supp.orted by the fact that pre-treatment of mycelium with non-labelled fungicide prevented subsequent uptake of [^{14}C]fenarimol.

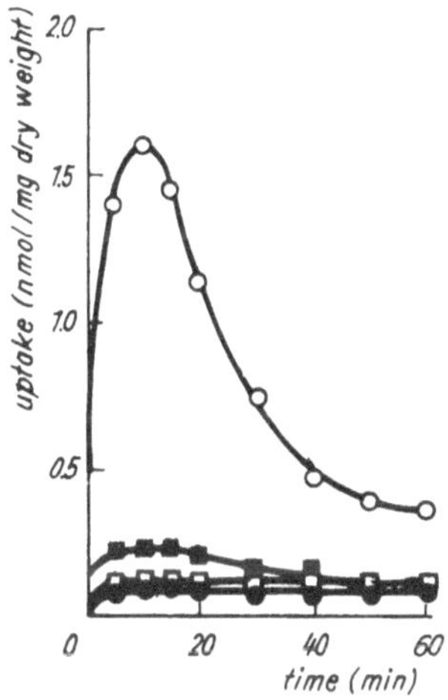

Fig. 2. Uptake of fenarimol (30 μM) by mycelium of wild-type strain 003 (○) and fenarimol-resistant mutants J146 (●), M193 (□), and R264 (■) of *Aspergillus nidulans* in KH_2PO_4 (25 mM) — $K_2HPO_4 \cdot 3H_2O$ (12.5 mM) buffer pH 6.0 containing 0.1 mM $CaCl_2 \cdot 2H_2O$ and 1% glucose (from: DE WAARD and VAN NISTELROOY, 1980)

Uptake by fenarimol-resistant mutants J146, M193, and R264 of *A. nidulans* which all possess the imaB gene for resistance, was relatively low and almost constant in time. In this case, uptake appeared to be considerably enhanced by low temperature, anaerobiosis, starvation of mycelium and incubation with respiratory inhibitors. Low uptake by these mutants can be attributed to an energy-dependent efflux activity for fenarimol which rather than being inducible like in the wild-type strain is most probably constitute. Upon inhibition of the efflux activity, net uptake resulted from the remaining passive influx, which in that case might be as high as in the wild-type strain.

The results suggested that wild-type and resistant strains only differ in 'efficiency' to excrete fenarimol from the mycelium. The efflux rate of resistant strains seems to be high enough to maintain a low level of uptake immediately from the moment of addition of fenarimol. In that case the fungicide might not be able to reach the target site at a concentration sufficient to exert its fungistatic action. The high efflux rate may be associated with an alteration in the regulation of the expression of the socalled imaB gene for resistance to imazalil which all strains tested have in common. The altered energy-dependent efflux in resistant mutants may also explain their cross-resistance or collateral sensitivity to other chemicals. The regulation of the energy-dependent efflux mechanism in the fenarimol-sensitive wild-type strain may be such that there is an initial lag before the steady-state being attained. The amount of fenarimol accumulated during this initial lag may be high enough to saturate the target site(s) in the fungus and consequently exert its fungistatic action (DE WAARD and VAN NISTELROOY [24]).

If the observed effects on the efflux activity of fenarimol should apply to other ergosterol synthesis inhibitors as well, they might also be of consequence for the practical efficacy of these fungicides against plant pathogenic fungi, provided that these have a similar efflux mechanism as *A. nidulans*. Practical conditions, such as low temperature, semi-anaerobiosis, lack of nutrients, may at times be so unfavourable for fungal growth, that efflux activity in both wild-type and resistant strains is impaired. Consequently, higher amounts of these fungicides might be taken up, and might thus enhance their efficacy to wild-type strains of pathogenic fungi or circumvent resistance mechanisms and thus eradicate resistant strains.

Concluding remarks

Clearly, the ergosterol biosynthesis inhibitors comprise a chemically heterogeneous group of systemic fungicides. This heterogeneity is even further amplified by the fact that almost all compounds involved occur as different stereoisomers, with inherent differences in solubility characteristics as well as in intrinsic fungitoxic activity. In view of this chemical variation it is hard to visualize a chemical 'common denominator' responsible for the biological activity evidently shared by these compounds. Also with respect to the 'target organisms' there is much variation. In itself, it is a remarkable phenomenon, that of taxonomically closely related fungal species one is resistant to all — or at least most of all — ergosterol biosynthesis inhibiting fungicides by nature, whereas the other is highly sensitive.

Most probably, at least two different mechanisms are involved, which determine whether or not the chemicals in question are toxic to a certain fungal species. In the first place, the intrinsic toxicity of a compound may be determined by its affinity to the target sites in the organism, of which in the case of ergosterol biosynthesis there might be at least several. Secondly, and superimposed on the first, the mechanism of uptake might be an important differentiating factor, allowing only certain ergosterol biosynthesis inhibitors to be taken up to toxic concentrations. Moreover, since active transport may be stereospecific, the rate of uptake, being the resultant of passive influx and active efflux, might vary depending on the stereoisomer present. Thus, when an ergosterol biosynthesis inhibiting fungicide be applied in practice, the prevalent stereoisomer present in the commercial formulation used as well as its chemical 'versatility' within the cell might further influence its actual fungitoxicity.

If in naturally sensitive fungal species resistance is induced, usually mutants arise with seriously impaired biological functions apparently allowing of growth on nutritionally rich media, but reducing the chances of survival in nature. The mutant strains are most probably constitutive with respect to overall efflux activity: not only ergosterol biosynthesis inhibiting fungicides but also cell metabolites are excreted at a rate exceeding that of the wild-type strain (DE WAARD and VAN NISTELROOY [23]). Resistance, thus, is accompanied with a serious disturbance of overall cell economy. Herein, the altered efflux mechanism seems crucial; therefore, its exact nature is currently being studied.

References

[1] BARUG, D., and A. KERKENAAR, Cross-resistance of UV-induced mutants of *Ustilago maydis* to various fungicides which interfere with ergosterol biosynthesis, Meded. Fac. Landbouwwet. Rijksuniv. Gent **41** (1979), 421–427

[2] BUCHENAUER, I. F., and H. R. HALL, Induced and natural tolerance of fenarimol (EL-222) in *Cladosporium cucumerinum* and *Venturia inaequalis*, Phytopathology **69** (1979), 914

[3] BUCHENAUER, H., Untersuchungen über Wirkungsmechanismus und Selektivität von Triadimefon, Fluotrimazol, Nuarimol, Fenarimol und Imazalil sowie über die Resistenz von Pilzen gegenüber diesen Fungiziden, Mitt. Biol. Bundesanstalt Land- und Forstw. Berlin-Dahlem **178** (1977), 157
[4] BUCHENAUER, H., Comparative studies on the antifungal activity of triadimefon, triadimenol, fenarimol, nuarimol, imazalil and fluotrimazole in vitro, Z. PflKrh. PflSch. **86** (1979), 341–354
[5] BUCHENAUER, H., Conversion of triadimefon into two diasteromers, triadimenol-I and triadimenol-II, by fungi and plants, Proc. IX Intern. Congr. Plant Pathol. Washington D. C., U.S.A. (1979), Abstr. No. 939
[6] BUCHENAUER, H., und E. RÖHNER, Zum Wirkungsmechanismus von Imazalil in Pilzen und Pflanzen, in H. LYR und C. POLTER (eds)/Systemfungizide, Akademie-Verlag, Berlin (1979), 175–185
[7] FUCHS, A., und C. A. DRANDARAEVSKI, Wirkungsbreite und Wirkungsgrad von Triforine in vitro und in vivo, Z. PflKrh. PflSch. **80** (1973), 403–417
[8] FUCHS, A., and C. A. DRANDAREVSKI, The likelihood of development of resistance to systemic fungicides which inhibit ergosterol biosynthesis. Neth. J. Plant Pathol. **82** (1976),85–87
[9] FUCHS, A., S. P. DE RUIG, J. M. VAN TUYL, and F. W. DE VRIES, Resistance to triforine: a nonexistent problem? Neth. J. Plant Pathol. **83S** (1977), 189–205
[10] FUCHS, A., and M. VIETS-VERWEIJ, Permanent and transient resistance to triarimol and triforine in some phytopathogenic fungi, Meded. Fac. Landbouwwet. Rijksuniv. Gent **40** (1975), 699–706
[11] GASZTONYI, M., and G. JOSEPOVITS, Biochemical and chemical factors of the selective antifungal effect of triforine, Acta Phytopathol. Acad. Sci. Hung. **10** (1975), 437–446
[12] GEORGOPOULOS, S. G., Development of fungal resistance to fungicides, in M. R. SIEGEL and H. D. SISLER (eds)—Antifungal Compounds, MARCEL DEKKER, New York (1977), Vol. II, 349–495
[13] HAMILTON-MILLER, J. M. T., Fungal sterols and the mode of action of the polyene antibiotics, Adv. appl. Microbiol. **17** (1974), 109–134
[14] JOSEPOVITS, G., and M. GASZTONYI, Biochemical and chemical factors of the selective antifungal effect of triforine. II. Isomerism and chemical breakdown of triforine, Acta Phytopathol. Acad. Sci. Hung. **11** (1976), 127–139
[15] OST, W., Triforine and its isomers, Proc. Symposium, Eigenschaften und Wirkungsverhalten von Triforine', Heidesheim, Germany (1976), 22–24
[16] POMMER, E.-H., und W. HIMMELE, Die Bekämpfung wichtiger Getreidekrankheiten mit Fenpropemorph, einem neuen Morpholinderivat, Meded. Fac. Landbouwwet. Rijksuniv. Gent **44** (1979), 499–510
[17] RAGSDALE, N. N., Inhibitors of lipid synthesis, in M. R. SIEGEL and H. D. SISLER (eds)/Antifungal Compounds, MARCEL DEKKER, New York (1977), Vol. II, 333–363
[18] RAGSDALE, N. N., and M. A. DE WAARD, Sterol patterns in fenarimol-susceptible and tolerant strains of *Aspergillus nidulans*, Proc. Amer. Phytopath. Soc. **4** (1977), 218
[19] SHERALD, J. L., and H. D. SISLER, Antifungal mode of action of triforine, Pestic. Biochem. Physiol. **5** (1975), 477–488
[20] SIEGEL, M. R., A. KERKENAAR, and A. KAARS SIJPESTEIJN, Antifungal activity of the systemic fungicide imazalil, Neth. J. Plant Pathol. **83S** (1977), 121–134
[21] TUYL, J. M. VAN, Genetics of fungal resistance to systemic fungicides, Meded Landbouwhogesch. Wageningen **77 2** (1977), 1–136
[22] WAARD, M. A. DE, and S. A. GIESKES, Characterization of fenarimol-resistant mutants of *Aspergillus nidulans*, Neth. J. Plant Pathol. **83S** (1977), 177–188
[23] WAARD, M. A. DE, and J. G. M. VAN NISTELROOY, Mechanism of resistance to fenarimol in *Aspergillus nidulans*, Pestic. Biochem. Physiol. **10** (1979), 219–229
[24] WAARD, M. A. DE, and J. G. M. VAN NISTELROOY, An energy-dependent efflux mechanism for fenarimol in a wild-type strain and fenarimol-resistant mutants of *Aspergillus nidulans*, Pestic. Biochem. Physiol. **11** (1980), in press
[25] WAARD, M. A. DE, and N. N. RAGSDALE, Fenarimol, a new systemic fungicide, in H. LYR und C. POLTER (eds/)Systemfungizide, Akademie-Verlag, Berlin (1979), 187–194

[26] Waard, M. A. de, and H. D. Sisler, Resistance to fenarimol in *Aspergillus nidulans*, Meded. Fac. Landbouwwet. Rijksuniv. Gent **41** (1976), 571–578
[27] Walmsley-Woodward, D. J., F. A. Laws, and W. J. Whittington, Studies on the tolerance of *Erysiphe graminis* f. sp. *hordei* to systemic fungicides, Ann. Appl. Biol. **92** (1979), 199–209

G. Würbach, H. Schubert und Erika Prater

Hautklinik der Medizinischen Akademie Erfurt, DDR

Allergische Reaktionen der Haut auf senfölabspaltende Verbindungen

Ausgehend von der bis ins Altertum zurückreichenden Bedeutung senfölführender Drogen in der Volksmedizin, die in einem breiten antimikrobiellen Wirkungsspektrum der Senföle (Isothiozyanate, ITZ) begründet sind, wurden in den Jahren 1955 bis 1975 insbesondere von Riesche u. a. [3–6] und Zsolnai [11] eine große Anzahl synthetischer Senfölbildner hergestellt und auf ihre antimikrobielle Wirkung untersucht. Die Verbindungen, welche unter bestimmten Bedingungen, so auch in Gegenwart von Mikroorganismen unter Abspaltung von Isothiozyanaten zerfallen, lassen sich in folgende 6 Gruppen einordnen: Perhydro-1,3,5-thiadiazin-2-thione; N-monosubstituierte Thiokarbamidsäureester (Thiourethane); N-monosubstituierte dithiokarbamidsaure Salze (Dithiokarbamate); N-monosubstituierte Dithiokarbamidsäureester (Dithiourethane); Dithiokarbamoylalkankarbonsäuren und die daraus herstellbaren 4-Oxothiazolidin-2-thione (Rhodamine).

Aufgrund der geringen Stabilität der meisten Verbindungen in Abhängigkeit von der Polarität und Basizität des Lösungsmittels (Würbach [9]) ist die praktische Anwendung auf wenige Substanzen beschränkt geblieben. So wird in der Humanmedizin das 3,5-Dibenzylperhydro-1,3,5-thiadiazin-2-thion unter den Bezeichnungen D 47, Dibenzthion und Sulbentin in den Präparaten Afungin® und Fungiplex® (BRD) als lokales Antimykotikum eingesetzt. In der Landwirtschaft sind es die Methylsenfölbildner 3,5-Dimethylpherhydro-1,3,5-thiadiazin-2-thion (D 35, Dazomet, Mylon®) und Natriummethyldithiokarbamat (Nematin®, Vapam®, Dicid®) die u. a. als Bodenfungizide Verwendung finden). Sowohl durch Behandlung mit Afungin-Präparaten als auch nach Umgang mit Nematinlösung kam es unter einer größeren Personenzahl zu Hautunverträglichkeitserscheinungen.

Zur Ermittlung des verantwortlichen Allergens wurden Untersuchungen mit Hilfe des üblichen Epikutantestes (Patchtest) durchgeführt, bei dem die Substanzen unter Oklusivbedingungen (Gothatest®) für 24 h zur Einwirkung kamen. Alle nach 24, 48 bzw. 72 h aufgetretenen Rötungen und stärkere Reaktionen vom Ekzemtyp wurden als positiv gewertet.

Da bei der Testung der Afunginbestandteile das Dibenzthion an mehreren Patienten Hautreaktionen gezeigt hatte, versuchten wir im Vergleich zu ähnlich gebauten Verbindungen und Spaltprodukten nachzuweisen, inwieweit der intakte heterozyklische Perhydrothiadiazin-Ring das eigentliche Hapten darstellt, oder ob bestimmte Substituenten für die Sensibilisierung verantwortlich zeichnen (Würbach u. Schubert [10]). Eine Auswahl der von uns ermittelten Testergebnisse ist in der Tabelle 1 zu-

Tabelle 1
Testergebnisse mit Benzyl- und β-Phenyläthyl-ITZ-Bildnern an Afungin-Allergikern; s. auch Zerfallsschema

Substanz	Chem. Konstitution R^1	R^2	abgespalt. Senföl	Testkonz. % Vaseline	Testergebnisse n. 48 h positiv/ Probandenzahl
Afungin®	—	—	Benzyl-ITZ	unverd.	5/5
Dibenzthion	$C_6H_5CH_2$	$C_6H_5CH_2$	Benzyl-ITZ	3,0	2/5
Benzylammonium-benzyldithiokarbamat II	$C_6H_5CH_2$	$C_6H_5CH_2$	Benzyl-ITZ	0,1	5/5
Benzyl-ITZ III	$C_6H_5CH_2$		Benzyl-ITZ	0,01	4/5
3-Benzyl-5-methyl-perhydro-1, 3, 5-thiadiazin-2-thion (D 4735) I	$C_6H_5CH_2$	CH_3	Benzyl-ITZ	1,0	5/5 6
3-β-Phenyläthyl-5-methylperhydro-1, 3, 5-thiadiazin-2-thion (D 14 735) I	$C_6H_5(CH_2)_2$	CH_3	β-Phenyl-äthyl-ITZ	1,0	0/5
Methylammonium-β-phenyläthyl-dithiokarbamat II	$C_6H_5(CH_2)_2$	CH_3	β-Phenyl-äthyl-ITZ	1,0	0/5
β-Phenyläthyl-ITZ	$C_6H_5(CH_2)_2$		β-Phenyl-äthyl-ITZ	0,01	0/5

sammengestellt. Die Dithiokarbamate (II) und Isothiozyanate (III) entstehen aus dem Perhydro-1,3,5-thiadiazin-2-thionen nach folgendem Zerfallsschema:

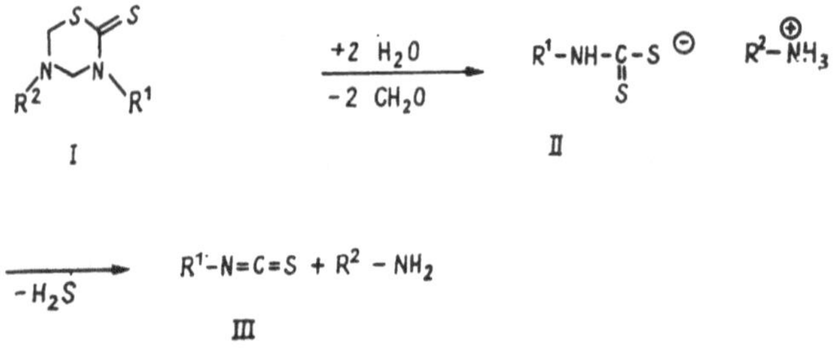

Aus der Tabelle 1 geht hervor, daß alle Benzylsenfölbildner, die an hautgesunden Kontrollpersonen in den angegebenen Konzentrationen keine Reaktion gezeigt hatten, bei Afungin-Allergikern Kontaktekzeme hervorriefen. Wie der Vergleich zwischen Dibenzthion, D 4735 und D 14735 zeigt, wird die Allergie nicht durch den intakten Perhydrothiadiazin-Ring, sondern durch Spaltprodukte hervorgerufen, von denen das Benzyl-ITZ als das eigentliche Hapten identifiziert werden konnte. Die allergisierende Wirkung ist jedoch nicht eine Gruppeneigenschaft der ITZ sondern von bestimmten Substituenten abhängig. Durch Austausch des Benzylrestes gegen den β-Phenyläthylrest wird die sensibilisierende Wirkung aufgehoben. Die Aussagen konnten sowohl durch immunologische Untersuchungen mit Hilfe des Lymphozytentransformationstestes (SCHUBERT u. a. [7]) als auch durch tierexperimentelle Sensibilisierungen (SCHUBERT u. a. [8]) untermauert werden.

Wir empfehlen deshalb einen Austausch des Dibenzthions in den Afungin-Präparaten gegen das 3-β-Phenyläthyl-5-methylperhydro-1,3,5-thiadiazin-2-thion (D 14735), das bei besserer antifungieller Wirksamkeit nur eine geringe Sensibilisierungspotenz aufweist.

Zu Hautunverträglichkeiten auf Nematin kam es besonders bei der Kartoffelselektierung mit Hilfe einer alkalischen Lösung des Natriummethyldithiokarbamats (I), aus dem im sauren pH-Bereich der Haut freies Methyl-ITZ (II) entstehen kann.

$$\underset{\text{I}}{CH_3NH\text{-}CS\text{-}S^{\ominus} \quad Na^{\oplus}} \xrightarrow[-H_2S]{+HO} \underset{\text{II}}{CH_3\text{-}N = C = S} + NaOH$$

Unter Anwendung des Epikutantestes führten wir vergleichende Untersuchungen sowohl mit Methyl- und Benzylisothiozyanatbildnern als auch mit N-disubstituierten Dithiokarbamaten durch. Die Ergebnisse sind auszugsweise in der Tabelle 2 zusammengestellt. Nach der Testung von 4 Patienten konnte auch in der Wiederholung das Natriummethyldithiokarbamat als die allergisierende Substanz identifiziert werden. Wie an einer Patientin sichtbar, dürfte auch hier analog der Afungin-Allergie das unter Testbedingungen leicht entstehende Methyl-ITZ das eigentliche Hapten darstellen. Während das Natriumäthyldithiokarbamat ebenfalls Hautreaktionen verursachte, reagierten die nematinempfindlichen Patienten in 3 von 4 Fällen nicht auf benzylsenfölabspaltende Verbindungen.

Die Kreuzvergleiche zwischen beiden Patientengruppen lassen 2 verschiedene Allergieformen erkennen, was auch von RICHTER u. a. [2] tierexperimentell bestätigt werden konnte. So wie bei der Afungin-Allergie der Benzylrest, scheinen bei der Nematin-Allergie

Tabelle 2
Testergebnisse an Nematin-Allergikern

Substanz	abgespalt. Senföl	Testkonz. % Vaseline	Testergebnis n. 48 h pos./Probandenzahl
Nematin-Lösung, 0,01% wässrig	Methyl-ITZ	unverdünnt	4/4
Natriummethyldithiokarbamat	Methyl-ITZ	0,1	3/4
Methyl-ITZ	Methyl-ITZ	0,01	1/1
3,5-Dimethylperhydro-1,3,5-thiadiazin-2-thion (D 35)	Methyl-ITZ	1,0	1/4
3-Methyl-5-benzylperhydro-1,3,5-thiadiazin-2-thion (D 3547)	Methyl-ITZ	1,0	1/4
Natriumäthyldithiokarbamat	Äthyl-ITZ	0,1	4/4
Dibenzthion	Benzyl-ITZ	3,0	0/4
Benzylammoniumbenzyldithiokarbamat	Benzyl-ITZ	0,1	1/4
Benzyl-ITZ	Benzyl-ITZ	0,01	0/4
3-Benzyl-5-methylperhydro-1,3,5-thiadiazin-2-thion (D 4735)	Benzyl-ITZ	1,0	1/4

die Methyl- und Äthylreste die allergenen Gruppen darzustellen, während die Isothiozyanatgruppierung aufgrund ihrer hohen Affinität zu Eiweißen für die Haptenkopplung zum Vollantigen verantwortlich sein dürfte.
Bei sehr hochgradiger Afungin- und Nematin-Allergie kann auch eine Kopplung der allergenen Gruppen über die Thiokarbamatstruktur ($>N-\underset{\underset{S}{\|}}{C^{+}}-$), die eine Elektronenlücke am Thiokarbonyl-C aufweist (Perkow [1]), nicht ausgeschlossen werden. Die bei wenigen dieser Patienten zusätzlich aufgetretenen Hautreaktionen auf Thioharnstoffe und disubstituierte Dithiokarbamate, die eine solche Struktur enthalten, sind in der Tabelle 3 zusammengestellt.

Tabelle 3
Hautreaktionen auf Thioharnstoffe und N-disubstituierte Dithiokarbamate

Substanz	Testkonz. % Vaseline	Allergieform	Testergebnis n. 48 h pos./ Probandenoahl
Methylbenzylthioharnstoff	1,0	Afungin-A.	1/10
Dibenzylthioharnstoff	1,0	Afungin-A.	2/9
Methylthioharnstoff	1,0	Nematin-A.	1/4
Dimethylthioharnstoff	1,0	Nematin-A.	1/4
Eisendimethyldithiokarbamat (Ferbam)	1,0	Nematin-A.	1/4
Zinkdimethyldithiokarbamat (Ziram)	1,0	Nematin-A.	1/4

Literatur

[1] Perkow, W., Konstitution und Wirkung biologisch aktiver Verbindungen. 15. Mitt.: Säureamide, Carbamate, Thiocarbamate, Harnstoffderivate, Arzneim.-Forsch. (Drug res.) **12** (1962), 1185–1191
[2] Richter, G., U. Heidelbach, I. Heidenbluth und H. Kadner, Untersuchungen zur gruppenallergischen Reaktionsbreite und Sensibilisierungspotenz von Benzylisothiozyanat, Berufsdermatosen **24** (1976), 147–151
[3] Rieche, A., G. Hilgetag, A. Martini und I. Kreyzy, Synthetische Senfölbildner III. Untersuchungen über Dithiokarbamidsäureester als mikrobizidwirksame Substanzen, Arch. Pharmaz. **296** (1963), 310–324
[4] Rieche, A., G. Hilgetag, A. Martini, O. Nejedly und I. Schlegel, Neue Verbindungen mit bakterizider, fungizider und das Wachstum von Viren hemmender Wirkung I., Arch. Pharmaz. **293** (1960), 957–967
[5] Rieche, A., G. Hilgetag, A. Martini und R. Philippson, Neue Verbindungen mit bakterizider, fungizider und das Wachstum von Viren hemmender Wirkung II., Arch. Pharmaz. **294** (1961), 201–209
[6] Rieche, A., D. Martin und W. Schade, Synthetische Senfölbildner V. Darstellung und antimikrobielle Eigenschaften wasserlöslicher, Senföle bildender Verbindungen, Arch. Pharmaz. **296** (1963), 770–784
[7] Schubert, H., G. Würbach und H.-D. Göring, Untersuchungen über die Afungin®-Allergie, 2. Mitt. Dermatol. Monatsschr. **163** (1977), 384–386
[8] Schubert, H., G. Würbach und V. Ziegler, Untersuchungen über die Afungin®-Allergie, 3. Mitt. Dermatol. Monatsschr. **165** (1979), 491–494 **3**
[9] Würbach, G., Untersuchungen über die Zerfallsbedingungen und den Zerfallsmechanismus synthetischer Senfölbildner, Inaug.-Diss. Jena, 1965
[10] Würbach, G. und H. Schubert, Untersuchungen über die Afungin®-Allergie, Dermatol. Monatsschr. **162** (1976), 317–322
[11] Zsolnai, T., Die antimikrobielle Wirkung von potentiellen Isothiozyanatbildern, 7. Mitt. Arzneim.-Forsch. (Drug res.) **25** (1975), 1336–1339

E. V. Yurina

Moscow State University, Moscow, USSR

Features of Peroxidase Isoenzyme Content of Rust Uredospores

The molecular heterogeneity of enzymes, determined by cell genome, essentialy increases the informative capacity of catalytic system and also increases adaptive possibilities organismic during changes of internal medium and environment. The biosynthesis of particular enzyme subunits induced in host plants by pathogen indicates to adaptive nature of isoenzymes [5, 7].
According the current ideas the basis of molecular mechanisms of isoenzyme production is the single substitution in nucleotide sequences of structural genes. Therefore the particular isoenzymes can be regarded as the informative units of metabolic and genetic levels of regulation. It may be concluded that isoenzymes are the most important mechanism of plant cells during their growth and development, adaptation to environmental changes and the metabolic during fungal and virus pathogenesis.
The data cited in the literature [4] indicate that the molecular heterogeneity of catalytically active proteins might provide the sufficiently reliable criterion for consideration of host-parasite compatibility. The nature of molecular mechanisms provided specialization of pathogenic organisms has been insufficiently studied. The investigations in this field are very important. The rust fungus with their high virulence and great adaptive flexibility are the suitable objects for such investigation [2].
The present report relates to the study of isoenzyme content of enzyme of oxidative metabolism-peroxidase. Peculiarities of this enzyme function in plant and fungal tissues determine their resistance, virulence and affinity to definite host plant.

Material and Methods

In experiments we used the uredospores of three rust species: *Puccinia graminis* Henn., *Puccinia dispersa* Erikss., *Puccinia coronifera* Kleb. and also three specialized forms of stem rust: *P. graminis* f. *tritici*, *P. graminis* f. *secalis*, *P. graminis* f. *avenae*. For estimation of the degree of complementary of parasite and host-plant enzyme systems we investigated simultaneously peroxidase isoenzyme composition in leaves of corresponding host plant (rye, oat, barley and wheat cultivars which differed in resistance).
Gel electrophoresis were for the separation multicomponent protein mixture [8] and immunochemical methods were used the elucidation specific features enzymic subunits [1]. In order to determine enzyme activity plant leaves and rust uredospores were broken by grinding in jasper mortar and buffer Tris-glycin 0,05 M, pH 8,5 with 17% sucrose, 0,1% ascorbic acid and 0,1% cysteine was added [10]. The procedure was carried out at 0°. The homogenate was centrifuged for 20 min at 40000 g. The supernatant was used for analysis. The protein concentration was determined by the method of Lowry et al. [9]. In order to reveal contents of isoenzymes of peroxidase the gels were stained according Pears [11].

Results and Discussion

Comparative investigation of soluble cytoplasmic proteins of uredospores from different species and specialized forms of rust showed qualitative and quantitative differences in molecular plurality of peroxidase pattern. Peroxidase zymograme for single species and formes of rust uredospores essentially are varied in number of subunits, concentration of protein bands and their mobility in electric field. A protein component is concentrated in middle of gel, while the crown rust proteins pattern showed that the most peroxidase activity are represented by fast migrating components with Rf 0.8—0.9 (Fig. 1). The similar differences observed in peroxidase pattern of uredospores special stem rust forms collected from wheat. Among three mentioned forms of the rust fungi the most isoenzymes were revealed in wheat form of rust than in rye-and oat-form: 8,5 and 3 subunits correspondingly.

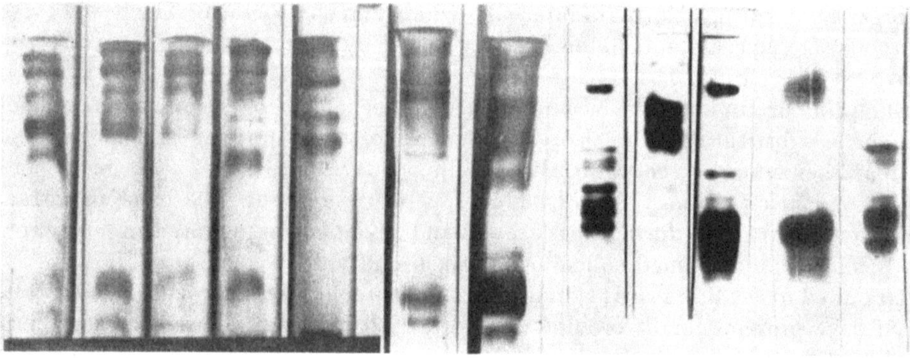

Fig. 1. Peroxidase zymogrames of different plant species (1), rust fungi species (2), specialized formes of *Puccinia graminis* (3)
Designation. wheat cultivars, M-Mironovskaya-808, A — Aurora, C — Caucasus, K — Khapli, B — Barley, R — Rye, O — Oat; *P. g.* — *Puccinia graminis*, *P. d.* — *Puccinia dispersa*, *P. c.* — *Puccinia coronifera*; *f. a.* — *P. graminis f. avenae*, *f. s.* — *P. graminis f. secalis*

Thus, these data indicate that physiologo-biochemical features of catalytically active cell proteins lie in basis of pathogenic microorganisms specialization. In this relation, it was interesting to compare peroxidase patterns of uredospores and host-plant. It has been shown that there is the characteristic peroxidase isoenzymes pattern for each species of plants and rust fungi. The similar type of isoenzymes can be detected during the comparison of plants and fungi zymograms. For instance the zymogrames of oat leaves peroxidase obviously showed high activity of isoenzyme with Rf 0.7—0.8. The most enzymes from crown rust und wheat stem rust, growing on oat, were also located in this part of gel. Analogous similarity are observed in zymograme of wheat cv. Khapli and in zymograme of brown rust, having one-type isoenzyme with Rf 0.3. Usually uninfected wheat cv. Khapli is high infected by brown rust and low infected by stem rust. As follows from Figure 1, the peroxidase zymogrames of stem rust uredospores does not almost show isoenzyme activity with Rf 0.3. Wheat, cv. Aurora and cv. Caucasus with high resistance to stem rust, have not this isoenzyme. The leaves of barley, with resistance to brown rust have not the isoenzyme with Rf 0.3.

We suggest from our experimental data in conjunction with literature data [6, 12] that the similarity of peroxidase for origin of compatible combination during rust infection. There are reports now that subunits of similar enzymes from host plant and pathogen can change and also form the hybride enzymes with different activities [4].

It has been accumulated considerable experimental materials revealing a specific role of catalytically active proteins as in formation of specific features of different pathogens as in processes of host-plant and pathogen interaction [7].

As a result of investigations it has been established than rust uredospores from different plants are nonhomogeneous and are distinguished in their aggression during the infection of other plants. The uredospores with different peroxidase isoenzymes patterns (Fig. 2), pathogenity, speed of germination on agar are formed in the course of reproduction on the same fungal race on different plants (Tab. 1). Fungal generation on the same cultivar caused the highest plant infection. The fungal passaging on another wheat cultivar resulted in considerable decrease of parasitic activity. Under the same spore leading the barley infection with uredospores from PPG-186 wheat and the PPG-186 infection with uredospores from barley plants were 7 and 12 times (622 and 93.61 and 5 pustules) less than the infections with uredospores from the same cultivars. These data indicate that using the periodic change of cultivars in agricultural crops can be considerably reduced the velocity of pathogen adaptation to resistant cultivars.

Serological studying of uredospores of the 15th race wheat stem rust has shown that uredospores of the same fungus race from wheat and barley plants are nonhomogeneous in their immunological reactions. These experimental results are represented in Figure 3 (A, B). Immunodiffusion was carried out in 1% agar "Difco" with antiserum to race 15th uredospores, collected from (well 6, in the center) against antigens race 15th uredospores, collected on barley (well 5) and on wheat (well 2—4). Antigen and anti-

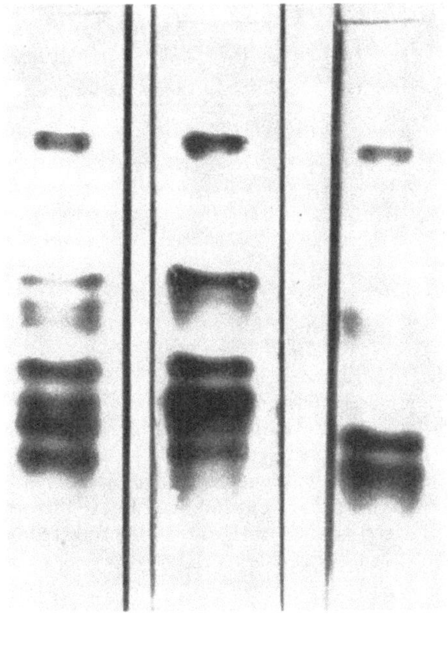

Fig. 2. Peroxidase zymogrames of uredospores from different plant species.
Designation: M — uredospores from wheat cv. Mironovskaya-808, K — Khapli, B — Barley cv

Table 1

The relative parasitic activity of Puccinia graminis uredospores *from different plants during the infection of wheat and barley spots*

host-plant	number of pustules per 10 PPG-plants	number of pustules per 10 barley plants	uredospores germination on agar %
wheat cultivars:			
PPG-186	61	93	85
Saratovskaya-29	37	188	84
Mironovskaya-808	35	167	89
Yubileinaya-50	25	180	63
Spelmar	28	57	58
Akme	26	62	61
Barley cultivar Vinner	5	622	51

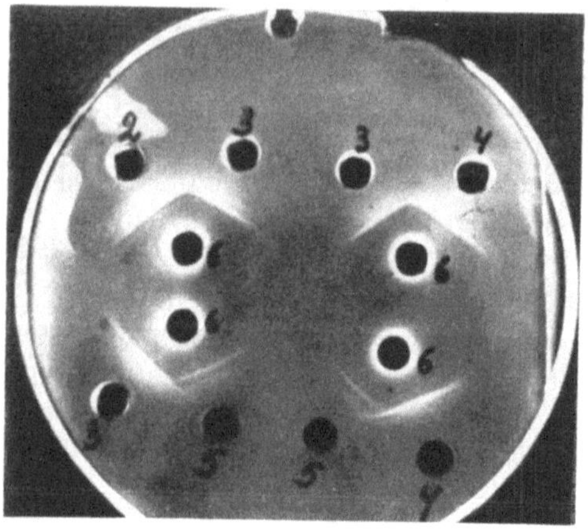

A

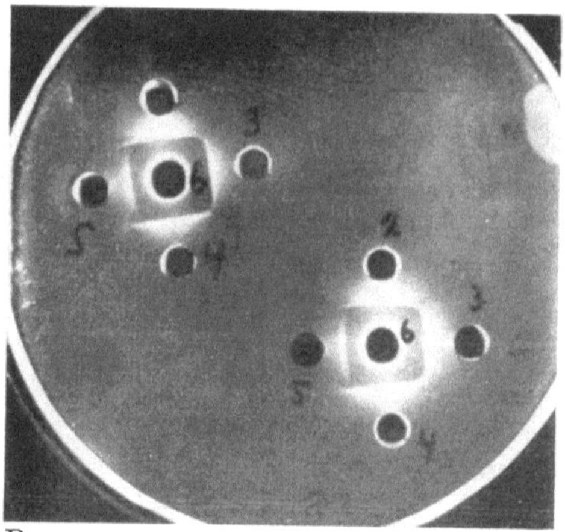

B

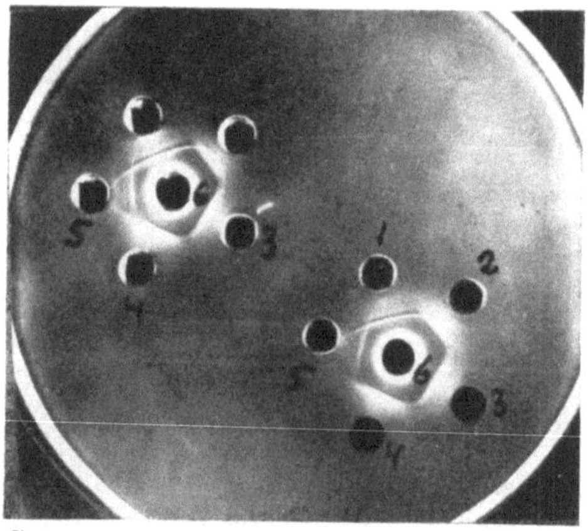

C

Fig. 3. Immunoprecipitation reactions of the 15th race uredospore antigenes from different plants with homological antiserum was carried in 1% agar (A, B, C).
Designation: In the center well (6) antiserum against the 15 race uredospores collected from wheat. Uredospore antigene collected from wheat (2—4) and barley (5)

body diffusing from wells into agar and after meeting gave as many precipitation lines as the number of antigen-antibody pairs in the system. About serological identity of comparable antigenes was indicated identity reaction. This type of reaction was revealed between wells 2—4 (Fig. 3). For example, there is precipitation against antigen of the 15th race uredospores from wheat and corresponding antiserum. Another type of reaction was observed when antigen of uredospores from wheat and barley were used. In this case full fusion precipitation lines was not observed. Precipitation line of the antigen of the 15th race uredospores from wheat in precipitation zone gives the process — "spur" with precipitation line of the same race, but collected from barley. The presence of such "spur" gives the reason to speak about the partial serological resemblance of comparable antigenes. This type of precipitation is revealed, when antiserum contained the antibodies to common determinants of comparable antigenes and to additional determinant of one of variants. The availability of common determinants leads to fusion of precipitation lines and the additional determinant one of the samples gaves "spur" against the antibody to them.

It thus appears that, uredospores of the same race of the fungus grown on wheat and barley plants are serologically nonhomogeneous. Under the influence of feeding plants it takes place the change of antigenic properties of parasitic fungus. There are definite reorganizations in cell protein components during adaptation to new host plant. Analysis of the data leads to a conclusion that features of oxidative isoenzyme patterns can be used for identification and determination of compatibility of partners during infection and also as a test for investigation of systemic effect of fungicides.

Literaturverzeichnis liegt nicht vor.

Autoren

FSC
www.fsc.org
MIX
Papier | Fördert gute Waldnutzung
FSC® C083411

Kolibri 360 GmbH
Ferdinand-Jühlke-Straße 7
99095 Erfurt, Deutschland
produktsicherheit@kolibri360.de